Bryophyte Biology

Bryophyte Biology provides a comprehensive yet succinct overview of morphology, systematics, ecology, and evolution of hornworts, liverworts, and mosses. A distinguished set of contributors provide state-of-the-art summaries of the most recent advances in bryology, with rich citation of the current literature. Revised classifications for the liverworts and mosses are presented that depart significantly from previous arrangements. These novel classifications reflect the results of recent phylogenetic analyses and include exhaustive lists of accepted genera with their familial placements. Accessible and well-illustrated overviews of morphology are provided for each group, with subsequent contributions focusing on current areas of active research, including developmental biology, molecular genetics, ecology, and microevolution. In addition to reviews of the most current literature, the chapters provide abundant reference to classical studies in bryology that will help those new to the topic to develop an historical perspective within which recent developments can be viewed.

A. JONATHAN SHAW is Associate Professor of Botany at Duke University, North Carolina, and Curator of the L. E. Anderson Bryophyte Herbarium. His research has covered monographic work, developmental anatomy, population genetics and, most recently, the molecular systematics and evolution of peat mosses. He teaches courses in evolutionary biology and speciation, as well as bryology.

BERNARD GOFFINET is Assistant Professor in the Department of Ecology and Evolutionary Biology at the University of Connecticut where he teaches bryology and plant morphology. His research focuses on the evolutionary biology and systematics of mosses, and he also maintains a strong interest in lichenology. He is the holder of the 1995 A. J. Sharp Prize, award by the American Bryological and Lichenological Society.

Bryophyte Biology

Edited by
A. JONATHAN SHAW
and BERNARD GOFFINET

PUBLISHED BY THE PRESS SYNDICATE OF THE UNIVERSITY OF CAMBRIDGE
The Pitt Building, Trumpington Street, Cambridge, United Kingdom

CAMBRIDGE UNIVERSITY PRESS
The Edinburgh Building, Cambridge CB2 2RU, UK
40 West 20th Street, New York, NY 10011–4211, USA
477 Williamstown Road, Port Melbourne, VIC 3207, Australia
Ruiz de Alarcón 13, 28014 Madrid, Spain
Dock House, The Waterfront, Cape Town 8001, South Africa

http://www.cambridge.org

First published 2000
Reprinted 2002

Printed in the United Kingdom at the University Press, Cambridge

Typeface TEFFLexicon 9/13 pt   *System* QuarkXPress® [SE]

A catalogue record for this book is available from the British Library

Library of Congress Cataloguing in Publication data
The biology of bryophytes / edited by A. Jonathan Shaw and Bernard Goffinet.
    p. cm.
  ISBN 0 521 660971 (hb)
    1. Bryophytes. I. Shaw, A. Jonathan (Arthur Jonathan) II. Goffinet, Bernard.
QK533.B48 2000
588–dc21   99-462301

ISBN 0 521 660971 1 hardback
ISBN 0 521 667941 1 paperback

Contents

Contributors

DR J. W. BATES
Department of Biology, Imperial College at Silwood Park, Ascot, Berkshire, SL5 7PY

DR W. R. BUCK
New York Botanical Garden, Bronx, NY 10458–5126, USA

DR M. L. CHRISTIANSON
Department of Ecology and Evolutionary Biology, University of Kansas, Lawrence, KS 66045, USA

DR D. COVE
Leeds Institute of Plant Biotechnology, University of Leeds, Leeds, LS2 9JT

DR B. CRANDALL-STOTLER
Department of Plant Biology, Southern Illinois University, Carbondale, IL 62901–6509, USA

DR B. GOFFINET
Department of Ecology and Evolutionary Biology, U-43 75 North Eagleville Road, University of Connecticut, Sorrs, CT 06268–3043, USA

DR R. MUES
Universität des Saarlandes, PO Box 151150, D-66041 Saarbrücken, Germany

DR K. P. O'NEILL
Nicholas School of the Environment, Duke University, Durham, NC 27708, USA

Dr T. Pócs
Department of Botany, Esterházy College, EGER, Pf.43, H-3301,
Hungary

Dr M. C. F. Proctor
Department of Biological Sciences, Hatherley Laboratories,
University of Exeter, Prince of Wales Road, Exeter, EX4 4PS

Dr K. S. Renzaglia
Department of Plant Biology, Southern Illinois University,
Carbondale, IL 62901–6509, USA

Dr A. J. Shaw
Department of Botany, Duke University, Durham, NC 27708, USA

Dr R. E. Stotler
Department of Plant Biology, Southern Illinois University,
Carbondale, IL 62901–6509, USA

Dr B. C. Tan
Department of Biological Sciences, National University of
Singapore, Singapore 119260

Dr K. C. Vaughn
USDA-ARS Southern Weed Science Laboratory, Stoneville, MS
38776, USA

Dr D. H. Vitt
Department of Biological Sciences and Devonian Botanic Garden,
University of Alberta, Edmonton, AB T6G 2E1, Canada

Preface

Interest in bryophytes has undergone a resurgence in the last decade. This renewed focus on the mosses, liverworts, and hornworts has converged from diverse quarters within the scientific community. With recent advances in DNA sequencing technology and analytical approaches to phylogeny reconstruction, systematists have made unprecedented progress toward reconstructing the "tree of life." One of the truly monumental events in the history of life was the origin of land plants, or Embryophytes. The bryophytes have long been considered a pivotal group positioned at or near the base of the embryophytes and a great deal of molecular work has recently been aimed at resolving relationships among the disparate groups of bryophytes, and their relationships to the tracheophyte clade (see chapter 4). At the same time, the utility of bryophytes, especially mosses, for analyses of plant function and development has been increasingly appreciated and capitalized upon (see chapter 7). Haploidy and structural simplicity among land plants gives the mosses "added value" for research in functional genomics and several species are presently being utilized as model systems (see chapter 5). The ecological importance of bryophytes has long been appreciated, but recent concerns about the implications of global climate change has focussed renewed attention on some bryophyte-dominated ecosystems, especially boreal peatlands (see chapters 10 and 11).

The idea for this volume came about from two divergent directions. Schofield's recent textbook of bryology[1] is no longer in print, and students in bryology classes have few other succinct options. In addition, growing attention to bryophytes by ecologists and molecular biologists

1. Schofield, W. B. (1985). *Introduction to Bryology*. New York and London: Macmillan.

raises the need for an accessible but inclusive reference on the biology of bryophytes. Fulfilling these rather disparate needs provided the stimulus for this book. Bryology students need a readable overview of bryophyte biology, and it is our opinion that students are well served by rich citation of both current and classic research. We envision that this volume will be especially appropriate for advanced undergraduate and graduate-level bryology students. For ecologists, geneticists, and other researchers who use bryophytes as model systems, we hope to provide a one-stop overview that provides entry into current literature from the spectrum of bryophyte research, as well as descriptions of basic biological characteristics. We tried to find the best compromise between these goals.

Mishler and Churchill[2] presented the first formal cladistic analysis of mosses, liverworts, and hornworts in relation to the tracheophyte clade that includes all other land plants. Their conclusion that the "bryophytes" comprise a paraphyletic basal grade rather than a monophyletic group has been supported by most recent molecular analyses (see chapter 4). What the three groups do have in common is a pleisomorphic life cycle that is nevertheless unique among land plants. The gametophytes of bryophytes are typically large and free-living, perennial, and photosynthetic, while the sporophytes remain attached to the gametophytes and are photosynthetic for a relatively short time. Although the bryophytes do not form a natural group in the sense of monophyly, their comparable life cycles promote a cohesive field of scientific study: bryology. The word bryophyte is used throughout this book in an informal sense in reference to those basal land plants that share the haploid-dominant dibiontic life cycle.

2. Mishler, B. D. & Churchill, S. P. (1984). A cladistic approach to the phylogeny of the "bryophytes." *Brittonia*, **36**, 406–24.

1

Anatomy, development and classification of hornworts

1.1. Introduction

Recently implicated as the oldest extant lineage of land plants, the anthocerotes hold many clues to the early diversification of terrestrial organisms (Malek *et al.* 1996, Garbary & Renzaglia 1998, Hedderson *et al.* 1998, Vaughn & Renzaglia 1998, Beckert *et al.* 1999, Nishiyama & Kato 1999, Renzaglia *et al.* 2000). Insights into adaptive strategies that enabled plants to survive during early land radiation and to persist through the millennia are gained through exploration of the morphology and reproductive biology of this ancient plant lineage (Renzaglia *et al.* in press). Such information also provides an essential foundation for future comparative studies among bryophytes and with basal groups of tracheophytes. In this chapter, we overview the current state of our knowledge on the morphology, ultrastructure, and developmental diversity within the anthocerotes. This information is evaluated from a comparative point of view in a broader context of relationships among streptophyte lineages. Throughout our discussion, we identify future lines of investigation that will provide significant new phylogenetic information on hornworts. Finally, we briefly review the prevalent ideas on the classification of anthocerotes.

1.2 Anatomy and development

The description that follows is intended to provide 1) an overview of the unifying morphological features of anthocerotes, 2) a brief survey of diversity in structure among hornwort taxa, and 3) a synthesis of published and unpublished data derived from recent ultrastructural studies.

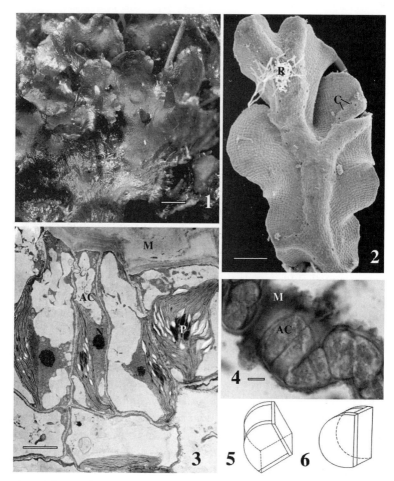

Fig. 1.1. *Phaeoceros laevis* subsp. *carolinianus* (Michx.) Prosk. Gametophyte with young
sporophytes seen as bumps (upper thalli) and as horns breaking through
protective involucres (right central thallus). Bar = 3.0 mm.

Fig. 1.2. *Dendroceros tubercularis* Hatt. SEM of ventral thallus showing swollen central
midrib and monostromatic wings. Pore-like mucilage clefts (C) occur in two
irregular rows on either side of the midrib and a tuft of rhizoids (R) is
positioned below the terminal bifurcation. Bar = 0.25 mm.

Fig. 1.3. *Phaeoceros laevis* subsp. *carolinianus*. Transmission electron microscope (TEM)
horizontal longitudinal section of growing notch overarched by mucilage (M).
The rectangular apical cell (AC) and surrounded derivatives are highly
vacuolated and contain a nucleus (N) in close association with a well-developed
chloroplast (P). Bar = 4.0 μm.

Fig. 1.4. *Dendroceros japonicus* Steph. Light microscope LM cross-section of growing
notch through hemidiscoid apical cell (AC) and immediate derivative
(indistinguishable from each other) surrounded by mucilage (M). This thallus
may be in the process of branching. The single layered wing extends to either
side. Bar = 10.0 μm.

Among land plants, hornworts possess a bewildering array of structural features that are algal-like (chloroplast structure and biochemistry), liver-wort-like (antheridial development and structure, apical cell architecture), moss-like (columella, stomates), and yet others that suggest affinities with seedless vascular plants (stomates, sunken gametangia, embryo development) (Campbell 1895, Bartlett 1928, Renzaglia 1978, Vaughn *et al.* 1992). These features will be examined at the cellular and tissue levels and the evolutionary significance of each will be considered.

The vegetative gametophyte of hornworts is a flattened thallus, with or without a thickened midrib (Figs. 1.1, 1.2). Growing regions that contain solitary apical cells and immediate derivatives typically are located in thallus notches and are covered by mucilage that is secreted by epidermal cells (Figs. 1.3, 1.4). Growth forms are correlated with apical cell geometry. The wedge-shaped apical cell of most taxa segments along four cutting faces: two lateral, one dorsal, and one ventral (Fig. 1.5). The resulting growth form tends to be orbicular and the thallus in cross-section gradually narrows from the center to lateral margins. In comparison, the hemidiscoid apical cell (Figs. 1.4, 1.6) of *Dendroceros* cuts along two lateral and one basal face and is responsible for producing a ribbon-shaped thallus with an enlarged midrib (Fig. 1.2). Aside from anthocerotes, wedge-shaped and, less commonly, hemidiscoid apical cells occur only in complex and simple thalloid liverworts (Crandall-Stotler 1980, Renzaglia 1982).

At the cellular level, hornworts typically contain solitary chloroplasts with central pyrenoids and channel thylakoids, features shared with algae but found in no other land plants (Duckett & Renzaglia 1988, Vaughn *et al.* 1992) (Figs. 1.7, 1.9, 1.10). Even the apical cell and immediate derivatives contain well-developed chloroplasts that are intimately associated with the nucleus (Fig. 1.3). Rubisco localizations in the pyrenoid and lack of grana end membranes (Figs. 1.10, 1.11) that characterize land plants may be viewed as plesiomorphies and further suggest ties with charophytes (Vaughn *et al.* 1990, 1992). Thylakoids traverse the pyrenoid and separate typically lens-shaped subunits, giving the appearance of

Fig. 1.5. Wedge-shaped apical cell characteristic of most hornworts. The two triangular lateral cutting faces, one rectangular dorsal cutting face, and one rectangular ventral cutting face produce a total of four derivatives in spiraled rotation. Modified from Renzaglia (1978).

Fig. 1.6. Hemidiscoid apical cell of *Dendroceros* with two semicircular lateral cutting faces and a single rectangular basal cutting face. Modified from Renzaglia (1978).

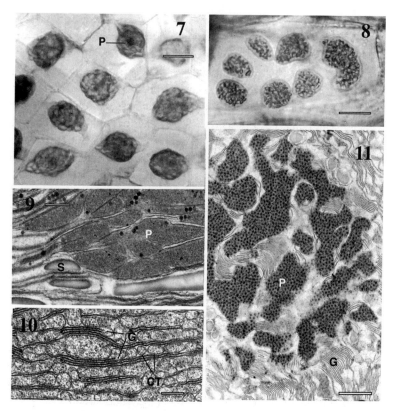

Fig. 1.7. *Phaeoceros laevis* subsp. *carolinianus*. Upper epidermal cells of gametophyte, each with single lens-shaped plastid containing abundant starch and central pyrenoid (P). Bar = 20.0 μm.

Fig. 1.8. *Megaceros aenigmaticus* Schust. Internal cell of thallus with seven starch-filled plastids that lack pyrenoids; the plastid on the right may be preparing for division. Bar = 10.0 μm.

Fig. 1.9. *Folioceros fuciformis* Bharadw. TEM of pyrenoid (P) consisting of lens-shaped subunits delimited by thylakoids and scattered pyrenoglobuli. Starch (S) surrounds the pyrenoid and narrow grana stacks traverse the plastid. Bar = 0.5 μm.

Fig. 1.10. *Folioceros appendiculatis* Haseg. TEM of plastid showing grana (G) that lack end membranes and associated channel thylakoids (CT). Bar = 0.2 μm.

Fig. 1.11. *Dendroceros tubercularis*. TEM of spherical pyrenoid (P) with irregularly shaped subunits containing uniform electron-dense inclusions. Thylakoids, including grana, interrupt the pyrenoid and stroma grana (G) lack end membranes. Bar = 0.5 μm.

"multiple pyrenoids" (Fig. 1.9). The shape of pyrenoid subunits and the existence/location of pyrenoid inclusions are taxonomically informative features of the chloroplast. For example, chloroplast structure in *Dendroceros* deviates from that of the "typical" hornwort in that the pyrenoid is spherical and contains irregularly-shaped subunits with regularly spaced electron-opaque inclusions (Fig. 1.11). Chloroplasts of *Megaceros* further diverge from the anthocerote norm in that they occur in multiples and lack a pyrenoid (Fig. 1.8). Chloroplasts in this genus may number as many as 14 per internal thallus cell (Burr 1969). As in other land plants, Rubisco is scattered amongst starch grains in the chloroplast stroma of *Megaceros*.

Cell division in all hornworts is monoplastidic and involves plastid division and morphogenetic migration that is tightly linked with nuclear division (Brown & Lemmon 1990, 1993). Spindle microtubules originate from an aggregation of electron-dense material at the poles, suggesting the vestige of algal-like centriolar centrosomes (Vaughn & Harper 1998). Further investigations into cell cycle and cytoskeletal proteins are required to specify homologies of this structure to the polar bodies of liverworts and to centrosomes of other eukaryotes.

The thickened thallus of the hornwort gametophyte lacks internal differentiation (Fig. 1.12), except for the occurrence of rather extensive schizogenous mucilage canals in species of *Anthoceros* (Fig. 1.13) and *Apoceros* (Schuster 1987). In some taxa, especially *Megaceros*, the epidermal cells are smaller than internal parenchyma cells (Fig. 1.12). Unlike in the sporophyte, all epidermal cells of the gametophyte contain chloroplasts (Fig. 1.15). Mucilage-filled cells are abundant and scattered amongst photosynthetic parenchyma in most taxa (Fig. 1.12). Band-like wall thickenings (Fig. 1.14) and primary pit fields may occur in thallus cells subtending archegonia and later the sporophyte foot (Leitgeb 1879, Proskauer 1960, Renzaglia 1978). Ultrastructural observations of these cells will enable an evaluation of their potential role in food transport. Vesicular–arbuscular endomycorrhizae are common in internal thallus cells of most taxa (Renzaglia 1978, Ligrone 1988). Rhizoids are unicellular, smooth and may have branched tips (Hasegawa 1983). They are typically ventral in position and occasionally they develop from the outer cell derived from a periclinal division of an epidermal cell (Fig. 1.14).

A distinctive feature of anthocerotes is the occurrence of apically-derived mucilage clefts (Figs. 1.2, 1.15) on the ventral thallus through which *Nostoc* enters the plant and becomes established as a colonial endosymbiont. Two cells that resemble guard cells surround the opening

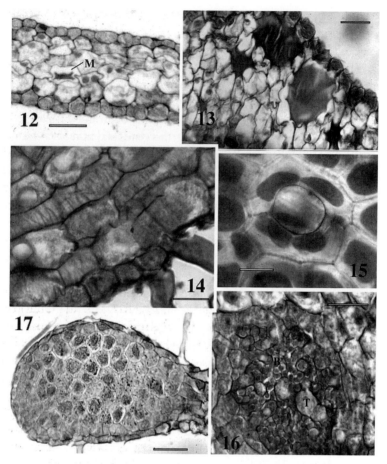

Fig. 1.12. *Megaceros aenigmaticus.* LM cross-section of undifferentiated, simple thallus. Epidermal cells are smaller than internal cells of which one is mucilage-filled (M). Bar = 25.0 μm.

Fig. 1.13. *Anthoceros punctatus* L. Numerous mucilage-containing schizogenous cavities below the upper epidermis of the gametophyte. Bar = 25.0 μm.

Fig. 1.14. *Dendroceros* sp. LM cross-section of the gametophyte thallus showing several internal cells with band-like wall thickenings which often subtend archegonia and later sporophytes. Rhizoids are unicellular and smooth and may originate from a surface cell after periclinal division of the epidermal cell (left rhizoid). Bar = 25.0 μm.

Fig. 1.15. *Megaceros aenigmaticus.* Surface view of mucilage cleft in ventral epidermis of gametophyte. Both cells contain recently divided plastid. Bar = 10.0 μm.

Fig. 1.16. *Phaeoceros laevis* subsp. *carolinianus.* LM section through a *Nostoc* colony. Cells of the hornwort thallus (T) penetrate the colony and are interspersed amongst the small, spherical cells of the cyanobacterium (B). Bar = 20.0 μm.

Fig. 1.17. *Anthoceros punctatus.* LM longitudinal section of tuber. This swollen thallus protuberance contains abundant food reserve. Bar = 50.0 μm.

which, once formed, lacks the ability to open and close. Although considered by some authors (e.g., Schuster 1992) homologous to the stomates in the sporophyte of some anthocerotes, this interpretation is likely inaccurate due to the function and ventral location of these clefts. As the *Nostoc* colony increases in size, the internal chamber in which the alga is housed increases in size and fills with mucilage secreted by the hornwort. Thallus outgrowths penetrate the algal colony (Fig. 1.16).

As in most bryophytes, asexual reproduction is widespread in anthocerotes. Indeed, taxa such as *Megaceros aenigmaticus* in which the male and female plants are geographically separated into different watersheds rely entirely on vegetative reproduction for dissemination and propagation (Renzaglia & McFarland 1999). Fragmentation, regenerant formation, and gemmae production have been reported in various taxa. Under adverse environmental conditions, some species of hornworts produce nutrient-filled tubers as perennating bodies (Goebel 1905, Parihar 1961, Renzaglia 1978) (Fig. 1.17).

Gametangia are produced along the dorsal thallus midline. Archegonia are exogenous, i.e., they develop from surface cells, and ultimately they are sunken in thallus tissue (Fig. 1.18). In addition to the central cells of the archegonium, the archegonial initial gives rise to a one- to two-layered venter (Fig. 1.19) and six rows of neck cells that slightly protrude from the thallus surface and are overarched by a layer of mucilage (Figs. 1.20, 1.21). Two to four cover cells cap the canal until the egg reaches maturity at which time they are dislodged from the neck (Fig. 1.20). Venter cells are smaller than the surrounding parenchyma; they are less vacuolated and contain a prominent nucleus with nucleolus and an associated flattened plastid (Fig. 1.19). The central cells of the archegonium typically consist of four to six neck canal cells and a ventral canal cell and egg (Fig. 1.18). The ventral canal cell and egg are similar in ultrastructure; both contain dense cytoplasm including abundant lipid reserve and a single elongated undifferentiated plastid that encircles the nucleus (Fig. 1.19). The ventral canal cell persists beyond degradation of the neck canal cells and disintegrates when the egg reaches maturity. Both cells are surrounded by callose.

Antheridia are referred to as endogenous because they develop from subepidermal cells and ultimately are positioned within internal thallus chambers (Fig. 1.22). In other embryophytes, antheridia develop from single epidermal cells. The difference in hornworts is that the antheridial initial is located at the base of a schizogenous antheridial chamber, and

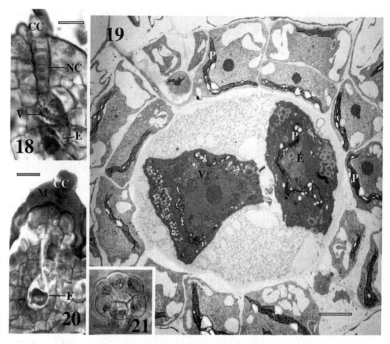

Fig. 1.18. *Phaeoceros laevis* subsp. *carolinianus*. Longitudinal section of an archegonium with two cover cells (CC), six neck canal cells (NC), ventral canal cell (V), and egg cell (E) containing nucleus. Bar = 20.0 μm.

Fig. 1.19. *Phaeoceros laevis* subsp. *carolinianus*. TEM oblique cross-section of venter of a nearly mature archegonium containing ventral canal cell (V) and egg cell (E); both are embedded in a callosic matrix and contain an elongated plastid near a large central nucleus (visible in ventral canal cell) and dense lipid-filled cytoplasm. The surrounding venter is one- or two-layered. Venter cells are small, each containing less-dense cytoplasm with small vacuoles and an elongated plastid (P) adjacent to the nucleus. Bar = 4.0 μm.

Fig. 1.20. *Dendroceros japonicus*. Longitudinal section of mature archegonium that projects from the dorsal thallus, is overarched by mucilage (M), and has discharged the cover cells (CC). The venter contains an egg cell (E). Bar = 20.0 μm.

Fig. 1.21. *Phaeoceros laevis* subsp. *carolinianus*. Surface view of mature archegonium containing six rows of neck cells each with a single prominent chloroplast. Bar = 20.0 μm.

not at the thallus surface. During hornwort evolution, there has been an apparent shift in developmental potential from epidermal (layer surrounding the external surface) to epithelial (layer surrounding an internal space) cells, both of which are surface cells that enclose tissue. However, development of the antheridium proper in hornworts resembles that of

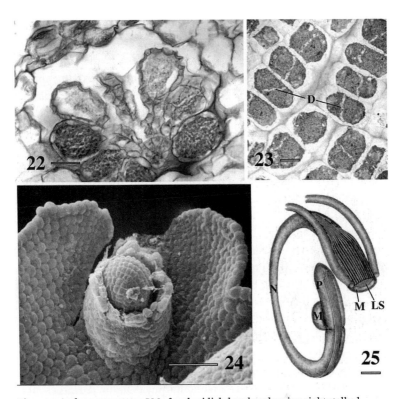

Fig. 1.22. *Anthoceros punctatus.* LM of antheridial chamber showing eight stalked antheridia. Bar = 20.0 μm.

Fig. 1.23. *Notothylas orbicularis* (Schwein.) Sull. TEM showing diagonal final mitotic division (D) that produces pairs of polygonal spermatids. Bar = 3.0 μm.

Fig. 1.24. *Dendroceros tubercularis.* SEM of dorsal thallus with ruptured, projecting chamber containing a single antheridium. Bar = 0.1 mm.

Fig. 1.25. *Phaeoceros laevis* subsp. *carolinianus.* Diagrammatic illustration of biflagellated sperm cell (modified from Carothers & Duckett 1980). The locomotory apparatus consists of two flagella that are inserted symmetrically into the cell anterior over a spline of 12 microtubules and an underlying anterior mitochondrion (M). A rim of remnant lamellar strip (LS) lies directly anterior to the spline microtubules. The cylindrical nucleus (N) with central constriction occupies most of the cell length and a round posterior mitochondrion (M) is positioned on the plastid (P) that terminates the cell. Bar = 0.5 μm.

other bryophytes, especially complex thalloid liverworts, in that the antheridial initial elongates without apical cell involvement and four primary spermatogones with eight peripheral jacket initials are produced in the formative stages of organogenesis. Thus, the designation of hornwort antheridia as endogenous refers only to the location of development and not to an inherently different developmental pathway from that in other bryophytes (Renzaglia *et al.* 2000).

Concomitant with antheridial morphogenesis is the growth and development of a two-layered chamber roof from the overlying epidermal initial. One to 25 antheridia (all derived from the same subepidermal cell) are ultimately enclosed in sunken chambers of the thallus (Figs. 1.22, 1.24). Thousands of minute spermatozoids are produced in each antheridium (Figs. 1.23, 1.25). When antheridia are mature, the plastids of the jacket layer typically have been converted to orange-colored chromoplasts (Duckett 1975). The roof of each antheridial chamber ruptures and the jacket cells dissociate, thus liberating the spermatozoids (Fig. 1.24).

Spermatogenesis provides clues to the phylogenetic history of hornworts (Renzaglia & Carothers 1986, Renzaglia & Duckett 1989, Garbary *et al.* 1993, Graham 1993, Vaughn & Renzaglia 1998). During spermiogenesis, pairs of bicentrioles arise de novo at the poles in the cell generation prior to the spermatid mother cell (Vaughn & Renzaglia 1998). Bicentrioles are diagnostic of archegoniates that produce biflagellated sperm cells but the timing of their origin in hornworts is earlier than in other taxa, where these organelles originate in the spermatid mother cell. Because green algal cells typically contain centrioles in all cell generations, this feature in hornworts is interpreted as a plesiomorphy (Vaughn & Harper 1998, Vaughn & Renzaglia 1998). As in *Coleochaete*, liverworts, and some pteridophytes, the final mitotic division in the spermatid mother cell is diagonal (Fig. 1.23). Spermatids develop in pairs and the mature spermatozoid is coiled, biflagellated, and symmetrical. Both flagella insert at the anterior extreme of the cell and are directed posteriorly. Spermatozoids are extremely small (approximately 3.0 μm in diameter) and only contain an anterior mitochondrion, a cylindrical nucleus with mid-constriction, and a posterior mitochondrion associated with a plastid containing one starch grain. Unlike spermatozoids of all other archegoniates which are sinistrally coiled, this cell exhibits a right-handed coil (Fig. 1.25). Because these cells are bilaterally symmetrical, as opposed to the bilateral asymmetry of motile sperm in other embryophytes, the direction of coiling may be inconsequential to swimming performance and thus was free to change during the extended evolutionary history of anthocerotes (Renzaglia *et al.* 2000).

The first division of the zygote is longitudinal and the endothecium of the embryo gives rise to a central columella in the sporophyte, if one exists (Renzaglia 1978). This is in contrast to most mosses and liverworts in which the zygote undergoes a transverse first division and the endothecium gives rise to sporogenous tissue (the notable exception is *Sphagnum*). The foot matures before the remaining histogenic regions. Growth of overarching gametophytic tissue occurs as the embryo develops, thus forming a protective involucre that in most taxa is ruptured with continued maturation of the sporophyte (Fig. 1.1). The involucre remains as a cylinder that surrounds the base of the sporophyte.

Multiple sporophytes are produced by each fertile thallus. At maturity, the sporophyte is differentiated into a bulbous foot embedded in gametophyte tissue and a prominent elongated cylindrical spore-bearing region which includes an epidermis, assimilative layer, sporogenous tissue, columella, and basal meristem (Figs. 1.26–1.29). This meristem is located above the foot and is responsible for the continuous production of sporogenous tissue throughout the growing season. No parallels of this developmental strategy are evident in any other embryophyte. Monosporangiate archegoniates have either generalized growth of the sporophyte (liverworts) or predominantly apical elongation (mosses), while polysporangiate land plants all exhibit apical growth (Kenrick & Crane 1997, Niklas 1997). Spore maturation occurs progressively from the base to the apex of the sporophyte. In *Notothylas*, the basal meristem functions for a limited period; the sporophyte remains small and it is frequently retained within the protective tissue of the gametophyte.

Stomates that resemble those of mosses and tracheophytes occur in the sporophyte of many taxa. Guard cells are characterized by inner (ventral) wall thickenings and apparently they are the only epidermal cells that contain plastids, specifically amyloplasts (Figs. 1.27, 1.28). These features suggest homology with stomates of other embryophytes. However, the lack of developmental, ultrastructural, and physiological studies on anthocerotes preclude an accurate determination of evolutionary origin of these phylogenetically significant structures. Epidermal cells are elongated and may contain transverse annular wall thickenings of unknown function (Fig. 1.27).

The details of the structure of the foot is species dependent and highly variable. For example, palisade-like epidermal cells surround the relatively small foot of *Anthoceros* while the massive foot of *Megaceros* contains thousands of small undifferentiated cells (Renzaglia 1978). In all cases, placental transfer cells that contain elaborate wall labyrinths are restricted to the gametophyte tissue, a feature that is shared with *Coleochaete* and rare in

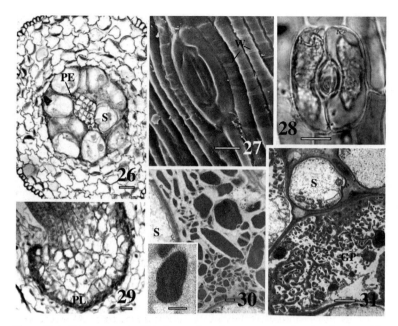

Fig. 1.26. *Anthoceros punctatus.* LM of sporophyte in cross-section showing from outside to inside: epidermis, assimilative layer, sporogenous zone with spore mother cells (S) and pseudoelaters (PE), and 16-celled columella. Bar = 30.0 μm.

Fig. 1.27. *Phaeoceros laevis* subsp. *carolinianus.* SEM of closed stomate in epidermis of sporophyte. Inner (ventral) walls of guard cells are thickened. The elongated surrounding epidermal cells contain band-like wall thickenings (W). Bar = 30.0 μm.

Fig. 1.28. *Phaeoceros laevis* subsp. *carolinianus.* LM of open stomate in sporophyte epidermis showing massive starch-filled plastids in guard cells. Bar = 30.0 μm.

Fig. 1.29. *Anthoceros punctatus.* LM of foot and basal meristem. Palisade-like epidermal cells (PL) of the foot adjacent to gametophyte. Bar = 30.0 μm.

Fig. 1.30. *Folioceros fuciformis.* TEM of protein crystals between sporophyte cells (S) that lack wall ingrowth and gametophyte (not visible) generations. Bar = 0.5 μm. Inset: Higher magnification showing substructure of protein crystal. Bar = 0.2 μm.

Fig. 1.31. *Folioceros appendiculatis.* TEM of gametophyte cells (GP) of the placenta with elaborate wall ingrowths adjacent to sporophyte cells (S) that lack wall ingrowths. Bar = 2.0 μm.

other bryophytes (Graham 1993, Ligrone *et al.* 1993). A distinctive feature of the hornwort placenta is the occurrence of abundant protein crystals between gametophyte and sporophyte cells in *Phaeoceros, Folioceros, Notothylas,* and some species of *Megaceros* (Ligrone *et al.* 1993, Vaughn & Hasegawa 1993) (Fig. 1.30).

An assimilative (photosynthetic) layer of variable thickness underlies the epidermis and the sporogenous tissue is situated between this layer

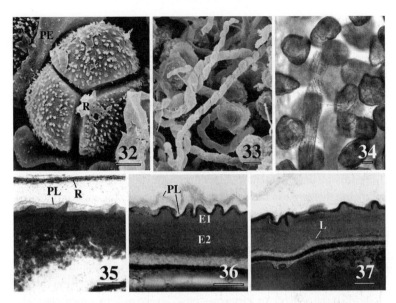

Fig. 1.32. *Phaeoceros laevis* subsp. *carolinianus*. SEM of spore tetrad surrounded by short, smooth pseudoelaters (PE) and still enclosed in the sporophyte. Note remnant spore mother cell wall (R) over the spore surfaces (compare with Fig. 1.35). Bar = 10.0 μm.

Fig. 1.33. *Megaceros flagellaris* Steph. SEM of spores and unicellular pseudoelaters with spiraled wall thickenings. Bar = 20.0 μm.

Fig. 1.34. *Dendroceros* sp. Precocious endosporic germination of spores still enclosed in the sporophyte and surrounding 16-celled columella. Bar = 20.0 μm.

Fig. 1.35. *Notothylas temperata* Haseg. TEM of developing spore enclosed in remnant spore mother cell wall (R). The spore wall consists of an outer exine and an inner exine of flocculent, electron-dense material. A perine-like layer (PL) is being deposited from surrounding spore mother cell wall material. Bar = 0.5 μm.

Fig. 1.36. *Notothylas temperata*. TEM of slightly older spore wall with further deposition of perine-like layer (PL) from spore mother cell wall material. Exine 1 (E1) and exine 2 (E2) overlie an electron-lucent layer and a dense basal layer. Bar = 0.5 μm.

Fig. 1.37. *Notothylas temperata*. TEM of nearly mature spore wall. Dark perine-like layer deposited from spore mother cell wall covers spore. One or two lamellae (L) lie between exine 2 and three inner layers: electron-lucent outer and inner layers, with an electron-dense layer between. Bar = 0.5 μm.

and the columella (Fig. 1.26). Pseudoelaters are interspersed among the spores and are multicellular or unicellular with smooth or spiraled wall thickenings (Figs. 1.32, 1.33). The central columella is persistent and usually consists of 16 cells in cross section (Figs. 1.26, 1.34) but may contain as many as 40 cells in *Megaceros*. Dehiscence typically occurs along two longitudinal lines that originate near the sporophyte tip. Pseudoelaters and the columella facilitate spore separation and assist in dispersal. Aside from preliminary studies of sporogenesis (Brown & Lemmon 1988) and investigations of the placenta (Ligrone *et al.* 1993), the hornwort sporophyte remains unexplored at the ultrastructural level.

Sporogenesis in hornworts resembles that in many other basal embryophytes in that meiosis is monoplastidic. Associated with monoplastidy in archegoniates, but not in green algae, is a unique quadripolar microtubule system that is organized at the plastids and predicts polarity of the two meiotic divisions (Brown & Lemmon 1997).

Spore wall development involves the production of a spore special wall following meiosis and the development of a primexine containing loose fibrils that provides a framework for spore wall deposition (Brown & Lemmon 1988). Initially, two exine layers are deposited: a thin outer layer (exine 1) and a thicker inner layer (exine 2) that forms through the deposition of flocculent electron-dense material (Fig. 1.35). In *Notothylas temperata*, subsequent wall layering involves the deposition of an electron-lucent layer, an electron-dense layer, and an inner electron-lucent zone (Figs. 1.36, 1.37). Curiously, during this development, a heretofore unseen layer of one or two lamellae often occurs at the base of exine 2 (Fig. 1.37). This lamellar zone likely represents compaction of fibrils from the primexine and as such is not homologous to the tripartite lamellae of mosses, liverworts, and pteridophytes.

During the final stages of spore wall development, a thin dark band of fibrous material is laid down on the outer spore surface. This layer is derived from deposition of remnant sporocyte wall and intrasporal septum (Figs. 1.32, 1.35–1.37). Thus, although it is of extrasporic origin, this covering is not a true perine which by definition derives from the inner sporangial wall. The details of sporogenesis and spore wall deposition are insufficiently known among hornwort taxa and these processes require further scrutiny (Brown & Lemmon 1988). Spores remain in tetrads until nearly mature and are dispersed individually.

Spore germination results in the production of a single gametophyte. In most taxa, germination is exosporic, resulting in a globose sporeling that produces an apical cell and flattens with continued development (Renzaglia 1978). Spores may overwinter or remain quiescent until favorable conditions for germination are encountered. In *Dendroceros*, an epiphytic taxon, spores are precocious and initially endosporic. Multicellular "spores" are released from the capsule and develop upon contact with the substrate (Fig. 1.34).

1.3 Anthocerote classification

Interrelationships among genera and delineation of generic boundaries remain problematic in hornworts. Cladistic analyses of morphological and ultrastructural data have been conducted and some of these phylogenetic hypotheses have been used to propose alternate classifications for anthocerotes (Hässel de Menéndez 1988, Hyvönen & Piippo 1993, Hasegawa 1994). Different concepts of classification are demonstrated in Table 1.1 wherein the four taxonomic schemes recognize from five to nine genera. In total, 11 genera of hornworts have been named, of which only six have gained wide recognition: *Anthoceros, Phaeoceros, Folioceros, Notothylas, Megaceros*, and *Dendroceros*. The remaining five genera are defined by anomalies of the spores and pseudoelaters or by specific features of the antheridia or thallus.

As evidenced by the lack of agreement on classification schemes, the hierarchical organization of hornwort taxa remains equivocal. This is in part due to the presence of gradations between taxa and hence poorly delimited generic boundaries and in part to the lack of fundamental developmental and biochemical data to evaluate the value and homology of characters and character states. Although diagnostic features such as endogenous antheridia and an intercalary meristem in the sporophyte are ubiquitous in hornworts, other critical characters such as chloroplast number and presence or absence of stomates and columella show considerable variation. To date, no molecular studies have included enough hornworts to approach questions relating to generic boundaries and familial limits within the group. Clearly, greater taxon sampling is necessary to evaluate generic boundaries and to resolve relationships among anthocerote taxa.

Two schools of thought predominate on evolutionary trends within the anthocerotes. It has been speculated that the simpler sporophyte of *Notothylas* represents a basal form and within the group there has been a progressive elaboration of the sporophyte. This concept was promoted by Mishler and Churchill (1984) and is supported by characters interpreted as retained plesiomorphies, many of which are shared with other bryophytes, i.e., indehiscent sporophyte, determinate sporophyte growth, poorly developed to absent columella, and poorly developed pseudoelaters (Graham 1993, Hyvönen & Piippo 1993, Hasegawa 1994). The more traditional view is that the sporophyte of *Notothylas* is highly specialized and derived through extensive evolutionary reduction (Campbell 1895,

Table 1.1. *Four alternative classifications of the hornworts*

Hässel de Menéndez (1988)
Anthocerotales Limpricht in Cohn
  Anthocerotaceae Dum.
    Anthoceros (Mich.) L.
    Sphaerosporoceros Hässel
  Notothyladaceae (Milde) Müll.
    Subfam. Notothyladoideae Grolle
      Notothylas Sull.
    Subfam. Phaeocerotoideae Hässel
      Phaeoceros Prosk.
Foliocerotales Hässel de Menéndez
  Foliocerotaceae Hässel
    Folioceros Bharadwaj
Leiosporocerotales Hässel de Menéndez
  Leiosporocerotaceae Hässel de Menéndez
    Leiosporoceros Hässel de Men.
Dendrocerotales Hässel de Menéndez
  Dendrocerotaceae (Milde) Hässel de Men.
    Dendroceros Nees
    Megaceros Campbell

Hyvönen & Piippo (1993)
Anthocerotales Limpricht in Cohn
  Anthocerotaceae Dum.
    Anthoceros (Mich.) L.
    Folioceros Bharadwaj
    Leiosporoceros Hässel de Men.
    Mesoceros Piippo
    Phaeoceros Prosk.
    Sphaerosporoceros Hässel de Men.
  Dendrocerotaceae (Milde) Hässel de Men.
    Dendroceros Nees
    Megaceros Campbell
Notothylales
  Notothyladaceae (Milde) Müll.
    Notothylas Sull.

Schuster (1992)
Anthocerotales Limpricht in Cohn
  Anthocerotaceae Dum.
    Subfam. Anthocerotoideae Dumort.
      [a]*Aspiromitus* Steph. emend. Rink
        Subgen. *Aspiromitus*
        Subgen. *Folioceros*
      [a]*Anthoceros* (Mich.) L.
        Subgen. *Anthoceros*
        Subgen. *Leiosporoceros*
    Subfam. Dendrocerotoideae Schuster
      Megaceros Campbell
        Subgen. *Megaceros*
        Subgen. *Notoceros* Schuster
      Dendroceros Nees
        Subgen. *Dendroceros*
        Subgen. *Apoceros* Schuster
    Subfam. Notothyladoideae Grolle
      Notothylas Sull.

Hasegawa (1994)
Anthocerotales Limpricht in Cohn
  Notothyladaceae (Milde) Müll.
    Notothylas Sull.
  Anthocerotaceae Dum.
    Subfam. Anthocerotoideae
      Anthoceros (Mich.) L.
        Subgen. *Anthoceros*
        Subgen. *Folioceros*
      Leiosporoceros Hässel de Men.
      Hattorioceros Hasegawa
      Phaeoceros Prosk.
    Subfam. Dendrocerotoideae
      Dendroceros Nees
        Subgen. *Dendroceros*
        Subgen. *Apoceros* Schuster
      Megaceros Campbell
      Notoceros (Schuster) Hasegawa

Note: [a] This author recognizes *Aspiromitus* in place of *Anthoceros* and *Anthoceros* in place of *Phaeoceros*

Proskauer 1960, Renzaglia 1978, Schofield 1985, Schuster 1992). This hypothesis is supported by recent morphological and molecular analyses (Renzaglia *et al.* 2000). The interpretation of directionality of evolutionary modifications within the sporophyte generation is central to the question of deep-seated relationships among land plants. If, indeed, the anthocerotes are the earliest divergent embryophytes, then the basalmost hornwort may represent the oldest extant genus of land plant.

1.4 Conspectus

Hornworts are unique among embryophytes in key morphogenetic characters. Diagnostic morphological features of the group includes chloroplast structure, endogenous antheridia, details of the microtubule organizing center during mitosis, sperm cell architecture, sporophyte growth from a basal meristem, placental transfer cells restricted to the gametophyte generation, and non-synchronized sporogenesis. A striking contrast of hornworts when compared with mosses and liverworts is the lack of organized external appendages. No leaves, scales, slime papillae, or superficial gametangia evolved in the group. Perhaps it was precisely this strategy that enabled hornworts to cope with desiccation and to facilitate sexual reproduction. That is to say that adaptive innovations in anthocerotes involved internalization of organs and processes. Sex organs and *Nostoc* colonies are embedded within the thallus and mucilage production and photosynthetic activity are localized in relatively undifferentiated thallus parenchyma. Small size and associated rapid life cycle, coupled with internal elaboration of structures, may be the key to the persistence of this relatively isolated taxon through the millions of years it has existed.

An understanding of episodically changing organismal characters is critical in determining character state homology and in clarifying the phylogenetic history of bryophytes. Information on the developmental anatomy and ultrastructure is lacking for phylogenetically significant structures and processes in hornworts, including stomates, mucilage clefts, cell cycle phenomena, embryology, and sporogenesis. Such comparative data are necessary to clarify homology and to elucidate directionality of character transformations. This information is essential to test hypotheses of adaptive evolution. By mapping well-described characters on robust phylogenetic trees derived from other data, especially molecular data, we may finally answer some of the more intriguing questions that relate to morphological innovations that accompanied early land invasion.

The hornworts are a small, morphologically distinct group of land plants that diversified early in land colonization. The radiation of bryophyte groups was likely an ancient and rapid process that involved high rates of phenotypic innovation followed by widespread decimation. With the concept that only limited lineages of plants survived the millions of years that followed land invasion, it becomes pertinent to examine thoroughly every aspect of the biology of these relatively simple land plants. At the cellular level, the hornworts exhibit symplesiomorphies with charophytes that are lacking in other embryophytes. These features alone, especially chloroplast ultrastructure and cell cycle characteristics, support hornworts as the basal-most land plant lineage (Renzaglia *et al.* 2000). Further insight on early land colonization strategies will be attain by continued investigation of the ultrastructure, morphogenesis, physiology, biochemistry, and phylogeny of this engaging plant group.

Acknowledgments

This study was supported by NSF grants DEB-9207626 and DEB-9527735. We thank Jessica Lucas for technical help and Drs John Bozzola, Steven Schmitt, and H. Dee Gates at the Center for Electron Microscopy, Southern Illinois University, for assistance and the use of the facility.

REFERENCES

Bartlett, E. M. (1928). The comparative study of the development of the sporophyte in the Anthocerotaceae, with special reference to the genus *Anthoceros*. *Annals of Botany*, **42**, 409–30.

Beckert, S., Steinhauser, S., Muhle, H. & Knoop, V. (1999). A molecular phylogeny of bryophytes based on nucleotide sequences of the mitochondrial sequences of the mitochondrial nad5 gene. *Plant Systematics and Evolution*, **218**, 179–92.

Brown, R. C. & Lemmon, B. E. (1988). Sporogenesis in bryophytes. *Advances in Bryology*, **3**, 159–223.

(1990). Monoplastidic cell division in lower land plants. *American Journal of Botany*, **77**, 559–71.

(1993). Diversity of cell division in simple land plants hold clues to evolution of the mitotic and cytokinetic apparatus in higher plants. *Memoirs of the Torrey Botanical Club*, **25**, 45–62.

(1997). The quadripolar microtubule system in lower land plants. *Journal of Plant Research*, **110**, 93–106.

Burr, F. A. (1969). Reduction in chloroplast number during gametophyte regeneration in *Megaceros flagellaris*. *Bryologist*, **72**, 200–9.

Campbell, D. H. (1895). *The Structure and Development of Mosses and Ferns (Archegoniatae)*. New York: Macmillan.

Carothers, Z. B. & Duckett, J. G. (1980). The bryophyte spermatozoid: a source of new phylogenetic information. *Bulletin of the Torrey Botanical Club*, **107**, 281–97.

Crandall-Stotler, B. J. (1980). Morphogenetic designs and a theory of bryophytes origins and divergence. *Bioscience*, **30**, 580–5.

Duckett, J. G. (1975). An ultrastructural sturdy of the differentiation of antheridial plastids in *Anthoceros laevis. Cytobiologie*, **10**, 432–48.

Duckett, J. G. & Renzaglia, K. S. (1988). Ultrastructure and development of plastids in bryophytes. *Advances in Bryology*, **3**, 33–93.

Garbary, D. J. & Renzaglia, K. S. (1998). Bryophyte phylogeny and the evolution of land plants: evidence from development and ultrastructure. In *Bryology for the Twenty-first Century*, ed. J. W. Bates, N. W. Ashton, & J. G. Duckett, pp. 45–63. Leeds: Maney and British Bryological Society.

Garbary, D. J., Renzaglia, K. S., & Duckett, J. G. (1993). The phylogeny of land plants: a cladistic analysis based on the male gametogenesis. *Plant Systematics and Evolution*, **188**, 237–69.

Graham, L. E. (1993). *Origin of Land Plants*. New York: John Wiley.

Goebel, K. (1905). *Organography of Plants, Especially of the Archegoniate and Spermatophyta. Part II. Special Organography*. Oxford: Clarendon Press.

Hasegawa, J. (1983). Taxonomical studies on Asian Anthocerotae. III. Asian species of *Megaceros. Journal of the Hattori Botanical Laboratory*, **54**, 227–40.

(1994). New classification of Anthocerotae. *Journal of the Hattori Botanical Laboratory*, **76**, 21–34.

Hässel de Menèndez, G. G. (1988). A proposal for a new classification of the genera within the Anthocerotophyta. *Journal of the Hattori Botanical Laboratory*, **64**, 71–86.

Hedderson, T. A., Chapman, R., & Cox, C. J. (1998). Bryophytes and the origins and diversification of land plants: new evidence from molecules. In *Bryology for the Twenty-first Century*, ed. J. W. Bates, N. W. Ashton, & J. G. Duckett, pp. 65–77. Leeds: Maney and British Bryological Society.

Hyvönen, J. & Piippo, S. (1993). Cladistic analysis of the hornworts (Anthocerotophyta). *Journal of the Hattori Botanical Laboratory*, **74**, 105–19.

Kenrick, P. & Crane, P. R. (1997). *The Origin and Early Diversification of Land Plants: A Cladistic Study*. Washington: Smithsonian Institution Press.

Leitgeb, H. (1879). *Untersuchungen über die Lebermoose*, vol. 5, *Die Anthoceroteen*. Graz: Leuschner & Lubensky.

Ligrone, R. (1988). Ultrastructure of a fungal endophyte in *Phaeoceros laevis* (L.) Prosk. (Anthocerotophyta). *Botanical Gazette*, **149**, 92–100.

Ligrone, R., Duckett, J. G., & Renzaglia, K. S. (1993). The gametophyte–sporophyte junction in land plants. *Advances in Botanical Research*, **19**, 231–317.

Malek, O., Lattig, K., Hiesel, R., Brennicke, A., & Knoop, V. (1996). RNA editing in bryophytes and a molecular phylogeny of land plants. *European Molecular Biology Organization Journal*, **15**, 1403–11.

Mishler, B. D. & Churchill, S. P. (1984). A cladistic approach to the phylogeny of the bryophytes. *Brittonia*, **36**, 406–24.

Niklas, K. J. (1997). *The Evolutionary Biology of Plants*. Chicago: University of Chicago Press.

Nishiyama, T. & Kato, M. (1999). Molecular phylogenetic analysis among bryophytes and tracheophytes based on combined data of plastid coded genes and the 18S rRNA gene. *Molecular Biology and Evolution*, **16**, 1027–36.

Parihar, N. S. (1961). *An Introduction to Embryophyta*, 4th edn. Allahabad: Central Book Depot.

Proskauer, J. (1960). Studies on Anthocerotales. VI. *Phytomorphology*, **10**, 1–19.

Renzaglia, K. S. (1978). A comparative morphology and developmental anatomy of the Anthocerotophyta. *Journal of the Hattori Botanical Laboratory*, **44**, 31–90.

 (1982). A comparative developmental investigation of the gametophyte generation in the Metzgeriales (Hepatophyta). *Bryophytorum Bibliotheca*, **24**.

Renzaglia, K. S. & Carothers, Z. B. (1986). Ultrastructural studies of spermatogenesis in the Anthocerotales. IV. The blepharoplast and mid-stage spermatid of *Notothylas*. *Journal of the Hattori Botanical Laboratory*, **60**, 97–104.

Renzaglia, K. S. & Duckett, J. G. (1989). Ultrastructural studies of spermatogenesis in the Anthocerotophyta. V. Nuclear metamorphosis and the posterior mitochondrion of *Notothylas orbicularis* and *Phaeoceros laevis*. *Protoplasma*, **151**, 137–50.

Renzaglia, K. S. & McFarland, K. D. (1999). Antheridial plants of *Megaceros aenigmaticus* in the Southern Appalachians: anatomy, ultrastructure and population distribution. *Haussknechtia Beiheft*, **9**, (RICLEF-GROLLE-Festschrift), 307–16.

Renzaglia, K. S., Duff, R. J., Nickrent, D. L., & Garbary, D. J. (2000). Vegetative and reproductive innovations of early land plants: implications for a unified phylogeny. *Philosophical Transactions of the Royal Society*, in press.

Schofield, W. B. (1985). *Introduction to Bryology*. New York: Macmillan.

Schuster, R. M. (1987). Preliminary studies on Anthocerotae. *Phytologia*, **63**, 193–200.

 (1992). *The Hepaticae and Anthocerotae of North America*, vol. V. Chicago: Field Museum of Natural History.

Vaughn, K. C. & Harper, D. I. (1998). Microtubule-organizing centers and nucleating sites in land plants. *International Review of Cytology*, **181**, 75–149.

Vaughn, K. C. & Hasegawa, J. (1993). Ultrastructural characteristics of the placental region of *Folioceros* and their taxonomic significance. *Bryologist,* **96**, 112–21.

Vaughn, K. C. & Renzaglia, K. S. (1998). Origin of bicentrioles in Anthocerote spermatogenous cells. In *Bryology for the Twenty-first Century*, ed. J. W. Bates, N. W. Ashton, & J. G. Duckett, pp. 189–203. Leeds: Maney and British Bryological Society.

Vaughn, K. C., Campbell, E. O., Hasegawa, J., Owen, H. A., & Renzaglia, K. S. (1990). The pyrenoid is the site of ribulose 1–5-bisphosphate carboxylase/oxygenase accumulation in the hornwort (Bryophyta: Anthocerotae) chloroplast. *Protoplasma*, **156**, 117–29.

Vaughn, K. C., Ligrone, R., Owen, H. A., Hasegawa, J., Campbell, E. O., Renzaglia, K. S., & Monge-Najera, J. (1992). The anthocerote chloroplast: a review. *New Phytologist*, **120**, 169–90.

2

Morphology and classification of the Marchantiophyta

2.1. Introduction

Like other bryophytes, liverworts are small, herbaceous plants of terrestrial ecosystems. They share, with the mosses and hornworts, a heteromorphic life cycle in which the sporophyte is comparatively short-lived and nutritionally dependent on the free-living gametophyte, but differ from both in numerous anatomical features as detailed by Crandall-Stotler (1984). Notable among these is that their sporophytes mature completely within the confines of gametophytic tissue, without differentiation of a meristematic zone, and always lack stomates and a columella. Gametophytes usually grow prostrate on their substrates and are of three fundamental types, 1) a leafy shoot system, 2) a simple thallus, or 3) a complex thallus, with air chambers. Traditionally, liverworts are subdivided into two major groups, the marchantioids and the jungermannioids, based somewhat on these growth forms. For example, complex thalloid organization is restricted to genera of the marchantioid group, while leafy shoot systems are the most common growth form in the jungermannioid group. Only simple thalloid morphologies are expressed in both marchantioid and jungermannioid taxa.

There are an estimated 6000 to 8000 species of liverworts, of which at least 85% are leafy jungermannioids (Schuster 1984*a*). In general, leafy liverworts possess fairly simple stems and two or three rows of unistratose, frequently divided leaves (Fig. 2.1). Some taxa are isophyllous, with all three rows of leaves transversely inserted, but more commonly, they are anisophyllous with a small row of transversely inserted ventral leaves, or amphigastria, and two rows of larger, obliquely inserted lateral leaves. The obliquity of the lateral leaves is due to apical cell tilt, with a dorsal tilt

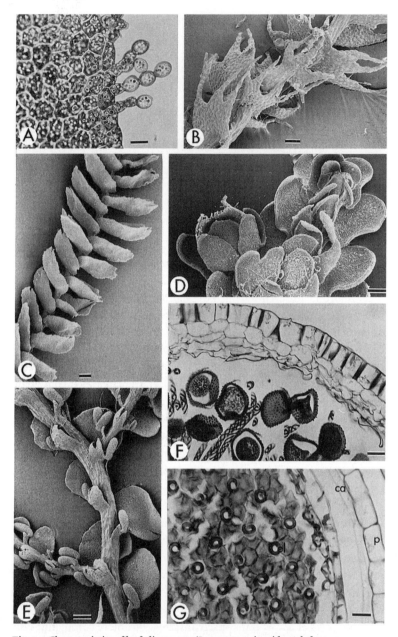

Fig. 2.1. Characteristics of leafy liverworts (Jungermanniopsida, subclass
Jungermanniidae). A. Blastic gemmae produced from cells of the leaf margin,
Scapania nemorea. Bar = 12 μm. B. *Chaetocolea palmata*, quadrifid leaves arranged
in three rows, with lateral leaves succubous and ventral leaves transverse,

producing a succubous insertion (Fig. 2.1B) and a ventral tilt, an incubous insertion (Fig. 2.1E). Simple thalloid morphologies gradate from almost leafy forms to thalli with distinct, multistratose midribs and unistratose wings to multistratose, undifferentiated, often strap-shaped thalli (Fig. 2.2). Ventral appendages may be present or absent, but when present are always in more than one row. In contrast to a simple thallus, a complex thallus is internally differentiated into an upper layer of air chambers, which open dorsally through air pores, and a lower parenchymatous storage zone. The ventral surface of the thallus usually bears two or more rows of leaf-like scales, which overlay horizontal files of pegged rhizoids. In leafy and simple thalloid liverworts, archegonia are borne either in apical clusters or on the dorsal surface of the thallus, respectively, and are typically associated with some type of leafy enclosure, such as a perianth or scales. In complex thalloid hepatics, in contrast, archegonia are produced on the undersurface of erect, highly specialized branches known as archegoniophores (Fig. 2.3).

Despite the heterogeneity of their gametophyte architectures, liverworts have traditionally been recognized to comprise a single natural unit, comparable in rank to mosses and hornworts (Grolle 1983, Schuster 1984a, Stotler & Crandall-Stotler 1977). As a consequence of their cladistic analysis of morphological characters, Kenrick and Crane (1997) propose that liverworts comprise the phylum, or division, Marchantiophyta, based on evidence of several shared sporophyte characters, as will be detailed below. Recently, however, Bopp and Capesius (1998) have concluded from an analysis of 18S rRNA gene sequences that liverworts do not form a natural unit, but are instead polyphyletic. They have even proposed a new classification (Capesius & Bopp 1997) in which the marchantioids have

Caption for **Fig. 2.1.** (*cont.*)
ventral view. Bar = 50 μm. C. *Mytilopsis albifrons*, leaves arranged in two rows, transverse to slightly incubous, equally bilobed and complicate, dorsal view. Bar = 100 μm. D. *Porella platyphylla*, with female branch on the left; lateral leaves complicate, unequally bilobed and incubous; ventral leaves smaller, undivided and transverse, ventral view. Bar = 250 μm. E. *Frullania duricaulis*, with exogenous, leaf modified type branches of the *Frullania*-type; lateral leaves three-parted with a dorsal lobe, inflated ventral lobule and stylus, incubous insertion; ventral leaves smaller, shortly bifid. Bar = 100 μm. F. *Porella platyphylla*, transverse section of the capsule, showing the multistratose wall, with I band thickenings in the outer wall cells, and elaters randomly dispersed among the spores. Bar = 30 μm. G. *Jubula pennsylvanica*, transverse section of the capsule, still surrounded by the calyptra (ca) and perianth (p), showing spore tetrads and vertically aligned elaters (el) confined by a bistratose capsule wall. Bar = 30 μm.

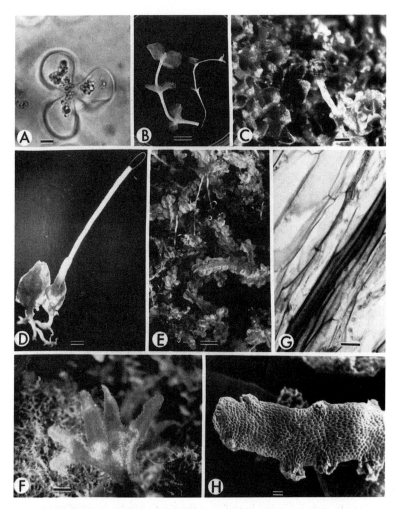

Fig. 2.2 Characteristics of simple thalloid liverworts (Jungermanniopsida, subclass
Metzgeriidae). A. *Radula obconica*, deeply furrowed sporocyte, diagnostic of all
Jungermanniopsida. Bar = 5 μm. B. *Phyllothallia nivicola*, a simple thalloid
taxon with opposite "leaves" and long internodes, dorsal view. Bar = 15 mm.
C. *Fossombronia foveolata*, illustrating an irregular capsule dehiscence pattern
(center). Bar = 1.5 mm. D. *Haplomitrium hookeri* var. *minutum*, showing a
subterranean stolon system, leafy shoots, and a mature sporophyte emerging
from the fleshy true calyptra. Bar = 2 mm. E. *Noteroclada confluens*, with
undivided, succubous "leaves," dorsal view. Bar = 5 mm. F. *Jensenia decipiens*, a
dendroid taxon in which the thallus possesses a central strand and marginal
teeth. Bar = 2 mm. G. *Symphyogyna brasiliensis*, longtudinal section through the
central strand of water-conducting cells. Bar = 20 μm. H. *Riccardia palmata*,
terminal branch of thallus, bearing several female branches, ventral view.
Bar = 100 μm.

been separated from other hepatics and established as an independent group, coordinate with all other bryophytes. To date, the question of how the two major lineages of liverworts are related to each other remains equivocal.

Additional molecular studies such as those of Hedderson *et al.* (1996, 1998) on the 18S rRNA sequence and Lewis *et al.* (1997) on the *rbcL* sequence resolve a single marchantioid/jungermannioid clade, but with only weak support, and in the latter case only when the data are weighted for transition/transversion bias. In an unweighted parsimony analysis of the *rbcL* data set, in fact, tree topology supports the hypothesis of Bopp and Capesius (1998). It is interesting, however, that branch lengths, compiled from 18S rRNA sequences, are significantly longer in the marchantioid clade than the jungermannioid clade (Bopp & Capesius 1998), while those compiled from *rbcL* sequences are significantly shorter in the marchantioid clade than in the jungermannioid clade (Lewis *et al.* 1997). Such lineage-specific branch length differences, of course, confound the phylogenetic inferences that can be deduced from these gene trees.

Morphological data, which can reflect episodic changes associated with major radiation events, such as those that occurred during early land plant evolution, are often more suitable to resolve deep-level phylogenetic relationships than are stochastic molecular sequences (Lewis *et al.* 1997). To address the questions raised by molecular studies, in this chapter we will review the significant diagnostic characters of liverworts, postulated to support their monophyly. We will then examine the diversity of these characters in a cladistic framework to identify potential phylogenetic trends and relationships among liverwort taxa. Finally, a classification scheme that will catalogue, or partition, the morphological diversity of hepatics into a hierarchy of phylogenetic units will be generated, based on this combination of intuitive and cladistic approaches.

2.2 Diagnostic characters of the Marchantiophyta

In contrast to the heterogeneity of gametophyte architectures, the sporophytes of liverworts are rather homogeneous in organization. In all liverworts the sporophyte develops entirely within the confines of gametophytic tissue (Fig. 2.4), derived either solely from the archegonium (= true calyptra), solely from the female gametophore (= solid perigynium), or from a combination of the two (= shoot calyptra). The extent to which the sporophyte is embedded in the gametophyte depends upon

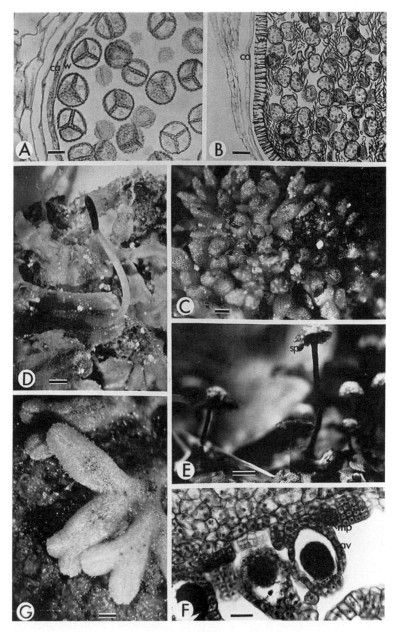

Fig. 2.3. Characteristics of complex thalloid liverworts (Marchantiopsida).
A. *Ricciocarpos natans*, longitudinal section of a capsule, still confined by the
calyptra (ca), showing the fragile, unistratose capsule wall (w) and unfurrowed
spore tetrads. Bar = 30 μm. B. *Conocephalum conicum*, longitudinal section of a

the activity of secondary meristems residing below the female inflorescence. For example, in the Lejeuneaceae as the embryo develops after fertilization, only cells of the stalked archegonium divide, and a true
calyptra is formed. In shoot calyptras and perigynia, a zone of meristematic cells just below the archegonial cluster becomes active after fertilization to form the tissues that surround the embryo either in part or
completely. If the meristematic zone is active for only a short period, as in
the case of a shoot calyptra, the upper part of the embryo will be surrounded by cells of archegonial venter origin, and the remnants of unfertilized archegonia will be elevated part way up the calyptra base. When a
perigynium is formed, on the other hand, many elements of the original
female inflorescence, including unfertilized archegonia, scales, and/or
bracts, will be scattered over the apex of the embedding structure. In
addition, there are usually other structures, associated with the female
inflorescence, that may enlarge after fertilization to surround the sporophyte, such as perianths (Fig. 2.1D), pseudoperianths (Fig. 2.4B), scales,
and paraphyllia. These structures enhance the water-retaining capacity of
the inflorescence, thereby increasing fertilization potential, and in the
case of perianths and pseudoperianths provide additional protection for
the developing sporophyte.

Although variation occurs in embryology, in all liverworts the sporophyte is determinant and excepting the Ricciales, is differentiated into a
foot, which forms a placental zone with the gametophyte, a seta, comprised of thin-walled, parenchymatous cells, and a sporangium or
capsule (Fig. 2.4). In almost all liverworts, after sporogenesis is com-
pleted, the fragile cells of the seta elongate up to 20 times their original

Caption for **Fig. 2.3.** (*cont.*)
capsule, still enclosed by the calyptra (ca), showing the unistratose capsule wall
(w), with transverse annular thickening bands, surrounding the mix of spores
and elaters. Bar = 30 μm. C. *Sphaerocarpos texanus*, bearing sporophytes within
bottle-shaped "involucres"; a small clump of male plants is indicated at *.
Bar = 1.5 mm. D. *Monoclea gottschei* subsp. *elongata*, with a mature sporophyte
emerging from its "involucre" at the apex of the thallus. Bar = 4 mm.
E. *Marchantia paleacea*, with a mature sporophyte (sp) emerging from the
calyptra, located on the underside of the archegonial receptacle that tops the
elongated archegoniophore; a very short antheridiophore can be seen just to
the left of this branch. Bar = 2 mm. F. *Marchantia polymorpha*, longitudinal
section through the archegonial receptacle after fertilization has taken place,
illustrating young sporophytes inside archegonial venters (av) and
marchantioid "pseudoperianths" (mp) enclosing each individual
archegonium. Bar = 20 μm. G. *Riccia hirta*, with embedded sporophytes
(darkened areas) near the middle of each thallus fork. Bar = 500 μm.

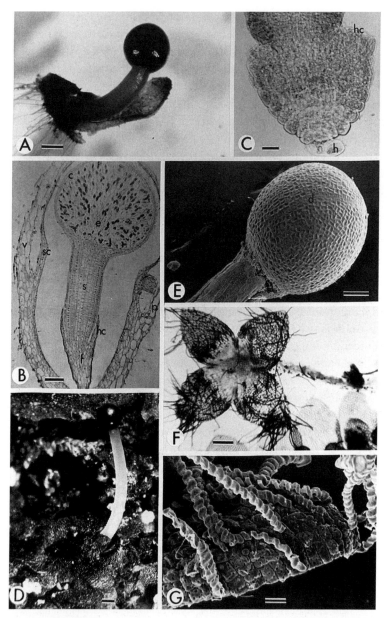

Fig. 2.4. The sporophyte generation, general structure. A. Intact sporophyte of *Pellia appalachiana*, prior to seta elongation; the shoot calyptra, a remnant of which lies to the right of the seta, has been cut away to show the sporophyte. Bar = 300 μm. B. Vertical longitudinal section of an immature sporophyte of *Pellia epiphylla*, showing the foot (f) with a well-developed haustorial collar (hc),

length, elevating the capsule up and out of the enclosing gametophytic tissues (Fig. 2.4D). This elongation process is auxin-mediated and can involve the synthesis of additional wall materials (Thomas & Doyle 1976). In marchantioid liverworts, seta elongation is abbreviated or absent in many taxa, but the structure of the unelongated seta is comparable to that of other liverworts. Capsule dehiscence and spore release occur shortly after seta elongation ceases, often within but a few hours of capsule emergence. In the vast majority of liverworts, capsule dehiscence occurs along differentiated dehiscence sutures (Fig. 2.4E). In the majority of taxa two such sutures extend longitudinally from near the capsule base on one side, over the capsule apex and down to the base on the other side, thereby dividing the capsule wall into four sectors, or valves. With drying, the cell walls between the two rows of suture cells tear, and the four valves bend backwards, releasing the mass of spores and elaters (Fig. 2.4F, G).

Although there is some variation in the size and shape of the sporophyte foot (Ligrone *et al.* 1993) and in the number and arrangement of cells in the seta (Schuster 1984*b*), most of the taxonomically informative variation of the sporophyte resides in the structure and correlated dehiscence properties of the capsule wall (Fig. 2.5). In all jungermannioids, except *Haplomitrium*, the capsule wall is comprised of two or more layers of cells, each of which typically displays a specific pattern of darkly pigmented wall thickenings (Fig. 2.5A, B). The thickenings are secondary wall deposits, laid down after expansion of the capsule wall cells is complete, concomitant with the late stages of sporogenesis and elater differentiation (Fig. 2.5C, D). Commonly, in the outer layer of wall cells, the thickenings are deposited as scattered I- or J-shaped bands on the longitudinal, radial

Caption for **Fig. 2.4.** (*cont.*)
unelongated seta (s) and capsule (c); lobed sporocytes and a basal elaterophore (e) are differentiated in the capsule; the sporophyte is enclosed by a shoot calyptra (sc), and a dorsally inserted, flap-like pseudoperianth (p); v = ventral thallus tissue. Bar = 150 μm. C. Intact sporophyte foot of *Paracromastigum bifidum*, showing the elongate, placental cells of the haustorial collar (hc) and the basal, two-celled haustorium (h). Bar = 30 μm. D. Sporophyte of *Pellia appalachiana*, after seta elongation; the torn shoot calyptra can be seen at the base of seta, just emerging from the pseudoperianth. Bar = 300 μm. E. Capsule of *Pellia epiphylla*, showing one of the two dehiscence sutures (d). Bar = 100 μm. F. Four-valved capsule dehiscence, illustrated by *Jubula pennsylvanica*; as characteristic of the Jubulineae, the elaters remain attached to the capsule valves after dehiscence. Bar = 30 μm. G. Inner surface of capsule wall in *Spruceanthus polymorphus*, showing a pitted or fenestrate pattern of thickenings. Bar = 15 μm.

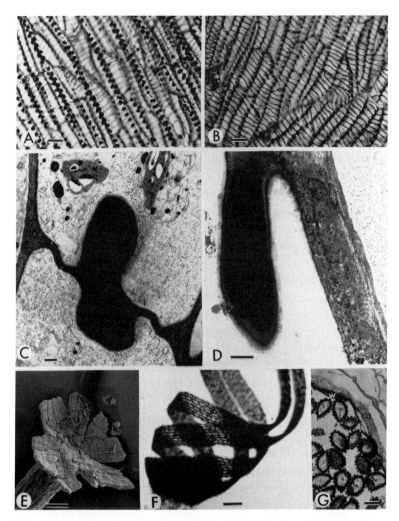

Fig. 2.5. The capsule wall and dehiscence patterns. A. Surface view of outer capsule wall cells in *Aneura maxima*, showing the nodular pattern produced by I-type thickenings. Bar = 30 μm. B. Surface view of inner capsule wall cells in *Aneura maxima*, showing the banded pattern produced by annular or semiannular thickenings. Bar = 30 μm. C. TEM of capsule wall cells of *Haplomitrium blumei*, showing deposition of the annular thickening bands. Bar = 1.3 μm. D. TEM of fully differentiated capsule wall cells of *Haplomitrium blumei* in which cell contents have been hydrolyzed. Bar = 1.0 μm. E. Dehisced capsule of *Fossombronia foveolata*, irregular pattern. Bar = 50 μm. F. Dehisced capsule of *Gyrothyra underwoodiana*, with four long, spirally twisted valves. Bar = 150 μm. G. Mature capsule of *Ricciocarpos natans*; the thin unistratose capsule wall (w) has degenerated, leaving the spores enclosed only by the calyptra (ca) and surrounding thallus tissue. Bar = 30 μm.

walls to produce a nodular pattern in surface view (Fig. 2.5A). Cells in the inner wall layers, in contrast, deposit annular or semiannular U-shaped thickenings that extend from the radial walls across the inner tangential wall, giving it a banded appearance in surface view (Fig. 2.5B). These multiple transverse bands make the inner wall layers more rigid than the outer, and consequently when the capsule opens, the separated valves bend outwards (Ingold 1939). In several taxa, including the North American endemic *Gyrothyra*, the valves are very long and spirally twisted (Fig. 2.5F). Neither the chemical nature of the thickening bands nor the mechanisms regulating their deposition are known. The fact that they are autofluorescent and have a homogeneous, osmiophilic appearance in TEM micrographs, however, suggests that they are composed of polyphenolics.

Within marchantioid liverworts, capsule walls are always unistratose and dehiscence only rarely occurs along four valves. In many taxa the capsule breaks apart into irregular plates of cells, a phenomenon also seen in some of the simple thalloid hepatics (Fig. 2.5E), while in others, e.g., the Aytoniaceae, dehiscence involves an apical operculum. In the Ricciales, the capsule wall actually deteriorates before the spores are mature, leaving them in a cavity lined by the calyptra (Fig. 2.5G); spore dispersal in this group requires thallus degeneration.

An additional, almost universal feature of liverwort sporophytes is the presence of elaters in the capsule. Among embryophytes, only the hornworts possess a somewhat similar structure. Generally regarded as nonhomologous, both the elaters of liverworts and the pseudoelaters of hornworts arise by the division of an archesporial cell into two sister cells, only one of which gives rise to sporocytes. Although the plane of this spore–elater division is transverse in hornworts and longitudinal in liverworts (Schuster 1984*b*), this difference may simply reflect constraints placed on the cells by the size and shape of the archesporium. We would agree with Kenrick and Crane (1997) that all sterile cells produced as sister cells to potential sporocytes, i.e., pseudoelaters, elaters, and nurse cells, should be considered as homologous structures. Despite their similar origins, liverwort elaters differ from the pseudoelaters of most hornworts in being always unicellular, and usually possessing spiral thickenings at maturity; i.e., evolutionary transformations of these initially homologous structures has progressed differently in the two groups.

The subsequent ontogeny of the sporocyte and elater initials also differs between liverworts and hornworts. In some Marchantiopsida, e.g., the Aytoniaceae, the division that produces the two initials is followed directly by sporocyte meiosis and elater differentiation, resulting in a 4:1 spore:elater ratio in the mature capsule. In most hepatics, however, the sporocyte initial divides several times prior to meiosis, while the elater initial remains undivided. As a consequence, in most hepatics, including many genera of the Marchantiopsida, spore:elater ratios are 8:1 or greater. In hornworts, in contrast, the sporocyte initial never divides before the onset of meiosis, but the elater initial does (Schuster 1984c) so the spore:elater ratio is less than 4:1, and in many cases approaches 1:1. With few exceptions, the absence of this division in the elater initial is a defining character of liverworts.

Among liverworts, *Pellia epiphylla* and *Conocephalum conicum*, both of which display endosporic, precocious spore germination, are reported to have spore:elater ratios less than 4:1 (Bischler 1998). *Conocephalum* is the only liverwort that produces spores in rhomboidal or linear rather than tetrahedral arrays through a unique process of cytoplasmic partitioning (Brown & Lemmon 1988, 1990). How this process might modify elater development is unknown. Likewise, whether the modified spore:elater ratio of *Pellia* involves secondary divisions of the elater initials, as in hornworts, is equivocal. In *Pellia* many of the elaters arise from a basal pad of sterile tissue, the elaterophore (Fig. 2.4B), and are, therefore, not homologous to the elaters that are sister cells to the sporocytes.

In the vast majority of hepatics, hygroscopically induced movements of the elaters help to break up the spore mass after the capsule opens (Ingold 1939). In the developing capsule, however, these sterile cells may serve as a dispersed tapetum (Crandall-Stotler 1984). Bartholomew-Began (1991) has shown that the immature elaters of *Haplomitrium* are packed with lipid globules and starch bodies, both of which disappear as elater thickenings are deposited. At the same time, the capsule lumen itself also contains numerous lipid droplets (see also Crandall-Stotler 1984). A similar nutritive function has often been postulated for the nurse cells of the Sphaerocarpales (Parihar 1961). Schuster (1992a: 799) has suggested that the nurse cells are not homologous to elaters because they are not formed as a consequence of a fixed spore–elater division, but instead seem to be sporocytes that fail to undergo meiosis. The studies of Doyle (1962) and Kelley and Doyle (1975), while demonstrating that the nurse cells are tapetal, do not solve the question of origin. Although variation in form occurs, the production of some type of dispersed sterile unicells in the archesporium appears to be a significant defining character of the Marchantiophyta as suggested by Mishler and Churchill (1984).

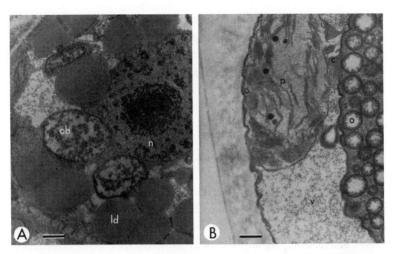

Fig. 2.6. Ultrastructure of liverwort oil bodies. A. Leaf cell of *Haplomitrium mnioides*, showing membrane-bound oil bodies of the small, homogeneous type (ob) as well as lipid droplets (ld) in the cytoplasm; n = nucleus. Bar = 1 μm. B. Leaf cell of *Radula obconica*, showing part of the single, large granular oil body that occupies much of the cell volume; c = cytoplasmic strand, o = oil droplet within the oil body, p = plastid, s = stroma or matrix of the oil body, v = vacuole. Bar = 500 nm.

Two additional synapomorphies of the liverwort clade, identified by Mishler and Churchill (1984), include the presence of lunularic acid as a growth inhibitor and the occurrence of oil bodies in the cells. Of these, the occurrence of oil bodies seems to be the least problematic (see also Kenrick & Crane 1997). The cladistic analysis of Garbary and Renzaglia (1998) also identifies several of the sporophyte characters described above as liverwort apomorphies. Oil bodies, which occur in 90% of liverwort species (Schuster 1966), are miscoded (see Garbary & Renzaglia 1998: table 2) and hence are not included in the list of apomorphies. Excepting this mistake, there is general agreement that the possession of oil bodies is a significant unique diagnostic character of the Marchantiophyta.

The liverwort oil body is, indeed, an intriguing organelle, found in no other group of streptophytes. In contrast to the lipid bodies of other embryophytes, which appear as simple osmiophilic spherules in the cytoplasm, the oil bodies of hepatics are true membrane-bound organelles (Fig. 2.6). They form either directly from the endoplasmic reticulum (Duckett & Ligrone 1995) or by fusion of dictyosome vesicles (Galatis *et al.* 1978a, Apostolakos & Galatis 1998) and contain a diversity of ethereal terpenoid oils, suspended in a carbohydrate- and/or protein-rich matrix (Müller 1905, Pihakaski 1972, Galatis *et al.* 1978*b*, Apostolakos & Galatis 1998). The enclosing membrane of the oil body resembles the tonoplast in having an asymmetrical, tripartite organization (Fig. 2.6). Oil bodies are absent from actively dividing cells, but are formed during early stages of cell maturation (Crandall-Stotler 1981). Frequently, differentiated cells will contain dispersed lipid globules in addition to the terpene-filled oil bodies (Fig. 2.6A). In other plants such globules are composed of triglycerides and serve as stored reserves, but this has not been verified in hepatics.

Variations that occur in oil body size, shape, color, number, distribution, and chemical composition are taxonomically informative characters of liverworts. Unfortunately, because of the volatility of the oils contained in them, they rapidly "disappear" in dried specimens. In fact, their morphology is often modified even during short-term storage in the dark so observations of oil body morphology must be conducted only on freshly collected samples. Ultrastructural evidence confirms that the oil body membrane and internal matrix remain intact for up to six weeks in dark-stored specimens, but the oil droplets within the matrix disappear within a few days (Crandall-Stotler, unpublished data).

Although many different oil body types have been described (Pfeffer 1874, Müller 1939, Gradstein *et al.* 1981, Schuster 1992*b*), their diversity can be partitioned among six categories, or states (Fig. 2.7). In about 10% of hepatics, oil bodies are absent, at least at the level of optical microscopy (Fig. 2.7A); whether these taxa also fail to elaborate the terpenoids which comprise oil bodies has never been tested. Schuster (1966) suggests that the absence of oil bodies in hepatics is a derived character, which has arisen independently in several families, including the Metzgeriaceae and Blasiaceae of the Metzgeriidae and the Cephaloziaceae and Lepidoziaceae of the Jungermanniidae. However, the basal positions of *Blasia*, *Sphaerocarpos*, *Geothallus*, *Riccia*, and *Ricciocarpos* in some analyses (see Fig. 2.8B below) suggest instead that oil bodies may not have been acquired until after the jungermannioid/marchantioid dichotomy and are not synapomorphic for liverworts. Phylogenetic reconstructions inferred from either *rbc*L (Lewis *et al.* 1997) or 18S rRNA (Bopp & Capesius 1998) gene sequences currently support Schuster's (1966) viewpoint, but the taxa included in these analyses may not be sufficient to test the alternative hypothesis. Whether synapomorphic or not, oil bodies are, nonetheless, diagnostic of most liverworts.

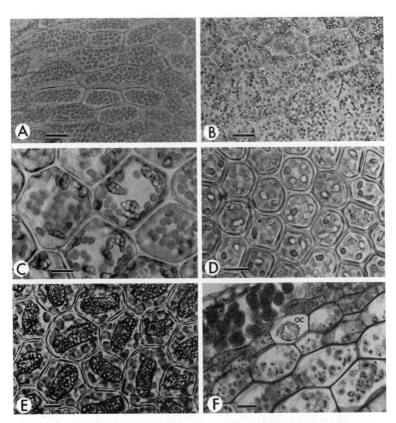

Fig. 2.7. Oil body diversity. A. *Blasia pusilla*, all cells lack oil bodies. Bar = 30 μm.
B. *Austrofossombronia peruviana*, small, homogeneous, shiny oil bodies. Bar = 30
μm. C. *Calypogeia muelleriana*, botryoidal oil bodies. Bar = 15 μm. D. *Frullania
asagrayana*, small, granular to segmented oil bodies. Bar = 10 μm.
E. *Leucolejeunea clypeata*, large segmented oil body, one per cell. Bar = 6 μm.
F. *Marchantia polymorpha*, transverse section of thallus to show idioblastic oil
cell (oc) in the parenchymatous zone. Bar = 10 μm.

In jungermannioid liverworts, oil bodies, when present, are typically
produced in all cells of both the sporophyte and gametophyte generations. Four oil body classes are distinguished, namely, oil bodies that are 1)
small, shiny and more or less homogeneous (Fig. 2.7B), usually very many
per cell; 2) of somewhat larger dimensions, shiny or sometimes pigmented and botryoidal or segmented (Fig. 2.7C), from few to many per
cell; 3) of variable size, with an opaque, granular to globular appearance
(Fig. 2.7D), few to many per cell; or 4) very large, often only one per cell,
opaque and appearing to be made up of many large globules (Fig. 2.7E).

In the marchantioids, oil bodies occur only in scattered idioblastic cells of the gametophyte and show little variation. These cells lack plastids and are almost filled by the single, large, granular to globular, brownish oil body (Fig. 2.7F). In contrast to the oil bodies of jungermannioid liverworts, the matrix of marchantioid oil bodies lacks proteins, but contains polysaccharides and polyphenols, which are absent in the former (Apostolakos & Galatis 1998). Though considered a diagnostic character of most complex thalloid hepatics, idioblastic oil cells of similar structure, called ocelli, also occur in the jungermannioid genus *Frullania* and in selected taxa of the Lejeuneaceae. In these taxa, however, the normal cells of the gametophyte also possess numerous, smaller oil bodies. Within the jungermannioid lineage, only *Treubia* resembles the complex thalloids in having oil bodies restricted to scattered oil cells, but in *Treubia* the oil cells also contain chloroplasts (Schuster & Scott 1966).

Various hypotheses have been formulated regarding oil body function, including suggestions that oil bodies deter herbivores and provide protection from cold and/or UV radiation (Schuster 1984b). There are no data, however, on oil body physiology; e.g., it is not known if the membranes contain V-type pumps like those of vacuoles or secretory vesicles, whether the proteins found in the matrix of jungermannioid liverworts are enzymatic or structural, or if there is active exchange of materials between the oil body and the cytoplasm after its formation. In the absence of such physiological data, speculations regarding the phylogenetic origin of this organelle are unfounded. The nearly ubiquitous occurrence of oil bodies in hepatics, however, at least suggests that this organelle arose early in liverwort phylogenesis and strongly supports the monophyly of the Marchantiophyta.

2.3 Brief history of liverwort classification

Liverwort nomenclature begins with *Species Plantarum* (Linnaeus 1753), in which 41 species were included in the single class of nonseed plants, the Cryptogamia. Excepting *Marchantia chenopoda* L., all of the hepatic binomials in Linnaeus (1753) reference the works of Micheli (1729) and/or Dillenius (1741), which were the sources of many of the names and diagnoses provided by Linnaeus. In *Species Plantarum* (1753) the leafy taxa of the comprehensive genus *Jungermannia* were listed first, followed by the simple thalloid taxa. The complex thalloids, *Targionia*, *Marchantia*, and *Riccia* (with one misplaced simple thalloid, *Blasia*, preceding *Riccia*), concluded the treatment. This arrangement was incorporated into many of the "natural" systems of classification that followed.

In the early 1800s the 25 species described by Linnaeus in *Jungermannia* were segregated among 22 liverwort genera. This partitioning was done independently by the contemporaries, Raddi (1808, 1818), Gray (1821), and Dumortier (1822), resulting, in some cases, in the establishment of three different generic names for the same taxon. As the number of genera increased, new systems of classification emerged. Of the various natural systems of plant classification that were proposed in the first half of the nineteenth century, that of Endlicher (1841) included the first comprehensive system for liverworts. Endlicher adopted the class name Hepaticae, which first appeared in the literature under "Acotyledones – Ordo III" in the system of de Jussieu (1789), and positioned the hornworts between the Ordo Ricciaceae and Ordo Targioniaceae. In addition, his system, which began with the complex thalloids, basically reversed the arrangement of Linnaeus (1753).

Synopsis Hepaticarum, the first world-wide treatment of liverworts, was published by Gottsche, Lindenberg, and Nees (1844–7) a few years later. Still recognized as a first-rate classification (Schuster 1966), the *Synopsis* created by these workers basically adopted the 1841 system of Endlicher but rearranged the taxa back to the sequence of the Linnaean system.

These alternative arrangements, in fact, reflect the two alternative views of liverwort evolution that have been debated by hepaticologists through the years. The "ascending" sequence of Endlicher (1841) hypothesizes that evolution involved a progressive elaboration of the sporophyte, proceeding from the highly simplified capsule of *Riccia* to the massive sporophytes of leafy liverworts like *Lepicolea*. The "descending" system of Gottsche, Lindenberg, and Nees (1844–7) postulates that evolution progressed in a reductive manner so that phylogenetically derived groups are structurally simple. Post-Darwinian classifications that proposed an ascending series basically accepted the antithetic theory of sporophyte origins (Bower 1890) while those arranged in a descending series reflected adherence to the homologous theory of sporophyte origins (Church 1919).

In a series of meticulous developmental studies, Leitgeb (1874–81) provided detailed anatomical descriptions of numerous hepatic taxa and demonstrated that the formation of archegonia in the leafy liverworts terminates further apical cell segmentation ("akrogyne") whereas in simple thalloid hepatics it does not ("anakrogyne"). His findings were quickly incorporated into classification systems, first in the much-improved classification system of Schiffner (1893, 1895) in which the categories Jungermanniales anakrogynae (simple thalloids) and Jungermanniales akrogynae (leafy liverworts) were established. Evans (1939) ultimately provided the formal names Jungermannineae and Metzgerineae with "Jungermanniales acrogynae and anacrogynae" in square brackets for these groups.

As expected, numerous "modified" liverwort classifications have been proposed over the years (see review in Schuster 1966). To summarize, in the early 1800s, a clearer concept of "genus" emerged and genera were sorted into appropriate groupings. Notable in that regard was the system proposed by Trevisan (1877) who established numerous tribes, most of which today equate to families. From the mid 1800s through the early 1900s, there was increased emphasis on better circumscription of families because of the explosive increase in numbers of named genera, especially exotics, as illustrated by Evans's (1939) classification. Since Evans's (1939) treatment, heterogenous families have been further partitioned, perhaps the most notable being the old Ptilidiaceae which has been split into no fewer than 13 families. The challenge that lies ahead is two-fold, first, rigorously to test family composition and second, to organize the families into phylogenetic groups. This task was begun by Schuster (1958, 1966, 1984*a*) who arranged families into suborders and by Schljakov (1972, 1975) who further aligned the suborders into orders and superorders. There are major differences in the intuitively derived hierarchical alignments presented by these two authors, and the circumscription of natural ranks within hepatics remains controversial.

2.4 Cladistic analysis of liverwort diversity

In contemporary systematics cladistic methods provide an objective means for reconstructing phylogenies. In these methods a selection of discrete, phylogenetically important characters are explicitly recorded in a data file for a suite of exemplar taxa, or operational taxonomic units. This file is then subjected to an analysis which seeks to group taxa on the basis of share-derived characters, and to reconstruct the most parsimonious pattern of taxonomic relationships. This pattern is presented as a branching diagram or cladogram. Although the cladogram does not explain the evolutionary trends and processes of a group, it does provide a hypothesis of relationships.

To assess systematic relationships and possible hierarchial rankings that are supported by morphological characters, we have conducted a cladistic analysis of 40 gametophyte and 21 sporophyte characters, distributed among 34 liverwort and four hornwort exemplars, along with *Takakia ceratophylla*. These exemplars include all of the taxa used in the molecular analyses of Hedderson *et al.* (1996), Capesius and Bopp (1997), and Lewis *et al.* (1997). The hornworts and *Takakia* were included as outgroup representatives. The data for this analysis have been compiled primarily from personal study of the exemplar taxa, but published references, including the comprehensive morphological treatments of Leitgeb (1874–81), Cavers (1911), Schuster (1966, 1984*b*), Renzaglia (1978, 1982), Crandall-Stotler (1981, 1984), Schofield and Hébant (1984), and Bischler (1998), have also been consulted. Additional taxon-specific references are indicated in Table 2.1. Other significant character-specific references are as follows: characters 2, 4, 6, 12, and 13 (Crandall 1969); character 14 (Pfeffer 1874, Müller 1939); characters 15, 16, 37, and 38 (Hébant 1977); characters 21 and 22 (Müller 1948*a*), characters 26, 27, and 28 (Knapp 1930); characters 32, 33, and 34 (Ligrone *et al.* 1993); characters 36 and 39 (Douin 1916); characters 41 and 42 (Müller 1948*b*); characters 48 and 50 (Neidhart 1979, Brown & Lemmon 1990); characters 52, 53, and 54 (Nehira 1984). Multistate characters were ordered only when ontogenetic evidence supported a transformation series (Table 2.2). Both unknown and inapplicable characters were coded as ?.

The data file (Table 2.3) was subjected to parsimony searches, using the heuristic search option of PAUP 3.1.1. Duplicate searches were performed in which starting trees were generated by stepwise addition, using the closest addition and random addition options, respectively. For random additions ten replication sequences were run. Regardless of the option chosen for generating starting trees, all searches employed the tree bisection–reconnection (TBR) algorithm for branch swapping, with steepest descent, MULPARS and COLLAPSE (branches of maximum length zero) options in effect. In all cases, the results of the duplicate searches were identical. Both midpoint and outgroup rooting options were applied, the former in analyses of the liverwort network alone, and the latter with three different outgroups specified. Since there is evidence that hornworts occupy a basal position among embryophytes (Malek *et al.* 1996, Garbary & Renzaglia 1998, Hedderson *et al.* 1998) and also share several characters with liverworts, they were initially designated as the outgroup taxa for a rooted analysis. To test Schuster's (1997) hypothesis that *Takakia* is closely related to the liverwort *Haplomitrium*,

Table 2.1. *List of exemplars, clades represented and significant taxon-specific references used to construct the data matrix*

Exemplars	Clade represented	References
Phaeoceros laevis (L.) Prosk.	Anthocerotales (outgroup)	Asthana & Srivastava 1991; Hasegawa 1984, 1988; Renzaglia 1978
Anthoceros punctatus L.	Anthocerotales (outgroup)	Hasegawa 1984, 1988; Renzaglia 1978
Megaceros vincentianus (Lehm. & Lindenb.) Campb.	Anthocerotales (outgroup)	Asthana & Srivastava 1991; Hasegawa 1988; Renzaglia 1978
Notothylas orbicularis (Schwein.) Sull.	Notothyladales (outgroup)	Hasegawa 1988; Lang 1907; Renzaglia 1978
Marchantia polymorpha L.	Marchantiales, Marchantiaceae	Bischler-Causse 1993; Bischler 1998
Asterella tenella (L.) P. Beauv.	Marchantiales, Aytoniaceae	Bischler 1998; Evans 1920
Reboulia hemisphaerica (L.) Raddi	Marchantiales, Aytoniaceae	Bischler 1998; Kachroo 1958; O'Hanlon 1930
Conocephalum conicum (L.) Dumort.	Marchantiales, Conocephalaceae	Bischler 1998; Cavers 1903a; Kobiyama & Crandall-Stotler 1999
Dumortiera hirsuta (Sw.) Nees	Marchantiales, Marchantiaceae	Bischler 1998; O'Hanlon 1934
Lunularia cruciata (L.) Lindb.	Marchantiales, Lunulariaceae	Bischler 1998; Saxton 1931
Ricciocarpos natans (L.) Corda	Ricciales, Ricciaceae	Bischler 1998; Kronestedt 1982
Riccia fluitans L.	Ricciales, Ricciaceae	Bischler 1998
Sphaerocarpos texanus Austin	Sphaerocarpales, Sphaerocarpaceae	Kelley & Doyle 1975
Geothallus tuberosus Campb.	Sphaerocarpales, Sphaerocarpaceae	Doyle 1962
Monoclea gottschei Lindb.	Monocleales, Monocleaceae	Campbell 1954; Cavers 1904; Johnson 1904
Pellia epiphylla (L.) Corda	Fossombroniales, Pelliaceae	Hutchinson 1915; Renzaglia 1982
Fossombronia foveolata Lindb.	Fossombroniales, Fossombroniaceae	Chalaud 1928; Haupt 1920; Renzaglia 1982
Makinoa crispata (Steph.) Miyake	Metzgeriales, Makinoaceae	Schuster 1964; Renzaglia 1982
Metzgeria furcata (L.) Dumort.	Metzgeriales, Metzgeriaceae	Kuwahara 1986, Renzaglia 1982
Pallavicinia lyellii (Hook.) Carruth.	Metzgeriales, Pallaviciniaceae	Haupt 1918; Perold 1993; Smith 1966; Renzaglia 1982
Petalophyllum ralfsii (Wilson) Nees & Gottsche	Fossombroniales, Fossombroniaceae	Cavers 1903b; Renzaglia 1982
Aneura pinguis (L.) Dumort.	Metzgeriales, Aneuraceae	Furuki 1991; Renzaglia 1982
Symphyogyna brasiliensis Nees & Mont.	Metzgeriales, Pallaviciniaceae	Evans 1927; Haupt 1943; Smith 1966
Blasia pusilla L.	Blasiales, Blasiaceae	Leitgeb 1874; Renzaglia 1982
Haplomitrium hookeri (Sm.) Nees	Haplomitriales, Haplomitriaceae	Bartholomew-Began 1991; Lilienfeld 1911

Haplomitrium mnioides (Lindb.) R. M. Schust.	Haplomitriales, Haplomitriaceae	Bartholomew-Began 1991
Treubia tasmanica R. M. Schust. & G. A. M. Scott	Treubiales, Treubiaceae	Schuster & Scott 1966; Renzaglia 1982
Bazzania trilobata (L.) Gray	Jungermanniales, Lepidoziaceae	Schuster 1969
Jungermannia leiantha Grolle	Jungermanniales, Jungermanniaceae	Schuster 1969
Calypogeia muelleriana (Schiffn.) Müll. Frib.	Jungermanniales, Calypogeiaceae	Schuster 1969
Herbertus pensilis (Taylor) Spruce	Jungermanniales, Herbertaceae	Fulford 1963
Jubula pennsylvanica (Steph.) A. Evans	Porellales, Jubulaceae	Crandall-Stotler & Guerke 1980
Lejeunea cavifolia (Ehrh.) Lindb.	Porellales, Lejeuneaceae	Schuster 1980
Frullania asagrayana Mont.	Porellales, Jubulaceae	Schuster 1992b
Lepidozia reptans (L.) Dumort.	Jungermanniales, Lepidoziaceae	Schuster 1969
Lophocolea heterophylla (Schrad.) Dumort.	Jungermanniales, Geocalycaceae	Schertler 1979
Porella pinnata L.	Porellales, Porellaceae	Schuster 1980
Scapania nemorea (L.) Grolle	Jungermanniales, Scapaniaceae	Zehr 1980
Takakia ceratophylla (Mitt.) Grolle	Takakiales (outgroup)	Crandall-Stotler & Bozzola 1988, Renzaglia *et al.* 1997, Schuster 1997
Liverwort ancestor	Ancestral states	Crandall-Stotler 1981

Table 2.2. *Characters and character state codes*

Character number	Character	Type	States
1	Gametophyte growth form	ordered	0 = nodal or "leafy"; 1 = costa and winged thallus; 2 = multistratose thallus
2	Apical cell geometry	unordered	0 = tetrahedral, dorsal; 1 = tetrahedral, ventral; 2 = lenticular; 3 = cuneate; 4 = hemidiscoid
3	Apical cell division plane	unordered	0 = parallel to apical cell wall; 1 = oblique to apical cell wall
4	Derivative first division	unordered	0 = periclinal; 1 = anticlinal
5	Phyllotaxy at the apex	unordered	0 = one-third; 1 = two-fifths; 2 = one-half
6	Apical cell tilt	unordered	0 = absent; 1 = dorsal; 2 = ventral
7	Ventral appendages, mature	unordered	0 = absent; 1 = one row; 2 = two rows; 3 = more than two rows
8	Ventral appendages, form	unordered	0 = short papillae; 1 = long cilia; 2 = foliose
9	Rhizoids	unordered	0 = absent; 1 = unicellular, smooth only; 2 = unicellular, dimorphic
10	Leaf-wing origin	unordered	0 = one central initial; 1 = anodic and kathodic initials; 2 = two acroscopic initials
11	Leaf morphology	unordered	0 = undivided; 1 = two-lobed, simple; 2 = two-lobed, complicate, dorsal; 3 = more than two-lobed; 4 = two-lobed, complicate, ventral
12	Lateral branching	unordered	0 = absent; 1 = dichotomous; 2 = axillary, moss type; 3 = terminal, leaf modified; 4 = terminal, leaf unmodified; 5 = collared, Lejeunea-type; 6 = endogenous; 7 = exogenous, intercalary
13	Ventral branching	unordered	0 = absent; 1 = terminal, leaf modified; 2 = exogenous, delayed; 3 = endogenous
14	Oil body occurrence	unordered	0 = absent; 1 = many, in all cells; 2 = one, in all cells; 3 = in idioblastic cells
15	Water "conducting" cells	ordered	0 = absent; 1 = thin-walled, simple perforation; 2 = thick-walled, simple perforation; 3 = thick-walled, pit fields
16	Food conducting cells	ordered	0 = absent; 1 = specialized parenchyma; 2 = deuters; 3 = leptoids
17	Sexual condition	unordered	0 = dioicous, dimorphic; 1 = dioicous, monomorphic; 2 = monoicous
18	Antheridial origin	unordered	0 = epidermal; 1 = subepidermal
19	Apical cell in antheridium	unordered	0 = absent; 1 = present
20	Division pattern, young antheridia	ordered	0 = four-celled; 1 = two-celled; 2 = one-celled
21	Antheridial stalk anatomy	unordered	0 = uniseriate; 1 = biseriate; 2 = four or more seriate

#	Character	States	Ordering
22	Antheridial jacket cells	0 = irregular, isodiametric; 1 = tiered, elongate	unordered
23	Perigonial position	0 = randomly scattered, dorsal; 1 = in bracts on main stem; 2 = in acrogynous clusters; 3 = embedded in receptacles; 4 = on specialized branches; 5 = in anacrogynous clusters; 6 = on antheridiophores	unordered
24	Perichaetial position	0 = randomly scattered, dorsal; 1 = anacrogynous, clustered; 2 = acrogynous on main stem; 3 = acrogynous on short branches; 4 = on archegoniophores	unordered
25	Number of archegonia per inflorescence	0 = one; 1 = more than one	unordered
26	Outer perichaetium	0 = absent; 1 = leaf-like scales; 2 = bracts and bracteoles; 3 = marchantioid "involucre"; 4 = hornwort involucre	unordered
27	Inner perichaetium	0 = absent; 1 = pseudoperianth; 2 = perianth; 3 = hollow perigynium; 4 = marchantioid pseudoperianth	unordered
28	Calyptra	0 = absent; 1 = true calyptra; 2 = shoot calyptra; 3 = vestigial, solid perigynium; 4 = epigonium	unordered
29	Archegonial neck	0 = four cell rows; 1 = five cell rows; 2 = six cell rows	ordered
30	Embryo type	0 = filamentous; 1 = octant; 2 = free nuclear	unordered
31	Embryo apical cell	0 = present; 1 = absent	unordered
32	Foot form	0 = small, bulbous; 1 = conical; 2 = collared, cup-shaped; 3 = irregular, with palisade layer; 4 = irregular, no palisade layer; 5 = absent	unordered
33	Transfer cells, gametophyte	0 = absent; 1 = present	unordered
34	Transfer cells, sporophyte	0 = absent; 1 = present	unordered
35	Seta development	0 = absent; 1 = from meristem; 2 = synchronous, elongation; 3 = synchronous, no elongation	unordered
36	Seta cell arrangement	0 = nonarticulate; 1 = articulate	unordered
37	Hydroids in sporophyte	0 = present; 1 = absent	unordered
38	Leptoids in sporophyte	0 = present; 1 = absent	unordered
39	Cruciate seta pattern	0 = absent; 1 = present	unordered
40	Sporangial development	0 = determinant; 1 = indeterminant	unordered
41	Capsule wall	0 = unistratose; 1 = bistratose; 2 = multistratose	unordered
42	Capsule dehiscence	0 = irregular; 1 = along one suture; 2 = into two valves; 3 = into four valves; 4 = apically with operculum; 5 = absent	unordered
43	Capsule stomates	0 = absent; 1 = present	unordered

Table 2.2. (cont.)

Character number	Character	Type	States
44	Spore:elater division ratio	unordered	0 = 4:1 (no division); 1 = >4:1 (smc divides); 2 = <4:1 (emc divides)
45	Elaters	unordered	0 = absent; 1 = present, unicellular; 2 = present, multicellular
46	Elater thickening bands	unordered	0 = absent; 1 = present
47	Elaterophores	unordered	0 = absent; 1 = basal; 2 = apical
48	Perine layer on spores	unordered	0 = absent; 1 = present
49	Columella	ordered	0 = absent; 1 = present, nonpersistent; 2 = present, persistent
50	Spore mother cells lobed	unordered	0 = absent; 1 = present
51	Gemmae	unordered	0 = absent; 1 = blastic; 2 = discoid; 3 = endogenous
52	Protonema	unordered	0 = filamentous, not heterotrichous; 1 = globose; 2 = flattened plate; 3 = long, cylindrical
53	Germ tube	unordered	0 = absent; 1 = present
54	Spore germination	unordered	0 = exosporic; 1 = endosporic
55	Paraphyses in perigonium	unordered	0 = absent; 1 = present
56	Paraphyses in perichaetium	unordered	0 = absent; 1 = present
57	Nostoc symbionts	unordered	0 = absent; 1 = endogenous; 2 = exogenous in domatia
58	Air chambers	unordered	0 = absent; 1 = Riccia-type; 2 = Reboulia-type; 3 = Marchantia-type
59	Air pores on thallus	unordered	0 = simple; 1 = compound
60	Schizog. mucilage cavities	unordered	0 = absent; 1 = present
61	Number of rhizoid furrows on female	unordered	0 = zero; 1 = one; 2 = two

analyses were also conducted in which it was designated as the rooting outgroup. In a final set of analyses, trees were rooted with a hypothetical ancestor, designed to reflect the intuitive hypotheses of ancestry postulated by Crandall-Stotler (1981, 1984) and Schuster (1984a).

Most of the analyses support the recognition of distinct jungermannioid and marchantioid lineages within the liverworts (Fig. 2.8). When considered in conjunction with the molecular data currently available for liverworts (Lewis *et al.* 1997, Bopp & Capesius 1998), they also suggest that these two lines diverged early in hepatic evolution and have had a long history of independent radiation. In the liverwort network, the two lineages are clearly defined as separate clades (Fig. 2.8A). The Sphaerocarpales are aligned as a basal group of marchantioids, and the midpoint taxa, *Blasia* and *Petalophyllum*, occur at the base of the jungermannioids. Except for the movement of the Ricciales to the base of the marchantioid lineage, this same topology is obtained in analyses in which hornworts are included as an outgroup (Fig. 2.8B). When *Takakia* is used to root the analysis, several changes in topology are evident (Fig. 2.8C). The simple thalloid liverworts are paraphyletic, with a *Pellia*/*Treubia* clade basal to the rest of the liverworts, *Petalophyllum* and *Blasia* basal to the marchantioids, and *Haplomitrium* and *Fossombronia* basal to the leafy liverworts. Relationships among the complex thalloid taxa are poorly resolved and there is weaker support for the basal nodes than is provided by a hornwort rooting. These results do not support Schuster's (1997) hypothesis that *Takakia* is closely related to *Haplomitrium*. Although *Takakia* shares some characters with select liverworts (Schuster 1997), in all molecular and morphological analyses that include both mosses and liverworts, it is always resolved as a member of the moss clade (Garbary *et al.* 1993, Garbary & Renzaglia 1998, Hedderson *et al.* 1998). In fact, when a subset of moss exemplars was designated as the outgroup in our analyses, resultant tree topology was identical to that obtained with *Takakia* alone. Existing evidence suggests that neither *Takakia,* nor the mosses in general, are appropriate outgroups for the liverworts.

Rooting the analysis with a hypothetical ancestor, in which the presumed ancestral state is coded for all characters (Table 2.3), results in some of the same groupings as those resolved with a hornwort rooting, but does not resolve the jungermannioids as a monophyletic group (Fig. 2.8D). The Sphaerocarpales are resolved as basal to the rest of the liverworts, and the tree is basically dichotomized into a "leafy, nodal" group that comprises *Petalophyllum, Fossombronia, Haplomitrium,* and all the true

Table 2.3. *Data matrix for all analyses; details of characters and their states are presented in Table 2.2*

Exemplars	1–10	11–20	21–30	31–40	41–50	51–61
Phaeoceros laevis	2 3 0 1 ? 0 0 ? 1 0	? 1 0 0 0 0 2 1 0 0	2 0 5 0 0 4 0 0 2 1	1 4 1 0 0 ? 1 1 ? 1	2 2 1 2 2 0 0 0 2 0	0 1 0 0 0 ? 1 0 ? 0 ?
Anthoceros punctatus	2 3 0 1 ? 0 0 ? 1 0	? 1 0 0 0 0 2 1 0 0	2 1 5 0 0 4 0 0 2 1	1 3 1 0 0 ? 1 1 ? 1	2 2 1 2 2 0 0 0 2 0	0 1 1 0 0 ? 1 0 ? 1 ?
Megaceros vincentianus	2 3 0 1 ? 0 0 ? 1 0	? 1 0 0 0 0 2 1 0 0	2 0 0 0 0 4 0 0 2 1	1 4 ? ? 0 ? 1 1 ? 1	2 2 0 2 2 1 0 0 2 0	0 1 0 0 0 ? 1 0 ? 0 ?
Notothylus orbicularis	2 3 0 1 ? 0 0 ? 1 0	? 1 0 0 0 0 2 1 0 0	2 0 5 0 0 4 0 0 2 1	1 4 1 0 0 ? 1 1 ? 0	2 0 0 2 1 0 0 0 1 0	0 1 0 0 0 ? 1 0 ? 0 ?
Marchantia polymorpha	2 3 0 1 ? 3 2 2 0	? 1 0 3 0 0 1 0 0 0	1 0 6 4 1 3 4 1 2 1	1 2 1 1 3 0 1 1 0 0	0 0 0 1 1 1 0 0 0 0	2 2 1 0 0 0 0 3 1 0 2
Asterella tenella	2 3 0 1 ? 2 2 2 0	? 1 2 3 0 1 2 0 0 0	? 0 3 4 1 3 4 1 2 1	1 0 ? ? 3 0 1 1 0 0	0 4 0 0 1 1 0 0 0 0	0 2 1 0 1 0 0 2 0 0 1
Reboulia hemisphaerica	2 3 0 1 ? 2 2 2 0	? 1 2 3 0 1 2 0 0 0	? 0 3 4 1 3 0 1 2 1	1 0 ? ? 3 0 1 1 0 0	0 4 0 0 1 1 0 0 0 0	0 2 1 0 1 0 0 2 0 0 1
Conocephalum conicum	2 3 0 1 ? 1 2 2 2 0	? 1 0 3 3 1 1 0 0 0	2 0 3 4 1 3 0 1 2 1	1 0 1 1 3 0 1 1 0 0	0 0 0 2 1 1 0 0 0 0	0 2 1 0 1 0 0 2 0 0 1
Dumortiera hirsuta	2 3 0 1 ? 1 2 2 2 0	? 1 2 3 0 0 2 0 0 0	? 0 6 4 1 3 0 1 2 1	1 1 1 2 0 1 1 0 0	0 3 0 1 1 1 0 0 0 0	0 1 0 1 0 0 0 3 0 1 1
Lunularia cruciata	2 3 0 1 ? 1 2 2 2 0	? 1 2 3 0 0 1 0 0 0	1 0 3 4 1 3 0 1 2 1	1 1 1 2 0 1 1 0 0	0 3 0 1 1 1 0 0 0 0	0 2 1 0 0 0 0 3 0 0 2
Ricciocarpos natans	2 3 0 1 ? 1 3 2 2 0	? 1 0 3 0 0 1 0 0 0	0 0 5 1 1 0 0 1 2 1	1 1 ? ? 2 0 1 1 0 0	0 5 0 ? 0 ? 0 0 0 0	2 2 1 0 0 0 0 3 0 0 0
Riccia fluitans	2 3 0 1 ? 1 1 2 2 0	? 1 0 3 0 0 2 0 0 0	0 0 5 1 1 0 0 1 2 1	1 5 ? ? 0 ? ? ? ? 0	0 5 0 ? 0 ? 0 0 0 0	0 2 1 0 0 0 0 2 0 0 ?
Sphaerocarpos texanus	1 3 0 1 ? 1 0 ? 1 0	? 1 0 0 0 0 0 0 0 0	1 0 0 0 0 0 4 1 2 0	1 0 1 1 3 1 1 1 1 0	0 5 0 ? 1 0 0 0 0 0	0 3 1 0 0 0 0 0 ? ? ?
Geothallus tuberosus	1 3 0 1 ? 1 0 ? 1 0	? 1 0 0 0 0 1 0 0 0	1 0 0 0 0 0 4 1 2 0	1 0 ? 3 1 1 1 1 0	0 5 0 ? 1 0 0 0 0 0	0 3 1 0 0 0 0 0 ? ? ?
Monoclea gottschei	2 3 0 1 ? 0 0 ? 1 0	? 1 0 3 0 0 0 0 0 0	2 0 3 1 1 3 0 1 2 2	1 1 1 2 0 1 1 0 0	0 1 0 1 1 1 0 0 0 0	0 2 0 0 1 1 0 0 ? 0 ?
Pellia epiphylla	2 4 0 1 ? 0 0 ? 1 0	? 1 0 1 0 0 2 0 0 1	2 0 5 2 1 0 1 2 1 0	1 1 1 2 0 1 1 0	2 3 0 2 1 1 1 0 0 1	0 1 0 1 1 1 0 0 ? 0 ?
Fossombronia foveolata	0 2 0 1 2 1 0 ? 1 0	0 4 0 1 0 0 2 0 0 1	2 1 0 0 0 1 2 1 0	1 0 1 1 2 0 1 1 0	1 0 0 1 1 1 0 0 0 1	0 1 0 1 1 1 0 0 ? 0 ?
Makinoa crispata	2 3 0 1 ? 0 3 1 1 0	? 1 0 1 0 0 0 0 0 1	2 0 3 1 1 1 0 3 1 0	1 0 1 1 2 0 1 1 0	1 2 0 1 1 1 0 0 1	0 1 0 0 0 0 0 0 ? 0 ?
Metzgeria furcata	1 2 0 1 ? 0 3 1 1 0	4 3 0 0 0 1 0 0 1	0 0 4 3 1 0 0 2 1 0	1 ? ? 2 0 1 1 0 0	1 3 0 1 1 0 0 0 1	0 1 0 0 0 0 0 0 ? 0 ?
Pallavicinia lyellii	1 2 0 1 ? ? 0 ? 1 0	? 4 2 1 2 0 0 0 0 1	0 0 4 3 1 0 0 2 1 0	1 ? ? ? 2 0 1 1 0 0	1 3 0 1 1 0 0 0 1	2 0 0 0 0 0 0 0 ? 0 ?
Petalophyllum ralfsii	1 1 0 1 ? 0 2 2 1 0	? 4 0 1 0 0 0 0 0 1	2 0 1 1 3 1 1 0	1 1 ? ? 2 0 1 1 0 0	2 0 0 1 1 1 0 0 0 1	0 1 0 0 0 0 0 0 ? 0 ?
Aneura pinguis	2 2 0 1 ? 0 0 ? 1 0	? 4 0 1 0 0 0 0 0 1	1 ? 4 3 1 1 0 2 1 0	1 1 0 0 2 0 1 1 0 0	1 3 0 1 1 1 2 0 0 1	0 ? 0 0 0 1 0 0 ? 0 ?
Symphyogyna brasiliensis	1 3 0 1 ? 0 0 ? 1 0	? 4 2 1 2 0 0 0 0 1	1 1 5 1 1 1 0 3 1 0	1 1 ? ? 2 0 1 1 0 0	2 3 0 1 1 0 0 0 1	2 1 0 0 0 0 0 0 ? 0 ?
Blasia pusilla	2 3 0 1 ? 0 2 2 1 0	? 1 0 0 0 0 0 0 0 ?	0 0 0 2 1 3 0 1 1 0	1 2 1 1 2 0 1 1 0 0	2 3 0 1 1 1 0 0 0 1	2 1 0 1 1 0 2 0 ? 0 ?
Haplomitrium hookeri	0 0 0 1 0 1 ? ? 0 0	0 7 2 1 1 0 0 0 0 2	2 0 1 0 1 2 0 1 0 0	1 1 1 1 2 0 1 1 0 0	0 1 0 1 1 1 0 0 0 1	0 1 0 0 1 1 0 0 0 1 ?
Haplomitrium mnioides	0 0 0 1 0 1 ? ? 0 0	0 7 2 1 1 0 0 0 0 2	2 0 2 2 1 2 0 2 0 0	1 1 ? ? 2 0 1 1 0 0	0 1 0 1 1 1 0 0 0 1	0 1 0 0 1 1 0 0 0 1 ?

```
Treubia tasmanica        0101210?10  1703001001  205111032?  11?2011110  2301110001  2????1100?
Bazzania trilobata       0101021211  3331001001  ??43122210  11?2011100  2301110001  01000000?
Jungermannia leiantha    0101210?11  0601002001  1012122210  1201201100  1301110001  11000000?
Calypogeia muelleriana   0101021211  0031002001  1143123210  12?2011100  1301110001  11000000?
Herbertus pensilis       0101001211  1631001001  1012122210  1?0201100   2301110001  01000100?
Jubula pennsylvanica     0101021211  43&8501002001  0142122110  11?2011100  1301110001  02010000?
Lejeunea cavifolia       0101021211  4501002001  0142022110  10?2111110  1301110001  02010000?
Frullania asagrayana     0101021211  4301001001  0043122110  10?2011100  1301110001  01010000?
Lepidozia reptans        0101021211  3331002001  ??43122210  11?2011100  2301110001  01000000?
Lophocolea heterophylla  0101011211  13&631002001  0012122210  1201201100  2301110001  01000000?
Porella pinnata          0101021211  4301000001  1043122210  11?2011100  2301110001  01010000?
Scapania nemorea         0101210?11  24&631001001  0012122210  1101201100  2301110001  11001000?
Takakia ceratophylla     0110102?02  32?0101011  205110042?0  101100120   210?0??110  0??20000?
Liverwort ancestor       1301?00?00  ?100001002  2000000120  1111201100  2000100000  03001100?
```

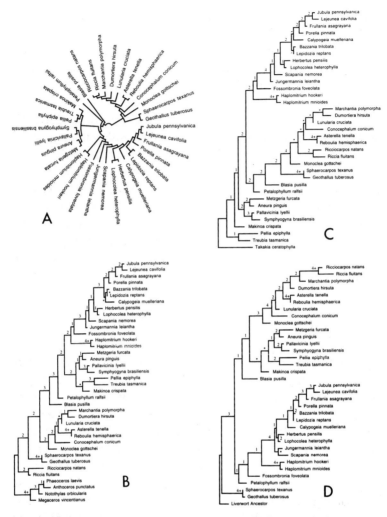

Fig. 2.8. Phylogenetic analyses. A. The unrooted liverwort network, one of 68 equally parsimonious trees, showing separation of the marchantioid (*Geothallus* through *Ricciocarpos*) and jungermannioid lineages (*Blasia* through *Jubula*); TL = 256, CI = 0.414, RI = 0.679. B. Outgroup rooting with the hornworts, phylogram of one of 30 equally parsimonious trees; numbers designate Bremer decay values, * denotes nodes not resolved in the strict consensus; TL = 290, CI = 0.403, RI = 0.697. C. Outgroup rooting with *Takakia*, phylogram of one of 35 equally parsimonious trees; numbers designate Bremer decay values, * denotes nodes not resolved in the strict consensus; TL = 274, CI = 0.412, RI = 0.671. D. Outgroup rooting with a hypothetical ancestor, phylogram of one of 191 equally parsimonious trees; numbers designate Bremer decay values, * denotes nodes not resolved in the strict consensus; TL = 266, CI = 0.402, RI = 0.674.

leafy liverwort taxa, and a thalloid group that consists of the rest of the simple thalloids and all of the complex thalloid forms. Of the rooted analyses, the hornwort rooting results in the intuitively most logical topology of liverwort relationships, except for the placement of the Ricciaceae at the base of the tree. This placement is incongruent with many other analyses which resolve this group as a derived lineage of the marchantioid clade (Lewis *et al.* 1997, Bopp & Capesius 1998, Garbary & Renzaglia 1998). For the most part, reductive evolutionary trends within lineages and a descending system of classification are supported by this analysis.

As a consequence of these and other phylogenetic studies (Churchill & Mishler 1984, Kenrick & Crane 1997, Lewis *et al.* 1997, Bischler 1998), we propose a classification scheme in which liverworts comprise the phylum (= division) Marchantiophyta, which is subdivided into two classes, the Marchantiopsida and the Jungermanniopsida. The morphological characters that define these classes and the hierarchial arrangements of taxa within each are detailed in the sections that follow.

2.5 Evolution and diversity in the class Marchantiopsida

The Marchantiopsida, comprising the Sphaerocarpidae and the Marchantiidae, are circumscribed by five morphological features, namely, a dorsal tilt to the apical cell (character 6), four primary androgonial cells (character 20), six rows of archegonial neck cells (character 29), a unistratose capsule wall (Fig. 2.3A, character 41), and unlobed spore mother cells (Fig. 2.3B, character 51). It should be noted that this latter character refers only to the shape of the sporocyte (Neidhart 1979), and not to internal cytoplasmic lobing associated with cytokinesis (Brown & Lemmon 1990). Consequently, we have interpreted the sporocytes of *Monoclea* as unlobed in agreement with Campbell (1954) and Bischler (1998) instead of lobed as in Garbary and Renzaglia (1998; see also Renzaglia *et al.* 1994: Fig. 15). Two characters of spermatid ultrastructure, namely a left-hand taper to the spline and a posterior notch in the lamellar strip, also seem to be diagnostic of the group, although few taxa have been sampled (Garbary *et al.* 1993).

To reflect its somewhat isolated position in the Marchantiopsida, we have segregated the Sphaerocarpales into a separate subclass, the Sphaerocarpidae. In this group the gametophyte thallus is differentiated into a midrib and unistratose wings, possesses only smooth rhizoids, and has no air chambers or ventral appendages. Idioblastic oil cells are developed only in the aquatic taxon *Riella* while the other two taxa lack oil bodies. Gametangia are scattered randomly on the thallus surface and are enclosed in bottle-shaped "involucres." Although these "involucres" are often considered unique to the Sphaerocarpidae (Schuster 1992a), they are homologous to perigonial and perichaetial enclosures of other marchantioids.

Leitgeb (1879) demonstrated that the male "involucre" in *Sphaerocarpos* is produced from a ring of epidermal initials that surround the antheridial initial (Leitgeb 1879: pl. VIII, Figs. 15 and 16). These initials divide in but a single plane, in synchrony with antheridial growth, to form a unistratose enclosure. In development this "involucre" is exactly like the flask-shaped antheridial cavity of the simple thalloid taxon *Noteroclada*. In *Riella* antheridial chambers are more like those of other members of the Marchantiopsida; i.e., the surrounding epidermal initials divide in more than a single plane and in advance of antheridial expansion, thereby more or less embedding the antheridium in a cavity in the lappet or dorsal wing. Some taxa of the Marchantiopsida, such as those of the Cleveaceae, combine these two developmental patterns, with early stages of cavity development involving multiple division planes, but late stages being restricted, so that a conspicuous unistratose, bottle-shaped antheridial "ostiole" extends above the antheridial cavity (see Bischler 1998: Fig. 9). Developing female involucres of *Sphaerocarpos* develop from cells of the archegonial stalk (see Leitgeb 1879: pl. VIII, Fig. 12) and are identical to the immature "pseudoperianths" of *Marchantia* (Fig. 2.3F) and *Asterella*, which Bischler (1998) also described as developing from archegonial stalks. The *Sphaerocarpos* involucre is not homologous to the marchantioid involucre, which is a multistratose enclosure produced by growth of the vegetative thallus around the entire perichaetium (Bischler 1998).

The remaining taxa of the Marchantiopsida, all of which have multistratose, thalloid gametophytes, comprise the subclass Marchantiidae. Both molecular and morphological data support this scheme (Lewis *et al.* 1997, Bischler 1998). Like *Sphaerocarpos*, *Monoclea* (Fig. 2.3D) blends characters of the Marchantiopsida with those of the jungermannioid lineage. *Monoclea* is more closely aligned to the rest of the complex thalloid taxa, however, and in analyses based on *rbcL* sequence data is even nested in the Marchantiaceae (Lewis *et al.* 1997). Major characters that differentiate *Monoclea* from the remaining members of the Marchantiidae are the absence of pegged rhizoids and ventral scales in the gametophyte and the occurrence of a massive seta, which fully elongates prior to dehiscence of the long, ellipsoidal capsule. The hood-like involucre around the archegonial cluster has been described as a *Pellia*-like pseudoperianth (Schuster 1984*a*), but developmental studies of Leitgeb (1877) and Johnson (1904) support its homology with the involucre of other Marchantiidae; i.e., it is multistratose and develops through continued growth of the thallus around the archegonia as they are being formed. The pseudoperianth of *Pellia*, in contrast, is unistratose and is derived from a ring or half-ring of epidermal cells surrounding the archegonial cluster. At present, *Monoclea* is the only bryophyte in which a free-nuclear stage precedes cell partitioning in the embryo (Ligrone *et al.* 1993). Based on a limited analysis of chloroplast $trnL_{UAA}$ sequences, Meissner *et al.* (1998) proposed establishment of the subclass Monocleidae for this taxon, but all other analyses support our recognition of *Monoclea* as distinct from the other Marchantiidae only at the ordinal level.

The hierarchial arrangement of the remaining taxa of the Marchantiidae is based on the morphological analysis of Bischler (1998). Both this analysis and a preliminary sequence analysis of the 26S rRNA gene (Boisselier-Dubayle *et al.* 1997) resolve two sister groups with contrasting evolutionary histories, which we recognize as the Marchantiales and Ricciales, respectively. Within the Marchantiales progressive morphological elaboration has occurred in the structures associated with the gametangia, culminating in the development of highly specialized gametangiophores in many of the lineages (Fig. 2.3E, F). In the Ricciales (Fig. 2.3G), in contrast, the trend has been to simplify both the gametophyte and the sporophyte (Bischler 1998). For the present, only the suborder Ricciineae is relegated to the Ricciales, while the Marchantiales comprises four suborders. The segregation of taxa among the suborders corresponds to the treatment of Bischler (1998), except for our placement of the Cyathodiaceae in the Corsiniineae rather than the Targioniineae. As intimated by Bischler (1998), there is only weak support for the placement of this family in the Targioniineae, while several characters link it to the Corsiniaceae.

2.6 Evolution and diversity of the Jungermanniopsida

The Jungermanniopsida are much more heterogeneous than the Marchantiopsida, with more than 300 genera currently recognized. Antheridial development involving one (*Haplomitrium*) or two primary androgonial cells, never four (character 20), embryos that are filamentous, never octants (character 30), capsule walls composed of two or more layers of cells, unistratose only in *Haplomitrium* (character 41), and lobed spore mother cells, or sporocytes, with often deeply furrowed walls (Fig. 2.2A, character 51) unambiguously distinguish the Jungermanniopsida from the Marchantiopsida. In 1990 Bartholomew-Began established the subclass Metzgeriidae to include *Haplomitrium* in the Haplomitriales (= Calobryales) and the Metzgeriales, as defined by Schuster (1972), leaving only the Jungermanniales in the subclass Jungermanniidae. Basically, in this system the Metzgeriidae comprise the simple thalloid or anacrogynous jungermannioids and the Jungermanniidae comprise the leafy or acrogynous jungermannioids.

In a comprehensive analysis of the morphological diversity of the Metzgeriidae (Crandall-Stotler *et al.* 1997, manuscript in preparation), five monophyletic clades are resolved within this subclass, but the branching order of the clades is ambiguous. Phylogenetic inferences drawn from ongoing analyses of molecular sequence data are also ambiguous, especially as regards the relationships of *Blasia, Treubia*, and *Haplomitrium*. The inability to resolve relationships with morphology alone is due in large part to the tremendous diversity of character expressions within the group. For example, gametophytes can be leafy, with nodal organization (Fig. 2.2B–E), or thalloid with a costa and unistratose wing (Fig. 2F, G), or thalloid without a costa (Fig. 2.2H). The group expresses five apical cell geometries (character 2), both acrogynous and anacrogynous perichaetia (character 24), and several types of capsule dehiscence (character 42). In both the leafy *Haplomitrium* (Fig. 2.2D) and the thalloid *Pallavicinia* clades, strands of hydrolyzed water-conducting cells are differentiated (Fig. 2.2G). Aside from the four characters that define the Metzgeriidae as part of the Jungermanniopsida, only character 10 is invariable within the group; i.e., all taxa, whether leafy or thalloid, form a central wedge-shaped leaf/thallus initial during early merophyte ontogeny (Renzaglia 1982).

The five lineages within the Metzgeriidae (Crandall-Stotler *et al.* 1997) are recognized in our classification scheme as orders. The hypothesis that the Metzgeriineae of the Metzgeriales are the most derived taxa of the subclass (Schuster 1984a, 1992a) is supported by our analyses, as are also the placements of the Blasiales and Fossombroniales near the base of the Jungermanniopsida. It is plausible, as Schuster (1992a) maintains, that the Metzgeriidae are the extant remnants of a much larger, ancient radiation. At least, the Devonian fossil *Pallaviciniites* is superficially similar to extant *Jensenia* (Fig. 2.2F) of the Pallaviciniaceae. A more refined resolution of relationships within this ancient group will require an additional suite of ontogenetic and molecular characters (Garbary & Renzaglia 1998).

In all of our analyses the Jungermanniidae, or leafy liverworts, are resolved as a monophyletic, crown group of the Jungermanniopsida. Characters that distinguish this largest group of hepatics from the Metzgeriidae include the following: apical cells are tetrahedral, with the third face ventral, in all taxa except *Pleurozia* (character 2), ventral appendages, when present, are produced in a single row (character 7) and are foliose (character 8), leaves are formed from two initials, that are anodic and kathodic in orientation (character 10), endogenous branches are common (characters 12 and 13), perichaetia are always acrogynous, terminating either the main stem or a branch (character 24), bracts/bracteoles and perianths are associated with the perichaetium (characters 26 and 27), the archegonial neck consists of five rows of neck cells (character 29), the capsule always dehisces into four valves (character 42) and gemmae, when present, are one- to two-celled and develop in a blastic manner (Fig. 2.1A, character 52).

In the absence of a comprehensive phylogenetic analysis of this subclass, we have relied heavily on the untested interpretations of Schuster (1984*a*) in developing the classification of the Jungermanniidae. The five orders proposed basically group suborders according to the illustration presented as Table II in Schuster (1972). Thus, the Lepicoleales includes the three suborders that possess massive sporophytes enclosed in solid perigynia or coelocaules and the suborder Ptilidiineae, which is hypothesized by Schuster (1972) to be ancestral to this group. The placement of *Mastigophora* in the Lepicoleineae by Grolle (1983), but in the Ptilidiineae by Schuster (1984*a*) indeed is suggestive of a link between these two suborders. The Jungermanniales comprises seven suborders in which a perianth and associated shoot calyptra are typical, or if a perigynium is elaborated, it is hollow. The elevation of the Pleuroziineae and Radulineae to ordinal rank is predicative of the isolated position of *Pleurozia* and *Radula*, respectively, and the establishment of the Porellales recognizes the highly specialized and derived features of this most advanced group of liverworts. Our recognition of two suborders within the Porellales, namely, the Porellineae, including only the Porellaceae, and the Jubulineae, to include the Goebeliellaceae, Jubulaceae, and Lejeuneaceae, is a major departure from other classification schemes. Both sequence data from the *rbcL* gene (Lewis *et al*. 1997) and our analyses of morphological data suggest that there is significant divergence between the Porellaceae and the other families of the Porellales. In the Porellaceae, for example, the leaves are fundamentally two-parted, with the lobules never forming water sacs, perianths are unbeaked, the sporophyte is enclosed by a well-developed shoot calyptra, the capsule wall is multistratose, and the elaters are free and scattered at random through the spore mass (Fig. 2.1F). In both the Jubulaceae and Lejeuneaceae, in contrast, leaves are often three-parted, consisting of a lobe, variously inflated water sac, and stylus or papilla, the perianths are beaked, the sporophyte is enclosed in a stalked, true calyptra, the capsule wall is bistratose, and the elaters are vertically aligned and attached to the valve apices (Fig. 2.1G). The placement of the Goebeliellaceae in the Jubulineae is provisional and is based on its three-parted leaves.

2.7 Classification of the Marchantiophyta

Classification schemes are hypotheses of taxonomic and phylogenetic relationships among groups of organisms. The classification scheme presented below reflects our current state of understanding of liverwort diversity and is based primarily on morphological evidence. As with all hypotheses, it is meant to be tested. Certainly, as new ontogenetic and molecular data are generated, the hypotheses presented herein will be scrutinized and refined. Latin diagnoses and other nomenclatural notes that are relevant to the classification are appended as annotations.

PHYLUM (DIVISION): MARCHANTIOPHYTA[1]

CLASS: MARCHANTIOPSIDA Stotler & Stotl.-Crand., Bryologist 80: 426. 1977.

SUBCLASS: SPHAEROCARPIDAE Stotler & Stotl.-Crand., Bryologist 80: 426. 1977.

ORDER: SPHAEROCARPALES Cavers, New Phytol. 9: 81. 1910.

SUBORDER: Sphaerocarpineae R. M. Schust. ex Stotler & Stotl.-Crand.[2]

Sphaerocarpaceae (Dumort.) Heeg, Verh. K. K. Zool.-Bot. Ges. Wien 41: 573. 1891. *Sphaerocarpos* Boehm., *Geothallus* Campb.

SUBORDER: Riellineae R. M. Schust. ex Stotler & Stotl.-Crand.[3]

Riellaceae Engl., Syll. Pflanzenfam. Grosse Ausgabe. p. 45. 1892. *Riella* Mont.

SUBCLASS: MARCHANTIIDAE Engl. [Unterklasse "Marchantiales"] in
Engler & Prantl, Naturl. Pflanzenfam. I, 3. I: 1. 1893.
ORDER: MONOCLEALES R. M. Schust., J. Hattori Bot. Lab. 26: 296. 1963.
    Monocleaceae (Nees) A. B. Frank in Leunis, Syn. Pflanzenk. (ed. 2) p.
    1556. 1877. *Monoclea* Hook.
ORDER: MARCHANTIALES Limpr. in Cohn, Krypt.-Fl. Schlesien 1: 239, 336.
1876[1877].
  SUBORDER: Marchantiineae Buch ex Schljakov, Bot. Zhurn. 57: 507. 1972.
    Aytoniaceae Cavers, New Phytol. 10: 42. 1911. *Asterella* P. Beauv.,
      Cryptomitrium Austin ex Underw., *Mannia* Opiz nom. cons.,
      Plagiochasma Lehm. & Lindenb. nom. cons., *Reboulia* Raddi nom.
      cons.
    Wiesnerellaceae Inoue, Ill. Jap. Hep. 2: 192. 1976. *Wiesnerella* Schiffn.
    Conocephalaceae Müll. Frib. ex Grolle, J. Bryol. 7: 207. 1972.
      Conocephalum Hill nom. cons.
    Lunulariaceae Klinggr., Höh. Crypt. Preuss. p. 9. 1858. *Lunularia*
      Adans.
    Marchantiaceae (Bisch.) Lindl., Nat. Syst. Bot. (ed. 2) p. 412. 1836.
      Bucegia Radian, *Dumortiera* Nees, *Marchantia* L., *Neohodgsonia* Perss.,
      Preissia Corda
    Monosoleniaceae Inoue, Bull. Natl Sci. Mus. Ser. 2, 9: 117. 1966.
      Monosolenium Griff., *Peltolepis* Lindb.
    Cleveaceae Cavers, New Phytol. 10: 42. 1911. *Athalamia* Falconer, *Sauteria*
      Nees
    Exormothecaceae Müll. Frib. ex Grolle, J. Bryol. 7: 208. 1972.
      Aitchisoniella Kashyap, *Exormotheca* Mitt., *Stephensoniella* Kashyap
  SUBORDER: Corsiniineae R. M. Schust. ex Schljakov, Bot. Zhurn. 57: 507.
  1972.
    Cyathodiaceae (Grolle) Stotler & Stotl.-Crand., stat. et fam. nov.[4]
      Cyathodium Kunze
    Corsiniaceae Engl., Syll. Pflanzenfam. Grosse Ausgabe. p. 44. 1892.
      Corsinia Raddi, *Cronisia* Berk.
  SUBORDER: Monocarpineae R. M. Schust. "Carrpinae," J. Hattori Bot. Lab.
  26: 299. 1963.
    Monocarpaceae D. J. Carr ex Schelpe, J. South African Bot. 35: 110.
    1969. *Monocarpus* D. J. Carr
  SUBORDER: Targioniineae R. M. Schust. ex Schljakov, Bot. Zhurn. 57: 507.
  1972.
    Targioniaceae Dumort., Anal. Fam. Pl. p. 68, 70. 1829. *Targionia* L.

ORDER: RICCIALES Schljakov emend. Stotler & Stotl.-Crand.[5]

Oxymitraceae Müll. Frib. ex Grolle, J. Bryol. 7: 215. 1972. *Oxymitra* Bisch. ex Lindenb.

Ricciaceae Reichenb., Bot. Damen. p. 255. 1828. *Riccia* L., *Ricciocarpos* Corda

CLASS: JUNGERMANNIOPSIDA Stotler & Stotl.-Crand., Bryologist 80: 425. 1977.

SUBCLASS: METZGERIIDAE Barthol.-Began, Phytologia 69: 465. 1990.

ORDER: HAPLOMITRIALES H. Buch ex Schljakov, Bot. Zhurn. 57: 499. 1972.

Haplomitriaceae Dedecek, Arch. Prír. Proskoumání Cech 5: 71. 1884. *Haplomitrium* Nees nom. cons.

ORDER: BLASIALES (R. M. Schust.) Stotler & Stotl.-Crand., stat. et ordo nov.[6]

Blasiaceae H. Klinggr., Höh. Crypt. Preuss. p. 14. 1858. *Blasia* L., *Cavicularia* Steph.

ORDER: TREUBIALES Schljakov emend. Stotler & Stotl.-Crand.[7]

SUBORDER: Treubiineae R. M. Schust. ex Stotler & Stotl.-Crand.[8]

Treubiaceae Verd., Man. Bryol. p. 427. 1932. *Apotreubia* S. Hatt. & Mizut., *Treubia* K. I. Goebel nom. cons.

SUBORDER: Phyllothalliineae R. M. Schust., Trans. Brit. Bryol. Soc. 5: 283. 1967.

Phyllothalliaceae E. A. Hodgs., Trans. Roy. Soc. New Zealand, Bot. 2: 247. 1964. *Phyllothallia* E. A. Hodgs.

ORDER: FOSSOMBRONIALES Schljakov emend. Stotler & Stotl.-Crand.[9]

SUBORDER: Fossombroniineae R. M. Schust. ex Stotler & Stotl.-Crand.[10]

Fossombroniaceae Hazsl. nom. cons., Magyar Birodalom Moh-Flórája. p. 20. 1885. *Austrofossombronia* R. M. Schust., *Fossombronia* Raddi, *Petalophyllum* Nees & Gottsche ex Lehm., *Sewardiella* Kashyap

Allisoniaceae (R. M. Schust. ex Grolle) Schljakov, [Liverworts, morphology, phylogeny, classification] p. 119. 1975. *Allisonia* Herzog, *Calycularia* Mitt.

SUBORDER: Pelliineae R. M. Schust. ex Schljakov emend Stotler & Stotl.-Crand.[11]

Pelliaceae H. Klinggr., Höh. Crypt. Preuss. p. 13. 1858. *Noteroclada* Taylor ex Hook. & Wils., *Pellia* Raddi nom. cons.

Sandeothallaceae R. M. Schust., New Man. Bryol. p. 951. 1984. *Sandeothallus* R. M. Schust.

ORDER: METZGERIALES Schljakov emend. Stotler & Stotl.-Crand.[12]

SUBORDER: Pallaviciniineae R. M. Schust. emend. Stotler & Stotl.-Crand.[13]

Pallaviciniaceae Mig., Krypt.-Fl. Deutschl., Moose. p. 423. 1904.
Greeneothallus Hässel, *Hattorianthus* R. M. Schust. & Inoue, *Jensenia*
Lindb., *Moerckia* Gottsche, *Pallavicinia* Gray nom. cons., *Podomitrium*
Mitt., *Seppeltia* Grolle, *Symphyogyna* Nees & Mont., *Symphyogynopsis*
Grolle, *Xenothallus* R. M. Schust.

Makinoaceae Nakai, [Ord. Fam. Trib. Nov.] p. 201. 1943. *Makinoa*
Miyake, *Verdoornia* R. M. Schust.

Hymenophytaceae R. M. Schust., J. Hattori Bot. Lab. 26: 296. 1963.
Hymenophyton Dumort.

SUBORDER: Metzgeriineae R. M. Schust. ex Schljakov, Bot. Zhurn. 57: 501.
1972.

Aneuraceae H. Klinggr., Höh. Crypt. Preuss. p. 11. 1858. *Aneura*
Dumort., *Cryptothallus* Malmb., *Lobatoriccardia* (Mizut. & S. Hatt.)
Furuki, *Riccardia* Gray nom. cons.

Mizutaniaceae Furuki & Z. Iwats., J. Hattori Bot. Lab. 67: 291. 1989.
Mizutania Furuki & Z. Iwats.

Vandiemeniaceae Hewson, J. Hattori Bot. Lab. 52: 163. 1982.
Vandiemenia Hewson

Metzgeriaceae H. Klinggr., Höh. Crypt. Preuss. p. 10. 1858. *Apometzgeria*
Kuwah., *Austrometzgeria* Kuwah., *Metzgeria* Raddi, *Steereella* Kuwah.

SUBCLASS: JUNGERMANNIIDAE Engl. emend. Stotler & Stotl.-Crand.[14]

ORDER: LEPICOLEALES Stotler & Crand.-Stotl., ordo nov.[15]

SUBORDER: Ptilidiineae R. M. Schust., J. Hattori Bot. Lab. 26: 227. 1963.

Ptilidiaceae H. Klinggr., Höh. Crypt. Preuss. p. 37. 1858. *Ptilidium* Nees

Mastigophoraceae R. M. Schust., J. Hattori Bot. Lab. 36: 345. 1972.
Dendromastigophora R. M. Schust., *Mastigophora* Nees nom. cons.

Chaetophyllopsidaceae R. M. Schust., J. Hattori Bot. Lab. 23: 68.
1960[1961]. *Chaetophyllopsis* R. M. Schust., *Herzogianthus* R. M. Schust.

SUBORDER: Lepicoleineae R. M. Schust., J. Hattori Bot. Lab. 36: 336, 338.
1972[1973].

Vetaformataceae Fulford & J. Taylor, Mem. New York Bot. Gard. 11: 27.
1963. *Vetaforma* Fulford & J. Taylor

Lepicoleaceae R. M. Schust., Nova Hedwigia 5: 27. 1963. *Lepicolea*
Dumort.

SUBORDER: Perssoniellineae R. M. Schust., J. Hattori Bot. Lab. 26:
229–230. 1963.

Schistochilaceae H. Buch, Commentat. Biol. 3: 9. 1928. *Gottschea* Nees
ex Mont., *Pachyschistochila* R. M. Schust. & J. J. Engel, *Paraschistochila* R.
M. Schust., *Pleurocladopsis* R. M. Schust., *Schistochila* Dumort.

Perssoniellaceae R. M. Schust. ex Grolle, J. Bryol. 7: 216. 1972.
Perssoniella Herzog

SUBORDER: Lepidolaenineae R. M. Schust., J. Hattori Bot. Lab. 36: 345.
1972[1973].

Trichocoleaceae Nakai, [Ord. Fam. Trib. Nov.] p. 201. 1943. *Eotrichocolea*
R. M. Schust., *Trichocolea* Dumort. nom. cons. [*Leiomitra* Lindb. =
Trichocolea]

Lepidolaenaceae Nakai, [Ord. Fam. Trib. Nov.] p. 200. 1943.
Gackstroemia Trevis., *Lepidogyna* R. M. Schust., *Lepidolaena* Dumort.,
Trichocoleopsis S. Okamura

Neotrichocoleaceae Inoue, Ill. Jap. Hepat. 1: 176. 1974. *Neotrichocolea* S.
Hatt.

Jubulopsidaceae (Hamlin) R. M. Schust., Phytologia 56: 68. 1984.
Jubulopsis R. M. Schust.

ORDER: JUNGERMANNIALES H. Klinggr. emend. Stotler & Stotl.-Crand.[16]

SUBORDER: Herbertineae R. M. Schust., J. Hattori Bot. Lab. 26: 232. 1963.

Herbertaceae Müll. Frib. ex Fulford & Hatcher, Bryologist 61: 284.
1958. *Herbertus* Gray, *Triandrophyllum* Fulford & Hatcher

Pseudolepicoleaceae Fulford & J. Taylor, Nova Hedwigia 1: 411. 1960.
Archeophylla R. M. Schust., *Blepharostoma* (Dumort.) Dumort.,
Chaetocolea Spruce, *Herzogiaria* Fulford ex Hässel, *Isophyllaria* E. A.
Hodgs. & Allison, *Pseudolepicolea* Fulford & J. Taylor [*Archeochaete* R. M.
Schust. and *Lophochaete* R. M. Schust. = *Pseudolepicolea*], *Temnoma* Mitt.

Trichotemnomataceae R. M. Schust., J. Hattori Bot. Lab. 36: 340. 1973.
Trichotemnoma R. M. Schust.

Grolleaceae Solari ex R. M. Schust., Phytologia 56: 66. 1984. *Grollea*
R. M. Schust.

SUBORDER: Balantiopsidineae R. M. Schust., J. Hattori Bot. Lab. 36: 353.
1972[1973].

Balantiopsidaceae H. Buch, Thüring. Bot. Ges. 1: 23. 1955. *Anisotachis*
R. M. Schust., *Austroscyphus* R. M. Schust., *Balantiopsis* Mitt.,
Eoisotachis R. M. Schust., *Hypoisotachis* (R. M. Schust.) J. J. Engel &
Merr., *Isotachis* Mitt., *Neesioscyphus* Grolle, *Ruizanthus* R. M. Schust.

SUBORDER: Lophocoleineae Schljakov, Bot. Zhurn. 57: 504. 1972 (=
Geocalycineae R. M. Schust., 1972[1973]).

Geocalycaceae H. Klinggr., Höh. Crypt. Preuss. p. 34. 1858 [incl.
Lophocoleaceae (Jörg.) Vanden Berghen in Robyns, Fl. Gén. Belgique,
Bryophytes 1: 208. 1956]. *Campanocolea* R. M. Schust., *Chiloscyphus*
Corda nom. cons., *Clasmatocolea* Spruce, *Conoscyphus* Mitt.,
Evansianthus R. M. Schust. & J. J. Engel, *Geocalyx* Nees, *Harpanthus*

Nees, *Hepatostolonophora* J. J. Engel & R. M. Schust., *Heteroscyphus*
Schiffn. nom. cons., *Lamellocolea* J. J. Engel, *Leptophyllopsis* R. M.
Schust., *Leptoscyphopsis* R. M. Schust., *Leptoscyphus* Mitt., *Lophocolea*
(Dumort.) Dumort., *Pachyglossa* Herzog & Grolle, *Pedinophyllopsis* R.
M. Schust. & Inoue, *Perdusenia* Hässel, *Pigafettoa* C. Massal., *Platycaulis*
R. M. Schust., *Pseudolophocolea* R. M. Schust. & J. J. Engel, *Saccogyna*
Dumort. nom. cons., *Saccogynidium* Grolle, *Stolonivector* J. J. Engel,
Tetracymbaliella Grolle, *Xenocephalozia* R. M. Schust.
Gyrothyraceae R. M. Schust., Trans. Brit. Bryol. Soc. 6: 87. 1970.
Gyrothyra M. Howe
Plagiochilaceae (Jörg.) Müll. Frib. & Herzog in Müller, Leberm. Eur. p.
877. 1956. *Acrochila* R. M. Schust., *Pedinophyllum* (Lindb.) Lindb.,
Plagiochila (Dumort.) Dumort. nom. cons., *Plagiochilidium* Herzog,
Plagiochilion S. Hatt., *Steereochila* Inoue, *Szweykowskia* Gradst. & Reiner-
Drehw., *Xenochila* R. M. Schust.
Acrobolbaceae E. A. Hodgs., Rec. Domin. Mus. 4: 177. 1962. *Acrobolbus*
Nees, *Austrolophozia* R. M. Schust., *Enigmella* G. A. M. Scott & K. G.
Beckm., *Goebelobryum* Grolle, *Lethocolea* Mitt. nom. cons., *Marsupidium*
Mitt., *Tylimanthus* Mitt.
Arnelliaceae Nakai, [Ord. Fam. Trib. Nov.] p. 200. 1943. *Arnellia* Lindb.,
Gongylanthus Nees, *Southbya* Spruce
SUBORDER: Lepidoziineae R. M. Schust., J. Hattori Bot. Lab. 26: 228. 1963.
Phycolepidoziaceae R. M. Schust., Bull. Torrey Bot. Club 93: 442. 1967.
Phycolepidozia R. M. Schust.
Calypogeiaceae (Müll. Frib.) Arnell in Holmberg, Skand. Fl. 2a. p. 189.
1928. *Calypogeia* Raddi nom. cons., *Eocalypogeia* (R. M. Schust.) R. M.
Schust., *Metacalypogeia* (S. Hatt.) Inoue
Lepidoziaceae Limpr. in Cohn, Krypt.-Fl. Schlesien 1: 310. 1876[1877].
Acromastigum A. Evans, *Arachniopsis* Spruce, *Austrolembidium* Hässel,
Bazzania Gray nom. cons., *Chloranthelia* R. M. Schust., *Dendrobazzania*
R. M. Schust. & W. B. Schofield, *Drucella* E. A. Hodgs., *Hyalolepidozia* S.
W. Arnell ex Grolle, *Hygrolembidium* R. M. Schust., *Isolembidium* R. M.
Schust., *Kurzia* G. Martins, *Lembidium* Mitt. nom. cons., *Lepidozia*
(Dumort.) Dumort. nom. cons., *Mastigopelma* Mitt., *Megalembidium*
R. M. Schust., *Micropterygium* Lindenb., Nees & Gottsche, *Mytilopsis*
Spruce, *Neogrollea* E. A. Hodgs., *Odontoseries* Fulford, *Paracromastigum*
Fulford & J. Taylor, *Protocephalozia* (Spruce) K. I. Goebel, *Pseudocephalozia*
R. M. Schust., *Psiloclada* Mitt., *Pteropsiella* Spruce, *Telaranea* Spruce ex
Schiffn., *Zoopsidella* R. M. Schust., *Zoopsis* Hook. f. ex Gottsche, Lindenb.
& Nees

SUBORDER: Cephaloziineae Schljakov, Bot. Zhurn. 57: 503. 1972
(=Cephaloziineae R. M. Schust., 1972[1973]).

Cephaloziaceae Mig., Krypt.-Fl. Deutschl., Moose. p. 465. 1904.
Alobiella (Spruce) Schiffn., *Alobiellopsis* R. M. Schust., *Anomoclada*
Spruce, *Cephalozia* (Dumort.) Dumort., *Cladopodiella* H. Buch,
Haesselia Grolle & Gradst., *Hygrobiella* Spruce, *Iwatsukia* N. Kitag.,
Metahygrobiella R. M. Schust., *Nowellia* Mitt., *Odontoschisma* (Dumort.)
Dumort., *Pleurocladula* Grolle, *Schiffneria* Steph., *Schofieldia* J. D.
Godfrey, *Trabacellula* Fulford

Cephaloziellaceae Douin, Bull. Soc. Bot. France, Mém. 29: 1, 5, 13.
1920. *Allisoniella* E. A. Hodgs., *Amphicephalozia* R. M. Schust.,
Cephalojonesia Grolle, *Cephalomitrion* R. M. Schust., *Cephaloziella*
(Spruce) Schiffn. nom. cons., *Cephaloziopsis* (Spruce) Schiffn.,
Cylindrocolea R. M. Schust., *Kymatocalyx* Herzog, *Stenorrhipis*
Herzog

Jackiellaceae R. M. Schust., J. Hattori Bot. Lab. 36: 395. 1973. *Jackiella*
Schiffn.

Adelanthaceae (Jörg.) Grolle, J. Hattori Bot. Lab. 35: 327. 1972.
Adelanthus Mitt. nom. cons., *Wettsteinia* Schiffn.

SUBORDER: Antheliineae R. M. Schust. ex Stotler & Stotl.-Crand.[17]

Antheliaceae R. M. Schust., J. Hattori Bot. Lab. 26: 236. 1963. *Anthelia*
(Dumort.) Dumort.

SUBORDER: Brevianthineae J. J. Engel & R. M. Schust., Bryologist 85: 382.
1982[1983].

Brevianthaceae J. J. Engel & R. M. Schust., Phytologia 47: 317. 1981.
Brevianthus J. J. Engel & R. M. Schust.

Chonecoleaceae R. M. Schust. ex Grolle, J. Bryol. 7: 206. 1972.
Chonecolea Grolle

SUBORDER: Jungermanniineae R. M. Schust. ex Stotler & Stotl.-Crand.[18]

Jungermanniaceae Rchb., Bot. Damen. p. 256. 1828 [incl.
Lophoziaceae Cavers, New Phytol. 9: 293. 1910]. *Anastrepta* (Lindb.)
Schiffn., *Anastrophyllum* (Spruce) Steph., *Andrewsianthus* R. M. Schust.,
Anomacaulis (R. M. Schust.) R. M. Schust. ex Grolle, *Barbilophozia*
Loeske, *Bragginsella* R. M. Schust., *Cephalolobus* R. M. Schust.,
Chandonanthus Mitt., *Cryptochila* R. M. Schust., *Cryptocolea* R. M.
Schust., *Cryptocoleopsis* Amak., *Denotarisia* Grolle, *Diplocolea* Amak.,
Gerhildiella Grolle, *Gottschelia* Grolle, *Gymnocolea* (Dumort.) Dumort.,
Gymnocoleopsis (R. M. Schust.) R. M. Schust., *Hattoria* R. M. Schust.,
Horikawaella S. Hatt. & Amakawa, *Jamesoniella* (Spruce) F. Lees,

Jungermannia L., *Lophonardia* R. M. Schust., *Lophozia* (Dumort.)
Dumort., *Mylia* Gray nom. cons., *Nardia* Gray nom. cons., *Nothostrepta*
R. M. Schust., *Notoscyphus* Mitt., *Pisanoa* Hässel, *Protosyzygiella* (Inoue)
R. M. Schust., *Pseudocephaloziella* R. M. Schust., *Rhodoplagiochila* R. M.
Schust., *Roivainenia* Perss., *Scaphophyllum* Inoue, *Sphenolobopsis* R. M.
Schust. & N. Kitag., *Syzygiella* Spruce, *Tetralophozia* (R. M. Schust.)
Schljakov, *Tritomaria* Schiffn. ex Loeske, *Vanaea* (Inoue & Gradst.)
Inoue & Gradst.

Mesoptychiaceae Inoue & Steere, Bull. Natl Sci. Mus., Tokyo, B 1: 62.
1975. *Mesoptychia* (Lindb.) A. Evans

Gymnomitriaceae H. Klinggr., Höh. Crypt. Preuss. p. 16. 1858.
Acrolophozia R. M. Schust., *Apomarsupella* R. M. Schust., *Eremonotus*
Lindb. & Kaal. ex Pearson, *Gymnomitrion* Corda nom. cons.,
Herzogobryum Grolle, *Marsupella* Dumort., *Nanomarsupella* (R. M.
Schust.) R. M. Schust., *Nothogymnomitrion* R. M. Schust.,
Paramomitrion R. M. Schust., *Poeltia* Grolle, *Prasanthus* Lindb.,
Stephaniella J. B. Jack, *Stephaniellidium* S. Winkl. ex Grolle

Scapaniaceae Mig., Krypt.-Fl. Deutschl., Moose. p. 479. 1904 [incl.
Blepharidophyllaceae (R. M. Schust.) R. M. Schust., nom. inval. Syst.
Assoc. Spec. Vol. 14: 74. "1979" 1980]. *Blepharidophyllum* Ångstr.
[*Clandarium* (Grolle) R. M. Schust. = *Blepharidophyllum*], *Delavayella*
Steph., *Diplophyllum* (Dumort.) Dumort. nom. cons., *Douinia* (C. N.
Jensen) H. Buch, *Krunodiplophyllum* Grolle, *Macrodiplophyllum* (H.
Buch) Perss., *Scapania* (Dumort.) Dumort. nom. cons., *Scapaniella* H.
Buch

ORDER: PORELLALES (R. M. Schust.) Schljakov emend. Stotler & Stotl.-
Crand.[19]

SUBORDER: Porellineae R. M. Schust., J. Hattori Bot. Lab. 26: 229. 1963.
Porellaceae Cavers nom. cons., New Phytol. 9: 292. 1910. *Ascidiota* C.
Massal., *Macvicaria* W. E. Nicholson, *Porella* L.

SUBORDER: Jubulineae (Spruce) Müll. Frib., Lebermoose 1: 403.
1909.
Goebeliellaceae Verd., Man. Bryol. p. 425. 1932. *Goebeliella* Steph.
Jubulaceae H. Klinggr., Höh. Crypt. Preuss. p. 40. 1858. *Frullania*
Raddi, *Jubula* Dumort. nom. cons., [*Amphijubula* R. M. Schust.,
Neohattoria Kamim., *Schusterella* S. Hatt., Sharp & Mizut., and *Steerea* S.
Hatt. & Kamim. = *Frullania*]
Bryopteridaceae Stotler, Bryophyt. Biblioth. 3: 57–8. 1974. *Bryopteris*
(Nees) Lindenb.

Lejeuneaceae Casares-Gil nom. cons., Fl. Ibér. Brióf. 1: 703. 1919.
Acanthocoleus R. M. Schust., *Acantholejeunea* (R. M. Schust.) R. M.
Schust., *Acrolejeunea* (Spruce) Schiffn. nom. cons., *Amphilejeunea* R. M.
Schust., *Anoplolejeunea* (Spruce) Schiffn., *Aphanolejeunea* A. Evans,
Aphanotropis Herzog, *Archilejeunea* (Spruce) Schiffn., *Aureolejeunea*
R. M. Schust., *Austrolejeunea* (R. M. Schust.) R. M. Schust.,
Blepharolejeunea S. W. Arnell, *Brachiolejeunea* (Spruce) Schiffn.,
Bromeliophila R. M. Schust., *Calatholejeunea* K. I. Goebel, *Capillolejeunea*
S. W. Arnell, *Caudalejeunea* (Spruce) Schiffn., *Cephalantholejeunea* R. M.
Schust., *Cephalolejeunea* Mizut., *Ceratolejeunea* (Spruce) Schiffn.,
Cheilolejeunea (Spruce) Schiffn., *Cladolejeunea* Zwick., *Cololejeunea*
(Spruce) Schiffn., *Colura* (Dumort.) Dumort., *Crossotolejeunea* (Spruce)
Schiffn., *Cyclolejeunea* A. Evans, *Cyrtolejeunea* A. Evans, *Cystolejeunea* A.
Evans, *Dactylolejeunea* R. M. Schust., *Dactylophorella* R. M. Schust.,
Dendrolejeunea (Spruce) Lacout., *Dicladolejeunea* R. M. Schust.,
Dicranolejeunea (Spruce) Schiffn., *Diplasiolejeunea* (Spruce) Schiffn.,
Drepanolejeunea (Spruce) Schiffn., *Echinocolea* R. M. Schust.,
Echinolejeunea R. M. Schust., *Evansiolejeunea* Vanden Berghen,
Frullanoides Raddi, *Fulfordianthus* Gradst., *Haplolejeunea* Grolle,
Harpalejeunea (Spruce) Schiffn., *Hattoriolejeunea* Mizut., *Kymatolejeunea*
Grolle, *Leiolejeunea* A. Evans, *Lejeunea* Lib. nom. cons., *Lepidolejeunea*
R. M. Schust., *Leptolejeunea* (Spruce) Schiffn., *Leucolejeunea* A. Evans,
Lindigianthus Kruijt & Gradst., *Lopholejeunea* (Spruce) Schiffn. nom.
cons., *Luteolejeunea* Piippo, *Macrocolura* R. M. Schust., *Macrolejeunea*
(Spruce) Schiffn., *Marchesinia* Gray nom. cons., *Mastigolejeunea*
(Spruce) Schiffn., *Metalejeunea* Grolle, *Metzgeriopsis* K. I. Goebel,
Microlejeunea Steph., *Myriocolea* Spruce, *Myriocoleopsis* Schiffn.,
Nephelolejeunea Grolle, *Neurolejeunea* (Spruce) Schiffn., *Nipponolejeunea*
S. Hatt., *Odontolejeunea* (Spruce) Schiffn., *Omphalanthus* Lindenb. &
Nees, *Oryzolejeunea* (R. M. Schust.) R. M. Schust., *Otolejeunea* Grolle &
Tixier, *Phaeolejeunea* Mizut., *Physantholejeunea* R. M. Schust.,
Pictolejeunea Grolle, *Pluvianthus* R. M. Schust. & Schäfer-Verwimp,
Potamolejeunea (Spruce) Lacout., *Prionolejeunea* (Spruce) Schiffn.,
Ptychanthus Nees, *Pycnolejeunea* (Spruce) Schiffn., *Rectolejeunea* A.
Evans, *Rhaphidolejeunea* Herzog, *Schiffneriolejeunea* Verd.,
Schusterolejeunea Grolle, *Siphonolejeunea* Herzog, *Sphaerolejeunea*
Herzog, *Spruceanthus* Verd., *Stenolejeunea* R. M. Schust., *Stictolejeunea*
(Spruce) Schiffn., *Symbiezidium* Trevis., *Taxilejeunea* (Spruce) Schiffn.
nom. cons., *Thysananthus* Lindenb., *Trachylejeunea* (Spruce) Schiffn.

nom. cons., *Trocholejeunea* Schiffn., *Tuyamaella* S. Hatt., *Tuzibeanthus* S. Hatt., *Verdoornianthus* Gradst., *Vitalianthus* R. M. Schust. & Giancotti
ORDER: RADULALES (R. M. Schust.) Stotler & Stotl.-Crand., stat. et ordo nov.[20]

> **Radulaceae** (Dumort.) Müll. Frib., Lebermoose 1: 404. 1909. *Radula* Dumort. nom. cons.

ORDER: PLEUROZIALES (R. M. Schust.) Schljakov, Bot. Zhurn. 57: 505. 1972.

> **Pleuroziaceae** (Schiffn.) Müll. Frib., Lebermoose 1: 404. 1909. *Pleurozia* Dumort. [*Eopleurozia* R. M. Schust. = *Pleurozia*]

Annotations to the classification

1. Since *Marchantia* is the most universally recognized liverwort genus, it is utilized as the prefix for the phylum name. Latin validation is made with reference to that for the Hepaticophyta "Hepatophyta" in Stotler and Crandall-Stotler (Bryologist 80: 425. 1977). This phylum is exclusive of the mosses and hornworts.

2. Schuster (1984*a*, 1992*a*) incorrectly cites Cavers as the author for the Sphaerocarpineae. We here provide the following Latin diagnosis for validation of this suborder: *Plantae thallosae, gametangiis in involucris ampulliformibus. Cellulae olei deest.*

3. Schuster (Bryologist 61:38. 1958) published the Riellineae *sine descr. lat.* Reference to the diagnosis of the Riellales Schljakov (Bot. Zhurn. 57: 506. 1972) may serve to validate the suborder Riellineae as it has the same circumscription as Schljakov's order.

4. Müller (1940) published the Cyathodiaceae in Rabenhorst, Krypt.-Fl. Deutschl. (ed. 2), 6 (Ergänzungsb.) p. 182, but failed to provide a Latin diagnosis or description. The Cyathodioideae Grolle is elevated to family rank without change in circumscription; the Cyathodiaceae is here validated with reference to the Latin diagnosis of the subfamily by Grolle (J. Bryology 7: 208. 1972).

5. The Ricciales Schljakov (Bot. Zhurn. 57: 499. 1972) included the Monocarpineae and Corsiniineae as well as the Ricciineae. We here restrict this order to include only the Ricciaceae and the Oxymitraceae, the Ricciineae sensu Buch (Suomen Maksasammalet. p. 39. 1936).

6. The Blasiineae R. M. Schust. is elevated to ordinal rank without change in circumscription. The Blasiales is here validated with reference to the Latin diagnosis of the suborder by Schuster (Phytologia 56: 66. 1984).

7. Schljakov (Bot. Zhurn. 57: 499. 1972) restricted the Treubiales to include only the Treubiineae, placing the Phyllothalliineae in the order Phyllothalliales. We emend the Treubiales to include both of these suborders.

8. The Treubiineae was published by Schuster in 1964 (J. Hattori Bot. Lab. 27: 184) but was never validated. The Latin diagnosis of the Treubiales by Schljakov (Bot. Zhurn. 57: 499. 1972) may serve as the diagnosis of the suborder as it has the same circumscription as Schljakov's order. We do not, however, credit Schljakov as the validating author since he did not recognize the rank Treubiineae.

9. The Fossombroniales was restricted by Schljakov (Bot. Zhurn. 57: 500. 1972) to include only the Fossombroniaceae. We emend this order to include the suborders Fossombroniineae and Pelliineae.

10. Schuster proposed a narrowly circumscribed Fossombroniineae in 1964. Later, in a footnote (Schuster 1992*b*) he referenced a 1987 publication for the Latin diagnosis, but

this citation dealt only with hornworts. We expand the Fossombroniineae to include the Fossombroniaceae and the Allisoniaceae and validate the suborder with the following Latin diagnosis: *Plantae foliosae vel thallosae. Capsulae sphaericae, dehiscentiis irregularibus. Germinatio exospora.*

11. The Pelliineae R. M. Schust. was provided with a Latin diagnosis by Schljakov (Bot. Zhurn. 57: 500. 1972). While this order sensu Schuster (1964, 1984*a*, 1992*a*) contained the Pelliaceae and the Allisoniaceae, Schljakov (1972, 1975) broadened it to include also the Blasiaceae, Pallaviciniaceae, and Makinoaceae. We emend and restrict this suborder to include only the Pelliaceae and the Sandeothallaceae.

12. Since Chalaud (1930), the Metzgeriales has been equated with the "Jungermanniaceae anakrogynae" of Leitgeb and Schiffner, or to the whole of simple thalloid liverworts. Our concept is more comparable to that of Schljakov who validated the order in 1972 (Bot. Zhurn. 57: 500) although we restrict it to include only the suborders Pallaviciniineae and Metzgeriineae. Schljakov (1972, 1975) included the Pelliineae sensu lato (within which he reduced the Pallaviciniineae) and the Metzgeriineae.

13. Schuster erected the Pallaviciniineae (Phytologia 56: 65. 1984) to accommodate the Pallaviciniaceae, the Sandeothallaceae, and the Makinoaceae. We emend this suborder to encompass the Pallavicinaceae and Makinoaceae but transfer the Sandothallaceae to the Pelliineae. In addition, we merge the suborder Hymenophytineae R. M. Schust. (Phytologia 56: 66. 1984), accepted here at the diminished rank of family.

14. Engler established the Jungermanniidae (Unterklasse "Jungermanniales," in A. Engler & K. Prantl, Naturl. Pflanzenfam. I, 3. I: 1. 1893) to include the Jungermanniaceae anakrogynae and Jungermanniaceae akrogynae. Following segregation of the anakrogynae or simple thalloids into the Metzgeriidae (Bartholomew-Began 1990) this subclass has been restricted to the acrogynae or leafy hepatics, exclusive of the Haplomitriales and, of course, the Takakiophytes. The Jungermanniidae is basically similar to most authors' concept of the old Jungermanniales.

15. Lepicoleales Stotler & Stotl.-Crand., ordo nov. *Plerumque, folia dissecta, triseriata. Plantae perigyniis succulentis, perianthiis reductis vel deest (praeter Ptilidiineae). Setae grandes, capsulae parietibus multistratosis.*

16. With the exception of Schljakov (1972, 1975) hepaticologists have placed virtually all leafy liverworts into the Jungermanniales, validated by Klinggräff (Höh. Crypt. Preuss. p. 16. 1858). This order is here much narrowed in concept and accommodates only a portion of the leafies.

17. In order to validate the problematic suborder Antheliineae, we here reference the Latin description for the single family that it includes, the Antheliaceae (Schuster, J. Hattori Bot. Lab. 26: 236. 1963).

18. The Jungermanniineae R. M. Schust., nom. invalid (Amer. Midl. Naturalist 49: 302. 1953; emend. Schuster, Syst. Assoc. Special Vol. 14: 74. 1979[1980]) has the same circumscription as the Lophoziineae Schljakov (Bot. Zhurn. 57: 504. 1972). Although the Lophoziineae is a legitimate name, the correct name for this suborder must be the Jungermanniineae in compliance with Article 16.1 of the I.C.B.N. (Greuter *et al.*, Regnum Veg. 131: 25. 1994). Reference is here made to the Latin diagnosis of the Lophoziineae Schljakov (Bot. Zhurn. 57: 504. 1972) to validate the Jungermanniineae R. M. Schust. ex Stotler & Stotl.-Crand.

19. The Porellales sensu Schljakov (Bot. Zhurn. 57: 505. 1972) included both the Porellineae and the Radulineae. We accept this order to comprise the Porellineae along

with the Jubulineae but sequester and elevate the Radulineae into the Radulales. 20. The Radulineae R. M. Schust. is elevated to ordinal rank without change in circumscription. The order Radulales is here validated with reference to the Latin diagnosis of the suborder (Schuster, J. Hattori Bot. Lab. 26: 229. 1963).

Acknowledgments

This work has been supported by NSF grant DEB-9521883, which is gratefully acknowledged. Thanks is given to John McNeill for clarification and helpful suggestions in regards to nomenclatural problems. We also acknowledge the Green Plant Phylogeny Research Coordination Group for stimulating workshops and discussions (USDA grant 94–37105–0713).

REFERENCES

Apostolakos, P. & Galatis, B. (1998). Microtubules and gametophyte morphogenesis in the liverwort *Marchantia paleacea* Bert. In *Bryology for the Twenty-first Century*, ed. J. W. Bates, N. W. Ashton, & J. G. Duckett, pp. 205–21. Leeds: Maney and British Bryological Society.

Asthana, A. K. & Srivastava, S. C. (1991). Indian hornworts (A taxonomic study). *Bryophytorum Bibliotheca*, **42**, 1–159 + 49 plates.

Bartholomew-Began, S. E. (1990). Classification of the Haplomitriales and Metzgeriales into the subclass Metzgeriidae, subclass nov. (Hepatophyta, Jungermanniopsida). *Phytologia*, **69**, 464–6.

(1991). A morphogenetic re-evaluation of *Haplomitrium* Nees (Hepatophyta). *Bryophytorum Bibliotheca*, **41**, 1–297 + 508 figures.

Bischler-Causse, H. (1993). *Marchantia* L. The European and African taxa. *Bryophytorum Bibliotheca*, **45**, 1–129.

Bischler, H. (1998). Systematics and evolution of the genera of the Marchantiales. *Bryophytorum Bibliotheca*, **51**, 1–201.

Boisselier-Dubayle, M.-C., Lambourdière, J., Leclerc, M.-C., & Bischler, H. (1997). Phylogenetic relationships in the Marchantiales (Hepaticae). Apparent incongruence between morphological and molecular data. *Compte Rendu de l'Académie des Sciences, Paris, Série 3, Sciences de la Vie*, **320**, 1013–20.

Bopp, M. & Capesius, I. (1998). A molecular approach to bryophyte systematics. In *Bryology for the Twenty-first Century*, ed. J. W. Bates, N. W. Ashton, & J. G. Duckett, pp. 79–88. Leeds: Maney and British Bryological Society.

Bower, F. O. (1890). On antithetic as distinct from homologous alternation of generations in plants. *Annals of Botany*, **4**, 347–70.

Brown, R. C. & Lemmon, B. E. (1988). Cytokinesis occurs at boundaries of domains delimited by nuclear-based microtubules in sporocytes of *Conocephalum conicum* (Bryophyta). *Cell Motility and the Cytoskeleton*, **11**, 139–46.

(1990). Sporogenesis in bryophytes. In *Microspores. Evolution and Ontogeny*, ed. S. Blackmore & R. B. Knox, pp. 55–94. London: Academic Press.

Buch, H. (1936). *Suomen Maksasammalet*. Helsinki: Kustannusosakeyhtiä.

Campbell, E. O. (1954). The structure and development of *Monoclea forsteri*, Hook. *Transactions of the Royal Society of New Zealand*, **82**, 237–48.

Capesius, I. & Bopp, M. (1997). New classification of liverworts based on molecular and morphological data. *Plant Systematics and Evolution*, **207**, 87–97.

Cavers, F. (1903a). Explosive discharge of antherozoids in *Fegatella conica*. *Annals of Botany*, **17**, 270–4.

(1903b). Notes on Yorkshire Bryophytes. I. *Petalophyllum ralfsii*. *Naturalist*, **1903**, 327–44.

(1904). On the structure and development of *Monoclea forsteri* Hook. *Revue Bryologique*, **21**, 69–80.

(1911). The inter-relationships of the Bryophyta. *New Phytologist Reprint*, **4**, 1–203.

Chalaud, G. (1928). *Le Cycle évolutif de Fossombronia pusilla* Dum. Paris: Librairie Générale de l'Enseignement.

(1930). Les derniers stades de la spermatogénèse chez les Hépatiques. *Annales Bryologici*, **3**, 41–50.

Church, A. H. (1919). Thalassiophyta and the subaerial transmigration. *Botanical Memoirs*, **3**, 1–95.

Crandall, B. J. (1969). Morphology and development of branches in the leafy Hepaticae. *Beihefte zur Nova Hedwigia*, **30**, 1–261.

Crandall-Stotler, B. (1981). Morphology/anatomy of hepatics and anthocerotes. *Advances in Bryology*, **1**, 315–98.

(1984). Musci, hepatics and anthocerotes – an essay on analogues. In *New Manual of Bryology*, ed. R. M. Schuster, pp. 1093–129. Nichinan: Hattori Botanical Laboratory.

Crandall-Stotler, B. & Bozzola, J. J. (1988). Fine structure of the meristematic cells of *Takakia lepidozioides* Hatt. et H. Inoue (Takakiophyta). *Journal of the Hattori Botanical Laboratory*, **64**, 197–218.

Crandall-Stotler, B. & Guerke, W. R. (1980). Developmental anatomy of *Jubula* Dum. (Hepaticae). *Bryologist*, **83**, 179–201.

Crandall-Stotler, B., Stotler, R. E., & Mishler, B. D. (1997). Phylogenetic relationships within the Metzgeriidae (simple thalloid liverworts) as inferred from morphological characters. *American Journal of Botany*, **84**(6 Supplement), 3–4 [abstract].

Dillenius, J. J. (1741). *Historia Muscorum*. Oxford: Theatro Sheldoniano.

Douin, C. (1916). Le pédicelle de la capsule des hépatiques. *Revue Générale de Botanique*, **28**, 129–32.

Doyle, W. T. (1962). The morphology and affinities of the liverwort *Geothallus*. *University of California Publications in Botany*, **33**, 185–267.

Duckett, J. G. & Ligrone, R. (1995). The formation of catenate foliar gemmae and the origin of the oil bodies in the liverwort *Odontoschisma denudatum* (Mart.) Dum. (Jungermanniales): a light and electron microscope study. *Annals of Botany*, **76**, 406–19.

Dumortier, B. C. (1822). *Commentationes Botanicae*. Tournay: Ch. Casterman-Dien.

Endlicher, S. (1841). *Enchiridion Botanicum*. Leipzig: W. Engelmann.

Engler, A. (1893). Embryophyta Zoidiogama (Archegoniateae). In *Die natürlichen Pflanzenfamilien*, vol. 1, ed. A. Engler & K. Prantl, pp. 1–2. Leipzig: W. Engelmann.

Evans, A. W. (1920). The North American species of *Asterella*. *Contributions from the United States National Herbarium*, **20**, 247–312 + vii–viii.

(1927). A further study of the American species of *Symphyogyna*. *Transactions of the Connecticut Academy of Arts and Sciences*, **28**, 295–354 + plate I.

(1939). The classification of the Hepaticae. *Botanical Review*, **5**, 49–96.

Fulford, M. H. (1963). Manual of the leafy Hepaticae of Latin America – Part I. *Memoirs of the New York Botanical Garden*, **11**, 1–172.

Furuki, T. (1991). A taxonomical revision of the Aneuraceae (Hepaticae) of Japan. *Journal of the Hattori Botanical Laboratory*, **70**, 293–397.

Galatis, B., Apostolakos, P., & Katsaros, C. (1978*a*). Ultrastructural studies on the oil bodies of *Marchantia paleacea* Bert. I. Early stages of oil-body cell differentiation: origination of the oil body. *Canadian Journal of Botany*, **56**, 2252–67.

Galatis, B., Katsaros, C., & Apostolakos, P. (1978*b*). Ultrastructural studies on the oil bodies of *Marchantia paleacea* Bert. II. Advanced stages of oil-body cell differention: synthesis of lipophilic material. *Canadian Journal of Botany*, **56**, 2268–85.

Garbary, D. J. & Renzaglia, K. S. (1998). Bryophyte phylogeny and the evolution of land plants: evidence from development and ultrastructure. In *Bryology for the Twenty-first Century*, ed. J. W. Bates, N. W. Ashton, & J. G. Duckett, pp. 45–63. Leeds: Maney and British Bryological Society.

Garbary, D. J., Renzaglia, K. S., & Duckett, J. G. (1993). The phylogeny of land plants: a cladistic analysis based on male gametogenesis. *Plant Systematics and Evolution*, **188**, 237–69.

Gottsche, C. M., Lindenberg, J. B. G., & Nees von Esenbeck, C. G. (1844–7). *Synopsis Hepaticarum*. Hamburg: Meissner.

Gradstein, S. R, Matsuda, R., & Asakawa, Y. (1981). Studies on Colombian cryptogams XIII. Oil bodies and terpenoids in Lejeuneaceae and other selected Hepaticae. *Journal of the Hattori Botanical Laboratory*, **50**, 231–48.

Gray, S. F. (1821). *A Natural Arrangement of British Plants*. London: Baldwin, Cradock & Joy.

Grolle, R. (1972). Die Namen der Familien und Unterfamilien der Lebermoose (Hepaticopsida). *Journal of Bryology*, **7**, 201–36.

(1983). Nomina generica Hepaticarum; references, types and synonymies. *Acta Botanica Fennica*, **121**, 1–62.

Hasegawa, J. (1984). Taxonomical studies on Asian Anthocerotae IV. A revision of the gencra *Anthoceros*, *Phaeoceros* and *Folioceros* in Japan. *Journal of the Hattori Botanical Laboratory*, **57**, 241–72.

(1988). A proposal for a new system of the Anthocerotae, with a revision of the genera. *Journal of the Hattori Botanical Laboratory*, **64**, 87–95.

Haupt, A. W. (1918). A morphological study of *Pallavicinia lyellii*. *Botanical Gazette*, **66**, 524–33.

(1920). Life history of *Fossombronia cristula*. *Botanical Gazette*, **69**, 318–31.

(1943). Structure and development of *Symphyogyna brasiliensis*. *Botanical Gazette*, **105**, 193–201.

Hébant, C. (1977). The conducting tissues of bryophytes. *Bryophytorum Bibliotheca*, **10**, 1–157 + 80 plates.

Hedderson, T. A., Chapman, R. L., & Rootes, W. L. (1996). Phylogenetic relationships of bryophytes inferred from nuclear-encoded rRNA gene sequences. *Plant Systematics and Evolution*, **200**, 213–24.

Hedderson, T. A., Chapman, R., & Cox, C. J. (1998). Bryophytes and the origins and diversification of land plants: new evidence from molecules. In *Bryology for the Twenty-first Century*, ed. J. W. Bates, N. W. Ashton, & J. G. Duckett, pp. 205–21. Leeds: Maney and British Bryological Society.

Hutchinson, A. H. (1915). Gametophyte of *Pellia epiphylla*. *Botanical Gazette*, **60**, 134–43.

Ingold, C. T. (1939). *Spore Discharge in Land Plants*. Oxford: Clarendon Press.

Johnson, D. S. (1904). The development and relationship of *Monoclea*. *Botanical Gazette*, **38**, 185–205 + plates XVI and XVII.

Jussieu, A. L. de (1789). *Genera Plantarum*. Paris: Viduam Herissant.

Kachroo, P. (1958). Morphology of Rebouliaceae. III. Development of sex organs, sporogonium and interrelationships of the various genera. *Journal of the Hattori Botanical Laboratory*, **19**, 1–24.

Kelley, C. B. & Doyle, W. T. (1975). Differentiation of intracapsular cells in the sporophyte of *Sphaerocarpos donnellii*. *American Journal of Botany*, **62**, 547–59.

Kenrick, P. & Crane, P. R. (1997). *The Origin and Early Diversification of Land Plants: A Cladistic Study*. Washington: Smithsonian Institution Press.

Klinggräff, H. von (1858). *Die höheren Cryptogamen Preussens*. Königsberg: Wilhelm Koch.

Knapp, E. (1930). Untersuchungen über die Hüllorgane um Archegonien und Sporogonien der akrogynen Jungermanniaceen. *Botanische Abhandlungen*, **16**, 1–168.

Kobiyama, Y. & Crandall-Stotler, B. (1999). Studies of specialized pitted parenchyma cells of the liverwort *Conocephalum* Hill and their phylogenetic implications. *International Journal of Plant Sciences*, **160**, 351–70.

Kronestedt, E. (1982). Anatomy of *Ricciocarpus natans*, with emphasis on fine structure. *Nordic Journal of Botany*, **2**, 353–67.

Kuwahara, Y. (1986). The Metzgeriaceae of the neotropics. *Bryophytorum Bibliotheca*, **28**, 1–254.

Lang, W. H. (1907). On the sporogonium of *Notothylas*. *Annals of Botany*, **21**, 201–10.

Leitgeb, H. (1874–81). *Untersuchungen über die Lebermoose*. I–VI; I. *Blasia pusilla*, 1874; II. Die Foliosen Jungermannieen, 1875; III. Die Frondosen Jungermannieen, 1877; IV. Die Riccieen, 1879; V. Die Anthoceroteen, 1879; VI. Die Marchantieen, 1881. Vols. I–III, Jena: O. Deistungs Buchhandlung; Vols. IV–VI, Graz: Leuschner & Lubensky.

Lewis, L. A., Mishler, B. D., & Vilgalys, R. (1997). Phylogenetic relationships of the liverworts (Hepaticae), a basal embryophyte lineage, inferred from nucleotide sequence data of the chloroplast gene *rbcL*. *Molecular Phylogenetics and Evolution*, **7**, 377–93.

Ligrone, R., Duckett, J. G., & Renzaglia, K. S. (1993). The gametophyte–sporophyte junction in land plants. *Advances in Botanical Research*, **19**, 232–317.

Lilienfeld, F. (1911). Beiträge zur Kenntnis der Art *Haplomitrium hookeri*. *Bulletin International de l'Académie des Sciences de Cracovie*, **1911**, 315–99.

Linnaeus, C. (1753). *Species Plantarum*. Stockholm: Laurentii Salvii.

Malek, O., Lättig, K., Hiesel, R., Brennicke, A., & Knoop, V. (1996). RNA editing in bryophytes and a molecular phylogeny of land plants. *European Molecular Biology Organization Journal*, **15**, 1403–11.

Meissner, K., Frahm, J.-P., Stech, M., & Frey, W. (1998). Molecular divergence patterns and infrageneric relationship of *Monoclea* (Monocleales, Hepaticae). *Nova Hedwigia*, **67**, 289–302.

Micheli, P. A. (1729). *Nova Plantarum Genera*. Florence: Bernardi Paperinii.

Mishler, B. D. & Churchill, S. P. (1984). A cladistic approach to the phylogeny of the "bryophytes." *Brittonia*, **36**, 406–24.

Müller, K. (1905). Beitrag zur Zenntnis der ätherischen Öle bei Lebermoosen. *Zeitschrift für physiologischen Chemie*, **45**, 299–319.

(1939). Untersuchungen über die Ölkörper der Lebermoose. *Berichte der deutschen botanischen Gesellschaft*, **57**, 325–70.

(1940). Die Lebermoose. In *Dr. L. Rabenhorst's Kryptogamen-Flora von Deutschland, Oesterreich und der Schweiz*, 2nd edn, vol. 6, ed. L. Rabenhorst, pp. 161–320. Leipzig: Eduard Kummer.

(1948*a*). Morphologische und anatomische Untersuchungen an Antheridien beblätter Jungermannien. *Botaniska Notiser*, **1948**, 71–80.

(1948*b*). Der systematische Wert von Sporophytenmerkmalen bei den beblättern Lebermoosen. *Svensk Botanisk Tidskrift*, **42**, 1–16.

Nehira, K. (1984). Spore germination, protonema development and sporeling development. In *New Manual of Bryology*, ed. R. M. Schuster, pp. 343–85. Nichinan: Hattori Botanical Laboratory.

Neidhart, H. V. (1979[1980]). Comparative studies of sporogenesis in bryophytes. In *Bryophyte Systematics*, ed. G. C. S. Clarke & J. G. Duckett, pp. 251–80. London: Academic Press.

O'Hanlon, M. E. (1930). Gametophyte development in *Reboulia hemisphaerica*. *American Journal of Botany*, **17**, 765–9.

(1934). Comparative morphology of *Dumortiera*. *Botanical Gazette*, **96**, 154–64.

Parihar, N. S. (1961). *An Introduction to Embryophyta*, 3rd. edn, vol. 1, *Bryophyta*. Allahabad: Central Book Depot.

Perold, S. M. (1993). The hepatics, *Symphyogyna podophylla* and *Pallavicinia lyellii* (Pallaviciniaceae) in southern Africa. *Bothalia*, **23**, 15–23.

Pfeffer, W. (1874). Die Ölkörper der Lebermoose. *Flora*, **57**, 2–6, 17–27, 33–43.

Pihakaski, K. (1972). Histochemical studies on the oil bodies of two liverworts, *Pellia epiphylla* and *Bazzania trilobata*. *Acta Botanica Fennica*, **9**, 65–76.

Raddi, G. (1808). Di alcune specie nuove e rare di piante crittogame ritrovate nei contorni di Firenze. *Atti dell' Accademia delle Scienze di Siena detta de' Fisiocritici*, **9**, 230–40.

(1818). Jungermanniografia Etrusca. Preprinted from: *Memorie di Matematica e di Fisica della Società Italiana delle Scienze Residente in Modena*, **18**, 14–56.

Renzaglia, K. S. (1978). A comparative morphology and developmental anatomy of the Anthocerotophyta. *Journal of the Hattori Botanical Laboratory*, **44**, 31–90.

(1982). A comparative developmental investigation of the gametophyte generation in the Metzgeriales (Hepatophyta). *Bryophytorum Bibliotheca*, **24**, 1–253.

Renzaglia, K. S., Brown, R. C., Lemmon, B. E., Duckett, J. G., & Ligrone, R. (1994). Occurrence and phylogenetic significance of monoplastidic meiosis in liverworts. *Canadian Journal of Botany*, **72**, 65–72.

Renzaglia, K. S., McFarland, K. D., & Smith, D. K. (1997). Anatomy and ultrastructure of the sporophyte of *Takakia ceratophylla* (Bryophyta). *American Journal of Botany*, **84**, 1337–50.

Saxton, W. T. (1931). The life-history of *Lunularia cruciata* (L.) Dum., with special reference to the archegoniophore and sporophyte. *Transactions of the Royal Society of South Africa*, **19**, 259–68.

Schertler, M. M. (1979). Development of the archegonium and embryo in *Lophocolea heterophylla*. *Bryologist*, **82**, 576–82.

Schiffner, V. (1893). Hepaticae (Lebermoose). In *Die natürlichen Pflanzenfamilien*, vol. 1, ed. A. Engler & K. Prantl, pp. 3–96. Leipzig: W. Engelmann.

(1895). Hepaticae (Lebermoose). In *Die natürlichen Pflanzenfamilien*, vol. 1, ed. A. Engler & K. Prantl, pp. 97–114. Leipzig: W. Engelmann.

Schljakov, R. N. (1972). On the higher taxa of liverworts – class Hepaticae s. str. *Botanicheskii Zhurnal*, **57**, 496–508. [in Russian]

(1975). *Liverworts: Morphology, Phylogeny, Classification*. Leningrad: Akademia Nauk S.S.S.R. [in Russian]

Schofield, W. B. & Hébant, C. (1984). The morphology and anatomy of the moss gametophore. In *New Manual of Bryology*, ed. R. M. Schuster, pp. 627–57. Nichinan: Hattori Botanical Laboratory.

Schuster, R. M. (1958). Keys to the orders, families and genera of Hepaticae of America north of Mexico. *Bryologist*, **61**, 1–66.

(1963). Studies on antipodal Hepaticae, I. Annotated keys to the genera of antipodal Hepaticae with special reference to New Zealand and Tasmania. *Journal of the Hattori Botanical Laboratory*, **26**, 185–309.

(1964). Studies on antipodal Hepaticae, IV. Metzgeriales. *Journal of the Hattori Botanical Laboratory*, **27**, 182–216.

(1966). *The Hepaticae and Anthocerotae of North America, East of the Hundredth Meridian*, vol. 1. New York: Columbia University Press.

(1969). *The Hepaticae and Anthocerotae of North America, East of the Hundredth Meridian*, vol. 2. New York: Columbia University Press.

(1972[1973]). Phylogenetic and taxonomic studies on Jungermanniidae. *Journal of the Hattori Botanical Laboratory*, **36**, 321–405.

(1980). *The Hepaticae and Anthocerotae of North America, East of the Hundredth Meridian*, vol. 4. New York: Columbia University Press.

(1984*a*). Evolution, phylogeny and classification of the Hepaticae. In *New Manual of Bryology*, ed. R. M. Schuster, pp. 892–1070. Nichinan: Hattori Botanical Laboratory.

(1984*b*). Comparative anatomy and morphology of the Hepaticae. In *New Manual of Bryology*, ed. R. M. Schuster, pp. 760–891. Nichinan: Hattori Botanical Laboratory.

(1984*c*). Morphology, phylogeny and classification of the Anthocerotae. In *New Manual of Bryology*, ed. R. M. Schuster, pp. 1071–92. Nichinan: Hattori Botanical Laboratory.

(1984*d*). Diagnoses of some new taxa of Hepaticae. *Phytologia*, **56**, 65–74.

(1992*a*). *The Hepaticae and Anthocerotae of North America, East of the Hundredth Meridian*, vol. 5. Chicago: Field Museum of Natural History.

(1992*b*). The oil-bodies of the Hepaticae. I. Introduction. *Journal of the Hattori Botanical Laboratory*, **72**, 151–62.

(1997). On *Takakia* and the phylogenetic relationships of the Takakiales. *Nova Hedwigia*, **64**, 281–310.

Schuster, R. M. & Scott, G. A. M. (1966). A study of the family Treubiaceae (Hepaticae; Metzgeriales). *Journal of the Hattori Botanical Laboratory*, **32**, 219–68.

Smith, J. L. (1966). The liverworts *Pallavicinia* and *Symphyogyna* and their conducting system. *University of California Publications in Botany*, **39**, 1–83.

Stotler, R. & Crandall-Stotler, B. (1977). A checklist of the liverworts and hornworts of North America. *Bryologist*, **80**, 405–28.

Thomas, R. T. & Doyle, W. T. (1976). Changes in the carbohydrate constituents of elongating *Lophocolea heterophylla* setae (Hepaticae). *American Journal of Botany*, **63**, 1054–9.

Trevisan, V. (1877). Schema di una nuova classificazione delle Epatiche. *Memorie del Reale Istituto Lombardo di Scienze e Lettere, Serie 3, Classe di Scienze Matematiche e Naturali*, **4**, 383–451.

Zehr, D. R. (1980). An assessment of variation in *Scapania nemorosa* and selected related species (Hepatophyta). *Bryophytorum Bibliotheca*, **15**, 1–140.

3

Morphology and classification of mosses

3.1 Introduction

Bryophytes, in the broad sense, are the second largest phylum of land plants, after the angiosperms, and inhabit every continent. Among the bryophytes, mosses (the Bryophyta sensu stricto) are the most speciose group, comprising approximately 10 000 or more species. They differ from liverworts and hornworts in a suite of characters (see chapters 1, 2), including macroscopic features such as a gametophyte composed of stems with undivided and often costate leaves that are typically arranged all around the axis, and a sporophyte terminated by a capsule that is elevated by the elongation of a seta prior to maturity, and whose mouth is lined by teeth involved in regulating the dispersal of spores. Mosses along with liverworts and hornworts represent the oldest lineages among extant land plants (chapter 5 in this volume). Evidence is mounting that mosses are currently at the highest level of diversity in their evolutionary history (e.g., Kürschner & Parolly 1999). This rather recent diversification of mosses is likely correlated with the advent of angiospermous forests which provide a wide array of habitats. This trend is best exemplified in tropical rainforests, where most of the diversity in mosses is found.

Although mosses are rather small organisms, their morphology is relatively complex. Macroscopic characters are limited and most taxonomic concepts rely on features of the cells composing individual tissues. Throughout their evolutionary history, mosses have undergone repeated morphological reduction and simplification (Frey 1981), often as a result of colonizing specialized, and particularly xeric or ephemeral habitats (Vitt 1981). Significant innovations or loss thereof occurred early in the

diversification of mosses, as revealed by the extreme morphological differentiation between the major groups (here recognized as classes, see
below).

The objective of this chapter is to offer a general overview of moss
morphology by describing characters and some of the most widely
encountered states. This brief characterization is followed by a classification of all moss genera that we currently recognize.

3.2 Synopsis of moss morphology

Here we present a general overview of the diversity of gametophytic and
sporophytic character-states. For more comprehensive treatments of
moss morphology, the reader is referred to Campbell (1895), Müller
(1895–8), Ruhland (1924), Schofield and Hébant (1984), Vitt (1984), and
Schofield (1985).

3.2.1 Gametophyte

The typical life history of mosses begins with the germination of spores.
The resulting filamentous, uniseriate, branched protonema grows vege-
tatively, and in most taxa is short-lived. (In rare cases germination may be
endosporic, with no obvious protonema.) The protonemata are differentiated into two kinds of filaments, the chloronemata and the caulonemata
(Duckett *et al.* 1998). The chloronemata are mainly involved in assimilation and propagation (via gemmae). The main role of the caulonema is
colonization. In a few cases (e.g., *Sphagnum*, *Andreaea*), the protonemata
are plate-like rather than filamentous (Anderson & Crosby 1965). Gametophores arise laterally from the caulonemata, and it is these leafy, gametophytic stems that are most conspicuous in the moss life cycle. The
gametophyte and sporophyte of mosses are covered by a surface layer at
least analogous to the cuticle of vascular plants, with compounds similar
to those of cuticular waxes of vascular plants (Holloway 1982, Cook &
Graham 1998). Stems may be erect, creeping, or pendent, varying from
less than 1 mm long in mature plants to well over 1 m (e.g., *Fontinalis* spp.).
Mosses are generally separated into two groups based upon the orientation and branching of the stems, and the position of gametangia. In acrocarps (Fig. 3.1) the stems are erect, not or minimally branched, and often
grow in tufts. The apical cell of the stem is eventually used to produce a
terminal gametangium. In some cases growth of the plant is resumed by
the development of subapical innovations which assume the function of

Fig. 3.1. *Polytrichum piliferum* (Polytrichaceae), as an example of an acrocarpous moss. Note the unbranched, erect stems, sometimes terminating in a sporophyte.

stem. The earliest mosses were probably acrocarpous. For the most part acrocarpous mosses have colonized terrestrial habitats, especially soil and rocks. Certainly they are common on the bases of trees, but for a few notable exceptions, e.g., the Calymperaceae, they have not been successful as epiphytes. The second group of mosses is the pleurocarps (Figs. 3.2, 3.3). The gametangia are produced laterally along the stems and thus the apical cell continues vegetative growth. These mosses have mostly creeping stems that have extensive lateral branching, resulting in mats. Consequently, virtually all pleurocarpous mosses are perennial, whereas acrocarps can be either annual or perennial. Although some lineages of pleurocarps have been successful colonizers of terrestrial habitats (especially in temperate climates, e.g., *Hypnum* and *Brachythecium*), in the tropics the bulk of the epiphytic mosses are pleurocarps. Growth forms of mosses have received considerable attention, especially in the ecological literature. Communities have often been characterized primarily by this feature alone. The earliest work in this field (Giesenhagen 1910) divided tropical bryophytes into cushions, turfs, and mats, and individual stems into unbranched dendroid, branched dendroid, feather, bracket, and hanging forms. More recent work has divided these various types more finely (e.g., Mägdefrau 1982). It should be emphasized that growth form is

Fig. 3.2. *Helodium blandowii* (Helodiaceae), as an example of a pleurocarpous moss. Note the well-branched stems. Gametangia and sporophytes would be formed laterally along the stem. Most pleurocarps are prostrate in habit, but this one grows erect in wet areas.

not necessarily correlated with systematic position (see also La Farge-England 1996).

Rhizoids

Mosses are typically attached to the substrate by rhizoids. Although lacking a conductive function, they are not lacking in taxonomic usefulness (Koponen 1982). Rhizoids are characterized by being typically branched, and with diagonal cross-walls between the uniseriately arranged cells. They may be variously colored or hyaline, smooth or papillose. Rhizoids may arise solely from certain parts of the stems (or rarely leaves), or they may cover extensive portions of the stems with a felt-like tomentum. Rhizoids most commonly are of the micronematal type, characterized by a slender diameter, sparse branching, and produced on stems

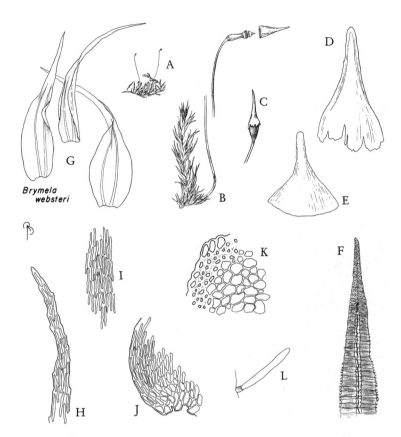

Fig. 3.3. *Brymela websteri* (Pilotrichaceae), as an example of a moss. A. Aspect, showing pleurocarpous habit. B. Short piece of stem with attached branch and sporophyte, the latter showing seta, capsule, operculum, and calyptra. C. Young capsule with calyptra. D. Calyptra. E. Operculum. F. Exostome tooth, showing median furrow. G. Leaves, showing strong double costa, a common feature of the Hookeriales. H. Leaf apex. I. Laminal cells at mid-leaf. J. Cells at base of leaf. K. Portion of stem cross-section. L. Axillary hair.

between leaves. In some groups (e.g., the Mniaceae) the rhizoids may be macronematal, with a larger diameter, extensive branching, and arising from around branch primordia.

Stems

Stem anatomy has been much used in differentiating taxa at all taxonomic levels (e.g., see Kawai & Ikeda 1970, Kawai 1971–91, Watanabe & Kawai 1975, Kawai *et al.* 1985, 1986, Kawai & Ochi 1987, Ron & Kawai 1991).

The external layer, or epidermis, is usually composed of relatively small, thick-walled cells (as seen in cross-section, Fig. 3.3K), but is sometimes differentiated as a hyalodermis of enlarged, thin-walled cells. Typically there is a layer of thick-walled cells subtending the epidermis, providing structural support for the moss. Internal to this layer is usually a layer of large, thin-walled parenchyma cells. Although considered "non-vascular" plants, mosses often have conducting tissue at the center of their stems. This central strand can be well developed, or completely lacking. Connections between the conducting tissues of the stem and that of the leaves occur in some Polytrichales (Hébant 1977). In most other mosses leaf traces are absent. Hydroids differ from analogous xylem in vascular plants by lacking lignin (Hébant 1977). Nevertheless, hydroids share several characteristics with tracheary elements (Schofield & Hébant 1984), which might suggest that they are derived from the same progenitor of vascular tissue.

Stems sometimes are variously ornamented with projections. The most conspicuous of these structures are paraphyllia. These filamentous or foliose, simple or branched structures are green and distributed over most areas of stems (and sometimes branches), especially on older parts. They are most common in pleurocarpous mosses (e.g., *Climacium*, *Hylocomium*, *Palustriella*). The paraphyllia provide additional photosynthetic area but their function is otherwise unknown. They may be involved in external water conductance. Rarely the epidermal cells of stems have small papillae (e.g., *Cyrto-hypnum pygmaeum*, *Heterocladium macounii*). In stems with branches, there are often other structures which are restricted to the periphery of branch primordia. These structures are called pseudo-paraphyllia and may be either filamentous or foliose. Although green, their function is most probably the protection of branch primordia, not photosynthetic enhancement. In recent years there has been considerable work on the origin of pseudoparaphyllia and some authors (Akiyama 1990, Akiyama & Nishimura 1993) have argued that there are two independent structures lumped under the single term: true pseudoparaphyllia (almost always filamentous) and "scaly leaves" which are little more than juvenile leaves. The question deserves a more detailed ontogenetic study to determine if the various structures originate from the apical cells of branches (presumably as in scaly leaves) or from the stem surface (presumably as in true pseudoparaphyllia).

In leaf axils is a structure that has garnered considerable taxonomic attention in recent years, axillary hairs (Fig. 3.3L). These uniseriate structures are apparently physiologically active when the leaves are juvenile (Ligrone 1986) and help keep the expanding leaves from dehydrating by secreting a mucilage. They vary in number from one to several per leaf axil. The basal cell or cells are often differentiated by being shorter and brownish. The apical cells, varying in number from one to over ten, are elongate and hyaline. Although often tedious to observe, they have proved very useful at different taxonomic levels (Hedenäs 1989*b*).

The architecture of the gametophyte can be described as modular (Mishler & De Luna 1991). The hierarchy of the modules is defined by the mode of branching. Two basic branching patterns occur in mosses: monopodial and sympodial (La Farge-England 1996). Sympodial branching occurs in perennial plants with axes of determinate growth. New branches develop from subapical meristems of the stem or primary module. These branches grow and function as new stems, sometimes bearing terminal sex organs. Sympodial branching thus reproduces a module of identical hierarchical level, and each module is of determinate growth. Sympodial branching is characteristic of all acrocarps and most epiphytic pleurocarps. In monopodial branching, the development of lateral branches is not determined by the cessation of apical meristematic activity. Lateral meristems develop and grow into branches, which are often morphologically and functionally distinct from the stems. Monopodial branching thus involves two distinct modules, here the stem and the branch. This is the standard pattern seen in most pleurocarps of the northern hemisphere (e.g., Hypnales sensu Brotherus 1925). It is typically associated with stable terrestrial habitats (see During 1979). Some mosses exhibit both modes of branching as they grow. In the Leucodontales (sensu Brotherus 1925, sub Isobryales) for example, which are mostly epiphytic, the creeping primary stem turns away from the substrate and the erect secondary stems often develop lateral branches. The stem will ultimately stop growing and, at some distance from the stem apex, a new primary stem arises to resume growth of the plant. This branching type is the most typical in epiphytic and to a lesser extent saxicolous mosses. Sympodial branching is likely the primitive condition in mosses. Monopodial branching is constant among true pleurocarps, but it has also arisen repeatedly among various acrocarpous lineages (La Farge-England 1996), and although pleurocarps often integrate both modes, they never branch following only a sympodial pattern.

Leaves

The apical cell of the stem is typically pyramidal with three cutting faces, but in a few taxa (e.g., *Fissidens*, *Distichium*), it has become lenticular and two-sided, resulting in distichously arranged leaves. In most taxa, though, phyllotaxy is spiral and leaves are arranged all around the stems (cf. Berthier 1972). Even in cases where the plants appear strongly flattened, the leaves still arise from all sides of the stems. As in sporophyte-dominant plants, leaves are the primary photosynthetic organ in the mosses. Moss leaves are typically formed heteroblastically, with the basal-most ones on a stem or branch significantly different than those formed later on the mature stems and branches. For the most part, leaves vary from lanceolate to ovate (Fig. 3.3G). Unlike in hepatics, moss leaves are almost never deeply divided into multiple lobes or filaments. The sole exception is *Takakia*, in which the leaves are divided into terete filamentous lobes. In some taxa of mosses, virtually all acrocarpous, leaves may be linear; in other mosses leaves may be orbicular. The leaf surface is most often flat or slightly concave. However, leaves may also be plicate (with longitudinal folds), undulate, or rugose. The leaf apex is often acute to acuminate, but it can also be obtuse, truncate, apiculate, or even retuse. The leaf insertion is mostly broad, with a wide attachment to the stem. Sometimes the leaf base may have variously developed auricles and sometimes the laminal cells on either side of the leaf extend down the stem as decurrencies. Almost all moss leaves are unistratose. Only in a few groups are multistratose leaves found, and these are mostly acrocarpous (e.g., Grimmiaceae, Diphysciaceae). The largest exception to leaves being unistratose is marginal thickening. Many mosses, in a large array of families, have thickened margins (e.g., *Fissidens* spp., *Racomitrium* spp., *Platylomella*). In some cases the cells of the margin are differentiated in shape, color, or wall thickness. Leaves with a border of hyaline, linear cells are not uncommon. When the margin is differentiated, it is often referred to as a limbidium. Limbate leaves presumably are an adaptation resulting in stronger structural integrity. Leaf margins are often not in the same plane as the rest of the lamina. Commonly the margins are bent under the leaf and then called recurved. When curled over the leaf, they are incurved. When the entire leaf is flat, the margins are referred to as plane. Most moss leaves are toothed on the margins. These teeth may be entire cells, cell ends, or lateral projections from cells. In almost all cases the teeth are larger toward the leaf apex, and smaller or absent toward the leaf base.

Most moss leaves have a costa (or midrib). In the acrocarps, a strong single costa is predominant. The costa originates from the base of the leaf (or rarely it is present above but absent below, e.g., some Dicnemonaceae). A single costa may extend beyond the laminal apex (excurrent), to the apex (percurrent), or almost anywhere below. The anatomy of the costa has been extensively used in some groups (especially the Pottiaceae) to differentiate taxa. Like in the stem, the external layers of the costa are often thick-walled, but sometimes are enlarged as a hyalodermis, either on one or both sides of the leaf. Across the middle of the costal cross-section there is sometimes a row of enlarged guide cells, essentially connecting the lamina on either side of the costa. Sometimes there are extensive areas of slender, very thick-walled cells (sclereids), that may be either below (abaxial to) the guide cells or on both sides of them. The costa may be short and double or essentially absent. This costal reduction seems to have occurred repeatedly in the pleurocarps, but is rare in acrocarps. The most unusual costal type is the strong double costa (Fig. 3.3G). This occurs commonly in the Hookeriales but is absent in acrocarps and rare in the Hypnales.

Some mosses, especially the Polytrichaceae and some Pottiaceae, have laminal lamellae. These structures, running parallel to the long axis of the leaf, are mostly unistratose, chlorophyllose sheets of cells. Often the terminal cell is differentiated from the subtending ones, and such features are used in taxonomic recognition. Typically the lamellae are situated over only the adaxial surface of the costa. Functionally, they are speculated to primarily increase photosynthetic area. Similar structures found in some Pottiaceae are uniseriate, chlorophyllose filaments. They presumably serve a similar function.

Laminal cells may vary in shape from linear to oblate (wider than long). Most acrocarpous mosses have cells that are relatively short or even isodiametric. On the other hand, most pleurocarps have elongate laminal cells (Fig. 3.3H, I). However, most cell shapes can be found in both groups. Oblate cells in the main part of the lamina are rare, and are seemingly correlated with extremely long-linear leaves with a narrow lamina (e.g., *Calymperes lonchophyllum*). The areolation of the Sphagnaceae is unique in the mosses. In this group, living, chlorophyllose cells (chlorocysts) form a reticulum that is enclosed by hyaline, hollow cells (hyalocysts). The leaves are still unistratose, but the green cells may be more exposed on either the inner or outer surface of the leaf, they may be equally exposed, or they may be entirely enclosed by the hyaline cells and not exposed on either surface. An extensive network of pores in the hyalocysts allows rapid uptake of water into the dead cells. The hyalocysts are typically reinforced with fibrils. A similar cell dimorphism occurs in the Leucobryaceae and some Calymperaceae, but here the chlorophyllose cells are embedded between two layers of leucocysts.

Typically cells are not uniform throughout the moss leaf, in addition to the possible differentiation of a limbidium. In leaves that have a relatively short apex, i.e., acute to obtuse or truncate, the apical laminal cells are usually shorter than those at midleaf (e.g., in *Platyhypnidium* and *Neckeropsis*). In almost all mosses the cells at the leaf insertion are differentiated, typically being longer and sometimes broader (Fig. 3.3J). In many groups the cells at the basal margins are differentiated. These so-called alar cells are typically subquadrate. They are sometimes colored or hyaline, and may be thick-walled or inflated and thin-walled. When there are leaf decurrencies, sometimes the cells of the decurrencies are similar to those of the alar regions, and sometimes they are different.

Laminal cells of mosses are sometimes ornamented, and this is more common in mosses adapted to dry habitats and is thus presumably adaptational in regard to water relations. The most common type of ornamentation is papillae. Although grouped under a single term, papillae are a heterogeneous assemblage of cuticular projections. They may be solid or hollow, they may be simple, C-shaped, coronate, or variously branched, and they range from scarcely visible to strongly projecting. Cells may have a single papilla, or may have multiple papillae in various arrangements, and the papillae may be only on one side of the leaf (most commonly the abaxial surface), or they may be on both sides. Papillae typically occur over the cell lumina, but sometimes (e.g., *Trachypus*) are positioned over the cell walls between the lumina. Until fairly recently another form of cell ornamentation was lumped into papillose, but is now standardly differentiated. In this case the upper end of the cell (or less often both ends, or even more rarely just the lower end of the cell) projects from the surface of the leaf. This is referred to as prorose or prorulose (depending on the degree of projection). Sometimes laminal cells are mammillose, with nipple-like projections. In this type of ornamentation, the lumen of the cell projects up into the projection, whereas in papillae the projection is entirely composed of wall material. Rarely cells have various other kinds of roughenings on their surfaces. When the upper laminal cells are ornamented, alar cells are almost always smooth.

Asexual propagula

Mosses have evolved numerous kinds of asexual reproductive structures (Correns 1899, Newton & Mishler 1994). When water is required for the transport of sperm to egg, and male and female plants may be quite distant (sometimes on different continents!), it is no surprise to find efficient mechanisms for vegetative dispersal. Most mosses can grow from leaf or stem fragments. Typically a leaf or stem fragment will first give rise to a protonematal stage prior to growth of a new leafy plant. Some taxa have developed specialized morphological strategies to take advantage of leaf fragmentation. In these taxa, widely dispersed across the mosses, the leaf apices are regularly broken off (typically referred to as fragile), either at the base of a specialized leaf apex (e.g, *Groutiella tomentosa*), or irregularly along cells lines (e.g., *Haplohymenium triste*). Sometimes almost the entire leaf can break into irregular laminal plates (e.g., *Zygodon fragilis*). Like in many other groups of plants and fungi, some mosses have developed a splash-cup mechanism for dispersing small gemmae which, at maturity, are loose within a cup (e.g., *Tetraphis*). Some mosses have evolved specialized gemmae in leaf axils (e.g., *Pohlia* spp.). These may be bulbil-like, or more elongate or vermicular; they may be single in leaf axils or clustered; they may be restricted to the apical portions of the stems, or throughout. Sometimes elongate, flagellate branches, usually in dense clusters, are formed in leaf axils and are easily shed (e.g., *Platygyrium*). These specialized branches have very reduced leaves. Sometimes specialized stems are produced solely for the production of asexual propagula, with leaves on these stems strongly reduced or absent (e.g., *Adelothecium*). Asexual reproductive structures are sometimes formed on leaves (e.g., *Calyptopogon*, *Calymperes*, *Syrrhopodon*). These often are in the form of few-celled gemmae or filaments. They may arise from specialized cells of the lamina, either along the margins, at the apex, or throughout, or they may arise from the costa, either on the abaxial or adaxial surface or both. Vegetative reproduction is not restricted to just the stems and leaves. Small bulbils are sometimes formed on rhizoids, presumably as perenniating structures (e.g., *Bryum* spp.). These may be of different colors and sizes in different taxa, and thus taxonomically useful (Arts 1989, 1994). In epiphytic or epiphyllous Hookeriales (esp. *Crossomitrium*), rhizoidal, filamentous asexual reproductive structures are not uncommon. Asexual reproductive structures are even known in some groups directly on the protonemata (e.g., Tetraphidales), but they are not known in the sporophyte generation. The diversity of asexual propagules is matched by a range of dispersal or abscission mechanisms for these gemmae, as well as by a variety of germination patterns (Duckett & Ligrone 1992).

Gametangia

The sex organs of mosses are archegonia (female) and antheridia (male). The flask-shaped archegonium contains a single egg cell in the venter, which is surmounted by an elongate neck. Numerous sperm cells are produced in the sack-like antheridium. The arrangement of gametangia in mosses is varied (Wyatt & Anderson 1984) and is often of taxonomic value. The two broad groups of sexuality are monoicous (with antheridia and archegonia on the same plant) and dioicous (with antheridia and archegonia on separate plants). In mosses sexuality terms typically have the ending -oicous whereas in vascular plants they typically end in -oecious, the latter being strictly applied to sporophytic or diploid sexuality. Within the monoicous arrangement of gametangia, the most common variations are autoicous, synoicous, paroicous, and heteroicous (= polyoicous). In the autoicous arrangement of gametangia the antheridia and archegonia are in separate inflorescences on the same plant. Within this broad grouping there is considerable variation, such as cladautoicous (with the male inflorescence terminal on a separate branch from the female), gonioautoicous (with the male inflorescence bud-like and axillary on the same stem or branch as the female), and rhizautoicous (with the male inflorescence on a short branch attached to the female stem by rhizoids and appearing to be a separate plant). In the synoicous condition the antheridia and archegonia are mixed within a single inflorescence, while in paroicous the antheridia and archegonia are in a single inflorescence but are not mixed, usually with the sex organs in different leaf axils. In plants with heteroicous sexuality, there is more than one form of monoicous inflorescence on the same plant (or species). In one form of dioicy, the male plants are dwarfed, consisting of little more than the male inflorescence and a few subtending, reduced leaves, and are epiphytic on the female plants, usually either on the leaves or tomentum of the stem. This is sometimes called a pseudautoicous or phyllodioicous condition.

In rare cases (e.g., *Splachnobryum*, *Takakia*), gametangia are naked in the axils of vegetative leaves. However, more typically the gametangia are surrounded and subtended by specialized leaves, usually differentiated from the vegetative leaves. The archegonia and their subtending leaves are referred to as perichaetia. The term is typically also used in synoicous inflorescences. Perichaetial leaves are commonly used taxonomic characters. They may become either larger or smaller from the outside to the inside of the inflorescence, and they typically resume growth after fertilization. The antheridia and their subtending leaves are called perigonia. The perigonial leaves or bracts are usually quite reduced and unlike the vegetative leaves, especially in autoicous taxa where the perigonia resemble axillary buds. Interspersed among the archegonia and antheridia within the inflorescences may be sterile hairs, referred to as paraphyses. Sometimes the apical cell of a paraphysis may be differentiated, e.g., in Funariaceae.

3.2.2 Sporophyte and associated gametophytic tissues

Seta

The leafless, unbranched sporophyte generation is intimately associated with the gametophyte. Although green for most of its development, the sporophyte is physiologically independent only shortly before sporogenesis (Paolillo & Bazzaz 1968). The basal-most part of the sporophyte is the foot, and it is embedded within the gametophyte. At the interface of the two generations is a transfer zone and the ultrastructural details of this area (cf. Ligrone & Gambardella 1988) have provided clues about the relationships of mosses with other land plants (e.g., Ligrone *et al.* 1993, Frey *et al.* 1996). The gametophytic tissue surrounding the foot is referred to as the vaginula. Traditional wisdom has the vaginula developing from the basal portions of the archegonium, but considering the occasional occurrence of archegonia and hairs on the vaginula, stem tissue seems also to be involved. The foot is the base of the seta.

The seta is the structure which elevates the sporangium, or spore capsule, and grows from an apical meristem. In most mosses the seta is elongate (Fig. 3.3A, B). However, in some reduced taxa (e.g., *Acaulon*, *Ephemerum*), the capsules remain immersed within the perichaetial leaves because the seta is greatly shortened. The seta is typically smooth, but in some mosses it is variously roughened (e.g., *Brachythecium* spp.) or even spinose (e.g., *Calyptrochaeta*). It is not uncommon for setae to be twisted and the pattern and direction of twisting may be taxonomically useful, e.g., in distinguishing subfamilies of Orthotrichaceae (Goffinet & Vitt 1998). There is usually a well-developed conducting system within the seta, even if such tissues are lacking in the gametophyte. Setae are mostly single per perichaetium, but polysety is known in a few groups (e.g., *Dicranum* spp.) where more than one archegonium per perichaetium is fertilized. In a few isolated groups (*Sphagnum* and *Andreaea*), the capsule is held aloft from the perichaetial leaves by an extension of gametophytic tissue, the pseudopodium. The foot of the capsule is embedded directly in this structure and there is no seta. In *Sphagnum*, the capsule is elevated at maturity, rather than long before sporogenesis as in most mosses.

Capsule

A single capsule develops atop the seta (Fig. 3.3B). In most mosses, capsules are stegocarpous (i.e., regularly dehiscent by shedding an operculum or lid), but in some groups, typically with immersed capsules, the capsules are cleistocarpous (i.e., without regular dehiscence), and spores are dispersed only upon decay and rupture of the capsule wall (Vitt 1981). Capsule shape varies greatly, from essentially globose to long–cylindric. Capsules may be symmetric, in which case the capsule is usually either erect or nutant (i.e., with the apex of the seta sharply bent up to 180° and the capsule nodding). Alternatively, capsules may be asymmetric. In these cases, as often found in pleurocarps and some acrocarps, the capsule itself is bent so that the mouth is oblique to the orientation of the rest of the capsule. Sometimes capsules are furrowed, either with eight or 16 ribs.

At the base of most capsules there is a differentiated neck, typically corresponding to the area below the spore sac within the capsule. In this area is where one most often finds stomata, although some groups of mosses lack stomata. The stomata are typically very similar to those in other land plants, i.e., with two guard cells around a pore. Rarely, there is a single doughnut-shaped guard cell (e.g., *Funaria*). In some groups they may be immersed in small pits (e.g., *Orthotrichum* spp.). In *Sphagnum* there are pairs of guard cells but no pore, and thus the structures are referred to as pseudostomata (Boudier 1988). In some mosses the neck of the capsule is greatly exaggerated (e.g., many Splachnaceae) while in others is it little more than a gradual tapering to the seta. The cells of the capsule wall are called exothecial cells. In a few groups these may be ornamented with various projections (e.g., *Symphyodon*), but for the most part they are smooth. However, the shape and size of the exothecial cells can be of taxonomic value. If a capsule is ribbed, the exothecial cells corresponding to the area of the ribs are usually narrower and thicker-walled. The cells directly below the mouth of the capsule are typically different than those of the rest of the capsule. In most mosses capsules dehisce by the loss of an operculum (Fig. 3.3B, E). The operculum itself may be almost flat to long–rostrate. At the mouth of the capsule there are sometimes cells specifically designed to aid in the dehiscence of the operculum, usually through swelling with water uptake. These cells, referred to as the annulus, may not be the result of a single origin. We define the annulus as differentiated cells which are deciduous, even if tardily so. In many groups there is a row of cells, usually differentiated in cell wall thickness, one to several cells high, directly intercalated between the exothecium and the operculum, which fragment at maturity. This type of annulus is typical of pleurocarps and some acrocarps. Another type of annulus found in acrocarps has the differentiated cells inside the capsule, at the exothecium–operculum junction. These are greatly swollen cells and, upon dehiscence, the annulus often falls as a complete unit (revoluble annulus). In an isolated group of mosses, *Andreaea*, *Andreaeobryum*, and *Takakia*, the capsule dehisces much like in hepatics, by longitudinal slits. In *Andreaea* there are typically four (rarely ten), long slits; in *Andreaeobryum* there are four to six, short irregular slits; and in *Takakia* there is a single spiral slit. It is not clear whether these modes of dehiscence are derived from a stegocarpous capsule, as suggested for *Andreaea* by Robinson and Shaw (1984).

In almost all mosses there is a column of sterile tissue running most of the length through the capsule. This structure is the columella. It has received little taxonomic attention but preliminary studies by the senior author indicate that it may well be a source of valuable systematic characters, in terms of shape (including cross-sectional) and length, especially in the acrocarps. In a few groups the apex of the columella is fused to the operculum (e.g., *Schistidium*). In this systylius condition, the columella eventually breaks free from its connection with the apex of the seta and falls with the operculum. In the Polytrichaceae the apex of the columella is expanded into a circular membrane, the epiphragm, which partially closes the capsule mouth after dehiscence. In *Sphagnum*, *Andreaea*, and *Andreaobryum* the columella is short, and the spore sac, rather than being cylindrical, is dome-shaped. Only in *Archidium* is a columella completely missing (Snider 1975).

Peristome

The capsule mouth of most stegocarpous mosses is lined by teeth, which together form the peristome (Fig. 3.3F). The peristome offers a variety of characters, some of which, such as the architecture and the pattern of cell

division in particular, are central to most modern higher-level classifications of mosses (e.g., Dixon 1932, Vitt 1984, Vitt *et al.* 1998). Much work has
been done in recent years on peristome architecture (e.g., Edwards 1979,
1984) and development (e.g., Shaw *et al.* 1987, 1989*a*, *b*, Shaw & Anderson
1988). Only a brief overview of the subject can be presented here. As in the
gametophyte, growth of the sporophyte is insured by the meristematic
activity of an apical cell. In the early stages of development (cf. Wenderoth
1931, Snider 1975, for a complete sequence of zygotic divisions) each cell
derived from the apical cell immediately undergoes an anticlinal division. By the time two derivatives have been produced and each has undergone a division, four cells are seen in transverse section below the apical
cell. Subsequent divisions lead to the differentiation of two fundamental
tissues, the endothecium and the amphithecium (Wenderoth 1931, Shaw
et al. 1987, 1989*a*, *b*). This organization is known as the fundamental
square (Goebel 1905). Each of the four merophytes will follow an identical
pattern of cell divisions. The endothecium will give rise to the columella
and sporogenous tissue in most mosses, except *Sphagnum* where the sporogenous tissue is of amphithecial origin. From the amphithecium are
derived the exothecial cells and their underlying parenchyma, which in
the apical portion of the capsule are differentiated into the peristome
forming layers. These innermost three layers of the amphithecium are
called the outer, the primary, and the inner peristomial layers (OPL, PPL,
and IPL, respectively), with the latter being adjacent to the endothecium
(Blomquist & Robertson 1941).

Edwards (1979) proposed the peristomial formula to describe the
numbers of cells in each of the three peristomial layers contributing to
the peristome. For ease of comparison, the peristomial formula uses the
number of cells in $\frac{1}{8}$ of the capsule circumference. For example, a 4:2:4
formula means that in a whole capsule there are 32, 16, and 32 cells in the
OPL, PPL, and IPL, respectively. Haplolepideous peristomes have a
formula of 4:2:3 and diplolepideous peristomes have formulae ranging
from 4:2:4 to 4:2:14. The peristome of *Funaria* has a formula of 4:2:4 (Fig.
3.4A, C). In addition, the peristome of *Funaria* has endostome segments
opposite the exostome teeth and all anticlinal walls in the IPL are aligned
with anticlinal walls of the PPL. In most other diplolepideous mosses,
where the endostome segments alternate with the exostome teeth, anticlinal walls of IPL cells are offset with regard to those in the PPL. Thus, the
4:2:4 formula of *Funaria* develops somewhat differently than the 4:2:4
formula of a *Pohlia* (Fig. 3.4D), for example.

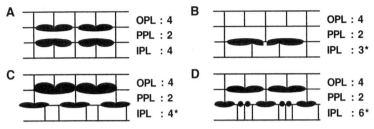

Fig. 3.4. Architecture of arthrodontous peristomes at maturity. One-eighth of the amphithecium is represented, with the corresponding peristomial formula indicating the number of cells found in one-eighth of each layer. Cell-wall thickening is represented in heavy black areas; fine lines correspond to walls that will be almost completely resorbed in the latest stages of development. A. Diplolepideous opposite peristome of *Funaria*. B. Haplolepideous peristome; note that each segment of the endostome is composed on its inner surface of one and a half cells. C. Diplolepideous alternate peristome as typically found in the Orthotrichaceae; for every two segments the inner peristomial layer (IPL) is composed of four cells at most, the inner surface of the segments bears a vertical line, although it is not median, or rarely so, depending on the degree of displacement of anticlinal walls during maturation, the outer peristomial layer (OPL) is thicker than the primary peristomial layer (PPL), and the exostome teeth are found paired rather than free. D. Typical diplolepideous alternate peristome (e.g., in the Bryidae and Hypnidae); note that the IPL has undergone additional divisions, leading to six cells per pair of segments, and to the development of cilia in between the segments; the inner surface of the segments never bears a median line, as all anticlinal IPL walls are displaced during maturation.

During the final stages of sporophyte maturation, additional material is deposited on the cell walls of the inner layers of the amphithecium. In the Polytrichopsida at maturity each tooth is composed of several elongate whole cells, and the peristome is designated as nematodontous (Fig. 3.5). In most other peristomate mosses, the cells are short and wide, and only periclinal walls are thickened while longitudinal anticlinal walls are resorbed so that at maturity the amphithecial layers appear as rows of teeth composed of cell wall remnants, i.e., the arthrodontous peristome (Figs. 3.6–3.8). Each tooth is composed of periclinal walls from two adjacent amphithecial cells, separated by their middle lamellae. The degree of thickening may vary between layers, thereby accounting for the hygroscopic nature of the peristome or how the peristome responds to changes in ambient moisture: a heavier outer thickening results in the teeth bending outward as they dry, while a thicker inner surfaces makes the peristome bend inward.

Arthrodontous peristomes can be classified into two general types:

Fig. 3.5. *Tetraphis pellucida* (Tetraphidaceae). Light micrograph of two of the four peristome teeth in the nematodontous peristome.

diplolepideous (Fig. 3.8) and haplolepideous (Figs. 3.6, 3.7). In the former, two concentric rows of teeth are developed around the capsule mouth, and the outer surface of the outer teeth bears a median line marking the junction of two rows of cells. By contrast, the outer surface of the haplolepideous tooth lacks such a line because it is made from a single column of cells. Although the diplolepideous peristome is typically composed of

Fig. 3.6. *Schistidium heterophyllum* (Grimmiaceae). Light micrograph of portion of the haplolepideous peristome.

two rows of teeth, an exostome and an endostome, while haplolepideous taxa typically bear only a single row (homologous to the endostome of the Diplolepideae), the terms refer solely to the number of cells composing the outer surface of the teeth. Since exostomes are lacking or strongly reduced in the Haplolepideae, it is clear that the terms compare two non-homologous characters, namely the outer surface of the exostome of the Diplolepideae with that of the endostome of the Haplolepideae.

The exostome is composed of 16 teeth. These can each be divided in

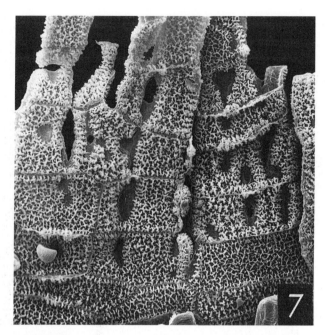

Fig. 3.7. *Tortula plinthobia* (Pottiaceae). SEM of portion of the haplolepideous peristome.

two or they can be fused into eight pairs. The thickening of the surfaces of the teeth or segments often follows a pattern that is characteristic for the family, genus or, more rarely, species. The inner surface of the exostome is typically not very ornamented, but commonly there are striking projections associated with the cell walls; this condition is referred to as trabeculate. In its most developed condition, the endostome of a diplolepideous taxon has a tall basal membrane, essentially a tube of tissue. Perched upon the basal membrane are 16 segments, most often alternating in position with the exostome teeth. These segments are often keeled and perforate and extend to about the same level as the apex of the exostome teeth. Alternating on the basal membrane with the segments, and thus typically opposite the exostome teeth, are groups of cilia. The presence of cilia is correlated with additional divisions in the IPL; they are always lacking in the Funariales and the Orthotrichales which consistently have 32 or fewer IPL cells, and present in many Bryales with 48 or more cells in the IPL. The cilia are mostly uniseriate structures and occur in groups of one to three or four. In some taxa, cilia bear attachments (then referred to as appendiculate) or are swollen at the

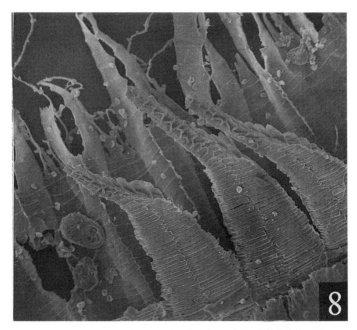

Fig. 3.8. *Pseudoscleropodium purum* (Brachytheciaceae). SEM of portion of the
diplolepideous peristome, showing median line on front surface of exostome
and projecting trabeculae on the back surface, and an endostome with a high
basal membrane, alternating, perforate segments, and cilia in groups of three.

nodes or joints between two cells of a column (nodulose). The peristome
of haplolepideous taxa may have individual, free teeth, sometimes
divided almost to their base into filaments, or sometimes the teeth are
fused into an extensive tube, with or without free filaments at its apex.
When selective pressures on the peristome are relaxed, or if selection
favors a reduced peristome, the endostome tends to be reduced prior to
the exostome.

When the capsule is young and not fully developed, there is a structure
of gametophytic tissue which encloses it. Near capsule maturity this structure, the calyptra, falls and frees the capsule, allowing dehiscence (Fig.
3.3B–D). In a few cases it stays attached to the apex of the seta. The calyptra
is formed from the expanded upper portions of the archegonium. There
are two basic forms of calyptrae, mitrate and cucullate. The mitrate calyptra is conic and undivided (similar to a bishop's miter) or equally lobed at
the base. A cucullate calyptra is split up one side, and is similar to a hood,
being longer on one side than the other. There seems to be a correlation

between obliquely rostrate opercula and cucullate calyptrae, and between erect–rostrate opercula and mitrate calyptrae. Calyptrae can be naked or hairy, smooth or roughened. Because of the limited variability found in calyptrae, convergent evolution has occurred repeatedly.

Spores

Moss spores are typically single-celled, spherical, and about 12–15 μm in diameter. In some cases the spores may be multicellular (e.g., Dicnemonaceae), typically associated with endosporic germination and relatively xeric environmental conditions. Some spores are bilaterally symmetrical or retain tetrad markings (or rarely remain in tetrads at maturity). Spores range in size from about 5 μm to well over 100 μm. For the most part moss spores are finely papillose, but range from essentially smooth to highly ridged, spinose, or reticulate. Strongly ornamented spores are often large and associated with cleistocarpous capsules. Spore coats are multilayered; their structure was reviewed by Mogensen (1983). Spore number per capsule is typically in the hundreds of thousands, but at the opposite extreme it may be as few as four (in *Archidium*).

In most stegocarpous mosses a peristome is attached around the circumference of the capsule mouth and is associated with spore dispersal. The peristome is typically attached to the spore sac and, when the peristome moves back and forth through hygroscopic movement, the spore sac is moved, keeping the spores from becoming compacted in the base of the capsule (Pfaehler 1904). However, the peristome has additional functions in spore dispersal as well. Through hygroscopic motion, the peristome either seals the mouth of the capsule or opens it. Some mosses open their capsule mouths under dry conditions (xerocastique), and others open their mouths when moist conditions prevail (hygrocastique) (Mueller & Neumann 1988). In diplolepideous mosses the endostome is not conspicuously hygroscopic, but in the Haplolepideae the hygroscopic function characteristic of the diplolepideous exostome is carried out by its peristome, the endostomial homolog (Crum 1977). Additionally, peristomes often act as filters, allowing spores to be released over extended periods of time. Peristomes may even be active spore dispersal agents by having spores get caught in the exostomial trabeculae and, through hygroscopic movement, flick the spores away from the capsule. Spore dispersal can be further enhanced by modifications of the capsule wall. Alternating bands of thick-walled and thin-walled exothecial cells result in the capsule contracting when dry, and thereby pushing spores up. In the Splachnaceae, differential thickening of walls is within individual cells. Here the transverse walls are thicker than the longitudinal ones and as a result, the capsule shrinks when it dries and the spore mass is pushed upward (Koponen 1990). In *Tetraplodon* the efficiency of this mechanism is improved by the presence of a pseudocolumella, which is located within the neck (or hypophysis) and pushes the whole spore sac up (Demidova & Filin 1994). Capsule contraction is also the mechanism by which *Scouleria* ejects its spores (Churchill 1985). In *Sphagnum*, the capsule wall bears pseudostomata which are arranged in a more or less median band and allow the capsule to contract quickly (Boudier 1988) leading to the dehiscence of the operculum and explosive spore dispersal. In valvate capsules, such as in *Andreaea*, the valves bend outward upon drying, thereby exposing the spore mass. Despite these various innovations, spore dispersal in most mosses is rather passive, relying on wind as the dispersing agent. The only exception is found in the Splachnaceae, where insects are the primary dispersal agent (Marino 1997). Such a dispersal mode is referred to entomophily (or more appropriately entomochory) and works by deceiving insects. Insects looking for fresh dung or cadavers on which to lay their eggs are attracted to the capsules of these mosses which emit volatile compounds through their stomata, the odor of which is reminiscent of decaying animal remains or dung. The deceived insect subsequently carries the sticky spores to a true nitrogenous substrate, upon which the spores germinate.

3.3 The classification of mosses

Throughout the lifetimes of most living bryologists classification systems have been proposed to reflect natural relationships. Of course through time our understanding of phylogeny and evolutionary processes has greatly changed. Nevertheless, through most of recent time, bryologists have attempted to understand how mosses are related, and then used this knowledge to construct classifications. Since the publication of Brotherus' treatment of the mosses in the second edition of Engler's *Die natürlichen Pflanzenfamilien* in 1924–5, moss classification has been greatly swayed by peristomial morphology and ornamentation. During preparation of this work, Brotherus was heavily influenced by Fleischer's monumental *Die Musci der Flora von Buitenzorg* (1904–23). Fleischer himself was influenced by the work of Philibert (1884–1902), who pioneered studies on peristome structure. Consequently, for most of the current century, details of the peristome were considered as valuable in understanding relationships as were flowers within the angiosperms. However, as noted above, peristomes are vital structures in spore dispersal and thus are as subject to environmental pressures as are any other part of the moss plant, and should therefore not be regarded of greater value for higher-level classification than gametophytic characters (Buck 1991). It has long been documented that the epiphytic condition is correlated with reductions in sporophytic characters (Mitten 1859). Capsules often become erect, peristomes become inserted below the capsule mouth, endostomial cilia are lost, the basal membrane becomes lower, eventually the endostome may be lost entirely, the inner surface of the exostome becomes less trabeculate, the ornamentation on the outer surface of the exostome becomes less, and in some extreme circumstances, the exostome is also lost. Nevertheless, there are certainly peristomial characters that remain valuable and diagnostic as taxonomic markers. In the diagnoses associated with the following classification, peristome characters are frequently used, and indeed seem to be extremely valuable. It is just the unquestioning use of peristome characters in the face of overwhelming contradictory evidence that should be challenged.

Among peristomial characters that are used in classification is the standard dichotomy of haplolepideous and diplolepideous. However, it appears as if the haplolepideous mosses arose from within the diplolepideous mosses and thus recognition of a formal grouping for all haplolepideous mosses leaves the diplolepideous mosses paraphyletic. Nevertheless, the haplolepideous mosses, on this single peristome character, are a natural group. Peristomial ornamentation has long been used for classification. For example, teeth with vertical pits characterize the Dicranaceae. Within the pleurocarps peristomial ornamentation has less value because of the repeated invasion by various lineages into epiphytic habitats with consequent sporophytic reductions. Thus the terrestrial taxa, most of which have striking horizontal striations on the external surface of the exostome, do not form a natural group, but rather an ecological one. Therefore, suites of correlated characters are needed to arrive at a substantiative classification.

Because of its more complex morphology, the gametophyte generation offers a bonanza of characters that can be used in classification (Hedenäs 1989a). Without going into great detail, characters that have been used for understanding relationships can be found thoughout the gametophore. These range from details of stem anatomy, to rhizoidal morphology and insertion, to structure of pseudoparaphyllia, to morphology of axillary hairs. These have been correlated with traditionally used gametophytic characters of leaf areolation and costal development.

Certainly there are still morphological characters that offer great promise in helping to better understand relationships that are currently not being used. Examples include columellar morphology and stem epidermal patterns in relationship to leaf insertion. Surely more will surface with increased observation. In recent years, though, molecular characters are having more influence on classifications than any other set of characters. Although these can be very powerful data sets, caution is needed. Single-gene classification systems are just as likely to be faulty as are any other type of single-character classification. This is easily observable when sequence data from different genetic loci reveal different phylogenies. They cannot all be correct. Evolution of a genetic locus or a plastid is not identical to that of the whole organism. Nevertheless, the use of sequence data cannot and should not be ignored. In groups of organisms with frequent morphological reduction, such as the mosses, insights from molecular data allows us to re-evaluate morphological data (Goffinet *et al.* 1998, Buck *et al.* in press).

3.4 A suggested classification of the mosses

The following classification is original in several aspects. Molecular data have been very influential (e.g., Goffinet *et al.* 1998, Cox & Hedderson 1999, Beckert *et al.* 1999, Buck *et al.* in press), but in the many cases where such data are lacking, morphology has been used exclusively. In some cases the morphological data have been interpreted differently than in the past. To date there are no cladistic analyses of morphological characters for mosses addressing the entire spectrum of families and orders. Undoubtedly, this classification, which is little more than a working hypothesis, will be a target. New information with implications for phylogeny and classification is being uncovered at a rapid pace. Thus, although this classification is our best estimate at the time of writing (September 1999), we are certain that in short order our own ideas will change. This is the nature of progress. The best that we can hope for is that some of our novel ideas will prove accurate and that the others will encourage bryologists to become more involved in this kind of basic systematic research. Please note that within the subclass Bryidae we have adopted superorders, following Reveal's (1998) suggestion to use the ending -*anae*, even though Reveal's specific proposal was defeated during the nomenclature session at the St Louis International Botanical Congress.

PHYLUM: BRYOPHYTA

CLASS: TAKAKIOPSIDA Leaves divided into terete filaments; capsules dehiscent by a single longitudinal spiral slit.

ORDER: TAKAKIALES

Takakiaceae S. Hatt. & Inoue. *Takakia* S. Hatt. & Inoue

CLASS: SPHAGNOPSIDA Branches usually in fascicles; leaves composed of a network of chlorophyllose and hyaline cells; setae lacking; capsules elevated on a pseudopodium; stomata lacking.

ORDER: SPHAGNALES Plants mostly branched, with branches in fascicles; stems with wood cylinder; leaves unistratose; antheridia subglobose; archegonia terminal on branches; capsules ovoid.

Sphagnaceae Dum. *Sphagnum* L.

ORDER: AMBUCHANANIALES Plants sparsely branched, with branches not in fascicles; stems without wood cylinder; leaves partially bistratose; antheridia oblong–cylindric; archegonia terminal on stems; capsules cylindrical.

Ambuchananiaceae Seppelt & H. A. Crum. *Ambuchanania* Seppelt & H. A. Crum

CLASS: ANDREAEOPSIDA Plants on acidic rocks, generally autoicous; cauline central strand absent; calyptrae small; capsules valvate, with four valves attached at apex; seta absent, pseudopodium present.

ORDER: ANDREAEALES

Andreaeaceae Dum. *Acroschisma* Lindl., *Andreaea* Hedw.

CLASS: ANDREAEOBRYOPSIDA Plants on calcareous rocks, dioicous; cauline central strand lacking; calyptrae large and covering whole capsule; capsules valvate, apex eroding and valves free when old; seta present.

ORDER: ANDREAEOBRYALES

Andreaeobryaceae Steere & B. M. Murray. *Andreaeobryum* Steere & B. M. Murray

CLASS: POLYTRICHOPSIDA Plants typically robust, dioicous; cauline central strand present; leaves costate, often with ventral lamellae on costa; capsules operculate; seta present; peristome nematodontous.

ORDER: TETRAPHIDALES Gametophore relatively small; stems not rhizomatous; costa slender, elamellate; capsules long–exserted.

Tetraphidaceae Schimp. Leaves unicostate; calyptrae small conic; capsule symmetric and erect, neck short; peristome single, of four erect teeth; *Tetraphis* Hedw. (Fig. 5), *Tetrodontium* Schwägr.

Oedipodiaceae Schimp. Leaves unicostate; calyptrae cucullate; capsule symmetric and erect, neck very long; capsules gymnostomous; *Oedipodium* Schwägr.

Buxbaumiaceae Schimp. Leaves ecostate; calyptrae cucullate or mitrate; capsule strongly asymmetric and horizontal, neck short; peristome double; *Buxbaumia* Hedw.

ORDER: POLYTRICHALES Gametophore typically large; stems typically rhizomatous; costa broad, with adaxial chlorophyllose lamellae; peristome mostly of (16)32–64 teeth.

Polytrichaceae Schwägr. *Alophosia* Card., *Atrichopsis* Card., *Atrichum* P. Beauv., *Bartramiopsis* Kindb., *Dawsonia* R. Br., *Dendroligotrichum* (Müll. Hal.) Broth., *Hebantia* G. L. S. Merr., *Itatiella* G. L. Sm., *Lyellia* R. Br., *Meiotrichum* (G. L. Sm.) G. L. S. Merr., *Notoligotrichum* G. L. Sm., *Oligotrichum* Lam. & DC., *Plagioracelopus* G. L. S. Merr., *Pogonatum* P. Beauv., *Polytrichadelphus* (Müll. Hal.) Mitt., *Polytrichastrum* G. L. Sm., *Polytrichum* Hedw. (Fig. 1), *Pseudatrichum* Reimers, *Pseudoracelopus* Broth., *Psilopilum* Brid., *Racelopodopsis* Thér., *Racelopus* Dozy & Molk., *Steereobryon* G. L. Sm.

CLASS: BRYOPSIDA Plants small to robust; leaves costate, typically lacking lamellae; capsules operculate; peristome arthrodontous.

SUBCLASS: DIPHYSCIIDAE Gametophore small, perennial; leaves costate; capsules immersed among long perichaetial leaves; peristome double.

ORDER: DIPHYSCIALES Plants perennial; leaves costate; capsules immersed among long perichaetial leaves.

Diphysciaceae M. Fleisch. *Diphyscium* D. Mohr, *Muscoflorschuetzia* Crosby, *Theriotia* Cardot

SUBCLASS: FUNARIIDAE Plants terricolous, acrocarpous; stem typically with central strand; annulus often well developed.

ORDER: TIMMIALES Plants robust; laminal cells mammillose; calyptrae cuculate, often persistent; capsules bent to pendent; cilia present.

Timmiaceae Schimp. *Timmia* Hedw.

ORDER: ENCALYPTALES Plants small to medium-size; laminal cells thick-walled, isodiametric above, rectangular and hyaline or reddish below; annulus not differentiated; calyptrae very large, enclosing the entire erect capsule.

Encalyptaceae Schimp. *Bryobrittonia* R. S. Williams, *Encalypta* Hedw.

ORDER: FUNARIALES Peristome lacking cilia, opposite, divisions in the IPL symmetric.

Funariaceae Schwägr. Protonema short-lived; costa well developed; laminal cells smooth and thin-walled; paraphyses with swollen apical cell; calyptrae smooth and naked; stomata with single guard cell; peristomes opposite, following a 4:2:4 pattern, or lacking; *Aphanorrhegma* Sull., *Brachymeniopsis* Broth., *Bryobeckettia* Fife, *Corynotheca* Ochyra, *Cygnicollum* Fife & Magill, *Entosthodon* Schwägr., *Funaria* Hedw., *Funariella* Sérgio, ×*Funariophyscomitrella* Wettst., *Loiseaubryum* Bizot, *Nanomitriella* E. B. Bartram, *Physcomitrella* Bruch & Schimp., *Physcomitrellopsis* Broth. & Wager, *Physcomitrium* (Brid.) Brid., *Pyramidula* Brid.

Disceliaceae Schimp. Protonemata persistent; costa weak to absent; calyptrae persistent below the urn; stomata lacking; peristomes reduced and opposite; *Discelium* Brid.

Gigaspermaceae Lindb. Gametophore united by subterranean rhizomatous aphyllous stem, cladocarpous; annulus not differentiated; capsules often immersed, gymnostomous; *Chamaebryum* Thér. & Dixon, *Costesia* Thér., *Gigaspermum* Lindb., *Lorentziella* Müll. Hal., *Neosharpiella* H. Rob. & Delgad., *Oedipodiella* Dixon

SUBCLASS: *DICRANIDAE* Peristome haplolepideous, with a formula of (4):2:3; exostome typically absent; late state division in the IPL asymmetric.

ORDER: GRIMMIALES Plants slender to robust; laminal cells with thick and often wavy walls; peristome of 16 entire or divided teeth.

Grimmiaceae Arn. Laminal cells with sinuose walls; leaves often terminated by hair-point; *Aligrimmia* R. S. Williams, *Coscinodon* Spreng., *Coscinodontella* R. S. Williams, *Dryptodon* Brid., *Grimmia* Hedw., *Indusiella* Broth. & Müll. Hal., *Jaffueliobryum* Thér., *Leucoperichaetium* Magill, *Racomitrium* Brid., *Schistidium* Bruch & Schimp. (Fig. 6)

Ptychomitriaceae Schimp. Laminal cells with straight walls; leaves lacking hair-points; *Campylostelium* Bruch & Schimp., *Glyphomitrium* Brid., *Ptychomitriopsis* Dixon, *Ptychomitrium* Fürnr.

Scouleriaceae S. P. Churchill. Plants blackish, acro- or cladocarpous, saxicolous in riparian habitats; calyptrae mitrate, smooth; annulus not differentiated; capsules urceolate to globose; *Scouleria* Hook., *Tridontium* Hook. f.

Drummondiaceae (Vitt) Goffinet, comb. nov. (Orthotrichaceae
subfam. Drummondioideae Vitt, Canad. J. Bot. 50: 1194. 1972). Stem
with central strand, cladocarpous; costa with differentiated adaxial
stereids; laminal cells thick-walled; peristome reduced; *Drummondia*
Hook.

ORDER: ARCHIDIALES Plants small, ephemeral; seta lacking; capsules
cleistocarpous, with less than 200 large spores (often 4–8); columella lacking.
   Archidiaceae Schimp. *Archidium* Brid.

ORDER: SELIGERIALES Plants generally saxicolous; alar cells differentiated;
capsules often systylious.
   Seligeriaceae Schimp. Plants small; leaves costate; costa homogenous
      in transverse section; peristome mostly deeply inserted, relatively
      well developed; spores small; *Blindia* Bruch & Schimp., *Brachydontium*
      Fürnr., *Hymenolomopsis* Thér., *Seligeria* Bruch & Schimp., *Trochobryum*
      Breidl. & Beck
   Wardiaceae W. H. Welch. Plants robust, rheophilic; leaves ecostate;
      peristome attached at mouth, reduced; spores large; *Wardia* Harv. &
      Hook.

ORDER: DICRANALES Plants small to large; laminal cells generally smooth;
alar cells often differentiated; peristome single, lacking basal membrane,
segments trabeculate and striate.
   Bryoxiphiaceae Besch. Leaves distichous with small dorsal extension
      (?) along costa; capsules gymnostomous; *Bryoxiphium* Mitt.
   Fissidentaceae Schimp. Leaves distichous and complanate, with
      vaginant lamellae; apical cell two-sided; *Fissidens* Hedw.
   Dicranaceae Schimp. Plants generally robust; cauline central strand
      present; leaves with well differentiated alar cells; peristome of 16 flat
      teeth divided in upper two-thirds, typically with vertically pitted
      outer surface; *Anisothecium* Mitt., *Aongstroemia* Bruch & Schimp.,
      Aongstroemiopsis M. Fleisch., *Arctoa* Bruch & Schimp., *Atractylocarpus*
      Mitt., *Braunfelsia* Paris, *Brothera* Müll. Hal., *Brotherobryum* M. Fleisch.,
      Bryohumbertia P. de la Varde & Thér., *Bryotestua* Thér. & P. de la Varde,
      Camptodontium Dusén, *Campylopodiella* Cardot, *Campylopodium* (Müll.
      Hal.) Besch., *Campylopus* Brid., *Chorisodontium* (Mitt.) Broth., *Cnestrum*
      I. Hagen, *Cryptodicranum* E. B. Bartram, *Dicranella* (Müll. Hal.)
      Schimp., *Dicranodontium* Bruch & Schimp., *Dicranoloma* (Renauld)
      Renauld, *Dicranum* Hedw., *Eucamptodontopsis* Broth., *Holomitriopsis* H.
      Rob., *Holomitrium* Brid., *Hygrodicranum* Cardot, *Kiaeria* I. Hagen,
      Kingiobryum H. Rob., *Leptotrichella* (Müll. Hal.) Lindb., *Leucoloma* Brid.,

Macrodictyum (Broth.) E. H. Hegew., *Mesotus* Mitt., *Microcampylopus*
(Müll. Hal.) Fleisch., *Mitrobryum* H. Rob., *Muscoherzogia* Ochyra,
Orthodicranum (Bruch & Schimp.) Loeske, *Paraleucobryum* (Limpr.)
Loeske, *Parisia* Broth., *Pilopogon* Brid., *Platyneuron* (Cardot) Broth.,
Pocsiella Bizot, *Polymerodon* Herzog, *Pseudephemerum* (Lindb.) I. Hagen,
Pseudochorisodontium (Broth.) C. H. Gao, Vitt, X. H. Fu & T. Cao,
Schliephackea Müll. Hal., *Sclerodontium* Schwägr., *Sphaerothecium*
Hampe, *Steyermarkiella* H. Rob., *Symblepharis* Mont., *Werneriobryum*
Herzog

Bruchiaceae Schimp. Alar cells not differentiated; capsules with
elongate necks; spores mostly with trilete markings, usually strongly
ornamented; *Bruchia* Schwägr., *Cladophascum* Sim, *Eobruchia* W. R.
Buck, *Pringleella* Cardot, *Trematodon* Michx.

Dicnemonaceae Broth. Plants robust, cladocarpous; central strand
generally absent; laminal cells elongate, thick-walled and porose;
perichaetial leaves differentiated and sheathing; calyptrae mitrate;
germination endosporous; *Dicnemon* Schwägr., *Eucamptodon* Mont.

Leucobryaceae Schimp. Plants robust, glaucous; cauline central strand
lacking; costa broad, occupying most of the leaf, with median
chlorophyllose cells and adaxial and abaxial layers of hyaline cells;
Cladopodanthus Dozy & Molk., *Leucobryum* Hampe, *Ochrobryum* Mitt.,
Schistomitrium Dozy & Molk.

Ditrichaceae Limpr. Plants slender; alar cells not differentiated;
peristome of 16 completely divided, terete teeth; *Astomiopsis* Müll.
Hal., *Austrophilibertiella* Ochyra, *Bryomanginia* Thér., *Ceratodon* Brid.,
Cheilothela Broth., *Chrysoblastella* R. S. Williams, *Cladastomum* Müll.
Hal., *Cleistocarpidium* Ochyra & Bednarek-Ochyra, *Crumuscus* W. R.
Buck & Snider, *Cygniella* H. A. Crum, *Distichium* Bruch & Schimp.,
Ditrichopsis Broth., *Ditrichum* Hampe, *Eccremidium* Hook. f. & Wilson,
Garckea Müll. Hal., *Kleioweisiopsis* Dixon, ×*Pleuriditrichum* A. L.
Andrews & F. J. Herm., *Pleuridium* Rabenh., *Rhamphidium* Mitt.,
Saelania Lindb., *Skottsbergia* Cardot, *Strombulidens* W. R. Buck, *Trichodon*
Schimp., *Tristichium* Müll. Hal., *Wilsoniella* Müll. Hal.

Rhabdoweisiaceae Limpr. Plants small to medium size; stem lacking
central strand; capsules ribbed, widest at mouth; *Amphidium* Schimp.,
Cynodontium Schimp., *Dichodontium* Schimp., *Dicranoweisia* Milde,
Holodontium (Mitt.) Broth., *Oncophorus* (Brid.) Brid., *Oreas* Brid.,
Oreoweisia (Bruch & Schimp.) De Not., *Pseudohyophila* Hilp.,
Rhabdoweisia Bruch & Schimp., *Verrucidens* Cardot

Rhachitheciaceae H. Rob. Laminal cells rectangular in lower half, short to isodiametric above; alar cells not differentiated; perichaetial leaves differentiated; capsules ribbed, rarely smooth; endostome teeth fused or not; IPL of 8 or 16 cells only (peristome formula: (4):2:2 or (4):2:1); *Hypnodontopsis* Z. Iwats. & Nog., *Jonesiobryum* B. H. Allen & Pursell, *Rhachitheciopsis* P. de la Varde, *Rhachithecium* Jolis, *Tisserantiella* P. de la Varde, *Uleastrum* W. R. Buck, *Zanderia* Goffinet

Erpodiaceae Broth. Plants cladocarpous; costa lacking; laminal cells often papillose; calyptrae mitrate; *Aulacopilum* Wilson, *Bryowijkia* Nog., *Erpodium* (Brid.) Brid., *Solmsiella* Müll. Hal., *Venturiella* Müll. Hal., *Wildia* Müll. Hal. & Broth.

Schistostegaceae Schimp. Gametophores dimorphic, small, annual, arising from persistent luminescent protonemata; leaves ecostate, distichous or in five rows; capsules globose, gymnostomous, lacking stomata and annulus; *Schistostega* D. Mohr

Eustichiaceae Broth. Leaves distichous; laminal cells quadrate and thick-walled; capsules ribbed; peristome of 16 teeth; *Eustichia* (Brid.) Brid.

Viridivelleraceae I. G. Stone. Protonemata persistent; stems producing gametangia and associated leaves only; capsules gymnostomous; *Viridivellus* I. G. Stone

ORDER: POTTIALES Plants minute to robust, generally orthotropic; upper laminal cells usually isodiametric and papillose; alar cells not differentiated; perichaetial leaves typically not differentiated; capsules erect; peristome typically papillose, not trabeculate.

Pottiaceae Schimp. Plants small to robust, primarily terrestrial; cauline central strand often present; leaves narrowly lanceolate to ligulate; laminal cells typically papillose; calyptrae cucullate, naked, smooth; peristome of 16 or 32 segments; *Acaulon* Müll. Hal., *Aloinella* Cardot, *Aloinia* Kindb., *Anoectangium* Schwägr., *Aschisma* Lindb., *Barbula* Hedw., *Bellibarbula* P. C. Chen, *Bryoceuthospora* H. A. Crum & L. E. Anderson, *Bryoerythrophyllum* P. C. Chen, *Calymperastrum* I. G. Stone, *Calyptopogon* (Mitt.) Broth., *Chenia* R. H. Zander, *Chionoloma* Dixon, *Crossidium* Jur., *Crumia* W. B. Schofield, *Dialytrichia* (Schimp.) Limpr., *Didymodon* Hedw., *Dolotortula* R. H. Zander, *Erythrophyllastrum* R. H. Zander, *Erythrophyllopsis* Broth., *Eucladium* Bruch & Schimp., *Ganguleea* R. H. Zander, *Gertrudiella* Broth., *Globulinella* Steere, *Goniomitrium* Hook.f. & Wilson, *Gymnostomiella* M. Fleisch., *Gymnostomum* Nees & Hornsch., *Gyroweisia* Schimp., *Hennediella* Paris, *Hilpertia* R. H. Zander,

Hymenostyliella E. B. Bartram, *Hymenostylium* Brid., *Hyophila* Brid.,
Hyophiladelphus (Müll. Hal.) R. H. Zander, *Hypodontium* Müll. Hal.,
Leptobarbula Schimp., *Leptodontiella* R. H. Zander & E. H. Hegew.,
Leptodontium (Müll. Hal.) Lindb., *Luisierella* Thér. & P. de la Varde,
Microbryum Schimp., *Microcrossidium* J. Guerra & M. J. Cano, *Mironia*
R. H. Zander, *Molendoa* Lindb., *Neophoenix* R. H. Zander & During,
Pachyneuropsis H. Mill., *Phascopsis* I. G. Stone, *Plaubelia* Brid., *Pleurochaete*
Lindb., *Pottiopsis* Blockeel & A. J. E. Sm., *Pseudocrossidium* R. S. Williams,
Pseudosymblepharis Broth., *Pterygoneurum* Jur., *Quaesticula* R. H. Zander,
Reimersia P. C. Chen, *Rhexophyllum* Herzog, *Sagenotortula* R. H. Zander,
Saitobryum R. H. Zander, *Sarconeurum* Bryhn, *Scopelophila* (Mitt.) Lindb.,
Splachnobryum Müll. Hal., *Stegonia* Venturi, *Stonea* R. H. Zander,
Streptocalypta Müll. Hal., *Streptopogon* Mitt., *Streptotrichum* Herzog,
Syntrichia Brid., *Teniolophora* W. D. Reese, *Tetracoscinodon* R. Br.,
Tetrapterum A. Jaeger, *Timmiella* (De Not.) Schimp., *Tortella* (Lindb.)
Limpr., *Tortula* Hedw. (Fig. 7), *Trachycarpidium* Broth., *Trachyodontium*
Steere, *Trichostomum* Bruch, *Triquetrella* Müll. Hal., *Tuerckheimia* Broth.,
Uleobryum Broth., *Weisiopsis* Broth., *Weissia* Hedw., *Weissiodicranum* W.
D. Reese, *Willia* Müll. Hal.

Ephemeraceae Schimp. Plants minute, developing from persistent
protonemata; capsules immersed, cleistocarpous or with poorly
differentiated operculum; columella disintegrating; *Ephemerum*
Schimp., *Micromitrium* Austin, *Nanomitriopsis* Cardot

Serpotortellaceae W. D. Reese & R. H. Zander. Plants robust,
cladocarpous, epiphytic; cauline central strand present; leaf margins
entire and unistratose; perichaetial leaves differentiated; peristome
well developed, reflexed when dry; *Serpotortella* Dixon

Calymperaceae Kindb. Plants epiphytic; stem lacking central strand;
leaves narrowly to broadly lanceolate; laminal cells papillose or
smooth; often with hyaline cancellinae on either side of costa at leaf
base; calyptrae persistent or not; peristome of 16 (rarely fused into 8)
segments, smooth, papillose or vertically striate; *Arthrocormus* Dozy &
Molk., *Calymperes* Sw., *Exodictyon* Cardot, *Exostratum* L. T. Ellis,
Leucophanes Brid., *Mitthyridium* H. Rob., *Octoblepharum* Hedw.,
Syrrhopodon Schwägr.

Cinclidotaceae K. Saito. Plants blackish and robust, generally
cladocarpous, riparian to aquatic; cauline central strand lacking;
leaves with thickened margins; capsules ribbed when dry; *Cinclidotus*
P. Beauv.

Bryobartramiaceae Sainsb. Plants very small, acrocarpous; calyptrae remaining attached to vaginula, persisting as an epigonium; capsules cleistocarpous; *Bryobartramia* Sainsb.

SUBCLASS: BRYIDAE Peristome double, of alternating teeth and segments; endostome ciliate; late stage division in the IPL asymmetric.

SUPERORDER: BRYANAE Plants acrocarpous or cladocarpous; pseudoparaphyllia generally lacking; leaves erect to spreading, lanceolate to ovate, mostly costate, costal anatomy mostly heterogeneous; laminal cells generally short.

ORDER: SPLACHNALES Laminal cells rhombic to elongate, typically smooth; capsules erect with differentiated neck; peristome single or double; cilia rudimentary or lacking.

Splachnaceae Grev. & Arn. Plants mostly coprophilous; laminal cells thin-walled, rhomboidal; annulus not differentiated; capsules erect, neck often differentiated into broad hypophysis; endostome fused to exostome or lacking; *Aplodon* R. Br., *Brachymitrion* Taylor, *Moseniella* Broth., *Splachnum* Hedw., *Tayloria* Hook., *Tetraplodon* Bruch & Schimp., *Voitia* Hornsch.

Meesiaceae Schimp. Capsules more or less horizontal, with elongated neck; annulus differentiated and revoluble; peristome double; exostome typically shorter then endostome; *Amblyodon* P. Beauv., *Leptobryum* (Bruch & Schimp.) Wilson, *Meesia* Hedw., *Neomeesia* Deguchi, *Paludella* Brid.

Catoscopiaceae Broth. Plants small, slender; leaves in three ranks; laminal cells quadrate and smooth; capsules black, asymmetric, horizontal; peristome double and reduced; *Catoscopium* Brid.

ORDER: ORTHOTRICHALES Plants medium-size, epiphytic or saxicolous; cauline central strand lacking; laminal cells typically papillose; capsules erect; peristome double or reduced; exostome recurved.

Orthotrichaceae Arn. Plants acrocarpous or cladocarpous; laminal cells mostly isodiametric, thick-walled; calyptrae typically plicate and hairy; capsules erect, rarely immersed, often ribbed; OPL thick and teeth recurved when dry; cilia lacking; *Bryomaltaea* Goffinet, *Cardotiella* Vitt, *Ceuthotheca* Lewinsky, *Codonoblepharon* Schwägr., *Desmotheca* Lindb., *Florschuetziella* Vitt, *Groutiella* Steere, *Leiomitrium* Mitt., *Leptodontiopsis* Broth., *Leratia* Broth. & Paris, *Macrocoma* (Müll. Hal.) Grout, *Macromitrium* Brid., *Matteria* Goffinet, *Muelleriella* Dusén, *Orthotrichum* Hedw., *Pentastichella* Müll. Hal., *Pleurorthotrichum* Broth., *Schlotheimia* Brid., *Sehnemobryum* Lewinsky-Haapasaari & Hedenäs,

Stoneobryum D. H. Norris & H. Rob., *Ulota* D. Mohr, *Zygodon* Hook. &
Taylor

ORDER: HEDWIGIALES Plants medium to robust, plagiotropic, acrocarpous
or cladocarpous; laminal cells thick-walled, papillose or smooth; capsules
gymnostomous and immersed.

Hedwigiaceae Schimp. Protonemata globular; laminal cells
pluripapillose; calyptrae smooth, and naked; *Braunia* Bruch &
Schimp., *Hedwigia* P. Beauv., *Hedwigidium* Bruch & Schimp.,
Pseudobraunia (Lesq. & James) Broth.

Helicophyllaceae Broth. Leaves dimorphic, with lateral leaves
strongly inrolled when dry, dorsal and ventral leaves reduced and
appressed; laminal cells smooth; *Helicophyllum* Brid.

Rhacocarpaceae Kindb. Leaves bordered by narrow cells; laminal cells
roughened; alar cells inflated; *Pararhacocarpus* Frahm, *Rhacocarpus*
Lindb.

ORDER: BRYALES Plants primarily terricolous; cauline central strand
present; laminal cells rhombic to elongate, smooth; annulus differentiated;
capsules pendent, neck differentiated; peristome double, typically well
developed and ciliate; exostome incurved.

Aulacomniaceae Schimp. Plants medium to robust; laminal cells
isodiametric, incrassate; capsules ribbed when dry; annulus
differentiated; peristome double, bryoid; *Aulacomnium* Schwägr.

Bartramiaceae Schwägr. Plants often robust; laminal cells
isodiametric, quadrate or rectangular, smooth or prorulose; annulus
typically undifferentiated; capsules subglose, erect or slightly curved,
typically ribbed; neck undifferentiated; *Anacolia* Schimp., *Bartramia*
Hedw., *Breutelia* (Bruch & Schimp.) Schimp., *Conostomum* Sw.,
Fleischerobryum Loeske, *Flowersia* D. G. Griffin & W. R. Buck, *Leiomela*
(Mitt.) Broth., *Philonotis* Brid., *Plagiopus* Brid., *Quathlamba* Magill

Orthodontiaceae (Broth.) Goffinet, comb. nov. (Bryaceae subfam.
Orthodontoideae Broth. in Engler, Die natürlichen
Pflanzenfamilien, ed. 2, 11: 347, 1924). Plants small to robust; laminal
cells elongate, lax; capsules often ribbed; annulus lacking;
endostomial membrane reduced or lacking; *Orthodontium* Wilson,
Orthodontopsis Ignatov & B. C. Tan

Bryaceae Schwägr. Plants erect, mostly unbranched, acrocarpous;
laminal cells mostly rhomboidal, smooth, thin-walled; costa single,
strong; capsules inclined to pendulous, smooth, with differentiated
neck; *Acidodontium* Schwägr., *Anomobryum* Schimp., *Brachymenium*

Schwägr., *Bryum* Hedw., *Mielichhobryum* J. P. Srivast., *Mniobryoides* Hörmann, *Perssonia* Bizot, *Plagiobryum* Lindb., *Pleurophascum* Lindb., *Rhodobryum* (Schimp.) Limpr., *Roellia* Kindb., *Rosulabryum* J. R. Spence

Phyllodrepaniaceae Crosby. Plants small; leaves complanate, in four rows; peristome single, of 16 segments; *Mniomalia* Müll. Hal., *Phyllodrepanium* Crosby.

Pseudoditrichaceae Steere & Z. Iwats. Plants very small; leaves ovate lanceolate; laminal cells thick-walled; capsules erect; peristome double; cilia lacking; *Pseudoditrichum* Steere & Z. Iwats.

Mniaceae Schwägr. Plants acro- or cladocarpous; leaves often bordered and often toothed; laminal cells thin-walled, rhomboidal to elongate; *Cinclidium* Sw., *Cyrtomnium* Holmen, *Epipterygium* Lindb., *Leucolepis* Lindb., *Mielichhoferia* Nees & Hornsch., *Mnium* Hedw., *Orthomnion* Wilson, *Plagiomnium* T. J. Kop., *Pohlia* Hedw., *Pseudobryum* (Kindb.) T. J. Kop., *Pseudopohlia* R. S. Williams, *Rhizomnium* (Broth.) T. J. Kop., *Schizymenium* Harv., *Synthetodontium* Cardot, *Trachycystis* T. J. Kop.

Leptostomataceae Schwägr. Plants forming dense mats; stems heavily tomentose; leaf margins entire, unbordered; annulus poorly differentiated to lacking; stomata cryptoporous; peristome strongly reduced; *Leptostomum* R. Br.

SUPERORDER: *RHIZOGONIANAE* Plants erect, generally from rhizomatous stem, acro-, clado-, or pleurocarpous; leaves oblong to lanceolate, costate, costal anatomy heterogeneous; laminal cells generally short and thick-walled.

ORDER: RHIZOGONIALES

Hypnodendraceae Broth. Plants robust, rhizomatous and stipitate; secondary stems erect, frondose to dendroid; laminal marginal cells differentiated or not, unistratose, often toothed; capsules often ribbed when dry; annulus differentiated; peristome well developed, ciliate; *Hypnodendron* (Müll. Hal.) Mitt.

Rhizogoniaceae Broth. Plants medium to large; cauline central strand present; marginal laminal cells often differentiated, bi- or multistratose, often toothed; costa typically toothed above; annulus differentiated; capsules generally smooth; peristome typically well developed, ciliate, or reduced to endostome or exostome; *Cryptopodium* Brid., *Goniobryum* Lindb., *Hymenodon* Hook.f. & Wilson, *Hymenodontopsis* Herzog, *Leptotheca* Schwägr., *Mesochaete* Lindb., *Pyrrhobryum* Mitt., *Rhizogonium* Brid.

Calomniaceae Kindb. Plants small, arising from persistent protonema; leaves tristichous and dimorphic, with ventral ones much smaller; perichaetial leaves highly differentiated; capsules gymnostomous; *Calomnion* Hook. f. & Wilson

Mitteniaceae Broth. Plants small, with luminescent protonemata; cauline central strand lacking; leaves complanate, decurrent; perichaetia polysetous; peristome double, outer row homologous to bryoid endostome; *Mittenia* Lindb.

Cyrtopodaceae M. Fleisch. Plants robust and large, with rhizomatous stem; secondary stems erect, frondose to dendroid; lamina bistratose above; margin unistratose, toothed or entire; costa toothed above; capsules smooth; peristome double or reduced, eciliate; *Bescherellia* Duby, *Cyrtopodendron* M. Fleisch., *Cyrtopus* (Brid.) Hook. f.

Spiridentaceae Kindb. Plants robust and large, with rhizomatous stem; secondary stems erect, sparsely branched; lamina bi- or pluristratose; margin typically toothed; costa toothed above; capsules smooth, immersed or exserted; peristome double, teeth and segments long, eciliate; *Franciella* Thér., *Spiridens* Nees

Pterobryellaceae (Broth.) W. R. Buck & Vitt. Plants robust and large, stipitate from rhizomatous stem, frondose to dendroid; cauline central strand lacking; capsules short oval; annulus differentiated; peristome double, with long teeth and segments but reduced cilia; *Pterobryella* (Müll. Hal.) A. Jaeger

Racopilaceae Kindb. Stems plagiotropic; leaves dimorphic with dorsal ones reduced; costa excurrent; capsules long–exserted; peristome double, well developed; *Powellia* Mitt., *Racopilum* P. Beauv.

SUPERORDER: HYPNANAE Plants pleurocarpous, typically freely branching; pseudoparaphyllia usually present; leaves mostly ovate, costate or not, costal anatomy homogeneous; laminal cells generally elongate.

ORDER: HOOKERIALES Alar cells mostly not differentiated; opercula mostly rostrate; exostome teeth often furrowed.

SUBORDER: Hookeriineae Plants mostly medium-sized to slender, never (?) phyllodioicous; stems monopodially branched; leaves seldom plicate; laminal cells often large and lax, thin-walled; costae often strong and double, sometimes single or short and double; capsules mostly smooth; exothecial cells often collenchymatous; opercula often rostrate; exostome teeth often furrowed; endostomial segments with baffle-like cross-walls; calyptrae often mitrate.

Hookeriaceae Schimp. Stems with central strand; pseudoparaphyllia
filamentous or absent; laminal cells large and lax; costa usually short
and double; calyptrae multistratose at middle, naked; *Achrophyllum* Vitt
& Crosby, *Cyathophorella* (Broth.) M. Fleisch., *Cyathophorum* P. Beauv.,
Dendrocyathophorum Dixon, *Hookeria* J. E. Sm., *Schimperobryum* Margad.

Leucomiaceae Broth. Stems lacking central strand; pseudoparaphyllia
absent; laminal cells linear, lax; costa lacking; calyptrae cucullate;
Leucomium Mitt., *Philophyllum* Müll. Hal., *Rhynchostegiopsis* Müll. Hal.

Daltoniaceae Schimp. Stems lacking central strand; pseudoparaphyllia
absent; laminal cells oval to long–hexagonal, differentiated at leaf
margins; costa single; calyptrae unistratose at middle, fringed at
base, usually naked; *Calyptrochaeta* Desv., *Crosbya* Vitt, *Daltonia* Hook.
& Taylor, *Distichophyllidium* M. Fleisch., *Distichophyllum* Dozy &
Molk., *Ephemeropsis* K. I. Goebel, *Leskeodon* Broth., *Leskeodontopsis*
Zanten, *Metadistichophyllum* Nog. & Z. Iwats.

Adelotheciaceae W. R. Buck. Stems lacking central strand;
pseudoparaphyllia filamentous; laminal cells isodiametric, not
differentiated at leaf margins; costa single; calyptrae unistatose at
middle, not fringed at base, densely hairy; *Adelothecium* Mitt.,
Bryobrothera Thér.

Pilotrichaceae Kindb. Stems lacking central strand; pseudoparaphyllia
none or foliose; laminal cells various; costa strong and double, or
short and double; calyptrae unistratose at middle, usually hairy;
Actinodontium Schwägr., *Amblytropis* (Mitt.) Broth., *Beeveria* Fife,
Brymela Crosby & B. H. Allen (Fig. 3), *Callicostella* (Müll. Hal.) Mitt.,
Callicostellopsis Broth., *Crossomitrium* Müll. Hal., *Cyclodictyon* Mitt.,
Diploneuron E. B. Bartram, *Helicoblepharum* (Mitt.) Broth., *Hemiragis*
(Brid.) Besch., *Hookeriopsis* (Besch.) A. Jaeger, *Hypnella* (Müll. Hal.) A.
Jaeger, *Lepidopilidium* (Müll. Hal.) Broth., *Lepidopilum* (Brid.) Brid.,
Pilotrichidium Besch., *Pilotrichum* P. Beauv., *Sauloma* (Hook. f. & Wilson)
Mitt., *Stenodesmus* (Mitt.) A. Jaeger, *Stenodictyon* (Mitt.) A. Jaeger,
Tetrastichium (Mitt.) Cardot, *Thamniopsis* (Mitt.) M. Fleisch.,
Trachyxiphium W. R. Buck, *Vesiculariopsis* Broth.

Hypopterygiaceae Mitt. Plants dendroid; amphigastria differentiated;
leaves often limbate; costa single; laminal cells short, mostly smooth;
alar cells not differentiated; exostome teeth not furrowed;
endostomial segments lacking baffle-like cross-walls;
Canalohypopterygium W. Frey & Schaepe, *Catharomnion* Hook. f. &
Wilson, *Hypopterygium* Brid., *Lopidium* Hook. f. & Wilson

SUBORDER: Ptychomniieae Plants usually robust and turgid, often
phyllodioicous; stems sympodially branched, usually lacking a central
strand; leaves usually plicate, often strongly toothed; costae short and
double; laminal cells elongate, often thick-walled and porose; alar cells
often colored; capsules mostly ribbed; endostomial segments lacking
baffle-like cross-walls; calyptrae often cucullate.

> **Ptychomniaceae** M. Fleisch. Alar cells mostly little differentiated,
> except for color; capsules long–exserted, strongly eight-ribbed;
> calyptrae cucullate; spores isosporous; *Cladomnion* Hook. f. & Wilson,
> *Cladomniopsis* M. Fleisch., *Glyphothecium* Hampe, *Hampeella* Müll.
> Hal., *Ptychomnium* (Hook.f. & Wilson) Mitt., *Tetraphidopsis* Broth. &
> Dixon
>
> **Garovagliaceae** (M. Fleisch.) W. R. Buck & Vitt. Alar cells typically well
> differentiated in shape and color; capsules often immersed, smooth
> or somewhat sulcate; calyptrae mitrate or cucullate; spores
> anisosporous; *Endotrichellopsis* During, *Euptychium* Schimp.,
> *Garovaglia* Endl.

ORDER: HYPNALES Stems monopodially or sympodially branched; alar cells
often differentiated; opercula various, mostly not rostrate; exostome seldom
furrowed; calyptrae mostly cucullate, naked.

> **Rutenbergiaceae** (Broth.) M. Fleisch. Stems sympodially branched,
> lacking a central strand; secondary stems little branched; costa
> single; laminal cells prorulose; alar cells well differentiated; capsules
> immersed; calyptrae mitrate, hairy; *Neorutenbergia* Bizot & Pócs,
> *Pseudocryphaea* Broth., *Rutenbergia* Besch.
>
> **Trachylomataceae** (M. Fleisch.) W. R. Buck & Vitt. Stems sympodially
> branched; secondary stems stipitate frondose, complanate–foliate;
> alar cells weakly differentiated; asexual propagula of stem-borne,
> filamentous gemmae; exostome teeth pale, densely papillose;
> *Braithwaitea* Lindb., *Trachyloma* Brid.
>
> **Amblystegiaceae** G. Roth. Plants typically growing in moist areas;
> stems monopodially branched; costa often variable; laminal cells
> mostly short; alar cells not or weakly differentiated; setae relatively
> long in comparison to size of plants; capsules strongly curved and
> asymmetric; exostome teeth yellow-brown, cross-striolate;
> *Amblystegium* Schimp., *Donrichardsia* H. A. Crum & L. E. Anderson,
> *Gradsteinia* Ochyra, *Hygroamblystegium* Loeske, *Hypnobartlettia* Ochyra,
> *Koponenia* Ochyra, *Leptodictyum* (Schimp.) Warnst., *Limbella* (Müll.
> Hal.) Müll. Hal., *Ochyraea* Váňa, *Orthotheciella* (Müll. Hal.) Ochyra,

Platylomella A. L. Andrews, *Sciaromiella* Ochyra, *Sciaromiopsis* Broth., *Sinocalliergon* Sakurai, *Vittia* Ochyra

Cratoneuraceae Mönk. Stems monopodially branched, often wiry, often densely tomentose; paraphyllia present, foliose; stem and branch leaves differentiated; costa single; laminal cells relatively short, smooth or prorulose; alar cells well differentiated; exostome teeth yellow-brown, cross-striolate; *Cratoneuron* (Sull.) Spruce, *Sasaokaea* Broth.

Helodiaceae (M. Fleisch.) Ochyra. Stems monopodially branched; paraphyllia present, filamentous to narrowly foliose, the cells elongate, not papillose; costa single; laminal cells mostly prorulose; alar cells often well differentiated; exostome teeth cross-striolate; *Actinothuidium* (Besch.) Broth., *Bryochenea* C. H. Gao & K. C. Chang, *Helodium* Warnst. (Fig. 2), *Palustriella* Ochyra

Hylocomiaceae (Broth.) M. Fleisch. Plants mostly robust; stems monopodially or sympodially branched; paraphyllia often present; leaves often strongly toothed; costae often strong and double; laminal cells elongate, smooth or prorulose; alar cells weakly differentiated; setae very elongate; capsules typically curved and asymmetric; exostome teeth yellow- to red-brown, often with reticulate pattern; *Hylocomiastrum* Broth., *Hylocomium* Bruch & Schimp., *Leptocladiella* M. Fleisch., *Leptohymenium* Schwägr., *Loeskeobryum* Broth., *Macrothamnium* M. Fleisch., *Meteoriella* S. Okamura, *Neodolichomitra* Nog., *Orontobryum* M. Fleisch., *Pleurozium* Mitt., *Rhytidiadelphus* (Limpr.) Warnst., *Rhytidiopsis* Broth., *Schofieldiella* W. R. Buck

Rhytidiaceae Broth. Plants robust; stems monopodially branched; paraphyllia none; leaves plicate, rugose; costa single; laminal cells linear, strongly porose, prorulose; alar cells well differentiated; exostome teeth yellow-brown, cross-striolate; *Rhytidium* (Sull.) Kindb.

Leskeaceae Schimp. Plants terrestrial or epiphytic; stems monopodially branched, often terete–foliate; paraphyllia nonpapillose; leaves mostly short–acuminate; costa mostly single; laminal cells short, usually unipapillose; alar cells weakly differentiated; capsules curved and asymmetric when plants terrestrial but in epiphytes often erect; exostome striate in terrestrial taxa but in epiphytes often pale, weakly ornamented; endostome often reduced; *Claopodium* (Lesq. & James) Renauld & Cardot,

Dolichomitriopsis S. Okamura, *Fabronidium* Müll. Hal., *Haplocladium*
(Müll. Hal.) Müll. Hal., *Hylocomiopsis* Cardot, *Leptocladium* Broth.,
Leptopterigynandrum Müll. Hal., *Lescuraea* Bruch & Schimp., *Leskea*
Hedw., *Leskeadelphus* Herzog, *Leskeella* (Limpr.) Loeske, *Lindbergia*
Kindb., *Mamillariella* Laz., *Miyabea* Broth., *Okamuraea* Broth.,
Orthoamblystegium Dixon & Sakurai, *Pseudoleskea* Bruch & Schimp.,
Pseudoleskeella Kindb., *Pseudoleskeopsis* Broth., *Ptychodium* Schimp.,
Rigodiadelphus Dixon, *Schwetschkea* Müll. Hal.

Regmatodontaceae Broth. Plants epiphytic; stems monopodially
branched; paraphyllia absent; costa single; laminal cells short, smooth;
alar cells not or weakly, differentiated; capsules erect; exostome teeth
much shorter than endostome segments; *Regmatodon* Brid.

Pterigynandraceae Schimp. Plants terrestrial or epiphytic, mostly
relatively small; stems monopodially branched, mostly
terete–foliate; paraphyllia absent; costa short and double; laminal
cells short, prorulose; alar cells weakly differentiated; gemmae stem-
borne; setae smooth; capsules often erect; peristome often reduced;
Habrodon Schimp., *Heterocladium* Bruch & Schimp., *Iwatsukiella* W. R.
Buck & H. A. Crum, *Myurella* Bruch & Schimp., *Pterigynandrum* Hedw.,
Trachyphyllum A. Gepp

Rigodiaceae H. A. Crum. Plants terrestrial or weakly epiphytic, more
or less stipitate; stems monopodially branched; paraphyllia absent;
stipe, stem and branch leaves differentiated; costa single; laminal
cells short, smooth; alar cells not or weakly differentiated; setae
smooth; capsules curved and asymmetric; exostome teeth densely
cross-striolate; *Rigodium* Schwägr.

Thuidiaceae Schimp. Plants terrestrial; stems monopodially branched;
paraphyllia present, the cells papillose; stem and branch leaves
differentiated; costa single; laminal cells short, papillose; alar cells
not or weakly differentiated; setae often roughened; capsules
typically curved and asymmetric; exostome teeth densely cross-
striolate; calyptrae naked or sparsely hairy.; *Abietinella* Müll. Hal.,
Boulaya Cardot, *Cyrto-hypnum* (Hampe) Hampe & Lorentz, *Fauriella*
Besch., *Pelekium* Mitt., *Rauiella* Reimers, *Thuidiopsis* (Broth.) M.
Fleisch., *Thuidium* Bruch & Schimp.

Campyliaceae (Kanda) W. R. Buck, comb. nov. (Amblystegiaceae
subfam. Campylioideae Kanda, J. Sci. Hiroshima Univ., Ser. B, Div. 2,
Bot. 15: 250. 1975 [1976]). Plants typically growing in moist areas,
relatively robust; stems monopodially branched; costa mostly single;

laminal cells elongate; alar cells often differentiated; setae not especially long; capsules asymmetric; exostome teeth yellow-brown; *Anacamptodon* Brid., *Callialaria* Ochyra, *Calliergon* (Sull.) Kindb., *Campyliadelphus* (Kindb.) R. S. Chopra, *Campylium* (Sull.) Mitt., *Conardia* H. Rob., *Cratoneuropsis* (Broth.) M. Fleisch., *Drepanocladus* (Müll. Hal.) G. Roth, *Hamatocaulis* Hedenäs, *Hygrohypnum* Lindb., *Loeskypnum* H. K. G. Paul, *Pictus* C. C. Towns., *Pseudocalliergon* (Limpr.) Loeske, *Sanionia* Loeske, *Scorpidium* (Schimp.) Limpr., *Straminergon* Hedenäs, *Tomentypnum* Loeske, *Warnstorfia* Loeske

Brachytheciaceae G. Roth. Plants mostly growing in mesic woodlands, terrestrial; stems monopodially branched; leaves often plicate; costa single, often projecting as a small spine; laminal cells elongate; alar cells mostly weakly differentiated; setae sometimes roughened; capsules often relatively short, curved, asymetric; opercula conic to rostrate; exostome teeth mostly red-brown; calyptrae mostly naked; *Aerobryum* Dozy & Molk., *Aerolindigia* M. Menzel, *Bestia* Broth., *Brachythecium* Schimp., *Bryhnia* Kaurin, *Bryoandersonia* H. Rob., *Bryostreimannia* Ochyra, *Cirriphyllum* Grout, *Clasmatodon* Hook. f. & Wilson, *Cratoneurella* H. Rob., *Eriodon* Mont., *Eurhynchiella* M. Fleisch., *Eurhynchium* Bruch & Schimp., *Flabellidium* Herzog, *Homalotheciella* (Cardot) Broth., *Homalothecium* Schimp., *Isothecium* Brid., *Juratzkaeella* W. R. Buck, *Lindigia* Hampe, *Mandoniella* Herzog, *Meteoridium* (Müll. Hal.) Manuel, *Myuroclada* Besch., *Nobregaea* Hedenäs, *Palamocladium* Müll. Hal., *Plasteurhynchium* Broth., *Platyhypnidium* M. Fleisch., *Pseudopleuropus* Takaki, *Pseudoscleropodium* (Limpr.) M. Fleisch. (Fig. 8), *Puiggariopsis* M. Menzel, *Rhynchostegiella* (Schimp.) Limpr., *Rhynchostegium* Bruch & Schimp., *Rozea* Besch., *Schimperella* Thér., *Scleropodiopsis* Ignatov, *Scleropodium* Bruch & Schimp., *Scorpiurium* Schimp., *Squamidium* (Müll. Hal.) Broth., *Steerecleus* H. Rob., *Stenocarpidiopsis* M. Fleisch., *Trachybryum* (Broth.) W. B. Schofield, *Zelometeorium* Manuel

Stereophyllaceae (M. Fleisch.) W. R. Buck & Ireland. Plants terrestrial or epiphytic; stems monopodially branched; costa typically single; laminal cells elongate, mostly smooth, rarely unipapillose; alar cells differentiated, collenchymatous, extending across base of costa; setae smooth; capsules inclined to erect; exostome teeth cross-striolate to papillose; *Catagoniopsis* Broth., *Entodontopsis* Broth., *Eulacophyllum* W. R. Buck & Ireland, *Juratzkaea* Lorentz, *Pilosium* (Müll. Hal.) M. Fleisch., *Sciuroleskea* Broth., *Stenocarpidium* Müll. Hal., *Stereophyllum* Mitt.

Myriniaceae Schimp. Plants often epiphytic, small; stems monopodially branched; costa single, often slender; laminal cells elongate, smooth; alar cells weakly differentiated; capsules often erect; peristomes mostly variously reduced; calyptrae rarely hairy; *Austinia* Müll. Hal., *Helicodontium* Schwägr., *Macgregorella* E. B. Bartram, *Merrilliobryum* Broth., *Myrinia* Schimp., *Nematocladia* W. R. Buck

Fabroniaceae Schimp. Plants epiphytic, often small; stems monopodially branched, sometimes fragile; leaves mostly acuminate; costa single, slender; laminal cells short, smooth; alar cells mostly weakly differentiated; capsules typically erect; peristome often reduced; exostome teeth often paired; *Dimerodontium* Mitt., *Fabronia* Raddi, *Ischyrodon* Müll. Hal., *Levierella* Müll. Hal., *Rhizofabronia* (Broth.) M. Fleisch.

Meteoriaceae Kindb. Plants epiphytic, often pendent; stems monopodially branched, often very elongate; costa short and double or single; laminal cells mostly elongate, sometimes short, often variously papillose; alar cells not or weakly differentiated; setae often short, roughened; capsules often immersed, erect, symmetric; exostome teeth cross-striolate to papillose; endostome often reduced; calyptrae mitrate or cucullate, often hairy; *Aerobryidium* M. Fleisch., *Aerobryopsis* M. Fleisch., *Ancistrodes* Hampe, *Barbella* M. Fleisch., *Barbellopsis* Broth., *Chrysocladium* M. Fleisch., *Cryphaeophilum* M. Fleisch., *Cryptopapillaria* M. Menzel, *Diaphanodon* Renauld & Cardot, *Duthiella* Renauld, *Floribundaria* M. Fleisch., *Lepyrodontopsis* Broth., *Meteoriopsis* Broth., *Meteorium* (Brid.) Dozy & Molk., *Neodicladiella* W. R. Buck, *Neonoguchia* S. H. Lin, *Pseudospiridentopsis* (Broth.) M. Fleisch., *Pseudotrachypus* P. de la Varde & Thér., *Sinskea* W. R. Buck, *Toloxis* W. R. Buck, *Trachycladiella* (M. Fleisch.) M. Menzel & W. Schultze-Motel, *Trachypodopsis* M. Fleisch., *Trachypus* Reinw. & Hornsch.

Plagiotheciaceae (Broth.) M. Fleisch. Plants terrestrial; stems monopodially branched, mostly complanate–foliate; leaves decurrent; costa short and double or absent; laminal cells elongate, often strongly chlorophyllose; alar cells differentiated into the decurrencies; setae smooth; capsules often curved and asymmetric; peristome teeth mostly pale yellow; exostome typically cross-striolate below; endostome well developed; *Plagiothecium* Bruch & Schimp.

Fontinalaceae Schimp. Plants aquatic; stems sympodially branched; costa single or short and double (and then the leaves concave to carinate); capsules immersed or short–exserted; endostome forming a trellis; calyptrae mitrate or cucullate; *Brachelyma* Cardot, *Dichelyma* Myrin, *Fontinalis* Hedw.

Entodontaceae Kindb. Plants often epiphytic; stems monopodially branched; costa short and double or absent; laminal cells linear, smooth; alar cells subquadrate, numerous; capsules erect and symmetric, long–exserted; columella often exserted; peristome inserted below mouth of capsule; endostome mostly strongly reduced; *Entodon* Müll. Hal., *Erythrodontium* Hampe, *Mesonodon* Hampe, *Pylaisiobryum* Broth.

Climaciaceae Kindb. Plants dendroid; stems sympodially branched, with paraphyllia on stipe; leaves not decurrent; costa single; laminal cells relatively short, smooth; capsules erect and symmetric; *Climacium* F. Weber & D. Mohr

Pleuroziopsidaceae Ireland. Plants dendroid; stems sympodially branched, with longitudinal lamellae on stipe; leaves decurrent; costa single; laminal cells relatively short, smooth; capsules inclined and asymmetric; *Pleuroziopsis* E. Britton

Hypnaceae Schimp. Stems monopodially branched; pseudoparaphyllia foliose or rarely filamentous; paraphyllia none; leaves often falcate or homomallous; costa short and double (or absent); laminal cells mostly linear; capsules mostly inclined and asymmetric; exothecial cells usually not collenchymatous; opercula apiculate to short–rostrate; exostome teeth mostly cross-striolate; calyptrae mostly naked; *Andoa* Ochyra, *Bardunovia* Ignatov & Ochyra, *Breidleria* Loeske, *Bryocrumia* L. E. Anderson, *Bryosedgwickia* Cardot & Dixon, *Callicladium* H. A. Crum, *Calliergonella* Loeske, *Campylophyllum* (Schimp.) M. Fleisch., *Caribaeohypnum* Ando & Higuchi, *Chrysohypnum* (Hampe) Hampe, *Crepidophyllum* Herzog, *Ctenidiadelphus* M. Fleisch., *Ctenidium* (Schimp.) Mitt., *Cyathothecium* Dixon, *Ectropotheciella* M. Fleisch., *Ectropotheciopsis* (Broth.) M. Fleisch., *Ectropothecium* Mitt., *Elharveya* H. A. Crum, *Elmeriobryum* Broth., *Entodontella* M. Fleisch., *Eurohypnum* Ando, *Fallaciella* H. A. Crum, *Giraldiella* Müll. Hal., *Gollania* Broth., *Herzogiella* Broth., *Homomallium* (Schimp.) Loeske, *Hondaella* Dixon & Sakurai, *Horridohypnum* W. R. Buck, *Hyocomium* Bruch & Schimp., *Hypnum* Hedw., *Irelandia* W. R. Buck, *Isopterygiopsis* Z. Iwats., *Isopterygium* Mitt., *Leiodontium* Broth.,

Macrothamniella M. Fleisch., *Mahua* W. R. Buck, *Microctenidium* M.
Fleisch., *Mittenothamnium* Henn., *Nanothecium* Dixon & P. de la Varde,
Orthothecium Bruch & Schimp., *Phyllodon* Bruch & Schimp.,
Plagiotheciopsis Broth., *Platydictya* Berk., *Platygyriella* Cardot,
Platygyrium Bruch & Schimp., *Podperaea* Z. Iwats. & Glime,
Pseudotaxiphyllum Z. Iwats., *Ptilium* De Not., *Pylaisiella* Grout,
Rhacopilopsis Renauld & Cardot, *Rhizohypnella* M. Fleisch.,
Sclerohypnum Dixon, *Serpoleskea* (Limpr.) Loeske, *Stenotheciopsis* Broth.,
Stereodontopsis R. S. Williams, *Syringothecium* Mitt., *Taxiphyllopsis*
Higuchi & Deguchi, *Taxiphyllum* M. Fleisch., *Trachythecium*
M. Fleisch., *Tripterocladium* (Müll. Hal.) A. Jaeger, *Vesicularia* (Müll.
Hal.) Müll. Hal., *Wijkiella* Bizot & Lewinsky

Catagoniaceae W. R. Buck & Ireland. Stems monopodially branched;
pseudoparaphyllia filamentous; leaves conduplicate; costa short and
double or absent; laminal cells linear, smooth; alar cells not
differentiated; exostome teeth cross-striolate; *Catagonium* Broth.

Symphyodontaceae M. Fleisch. Stems monopodially branched;
laminal cells mostly prorulose; alar cells not or weakly differentiated;
setae mostly roughened; capsules symmetric, typically spinose;
exostome teeth papillose to cross-striolate; calyptrae cucullate or
mitrate; *Chaetomitriopsis* M. Fleisch., *Chaetomitrium* Dozy & Molk.,
Dimorphocladon Dixon, *Symphyodon* Mont.

Sematophyllaceae Broth. Stems monopodially branched; leaves often
golden green; costa short and double or none; laminal cells mostly
linear, smooth or papillose; alar cells well differentiated; exothecial
cells mostly collenchymatous; opercula mostly obliquely rostrate;
exostome teeth often furrowed, cross-striolate; *Acanthorrhynchium* M.
Fleisch., *Acritodon* H. Rob., *Acroporium* Mitt., *Allioniellopsis* Ochyra,
Aptychella (Broth.) Herzog, *Aptychopsis* (Broth.) M. Fleisch., *Brotherella*
M. Fleisch., *Chinostomum* Müll. Hal., *Clastobryella* M. Fleisch.,
Clastobryophilum M. Fleisch., *Clastobryopsis* M. Fleisch., *Clastobryum*
Dozy & Molk., *Colobodontium* Herzog, *Donnellia* Austin, *Foreauella*
Dixon & P. de la Varde, *Gammiella* Broth., *Hageniella* Broth.,
Heterophyllium (Schimp.) Kindb., *Isocladiella* Dixon, *Leptoischyrodon*
Dixon, *Macrohymenium* Müll. Hal., *Mastopoma* Cardot, *Meiotheciella* B.
C. Tan, W. B. Schofield & H. P. Ramsay, *Meiothecium* Mitt., *Papillidiopsis*
(Broth.) W. R. Buck & B. C. Tan, *Paranapiacabaea* W. R. Buck & Vital,
Potamium Mitt., *Pseudohypnella* (M. Fleisch.) Broth., *Pterogonidium*
Broth., *Pterogoniopsis* Müll. Hal., *Pylaisiadelpha* Cardot, *Pylaisiopsis*

(Broth.) Broth., *Radulina* W. R. Buck & B. C. Tan, *Rhaphidostichum* M. Fleisch., *Schraderella* Müll. Hal., *Schroeterella* Herzog, *Sematophyllum* Mitt., *Struckia* Müll. Hal., *Taxitheliella* Dixon, *Taxithelium* Mitt., *Timotimius* W. R. Buck, *Trichosteleum* Mitt., *Trismegistia* (Müll. Hal.) Müll. Hal., *Trolliella* Herzog, *Warburgiella* Müll. Hal., *Wijkia* H. A. Crum

Hydropogonaceae W. H. Welch. Plants aquatic; stems monopodially branched; costa short and double or none; laminal cells linear, smooth; alar cells weakly differentiated; capsules cladocarpous, immersed; endostome none; calyptrae mitrate; *Hydropogon* Brid., *Hydropogonella* Cardot

Myuriaceae M. Fleisch. Stems sympodially branched; secondary stems little or not branched; leaves mostly long–acuminate; costa short and double or none; laminal cells linear, smooth; alar cells well differentiated, mostly colored; capsules long–exserted, erect; exostome teeth reduced, smooth, often perforate; endostome rudimentary; *Eumyurium* Nog., *Myurium* Schimp., *Oedicladium* Mitt., *Palisadula* Toyama, *Piloecium* (Müll. Hal.) Broth.

Cryphaeaceae Schimp. Stems sympodially branched; secondary stems little or not branched; costa single; laminal cells short, smooth or sometimes prorulose; alar cells numerous; capsules immersed or seldom emergent; exostome teeth pale, papillose; endostome rudimentary to absent; calyptrae mitrate; *Cryphaea* D. Mohr, *Cryphidium* (Mitt.) A. Jaeger, *Cyptodon* (Broth.) M. Fleisch., *Cyptodontopsis* Dixon, *Dendroalsia* E. Britton, *Dendrocryphaea* Broth., *Dendropogonella* E. Britton, *Pilotrichopsis* Besch., *Schoenobryum* Dozy & Molk., *Sphaerotheciella* M. Fleisch.

Prionodontaceae Broth. Plants epiphytic; stems sympodially branched; axillary hairs as in *Breutelia* (Bartramiaceae); leaves usually plicate and with strongly toothed margins; costa single; laminal cells short, papillose; alar cells differentiated in large areas; capsules immersed to emergent; annulus revoluble; exostome teeth papillose; endostome segments united into a reticulum; *Prionodon* Müll. Hal.

Leucodontaceae Schimp. Plants mostly epiphytic; stems sympodially branched; secondary stems often not or little branched, mostly curled when dry; leaves rapidly spreading when moist, mostly plicate; costa short and double or none; laminal cells oval to linear, mostly smooth, rarely prorulose; alar cells numerous; capsules usually exserted; annulus not differentiated; exostome teeth pale, papillose;

endostome mostly rudimentary; spores often large and
anisosporous; *Antitrichia* Brid., *Dozya* Sande Lac., *Eoleucodon* H. A.
Mill. & H. Whittier, *Felipponea* Broth., *Leucodon* Schwägr., *Pterogonium*
Sw., *Scabridens* E. B. Bartram

Pterobryaceae Kindb. Plants mostly epiphytic; stems sympodially
branched; secondary stems often well branched, and thus stipitate;
stem and branch leaves often differentiated, branch leaves sometimes
five-seriate; costa mostly single, sometimes short and double or
absent; laminal cells mostly linear, mostly smooth, sometimes
prorulose; alar cells usually differentiated, often thick-walled and
colored; capsules mostly immersed; exostome teeth pale, often
smooth; endostome most rudimentary; calyptrae cucullate or
mitrate, often hairy; *Calyptothecium* Mitt., *Cryptogonium* (Müll. Hal.)
Hampe, *Henicodium* (Müll. Hal.) Kindb., *Hildebrandtiella* Müll. Hal.,
Horikawaea Nog., *Jaegerina* Müll. Hal., *Micralsopsis* W. R. Buck,
Muellerobryum M. Fleisch., *Neolindbergia* M. Fleisch., *Orthorrhynchidium*
Renauld & Cardot, *Orthostichidium* Dusén, *Orthostichopsis* Broth.,
Osterwaldiella Broth., *Penzigiella* M. Fleisch., *Pireella* Cardot,
Pseudopterobryum Broth., *Pterobryidium* Broth. & Watts, *Pterobryon*
Hornsch., *Pterobryopsis* M. Fleisch., *Pulchrinodus* B. H. Allen, *Renauldia*
Müll. Hal., *Rhabdodontium* Broth., *Spiridentopsis* Broth., *Symphysodon*
Dozy & Molk., *Symphysodontella* M. Fleisch.

Phyllogoniaceae Kindb. Plants epiphytic; stems sympodially
branched; secondary stems irregularly branched, strongly
complanate–foliate; leaves conduplicate, cucullate, auriculate; costa
short and double or absent; laminal cells linear, smooth; alar cells
differentiated in small groups; capsules immersed or shortly
exserted; exostome teeth pale, not or scarcely ornamented;
endostome rudimentary or absent; calyptrae cucullate or mitrate,
naked or hairy; *Phyllogonium* Brid.

Orthorrhynchiaceae S. H. Lin. Plants terrestrial; stems monopodially
branched; leaves conduplicate, cucullate; costa short and double or
absent; laminal cells linear, smooth; alar cells undifferentiated;
capsules short–exserted, erect; exostome teeth pale, unornamented;
endostome none; calyptrae mitrate, hairy; *Orthorrhynchium*
Reichardt

Lepyrodontaceae Broth. Plants terrestrial or epiphytic; stems
sympodially branched; secondary stems not or little branched; leaves
sometimes plicate; costa single and weak or short and double to

absent; laminal cells linear, smooth, thick-walled and porose; alar
cells few or scarcely differentiated; capsules long–exserted;
peristome usually only endostomial; *Dichelodontium* Broth., *Lepyrodon*
Hampe

Neckeraceae Schimp. Plants terrestrial or epiphytic; stems mostly
sympodially branched, sometimes monopodial; stipes sometimes
differentiated and plants then frondose; leaves mostly complanately
arranged; costa typically single, sometimes short and double;
laminal cells fusiform to linear, rarely shorter, mostly smooth, rarely
prorulose or papillose; alar cells mostly few or weakly differentiated;
capsules immersed (mostly in epiphytes) to long–exserted (mostly in
terrestrial taxa); exostome teeth often pale, usually cross-striolate at
least at extreme base, papillose above; endostome often reduced;
calyptrae mostly cucullate; *Baldwiniella* M. Fleisch., *Bissetia* Broth.,
Bryolawtonia D. H. Norris & Enroth, *Caduciella* Enroth, *Crassiphyllum*
Ochyra, *Cryptoleptodon* Renauld & Cardot, *Curvicladium* Enroth,
Dixonia Horik. & Ando, *Dolichomitra* Broth., *Handeliobryum* Broth.,
Himantocladium (Mitt.) M. Fleisch., *Homalia* (Brid.) Bruch & Schimp.,
Homaliadelphus Dixon & P. de la Varde, *Homaliodendron* M. Fleisch.,
Hydrocryphaea Dixon, *Isodrepanium* (Mitt.) E. Britton, *Metaneckera*
Steere, *Neckera* Hedw., *Neckeropsis* Reichardt, *Neomacounia* Ireland,
Noguchiodendron Ninh & Pócs, *Pendulothecium* Enroth & S. He,
Pinnatella M. Fleisch., *Porothamnium* M. Fleisch., *Porotrichodendron* M.
Fleisch., *Porotrichopsis* Broth. & Herzog, *Porotrichum* (Brid.) Hampe,
Thamnobryum Nieuwl., *Touwia* Ochyra

Echinodiaceae Broth. Plants epipetric or less often on soil or bases of
trees; stems sympodially branched, wiry; secondary stems irregularly
branched; leaves mostly subulate; costa single, mostly excurrent;
laminal cells short, smooth; alar cells weakly differentiated; capsules
long–exserted, inclined to horizontal; exostome teeth reddish, cross-
striolate; endostome well developed; *Echinodium* Jur.

Leptodontaceae Schimp. Plants mostly epiphytic; stems sympodially
branched, often curled when dry; secondary stems irregularly
branched to bipinnate; costa typically single; laminal cells
isodiametric to long–hexagonal, smooth, unipapillose or prorulose;
alar cells numerous; capsules immersed to short–exserted; exostome
teeth pale, unornamented to spiculose; endostome rudimentary;
calyptrae hairy; *Alsia* Sull., *Forsstroemia* Lindb., *Leptodon* D. Mohr,
Taiwanobryum Nog.

Lembophyllaceae Broth. Plants often turgid; stems monopodially branched; leaves mostly strongly concave; costa mostly short and double (rarely single); laminal cells elongate, smooth; alar cells often somewhat differentiated; capsules mostly erect and immersed to short–exserted; endostome mostly reduced; calyptrae rarely mitrate, naked or hairy; *Acrocladium* Mitt., *Camptochaete* Reichardt, *Fifea* H. A. Crum, *Lembophyllum* Lindb., *Neobarbella* Nog., *Orthostichella* Müll. Hal., *Pilotrichella* (Müll. Hal.) Besch., *Weymouthia* Broth.

Anomodontaceae Kindb. Plants mostly epiphytic; stems sympodially or monopodially branched, secondary stems and/or branches often curled when dry, not complanate–foliate; paraphyllia none; leaves often acute to obtuse; costa single or short and double; laminal cells mostly short, papillose or prorulose; alar cells mostly poorly differentiated; capsules exserted, erect; exostome teeth pale to white, cross-striolate sometimes with overlying papillae to papillose; endostome often reduced; *Anomodon* Hook. & Taylor, *Bryonorrisia* L. R. Stark & W. R. Buck, *Chileobryon* Enroth, *Curviramea* H. A. Crum, *Haplohymenium* Dozy & Molk., *Herpetineuron* (Müll. Hal.) Cardot, *Schwetschkeopsis* Broth.

Theliaceae (Broth.) M. Fleisch. Plants terrestrial or on bases of trees; stems monopodially branched; paraphyllia present; leaves imbricate, little altered when moist, deltoid–ovate; costa single; laminal cells short, stoutly unipapillose; alar cells differentiated; capsules exserted, erect; exostome teeth white, smooth to papillose; endostome strongly reduced; *Thelia* Sull.

Microtheciellaceae H. A. Mill. & A. J. Harr. Plants epiphytic; stems monopodially branched?; costa single; laminal cells short, smooth; alar cells weakly differentiated;, capsules short–exserted, erect; exostome teeth truncate, reduced, weakly ornamented; endostome rudimentary; *Microtheciella* Dixon

Sorapillaceae M. Fleisch. Leaves distichous and complanate; capsules cladocarpous, immersed; peristome double, of 16 slender segments and 32 stout exostome knobs, cilia absent; *Sorapilla* Spruce & Mitt.

Acknowledgments

We particularly appreciate the unpublished sequence data generously provided by Cymon Cox. Thanks are also extended to Jon Shaw who also provided unpublished sequence data and made helpful comments on the manuscript and to Kevin Indoe who assisted with the digitization of the photographic figures.

REFERENCES

Akiyama, H. (1990). A morphological study of branch development in mosses with special reference to pseudoparaphyllia. *Botanical Magazine, Tokyo*, **103**, 269–82.

Akiyama, H. & Nishimura, N. (1993). Further studies on branch buds in mosses; "pseudoparaphyllia" and "scaly leaves". *Journal of Plant Research*, **106**, 101–8.

Anderson, L. E. & Crosby, M. R. (1965). The protonemata of *Sphagnum meridense* (Hampe) C. Muell. *Bryologist*, **68**, 47–54.

Arts, T. (1989). The occurrence of rhizoidal tubers in the genus *Campylopus*. *Lindbergia*, **15**, 60–4.

(1994). Rhizoidal tubers and protonematal gemmae in European *Ditrichum* species. *Journal of Bryology*, **18**, 43–61.

Beckert, S., Steinhauser, S., Muhle, H., & Knoop, V. (1999). A molecular phylogeny of bryophytes based on nucleotide sequences of the mitochondrial *nad*5 gene. *Plant Systematics and Evolution*, **218**, 179–92.

Berthier, J. (1972). Recherches sur la structure et le développement de l'apex du gamétophyte feuillé des mousses. *Revue Bryologique et Lichénologique*, **38**, 421–551.

Blomquist, H. L. & Robertson, L. L. (1941). The development of the peristome in *Aulacomnium heterostichum*. *Bulletin of the Torrey Botanical Club*, **68**, 569–84.

Boudier, P. (1988). Différenciation structurale de l'épiderme du sporogone chez *Sphagnum fimbriatum*. *Annales des Sciences Naturelles, Botanique, Séries 13*, **8**, 143–56.

Brotherus, V. F. (1924–25). Musci (Laubmoose). In *Die natürlichen Pflanzenfamilien*, 2nd edn, vols. 10–11, ed. A. Engler. Leipzig: Wilhelm Engelmann.

Buck, W. R. (1991). The basis for familial classification of pleurocarpous mosses. *Advances in Bryology*, **4**, 169–85.

Buck, W. R., Goffinet, B., & Shaw, A. J. (2000). Testing morphological concepts of orders of pleurocarpous mosses (Bryophyta) using phylogenetic reconstructions based on *trn*L–*trn*F and *rps*4 sequences. *Molecular Phylogenetics and Evolution*, in press.

Campbell, D. H. (1895). *The Structure and Development of the Mosses and Ferns*. London: Macmillan.

Churchill, S. P. (1985). The systematics and biogeography of *Scouleria* Hook. (Musci: Scouleriaceae). *Lindbergia*, **11**, 59–71.

Cook, M. E. & Graham, L. E. (1998). Structural similarities between surface layers of selected charophycean algae and bryophytes and the cuticles of vascular plants. *International Journal of Plant Sciences*, **159**, 780–7.

Correns, C. (1899). *Untersuchungen über die Vermehrung der Laubmoose durch Brutorgane und Stecklinge*. Jena: Fisher.

Cox, C. J. & Hedderson, A. J. (1999). Phylogenetic relationships among the ciliate arthrodontous mosses: evidence from chloroplast and nuclear DNA sequences. *Plant Systematics and Evolution*, **215**, 119–39.

Crum, H. A. (1977). *Meiothecium*, a new record for North America. *Bryologist*, **80**, 188–93.

Demidova, E. E. & Filin, V. R. (1994). False columella and spore release in *Tetraplodon angustatus* (Hedw.) Bruch et Schimp. in B.S.G. and *T. mnioides* (Hedw.) Bruch et Schimp. in B.S.G. (Musci: Splachnaceae). *Arctoa*, **3**, 1–6.

Dixon, H. N. (1932) Classification of mosses. In *Manual of Bryology*, ed. F. Verdoorn, pp. 397–412. The Hague: Martinus Nijhoff.

Duckett, J. G. & Ligrone, R. (1992). A survey of diaspore liberation mechanisms and germination patterns in mosses. *Journal of Bryology*, 17, 335–54.

Duckett, J. G., Schmid, A. M., & Ligrone, R. (1998). Protonemal morphogenesis. In *Bryology for the Twenty-first Century*, ed. J. W. Bates, N. W. Ashton & J. G. Duckett, pp. 223–46. Leeds: Maney and British Bryological Society.

During H. J. (1979). Life strategies of Bryophytes: A preliminary review. *Lindbergia*, 5, 2–18.

Edwards, S. R. (1979). Taxonomic implications of cell patterns in haplolepideous moss peristomes. In *Bryophyte systematics*, Systematics Association Special Volume 14, ed. G. C. S. Clarke & J. G. Duckett, pp. 317–46. New York: Academic Press.

(1984). Homologies and inter-relations of moss peristomes. In *New manual of bryology*, vol. 2. ed. R. M. Schuster, pp. 658–95. Nichinan: Hattori Botanical Laboratory.

Fleischer, M. (1904–23). *Die Musci der Flora von Buitenzorg (zugleich Laubmoosflora von Java)*, 4 vols. Leiden: Brill.

Frey, W. (1981). Morphologie und Anatomie der Laubmoose. *Advances in Bryology*, 1, 399–477.

Frey, W., Hofmann, M., & Hilger, H. H. (1996). The sporophyte–gametophyte junction in *Hymenophyton* and *Symphyogyna* (Metzgeriidae, Hepaticae): structure and phylogenetic implications. *Flora*, 191, 245–52.

Giesenhagen, K. (1910). Moostypen der Regenwälder. *Annales du Jardin Botanique de Buitenzorg*, Supplement 3(2), 711–90.

Goebel, K. (1905). *Organography of Plants especially of the Archegoniatae and Spermatophyta*, Part 2. Oxford: Clarendon Press.

Goffinet, B. & Vitt, D. H. (1998). Revised generic classification of the Orthotrichaceae based on a molecular phylogeny and comparative morphology. In *Bryology for the Twenty-first Century*, ed. J.W. Bates, N. W. Ashton & J. G. Duckett, pp. 143–59. Leeds: Maney and British Bryological Society.

Goffinet, B., Bayer, R. J., & Vitt, D. H. (1998). Circumscription and phylogeny of the Orthotrichales (Bryopsida) based on *rbc*L sequence analyses. *American Journal of Botany*, 85, 1324–37.

Goffinet, B., Shaw, A. J., Anderson, L. E, & Mishler, B. D. (2000). Peristome development in mosses in relation to systematics and evolution. V. Diplolepideae: Orthotrichaceae. *Bryologist*, 103, 581–94.

Hébant, C. (1977). The conducting tissues of bryophytes. *Bryophytorum Bibliotheca*, 10, 1–157.

Hedenäs, L. (1989a). Some neglected character distribution patterns among the pleurocarpous mosses. *Bryologist*, 92, 157–63.

(1989b). Axillary hairs in pleurocarpous mosses? A comparative study. *Lindbergia*, 15, 166–80.

Holloway, P. J. (1982). The chemical constitution of plant cutins. In *The Plant Cuticle*, ed. D. F. Cutler, K. L. Alvin, & C. E. Price, pp. 45–86. London: Academic Press.

Kawai, I. (1971a). Systematic studies on the conducting tissue of the gametophyte in Musci. (2) On the affinity regarding the inner structure of the stem in some species of Dicranaceae, Bartramiaceae, Entodontaceae and Fissidentaceae. *Annual Report of the Botanic Garden, Faculty of Science, Kanazawa University*, 4, 18–39.

(1971*b*). Systematic studies on the conducting tissue of the gametophyte in Musci. (3) On the affinity regarding the inner structure of the stem in some species of Thuidiaceae. *Science Reports of Kanazawa University*, **16**, 21–60.

(1971*c*). Systematic studies on the conducting tissue of the gametophyte in Musci. (4) On the affinity regarding the inner structure of the stem in some species of Mniaceae. *Science Reports of Kanazawa University*, **16**, 83–111.

(1976). Systematic studies on the conducting tissue of the gametophyte in Musci. (6) On the essential coordination among the anatomical characteristics of the stem in some species of Hypnaceae. *Science Reports of Kanazawa University*, **21**, 47–124.

(1977). Systematic studies on the conducting tissue of the gametophyte in Musci. (7) On the essential coordination among the anatomical characteristics of the stems in some species of Isobryales. *Science Reports of Kanazawa University*, **22**, 197–305.

(1978). Systematic studies on the conducting tissue of the gametophyte in Musci. (8) On the essential coordination among the anatomical characteristics of the stems in some species of Amblystegiaceae. *Science Reports of Kanazawa University*, **23**, 93–117.

(1979). Systematic studies on the conducting tissue of the gametophyte in Musci. (9) On regularity among anatomical characteristics of stems in some species of Dicranaceae. *Science Reports of Kanazawa University*, **24**, 13–43.

(1980*a*). Systematic studies on the conducting tissue of the gametophyte in Musci. (11) Anatomical characteristics of stems in some species of Leucobryaceae. *Science Reports of Kanazawa University*, **25**, 31–42.

(1980*b*). Systematic studies on the conducting tissue of the gametophyte in Musci. (12) Anatomical characteristics of stems in some species of Bartramiaceae. *Science Reports of Kanazawa University*, **26**, 31–50.

(1981). Systematic studies on the conducting tissue of the gametophyte in Musci. (10) Organization of the stem and its origin. *Hikobia Supplement*, **1**, 29–33.

(1989). Systematic studies on the conducting tissue of the gametophyte in Musci. (16) Relationships between the anatomical characteristics of the stem and the classification system. *Asian Journal of Plant Science*, **1**, 19–52.

(1991). Systematic studies on the conducting tissue of the gametophyte in Musci. (19) Relationships between the stem and seta in some species of Polytrichaceae, Bryaceae, Mniaceae, Bartramiaceae and Dicranaceae. *Science Reports of Kanazawa University*, **36**, 1–19.

Kawai, I. & Ikeda, K. (1970). Systematic studies on the conducting tissue of the gametophyte in Musci. (1) On the affinity regarding the conducting tissue of the stem in some species of Polytrichaceae. *Science Reports of Kanazawa University*, **15**, 71–98.

Kawai, I. & Ochi, H. (1987). Systematic studies on the conducting tissues of the gametophyte in Musci. (15) Relationships between the taxonomic system and anatomical characteristics of stems in some species of Bryaceae. *Science Reports of Kanazawa University*, **32**, 1–67.

Kawai, I., Yoshitake, S., & Yamazaki, M. (1985). Systematic studies on the conducting tissue of the gametophyte in Musci. (13) Anatomy of the stem through analysis of pigment deposition in *Polytrichum commune* Hedw. and *Pogonatum contortum* (Brid.) Lesq. *Science Reports of Kanazawa University*, **30**, 47–53.

Kawai, I., Yoshitake, S., & Yamamoto, E. (1986). Systematic studies on the conducting tissue of the gametophyte in Musci. (14) Anatomy of the stems of *Rhizogonium*, *Mnium* and *Fissidens*. *Science Reports of Kanazawa University*, **31**, 31–42.

Koponen, A. (1990). Entomophily in the Splachnaceae. *Botanical Journal of the Linnean Society*, **104**, 115–27.

Koponen, T. (1982). Rhizoid topography and branching patterns in moss taxonomy. *Beihefte zur Nova Hedwigia*, **71**, 95–99.

Kürschner, H. & Parolly, G. (1999). Pantropical epiphytic rain forest bryophyte communities: coeno-syntaxonomy and floristic–historical implications. *Phytocoenologia*, **29**, 1–52.

La Farge-England, C. (1996). Growth form, branching pattern, and perichaetial position in mosses: cladocarpy and pleurocarpy redefined. *Bryologist*, **99**, 170–86.

Ligrone, R. (1986). Structure, development and cytochemistry of muscilage-secreting hairs in the moss *Timmiella barbuloides* (Brid.) Moenk. *Annals of Botany*, **58**, 859–68.

Ligrone, R. & Gambardella, R. (1988). The sporophyte–gametophyte junction in bryophytes. *Advances in Bryology*, **3**, 225–274.

Ligrone, R., Duckett, J. G., & Renzaglia, K. S. (1993). The gametophyte–sporophyte junction in land plants. *Advances in Botanical Research*, **19**, 231–317.

Mägdefrau, K. (1982). Life-forms of bryophytes. In *Bryophyte Ecology*, ed. A. J. E. Smith, pp. 45–58. London: Chapman & Hall.

Marino, P. C. (1997). Competition, dispersal and coexistence of Splachnaceae in patchy habitats. *Advances in Bryology*, **6**, 241–63.

Mishler, B. D. & De Luna, E. (1991). The use of ontogenetic data in phylogenetic analyses of mosses. *Advances in Bryology*, **4**, 121–67.

Mitten, W. (1859). Musci Indiae Orientalis. An enumeration of the mosses of the East Indies. *Journal of the Proceedings of the Linnean Society*, Supplement to Botany, **1**, 1–171.

Mogensen, G. S. (1983). The spore. In *New Manual of Bryology*, vol. 1, ed. R. M. Schuster pp. 325–42. Nichinan: Hattori Botanical Laboratory.

Mueller, D. M. J. & Neumann, A. J. (1988). Peristome structure and the regulation of spore release in arthrodontous mosses. *Advances in Bryology*, **3**, 135–58.

Müller, C. (1895–8). Musci (Laubmoose). In *Die natürlichen Pflanzenfamilien*, vol. 1, ed. A. Engler and K. Prantl, pp. 142–202. Leipzig: Wilhelm Engelmann.

Newton, A. E. & Mishler, B. D. (1994). The evolutionary significance of asexual reproduction in mosses. *Journal of the Hattori Botanical Laboratory*, **76**, 127–45.

Paolillo, D. J. & Bazzaz, F. A. (1968). Photosynthesis in sporophytes of *Polytrichum* and *Funaria*. *Bryologist*, **71**, 335–43.

Pfaehler, A. (1904). Étude biologique et morphologique sur la dissémination des spores chez les mousses. *Bulletin de la Société Vaudoise des Sciences Naturelles*, **49**, 41–132 + plates VI–XIV.

Philibert, H. (1884–1902). De l'importance du péristome pour les affinités naturelles des mousses. *Revue Bryologique*, **11**, 49–52, 65–72 (1884); Études sur le péristome. *Revue Bryologique*, **11**, 80–7 (1884); **12**, 67–77, 81–5 (1885); **13**, 17–26, 81–6 (1886); **14**, 9–11, 81–90 (1887); **15**, 6–12, 24–8, 37–44, 50–60, 65–9, 90–3 (1888); **16**, 1–9, 39–44, 67–77 (1889); **17**, 8–12, 25–9, 39–42 (1890); **23**, 36–8, 41–56 (1896); **28**, 56–9, 127–130 (1901); **29**, 10–13 (1902).

Reveal, J. L. (1998). Seventeen proposals to amend the *Code* on suprageneric names. *Taxon*, **47**, 183–91.

Robinson, H. & Shaw, A. J. (1984). Considerations on the evolution of the moss operculum. *Bryologist*, **87**, 293–6.

Ron, E. & Kawai, I. (1990). Systematic studies on the conducting tissue of the gametophyte in Musci. (17) On the relationships between the stem and rhizome (forecast). *Annual Report of the Botanic Garden, Faculty of Science, Kanazawa University*, **13**, 15–18.

Ron, E. & Kawai, I. (1991). Systematic studies on the conducting tissue of the gametophyte in Musci. (18) On the relationships between the stem and rhizome. *Annual Report of the Botanic Garden, Faculty of Science, Kanazawa University*, **14**, 17–25.

Ruhland, W. (1924). Musci. Allgemeiner Teil. In *Die natürlichen Pflanzenfamilien*, 2nd ed. vol. 10, ed. A. Engler, pp. 1–100. Leipzig: Wilhelm Engelmann.

Schofield, W. B. (1985). *Introduction to Bryology*. New York: Macmillan.

Schofield, W. B. & Hébant, C. (1984). The morphology and anatomy of the moss gametophore. In *New Manual of Bryology*, vol. 2, ed. R. M. Schuster, pp. 627–57. Nichinan: Hattori Botanical Laboratory.

Schwartz, O. M. (1994). The development of the peristome-forming layers in the Funariaceae. *International Journal of Plant Sciences*, **155**, 640–57.

Shaw, J. & Anderson, L. E. (1988). Peristome development in mosses in relation to systematics and evolution. II. *Tetraphis pellucida* (Tetraphidaceae). *American Journal of Botany*, **75**, 1019–32.

Shaw, J., Anderson, L. E., & Mishler, B. D. (1987). Peristome development in mosses in relation to systematics and evolution. I. *Diphyscium foliosum* (Buxbaumiaceae). *Memoirs of the New York Botanical Garden*, **45**, 55–70.

(1989*a*). Peristome development in mosses in relation to systematics and evolution. III. *Funaria hygrometrica*, *Bryum pseudocapillare*, and *B. bicolor*. *Systematic Botany*, **14**, 24–36.

Shaw, J., Mishler, B. D., & Anderson, L. E. (1989*b*). Peristome development in mosses in relation to systematics and evolution. IV. Haplolepideae: Ditrichaceae and Dicranaceae. *Bryologist*, **92**, 314–25.

Snider, J. A. (1975). Sporophyte development in the genus *Archidium* (Musci). *Journal of the Hattori Botanical Laboratory*, **39**, 85–104.

Vitt, D. H. (1981). Adaptive modes of the moss sporophyte. *Bryologist*, **84**, 166–86.

(1984). Classification of the Bryopsida. In *New Manual of Bryology*, vol. 1, ed. R. M. Schuster, pp. 696–759. Nichinan: Hattori Botanical Laboratory.

Vitt, D. H., Goffinet, B., & Hedderson, T. (1998). The ordinal classification of mosses: questions and answers for 1990s. In *Bryology for the Twenty-first Century*, ed. J.W. Bates, N. W. Ashton, & J. G. Duckett, pp. 113–23. Leeds: Maney and British Bryological Society.

Watanabe, R. & Kawai, I. (1975). Systematic studies on the conducting tissue of the gametophyte in Musci. (5) What is expected of systematics regarding the inner structure of the stem in some species of Thuidiaceae. *Science Reports of Kanazawa University*, **20**, 21–76.

Wenderoth, H. (1931). Beiträge zur Kenntnis des Sporophyten von *Polytrichum juniperinum* Willdenow. *Planta*, **14**, 244–385.

Wyatt, R. & Anderson, L. E. (1984). Breeding systems in bryophytes. In *The Experimental Biology of Bryophytes*, Experimental Botany, vol. 19, ed. A. F. Dyer & J. G. Duckett, pp. 39–64. London: Academic Press.

4
_____

Origin and phylogenetic relationships of bryophytes

4.1 Introduction

The origin of a land flora during the Upper Ordovician–Lower Silurian border (± 440 millions years ago) represents a significant evolutionary event in the history of life. The subsequent evolution of a diverse autotrophic land flora created the conditions necessary for the diversification of a terrestrial heterotrophic fauna (Behrensmeyer *et al.* 1992). Today the land flora comprises roughly 300 000 species distributed among three major groups, the bryophytes, the pteridophytes, and the seed plants. Considering the evolutionary significance of land plants, as well as their dominance and thus their ecological importance in today's biosphere, it is not surprising that evolutionary biologists are investing much time and resources in understanding the transition to land, and the relationships between major lineages of terrestrial plants. Elucidating the evolution of early land plants may allow for a better understanding on how these plants overcame new obstacles encountered in a terrestrial habitat. Character innovations that coincide with exposure to new selective forces can then be examined in a phylogenetic context for their evolutionary significance or potential adaptive value (Knoll *et al.* 1984, Knoll & Niklas 1987).

Morphological, ontogenetic, and molecular characters suggest that land plants form a monophyletic group characterized by a series of synapomorphies: a multicellular sporophyte which is dependent upon maternal tissues at least during early developmental stages, sporic meiosis, spores with decay-resistant walls, presence of archegonia, antheridia enclosed in sterile jackets, apical meristematic cell with more than two cutting faces, coaxial bicentriolar replication of centrosomes, the orientation of lamellae in multilayered structures in sperm cells, and presence of preprophase bands at mitosis, conserved size and architecture of plastid genome (e.g., Mishler & Churchill 1984, Bremer 1985, Sluiman 1985, Bremer *et al.* 1987, Graham *et al.* 1991, Kenrick & Crane, 1991, 1997*a*, *b*, Garbary *et al.* 1993, Graham 1996). The origin of bryophytes needs therefore to be examined within the evolutionary history of all land plant lineages. Among extant embryophytes, bryophytes exhibit a series of characters reminiscent of the aquatic habitat, and thus appear as prime candidates for marking the transition to land. Resolving the relationships among three lineages of land plants, the bryophytes, pteridophytes, and seed plants, appears a priori straightforward. After all, there would only be three alternative hypotheses, depending on which of the three taxa is more basal in the evolutionary tree. Morphogenetic and molecular characters suggest, however, that extant land plants need to be accommodated in more than three natural groups. The monophyly of bryophytes and pteridophytes in particular has come under increased scrutiny. Furthermore, the exploration of sediments from the Lower Paleozoic has revealed the presence of a diverse flora, whose components may represent ancestors of extant and extinct lineages (Kenrick & Crane 1997*b*), suggesting that the earliest land plants may have given rise to multiple lineages. As more monophyletic lineages are identified, the number of possible evolutionary histories increases.

A phylogenetic tree that is not rooted merely represents the relative relationships among taxa composing the tree. Although polyphyly is revealed even in an unrooted tree, monophyly and paraphyly cannot be distinguished. Monophyly is determined by unique character innovations, shared by all descendants of the last ancestor who first acquired the trait. The phylogenetic significance of characters is determined in reference to the state found in the outgroup, i.e., a taxon who exhibits many and ideally all, ancestral character-states. Imperatively, the ancestor to land plants needs to be identified for the relationships among the lineages to be resolved.

Considering the complexity of these issues it is not surprising that a unified view of land plant evolution has not yet been reached. Nevertheless, remarkable progress has been achieved in recent decades in our understanding of the morphological, anatomical, ultrastructural, cytological, and molecular diversity of land plants (e.g., Gensel & Andrews 1987, Graham 1993, Stewart & Rothwell 1993, Taylor & Taylor 1993, Kenrick & Crane 1997*a, b*). The objective of this chapter is to examine the

evidence presented for resolving the origin of bryophytes and their relationships to early land plants. Specifically, I will 1) review the relevant
fossil record, 2) discuss the two fundamental theories of land plant evolution in terms of their implication for the affinities of bryophytes, 3) summarize the evidence for a charophycean origin of land plants, 4) examine
the implications of this hypothesis on the origin of bryophytes, 5) review
alternative phylogenetic hypotheses among lineages of bryophytes and
other land plants, and finally 6) address transformations of characters
based on a widely supported phylogenetic hypothesis.

4.2 The origin of bryophytes: a paleobotanical perspective

The degree at which plant remains are preserved in sediments depends,
all else being equal, on the resistance of the plant material to fossilization.
The ancestors to land plants, particularly if of freshwater origin, were
likely growing close to a terrestrial habitat, and thus exposed to temporal
water-level fluctuations. Species could overcome exposures to the aerial
environment by developing drought-resistant spores, that might have
remained dormant during the dry periods. The presence of sporopollenin
or sporopollenin-like compounds in spores may thus be seen as evidence
for the presence of land plants or their immediate ancestors at the time of
sedimentation (Kenrick & Crane 1997a). The oldest cryptospores (spores
whose corresponding sporophyte or sporangium is unknown) are present
in sediments of the Mid-Ordovician (Strother et al. 1996). Their discovery
is significant for two reasons: first, their persistence in the sediments
implies the presence of a decay-resistant polymer in the walls; and
second, the spores are arranged in permanent tetrads, a feature considered distinctive of land plants (Kenrick & Crane 1997a), or of their algal
precursor(s) (Johnson 1985). Other forms such as monads, or dyads, either
naked or surrounded by a thin envelope, compose early cryptosporic
assemblages. The significance of spore types other than tetrads is ambiguous, as their origin is unclear. Hemsley (1994) argues that these types (e.g.,
naked or membrane-bound monads, and dyads) can result from differences in sporopollenin deposition relative to meiotic divisions. Whether
these alternative spore types may be produced by individual taxa or are
each taxon-specific, and thus represent traces of distinct lineages of early
land plants that soon became extinct (Gray 1985), remains therefore open
to debate. Some cryptosporic tetrads are reminiscent of bryophytes, such

as the Sphaerocarpales (Gray 1985), whose spores are retained as tetrads within an envelope (Schofield 1985). Similarities in the ultrastructure of the spore wall between the dyad *Dyadospora* sp. and sphaerocarpalean liverworts (Taylor 1996) lent credence to the hypothesis that at least some of the early land plants have affinities with the liverwort lineage. Comparative spore ultrastructure studies offer new insights into the diversity and affinities of these microfossils, but these investigations need to be extended for them to offer reliable evidence (Edwards *et al.* 1998).

Trilete spores appear in the Lower Silurian and become increasingly abundant (Edwards *et al.* 1998), while dyads disappear. This shift in spore dominance may reflect important changes in the composition of the land flora (Gray & Boucot 1977, Steemans 1999). With regard to the origin of bryophytes, their occurrence is consistent with, but not necessarily indicative of, the presence of taxa of bryophyte affinities, since trilete marks occur on spores of representatives of hornworts, liverworts, and mosses, as well as of various other land plants (see Gray & Boucot 1977, Taylor 1982).

A minute coalified fossil from the Lower Devonian described by Edwards *et al.* (1995) is composed of a short axis terminated by two cup-shaped outgrowths that bear spore tetrads on their inner surface. These terminal structures could be interpreted as sporangia. Affinities with hepatics were proposed on the basis of similarities between some associated tubes and rhizoids or conducting cells, and the tetrahedral arrangement of the spores. The naked tetrads of spores (tetrads not enclosed within an envelope) resemble the tetrad *Tetrahedraletes medinensis* from the Ordovician, and could be the first link between Ordovician spores and their sporangium (Edwards *et al.* 1995). Whether the taxon from the Devonian is allied to the liverworts needs to be reconsidered, since the interpretation of the tubes associated with the sporangium was subsequently revised, and directed toward a fungal origin (Edwards *et al.* 1996).

Other microfossils from Ordovician to Silurian strata consist of cuticle fragments. Although they are now considered to be derived from land plants, their affinities remain ambiguous (Edwards *et al.* 1998). Cook and Graham (1998) suggest indeed that Charophyceae and bryophytes have surface layers similar to the cuticle of vascular plants. Kroken *et al.* (1996) subjected sporangial epidermis from *Sphagnum* and liverworts to acid hydrolysis, a treatment routinely used for retrieving fossil material from the sediments, and observed that these tissues resisted acetolysis. The scraps of bryophyte cell-remains appeared reminiscent of those obtained from deposits dating to the Ordovician (Kroken *et al.* 1996). The size and shape of the cells and their occurrence with spores led Kroken *et al.* (1996) to consider that at least some of the fragments from the Ordovician or Devonian could be homologous to sporophytic tissues of early bryophytes. Kodner and Graham (1999) recently compared acid-hydrolyzed remains of extant bryophytes with Silurian microfossils. Their observation, based on morphometric analyses, that tubes recovered from Silurian sediments are similar to "treated" calyptrae hairs of *Polytrichum* supports the view that bryophytes compose at least part of the assemblages of fossilized Silurian plants.

Megafossils from the Mid to Upper Silurian exhibit no affinities to bryophytes and appear related to vascular plants, although often with uncertain relationships (Kenrick & Crane 1997*a*). One exception may be the Upper Silurian *Tortilicaulis transwalliensis*. This plant has elliptic to fusiform sporangia terminating unbranched, short, naked, and twisted axes reminiscent of moss sporophytes. The poor state of preservation precludes, however, retrieving any anatomical evidence in support of a relationship to bryophytes (Edwards *et al.* 1998). *Tortilicaulis offaeus* has isotomous branching (Edwards *et al.* 1994), but its affinities, even to *T. transwalliensis*, are ambiguous (Edwards *et al.* 1998). Deposits from the Lower Devonian of Norway and Belgium contain another plant, *Sporogonites*, consisting of long, erect, and unbranched axes bearing terminal sporangia (Halle 1916, 1936) and lacking evidence of vascular tissue (Andrews 1958). Its sporangium possibly dehisces by vertical slits (Halle 1936) as in extant hepatics and the moss *Takakia*. The lack of elaters precludes affinities to the former (Krassilov & Schuster 1984). Furthermore, *Sporogonites* has a columella-like projection (Halle 1936), a feature lacking in liverworts but present in hornworts, and among extant mosses in putative primitive lineages, namely Sphagnopsida, Andreaeaopsida, and Takakiopsida (Schofield 1985, Renzaglia *et al.* 1997). Andrews (1958) interpreted the gametophyte as thalloid, whereas Crandall-Stotler (1984) suggested that the "thallus" may actually represent a persistent protonema or even small leafy gametophores, a hypothesis consistent with an affinity to mosses. Cladistic analyses of sporophytic characters resulted in unresolved relationships of *Sporogonites*, which formed a polytomy with the polysporangiophytes and the moss *Polytrichum* (Kenrick & Crane 1997*b*). The systematic position of *Sporogonites* and thus its significance in the evolution of bryophytes remain ambiguous (Kenrick & Crane 1997*b*), and need to be addressed further.

The oldest unequivocal fossil bryophyte dates from the Upper Devonian. *Hepaticites devonicus* (Hueber 1961) consists of a gametophyte composed of a small terete and apparently plagiotropic axis with unicellular rhizoids, from which dichotomous erect winged fronds originate. The unistratose lamellae bear lateral teeth, and the costal region comprises narrow elongate putative conducting cells. The plant is so similar to extant taxa of *Pallavicinia* (Metzgeriales sensu Schuster 1984) and related genera that Schuster (1966) accommodated the species in the new genus *Pallavicinites*. Other fossils from the Silurian and Devonian have been considered allied to the bryophytes, but these hypotheses have recently been disputed (Krassilov & Schuster 1984). Several megafossils allied to the liverworts have been recovered from deposits of the Lower Carboniferous. Like *Pallavicinites*, *Blasiites*, *Treubiites*, and *Metzgeriothallus* are all assigned to the Metzgeriales (Krassilov & Schuster 1984).

The earliest occurrence of a moss in the fossil record could be from the Lower Carboniferous. *Muscites plumatus* is described based on a single, densely foliate, sterile stem (Thomas 1972). Considered of ambigous affinities to extant mosses, *Muscites* may actually be best considered a lycopsid (see Rowe in Bateman *et al.* 1998). Other fossils clearly allied to mosses, *Muscites polytrichaceus* and *M. bertrandi*, are from the Upper Carboniferous (Krassilov & Schuster 1984). Permian permineralized specimens described by Neuburg (1960) are not only the best preserved and most abundant fossil mosses, but also come from the most diverse assemblage of Paleozoic mosses.

The ornamentation of the Paleozoic spores *Streelispora* and *Aneurospora* has been compared to that of *Anthoceros*, and used as evidence to support the presence of bryophytes in the Silurian (Richardson 1985). Considering variation in the ornamentation, of both the fossil and extant taxa (Gray in Richardson 1985), such conclusions are weak, at the very least. The first unambiguous record of hornworts in the fossil record dates from the Cretaceous (Oostendorp 1987).

The megafossil record for bryophytes is poor (Lacey 1969), and the evidence for bryophytes occurring early in the evolution of land plants is based solely on interpretation of spore and tube features. Opinions on the affinities of these microfossils can be volatile (Edwards *et al.* 1995 vs. Edwards *et al.* 1996), and the lack of consensus (e.g., Johnson 1985 vs. Gray 1985) reflects our inability at present to attribute spores unambiguously to higher-level taxa (Banks 1975). Detailed studies of the ultrastructure need to be pursued, as one of only few possible avenues for retrieving informative characters from our largest source of plant remains from the Ordovician and Silurian. The similarities between these microfossils and extant bryophytes are, however, at least congruent with the hypothesis that the early land plants associated with these fossils are derived from embryophytes at a bryophyte level of organization as suggested by Gray (1985) and Gray *et al.* (1992, Edwards *et al.* 1998).

4.3 The antithetic and homologous theories of evolution of land plants

Bryophytes share with other land plants a suite of morphological and cytological character-states (see above). To invoke an origin of bryophytes independent of that of other embryophytes, maybe even from a distinct algal ancestor as proposed by Schuster (1977, 1979), Ando (1978), Stewart (1983), Sluiman (1985), Asakawa (1986), Crandall-Stotler (1986), and others, would require convergence in all these features. It appears thus far more likely that these innovations occurred only once, and consequently that embryophytes compose a monophyletic group (Smith 1986, Graham *et al.* 1991, Graham 1993). Unlike other land plants, bryophytes are character- ized by a sporophyte that is short-lived, and attached to the dominant gametophyte. The gametophyte is free-living, and the biflagellate male gametes (antherozoids) typically require a continuous film of water to reach the archegonia. These reproductive constraints point to an aquatic ancestry and thus to the possible significance of bryophytes in the phylog- eny of land plants, and particularly the origin of extant vascular plants. Whether bryophytes mark the transition to land, and represent the earli- est terrestrial embryophytes, or whether they arose by modification (i.e., reduction) from a more complex ancestor shared with tracheophytes, is a controversy initiated over a 100 years ago (Celakovsky 1874, Pringsheim 1878) that is still at the forefront of the debate on the origin of land plants (Qiu & Palmer 1999 and Bateman 1996 respectively). Both theories are inspired by the studies by Hofmeister (1851) on the relationships between life phases in cryptogams and gymnosperms. Hofmeister proposed that the life cycle of plants is composed of two phases, a gametophyte bearing sexual organs, and a sporophyte producing spores, and that these funda- mental phases are homologous among plants.

Celakovsky (1874) examined Hofmeister's observations in an evolu- tionary context, and his hypotheses were later developed into what is known as the antithetic or interpolation theory (Bower 1890, 1908). This theory proposes that land plants arose from an algal ancestor with a haplobiontic life cycle, composed of heteromorphic phases, a multicellular gametophyte that is dominant and a unicellular sporophyte, composed only of the zygote which undergoes meiosis. The life cycle of bryophytes would have been achieved by delaying meiosis in the zygote through the intercalation of mitotic divisions, which yielded a multicellular sporophyte. In bryophytes, the sporophyte would remain attached and at least in part nutritionally dependent on the gametophyte. Increases in mitotic divisions would allow for additional growth in body size of the sporophyte, for more complex tissue to develop and thereby give rise to early vascular plants. In this scenario, bryophytes or an organism allied to them represent the ancestor to the line that gave rise to the tracheophytes. Results from recent phylogenetic studies of green plants (e.g., Mishler & Churchill 1985, Bremer *et al.* 1987, Graham *et al.* 1991, Mishler *et al.* 1994) have been interpreted as supporting this evolutionary scenario (Kenrick & Crane 1997*b*, Qiu & Palmer 1999). The fossil record, however, does not provide unambiguous evidence in support of this hypothesis.

In contrast to the antithetic theory, the homologous theory (Pringsheim 1878) suggests that the ancestor to land plants had two isomorphic, independent and thus autotrophic phases. Since both phases shared a single genetic pool, there was a priori no reason for the phases to be morphologically distinct under identical environmental pressures. Pringsheim (1878) based his theory of homology of the alternating phases in land plants based on his earlier observation of apospory in mosses (Pringsheim 1876). The development of a gametophyte directly from sporophytic tissue, without the production of spores, and thus without meiotic reduction, suggested that both phases were potentially identical, i.e., homologous. Springer (1935) demonstrated the reverse process, apogamy in mosses, providing further support for the theory of homologous generations. Each of these phenomena was subsequently replicated and confirmed by Wettstein (1942). The discovery of tracheids in the gametophyte of *Psilotum* (Holloway 1939) and photosynthetic sporophytes in bryophytes lent further credence to this theory, although the occurrence of tracheids varies with the length of the axis (Bierhorst 1971), while the latter is also congruent with the antithetic theory.

The central point of the homologous theory is that from an ancestor with isomorphic gametophyte and sporophyte, two lines of evolution diverged. Both lineages are characterized by heteromorphic life cycles but differ in the nature of the dominant phase. In the first line, leading to the bryophytes, the sporophyte underwent reduction while the gametophyte became more complex. In contrast the pteridophytes would have arisen by amplification of the sporophyte at the expense of the gametophyte. The implications of this theory for the evolution of a land flora are thus far-reaching. Although proponents of this theory of transformation would still consider bryophytes and pteridophytes related (Zimmerman 1932), the latter would have arisen independently from the former. In other words bryophytes had no significance regarding the evolution of the Polysporangiophyta, since they would at best represent a sister group derived by reduction from an ancestor allied to the Polysporangiophyta.

It is evident that these two alternative theories are not fully incongruent. Indeed, both evolutionary scenarios likely comprise a stage characterized by more or less isomorphic phases or at least co-dominating phases. Indeed, even within the antithetic scenario, a direct transition from a gametophyte-dominated life cycle to a sporophyte-dominated one appears a priori unlikely (see Kenrick & Crane 1991, Kenrick 1994). A derivation of bryophytes from an ancestral type with isomorphic generations could thus be accommodated within the antithetic theory of evolution of land plants. Therefore, the verification of the antithetic theory of evolution does not rely on demonstrating that bryophytes mark the transition to land.

4.4 The ancestor to the embryophytes

Green plants form a natural evolutionary entity, the Chlorobiota, defined by a suite of morphological, ultrastructural, and biochemical characters (Kenrick & Crane 1997b). Among the green algae, alternations of isomorphic gametophytes and sporophytes are known only from the Ulvophyceae; all other lineages are consistently characterized by a haplobiontic life cycle in which the zygote undergoes meiosis (Graham & Wilcox 1999). Among the latter, the Charophyceae, a group of freshwater algae, are characterized by a phragmoplast type of cell division, which is defined by the right-angle orientation of microtubules to the plane of division. The significance of this fundamental difference between the Charophyceae and the remaining green algae is highlighted by the presence of this type of cell division in all land plants (Marchant & Pickett-Heaps 1973). Charophyceae and embryophytes also share the presence of a multilayered sheath in the cytoskeleton of the swimming sperm, a nuclear envelope that disperses during mitosis, the loss of rhizoplast and eye spot, and an orientation of the flagella to the right (see Melkonian 1982, Graham 1985, Mishler & Churchill 1985, Sluiman 1985). Another character indicative of a close relationship between Charophyceae and basal lineages of land plants is the occurrence of monoplastidic meiosis (Renzaglia *et al.* 1993). Affinities between the Charophyceae and land plants as evidenced by these ultrastructural features are congruent with the occurrence of the photorespiratory enzyme glycolate oxidase restricted to these groups (Frederick *et al.* 1973), the presence of introns in chloroplast genes tRNA[ALA] and tRNA[ILE] (Manhart & Palmer 1990), and the presence, in at least some Charophyceae, of the *tuf*A gene in the nuclear rather than in the chloroplast genome (Baldauf *et al.* 1990). Analyses of nucleotide sequence data corroborate further a close phylogenetic relationship of the Charophyceae and the land plants (Mishler *et al.* 1994, Friedl 1997, Bhattacharya & Medlin 1998, Bhattacharya *et al.* 1998). Together these green plants compose the Streptobionta, a group likely derived from a prasinophyte, allied to *Mesostigma* (Melkonian 1982, Melkonian & Surek 1995, Bhattacharya *et al.* 1998).

The Charophyceae are distributed among five orders (i.e., the Charales, Chlorokybales, Coleochaetales, Klebsormidiales, and Zygnematales). The relationships among these orders, as well as the monophyly of the class, remain ambiguous (see summary in Chapman *et al.* 1998). Within a paraphyletic Charophyceae, either the Charales or the Coelochaetales would form a sister group to the land plants (Graham *et al.* 1991). In either case, the phylogenetic hypothesis would be congruent with the antithetic theory of evolution of land plants (Kenrick & Crane 1997*b*), although here again, the fossil record does not provide any supporting evidence: charophycean algae are indeed predated by land plants, namely the Zosterophyllopsida (Graham 1996, Kenrick & Crane 1997*a*).

4.5 Relationships of bryophyte lineages

The fossil record offers valuable insights into the taxonomic diversity of the early land plant flora, yet fails to offer unambiguous evidence for the presence of bryophytes among early plant lineages. The overall lack of complexity and their haplo–diplobiontic life cycle suggest, however, that bryophytes compose a rather basal lineage of land plants. Except for these features, the liverworts, hornworts, and mosses seem to have little in common, raising doubts about close affinities among them (Steere 1969,

Crandall-Stotler 1980, 1984, 1986). Due to the lack of relevant fossil taxa, reconstruction of the relationships among bryophytes thus relies primarily on the analyses of characters of extant taxa.

To address the phylogenetic relationships of bryophytes we need first to examine how many lineages should be considered. The hornworts are generally regarded as a well-defined natural group (Mishler & Churchill 1985, Renzaglia & Vaughn, chapter 1 in this volume). Although mosses are generally considered to compose a monophyletic group, the affinities of *Takakia* have puzzled botanists ever since its discovery (see Schuster 1997). Crandall-Stotler (1986), for example, accommodated the isolated position of *Takakia* by erecting the Takakiophyta, based on morphogenetic, anatomical, and cytological features (see Crandall-Stotler & Bozzola 1988). Based on morphological characters of the gametophyte, Murray (1988) interpreted *Takakia* as representing a primitive group of mosses, a hypothesis congruent with morphological features of the sporophyte (Smith & Davison 1993). Schuster (1997) recognized that *Takakia* was unique, but only by the combination of characters it exhibited, arguing that most individual characters could be found in either mosses or liverworts. Schuster (1997) chose to retain *Takakia* with the liverworts, albeit in a basal position, serving as the "glue that holds the bryophytes together." Recently, several studies (Ligrone *et al.* 1993, Hedderson *et al.* 1996, Renzaglia *et al.* 1997) converged toward a hypothesis wherein *Takakia* belongs to the mosses.

Another alternative to the "three-lineage" concept of bryophytes was raised by Capesius (1995), based on analyses of nucleotide sequences of the nuclear gene coding for the RNA composing the small subunit of the ribosomes. In this and subsequent analyses (Kranz *et al.* 1995, Bopp & Capesius 1996, Capesius & Bopp 1997) the thalloid and leafy liverworts (i.e., the marchantioid and jungermannioid liverworts) are resolved as a paraphyletic group, with the leafy liverworts sister to the mosses, a hypothesis proposed earlier by Hori *et al.* (1985). Lewis *et al.* (1997) tested this hypothesis using sequences of the chloroplast gene *rbc*L. Although inferences under certain assumptions yielded topologies congruent with Capesius's hypothesis, analyses accommodating rate heterogeneity among codon positions resulted in a monophyletic lineage of liverworts. Analyses by Garbary *et al.* (1993) resolved liverworts as a paraphyletic group when morphological characters of the gametophyte were analyzed, or a monophyletic clade when sporophytic features were included. More recently Nishiyama and Kato (1999) combined sequences of the nuclear 18S and various chloroplast genes for addressing the phylogeny of land plants. Although their taxon sampling was extremely limited, including only a single exemplar for each liverwort clade, they concluded that liverworts form a monophyletic group, corroborating the results obtained by earlier workers (e.g., Mishler & Churchill 1985, Garbary *et al.* 1993, Samigullin *et al.* 1998, Beckert *et al.* 1999, Renzaglia *et al.* 2000). Morphological, ontogenetic, and ultrastructural characters also point to a single liverwort lineage (see Crandall-Stotler & Stotler, chapter 2 in this volume). Further study may be warranted, but in the light of the growing consensus in favor of the monophyly of hepatics, the following discussion will rely on this hypothesis.

Within a monophyletic Embryophyta, mosses, liverworts, and hornworts compose either a monophyletic lineage, defined by a common ancestor whose descendants do not include the Polysporangiophyta, or a paraphyletic group, with one or two lineages of bryophytes originating from an ancestor shared with the polysporangiophytes (Fig. 4.1). For a group of four taxa, 12 different relationships are possible. Most if not all of these have received some support from one or the other set of characters (see for example results from different analyses performed by Mishler *et al.* 1994). It is impossible to evaluate every phylogenetic hypothesis in terms of its implications for character evolution. Instead I will briefly present the major trends suggested by these alternative hypothesis and discuss the implications of the currently preferred topology in terms of the evolution of morphological characters focusing on those traditionally used to infer the phylogeny of land plants.

A monophyletic concept for bryophytes has gained little support in recent years (Fig. 4.1A). Except for the analyses of 5S nucleotide sequences (Hori *et al.* 1985) and data from sperm cells (Garbary *et al.* 1993, Renzaglia *et al.* 2000), this hypothesis has generally been rejected. Mishler and Churchill (1984) proposed the first formal phylogenetic hypothesis for the origin of bryophyte lineages. Based on morphological, physiological, and biochemical characters, they proposed that the bryophytes compose a paraphyletic group; among the descendants of their common ancestor were the vascular plants (Fig. 4.1E). Their phylogenetic scenario showed the liverworts as the most basal lineage among extant land plants, followed by the hornworts, and mosses forming a sister group to the tracheophytes. Support for this topology has been gained from analyses of data sets restricted to general morphology (Bremer 1985, Kenrick & Crane 1997*b*) or in combination with sequence data of the nuclear genes, 18S and

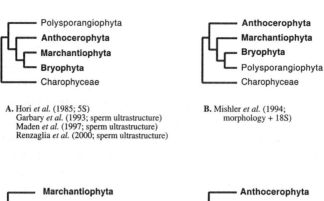

A. Hori *et al.* (1985; 5S)
   Garbary *et al.* (1993; sperm ultrastructure)
   Maden *et al.* (1997; sperm ultrastructure)
   Renzaglia *et al.* (2000; sperm ultrastructure)

B. Mishler *et al.* (1994;
   morphology + 18S)

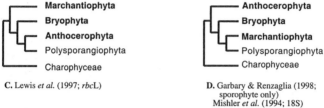

C. Lewis *et al.* (1997; *rbc*L)

D. Garbary & Renzaglia (1998;
   sporophyte only)
   Mishler *et al.* (1994; 18S)
   Malek *et al.* (1996; *cox*3)

E. Mishler & Churchill (1984; morphology),
   Bremer (1985; morphoogy)
   Mishler *et al.* (1994;
     morphology   26S &   18S)
   Kenrick & Crane (1997b; morphology)

F. Hedderson *et al.* (1996 & 1998; 18S)
   Crowe *et al.* (1997; *psb*A)
   Garbary & Renzaglia (1998; morphology)
   Nishiyama & Kato (1999; 18S+*rbc*L
     *psa*A, *psa*B, *psb*D, *rpo*C2)
   Duff & Nickrent (1999; 19S)
   Renzaglia *et al.* (2000; morphology and
     ontogeny)

Fig. 4.1. Summary of alternative phylogenetic relationships among extant lineages of
land plants, based on the assumption that land plants are composed of four
monophyletic lineages, i.e., hornworts, liverworts, mosses, and vascular
plants. Only studies including exemplars of all four lineages are considered
here.

26S (Mishler *et al.* 1994). A paraphyletic bryophyte assemblage with liverworts composing the most basal group (Fig. 4.1C) was also resolved by
analyzing *rbc*L sequence data (Lewis *et al.* 1997). Unlike morphological
characters (Fig. 4.1E), these analyses resulted in the hornworts being
sister to the tracheophytes (Fig. 4.1C). Except for small data sets analyzed

by Mishler *et al.* (1994), this result has found little additional congruent data.

Most other studies resolve bryophytes as paraphyletic, too, but differ in the position of hornworts, which compose the most basal lineage of land plants (Fig. 4.1B, D, F). The relationships within the sister group to the hornworts differ, however, among the studies. Reconstructions based on sporophytic characters suggest that liverworts share a common ancestor with polysporangiophytes (Garbary & Renzaglia 1998), a hypothesis congruent with topologies obtained based on 18S and *cox3* data (Mishler *et al.* 1994, Malek *et al.* 1996). Combination of 18S data with morphological characters for a reduced set of taxa results in a sister group relationship between mosses and polysporangiophytes (Mishler *et al.* 1994). The remaining studies converge toward a phylogenetic hypothesis wherein mosses and liverworts compose a monophyletic clade sister to the polysporangiophytes (Fig. 4.1F). This hypothesis was first proposed by Hedderson *et al.* (1996) on the basis of 18S sequence data analyses, and has subsequently been corroborated by various other data sets (Fig. 4.1F), including the latest morphological data set (Renzaglia *et al.* 2000), and a combination of sequences derived from all three genomes (Nickrent *et al.* 2000).

4.6 Implications for character transformations

This scenario of relationships among extant lineages of land plants is congruent with various morphological transformations, including the loss of pyrenoids in the chloroplast of the ancestor to the moss–liverwort–tracheophyte clade (Fig. 4.2). Other characters maximally consistent with this topology are putative innovations such as grana endmembranes, staggered flagella, and the loss of channeled thylakoid membranes in the sister taxon to the hornworts (Fig. 4.2; Renzaglia *et al.* 2000). This hypothesis is, however, incongruent with hypothesized transformations in other characters, implying multiple parallelisms. Qiu and Palmer (1999) recently examined the distribution of introns in three mitochondrial genes among land plants. These introns were lacking in the exons of liverworts and algae but were present and considered homologous among nearly all other land plants, thereby pointing to liverworts as the most basal group of land plants. Their assumption for homology of the introns across embryophytes was based in part on their identical location in the exon of these genes, and also on the relative high degree of similarity

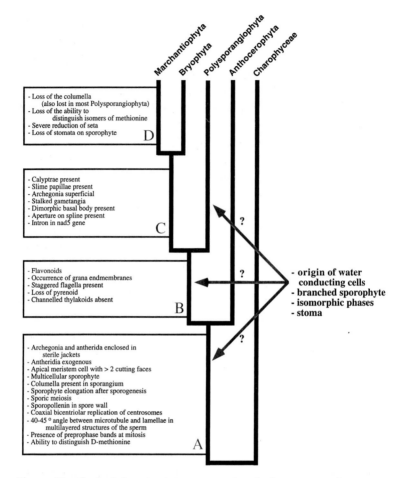

Fig. 4.2. Hypothesized character-state reconstructions for the ancestor to the embryophytes (branch A) and to clades composed of two or more lineages of extant land plants (branches B and C), based a phylogeny of embryophytes as suggested by most recent analyses of morphological and molecular characters (see Fig. 4.1F). Autapomorphic states defining individual lineages are not shown, except for states interpreted as reversals in the evolution of liverworts (branch D). Whether isomorphic phases and the ability of the sporophyte to branch were innovations that arose in the ancestor to the Polysporangiophyta only, as suggested by Kenrick and Crane (1991, 1997*b*), or in a lineage preceding the divergences between the moss + liverwort clade and the Polysporangiophyta, remains open to debate. Similarly the origin(s) of the stoma and of water-conducting cells remains unresolved.

between intron sequences of the hornworts and angiosperms. Similar introns have been recorded from another mitochondrial gene, but with a different distribution. A group I intron is present in the $nad5$ gene of all land plants except hornworts (Steinhauser et al. 1999). Such conflicts in phylogenetic signals between introns, traditionally considered good markers for cladogenic events, are not readily reconciled, and need further study.

The hypothesis of hornworts being basal is also in sharp contrast with previous phylogenetic hypotheses derived from morphological characters. If we examine which homology assumptions would be altered by fitting the morphological data onto the above phylogeny (Fig. 4.1F), the following observations can be made. A columella occurs in the sporangium of hornworts, mosses, and some basal polysporangiophytes (i.e., the Horneophytopsida sensu Kenrick & Crane 1997b). Its absence from other polysporangiophytes is best explained as resulting from an irreversible loss. If liverworts are confirmed as the most likely sister group to mosses, a similar explanation should be invoked to account for the universal lack of a columella in hepatics. Mishler and Churchill (1984) and Bremer (1985) considered the stoma a synapomorphy for a clade of land plants excluding the liverworts. Considering the phylogenetic hypothesis of Fig. 4.1F, stomata could have arisen once, and have been subsequently lost in liverworts. A homology assumption for stomata across land plants relies on the hypothesis that stomata are synplesiomorphic in all land plant lineages. Stomata are widespread among early polysporangiophytes (Kenrick & Crane, 1997b), but are lacking in basal mosses such as *Takakia*, the Andreaeopsida, Andreaeobryopsida, and Sphagnopsida (Renzaglia et al. 2000), as well as in putatively basal hornworts (Hyvönen & Piippo 1993). Evolutionary scenarios invoking multiple gains versus losses of stomata appear similarly parsimonious. If bryophytes arose through reduction from a polysporangiophyte ancestor (see below), the stoma could be considered to be of a single origin, and to have been lost, like the columella, in the ancestor to liverworts.

Mishler and Churchill (1984) interpreted the inability of liverworts to distinguish D-methionine from its L- isomer as further support for their basal position. However, as noted already by these authors, some mosses lack this ability, too. Considering that such losses have already occurred repeatedly, and assuming that the above inability is indeed universal among hepatics, a reversal early in the evolutionary history of liverworts cannot be excluded. A final biochemical feature that hornworts share with mosses and vascular embryophytes is the ability to conjugate the hormone auxin (IAA), whereas liverworts and charophycean algae lack this ability (Sztein *et al.* 1995). Again, one could, within the phylogenetic scenario proposed recently (Fig. 4.1F), invoke a loss of such ability in liverworts.

A close relationship between mosses and tracheophytes had been hypothesized in part, based on the assumption that the conducting cells in these groups are homologous (Scheirer 1980, Mishler & Churchill 1984, see also Niklas 1997). With mosses and liverworts sharing a common ancestor, and sister to the polysporangiophytes, this assumption may need to be somewhat re-evaluated. Conducting cells in liverworts have been documented in the Metzgeriales (Frey *et al.* 1996) and *Haplomitrium* (Hébant 1977), and have recently also been documented in *Conocephalum*, a marchantialean liverwort (Kobiyama & Crandall-Stotler 1999; and see chapter 3 in this volume). Based on the phylogenetic scenario adopted here, it is possible that the evolution of conducting cells was initiated in the ancestor to the moss–liverwort–polysporangiophyte clade, at least in the form of apoptosis, the genetically programmed death of the cells to improve water conduction (see Hébant 1977, Friedman & Cook 2000). Subsequent transformations, leading to the cell types recognized within this large clade of land plants (see Kenrick & Crane 1991), may, however, not be homologous among various lineages, as suggested by Frey *et al.* (1996) and Kobiyama and Crandall-Stotler (1999).

4.7 The origin of bryophytes

A charophycean ancestry of embryophytes suggest that the multicellular sporophyte of land plants arose de novo, as meiosis is delayed by interpolation of mitotic divisions. As mentioned above, this scenario, does not preclude the occurrence in the evolutionary history of land plants of a stage characterized by isomorphic haploid and diploid phases. A common ancestry between the moss–liverworts clade and the polysporangiophytes revives the question whether bryophytes could have evolved from a polysporangiophyte-like ancestor. If bryophytes, or maybe only the ancestor to the liverworts and mosses, evolved through reduction, the Polysporangiophyta sensu Kenrick and Crane (1997*b*) may not be monophyletic. The polysporangiophytes, a clade composed of the Horneophytopsida, *Aglaophyton*, Rhyniophytopsida, and the Tracheophyta (Kenrick & Crane 1997*b*), differ from bryophytes by three characters, namely 1) the presence of protoxylem (i.e., conducting cells that differentiate during the period of elongation), 2) an independent sporophyte that moreover is 3) branched and thus multisporangiate. Evidence for protoxylem differentiation is missing for four taxa composing the grade basal to the tracheophytes, being found only in *Horneophyton* (Kenrick & Crane 1997*b*). Perhaps more significant here, is the absence of protoxylem in *Stockmansella*, a member of the Rhyniopsida (Kenrick & Crane 1997*b*). Thus, even if a single origin for the differentiation of protoxylem is likely, subsequent losses of this character should not be excluded. If conducting cells are considered homologous among land plants (see above), the lack of differentiation of conducting cells during the elongation of the axis (Hébant 1977), may be the result of a loss in bryophytes.

The physiological independence of the sporophyte of polysporangiophytes is inferred based on the presence of roots or rhizomes. Sporophytes of bryophytes remain attached to the gametophyte, and thus never produce any roots, except for the actual foot, that allows for transfer of metabolites from the gametophyte. The sporophyte of bryophytes is, however, photosynthetic and at least in mosses nutritionally independent at the time sporogenesis is initiated (Paolillo & Bazzaz 1968, Wiencke & Schulz 1978). A plant body comprising the gametophyte and the sporophyte is restricted among extant land plants to bryophytes, but has also been hypothesized for some early polysporangiophytes (Rothwell 1995). In these taxa, sporophytes were likely dependent on the gametophyte, too, if only during the early stage of their development.

The only character that remains in Kenrick and Crane's (1997*b*) phylogenetic scenario, to differentiate bryophytes and polysporangiophytes, is the presence of branched sporophytes. This distinction is precisely at the core of the controversy regarding the origin of the sporophyte in bryophytes. Gametophytes of early land plants are poorly known, as gametangia are rare. Remy *et al.* (1993) recently described three gametophytes from the Lower Devonian and postulated a relationship with sporophytes representing members of a protracheophyte grade (sensu Kenrick & Crane 1991), that differ from the haploid phase primarily by their branched axes. Branched gametophytes are, however, known for *Sciadophyton*, also from the Lower Devonian. *Zosterophyllum*, its putative sporophyte, bears multiple sporangia, and is thus branched (Schweitzer 1983). The Zosterophyllopsida date back to the Upper Silurian (Kenrick & Crane 1997*a*). The reconstruction of *Zosterophyllum* would fit the *Cooksonia* model of growth, that differs from the bryophyte model in the occurrence of a branched sporophyte (Rothwell 1995). More elaborate models of other polysporan-giophytes differ in the determinate growth of the sporophyte (Rothwell 1995). It appears, thus, that taxa with branched sporophytes attached to the gametophyte occurred early in the evolution of land plants, and that they may have predated the origin of bryophytes. Whether bryophytes actually arose from such a cooksonioid ancestor (see Miller 1982) needs to be addressed further.

Considering that the gametophytes of early polysporangiophytes are terete (Remy *et al.* 1993), it is possible that the ancestor to the liverwort and moss clade had gametophytes that are radially symmetric. In the current context of land plant phylogeny, the hypothesis that the archetype to land plants was characterized by a thalloid haploid phase, as suggested by Mishler and Churchill (1985), may need to be reconsidered. Bilaterally symmetric gametophytes could indeed have arisen multiple times (e.g., hornworts, marchantioid liverworts, filicalean ferns; see Schuster 1979), an evolutionary trend consistent with the inference from simulations of apical development, that bilateral symmetry is a derived feature in land plants (Niklas 1979; see also Sattler 1998).

4.8 Conclusions

The relationships of bryophytes and their significance in the evolution of land plants have preoccupied biologists for more than a century. The view that land plants arose from a charophycean-like ancestor prevails based on phylogenetic analyses of the green plants. The multicellular sporo-phyte evolved through interpolation of mitotic divisions prior to sporo-genesis. Recent phylogenetic reconstructions of relationships among extant lineages of land plants converge toward a hypothesis wherein the hornworts compose the most basal group, with the mosses and liverworts forming a monophyletic clade sister to the polysporangiophytes. Charac-ter reconstructions remain equivocal in many cases, in part due to the ambiguous relationships of bryophytes to early polysporangiophytes. Recent paleobotanical studies reveal the existence among early land plants of taxa that are characterized by nearly isomorphic gametophytes and sporophytes. The interpretations of these fossils from the Devonian revived the hypothesis that bryophytes may have evolved from such forms, and thus, that the sporophyte of bryophytes is the result of reduc-tion, through the loss of the ability to branch, and to develop multiple sporangia.

Acknowledgments

This study was supported by NSF grant DEB-9806955 to A. J. Shaw and B. Goffinet. Comments by Jennifer Arrington and Jon Shaw on an earlier draft of the manuscript were appreciated. I also thank Karen Renzaglia and Daniel Nickrent for sharing manuscripts of yet unpublished studies with me.

REFERENCES

Ando, H. (1978). A view on the evolution of bryophytes. *Proceedings of the Bryological Society of Japan*, **2**, 80–3.

Andrews, H. N. (1958). Notes on Belgian specimens of *Sporogonites*. *Paleobotanist*, **7**, 85–9.

Asakawa, Y. (1986). Chemical relationships between algae, bryophytes and pteridophytes. *Journal of Bryology*, **14**, 59–70.

Baldauf, S. L., Manhart, J. R, & Palmer, J. D. (1990). Different fates of the chloroplast *tuf*A gene following its transfer to the nucleus in green algae. *Proceedings of the National Academy of Sciences USA*, **87**, 5317–21.

Banks, H. P. (1975). Early vascular plants: proof and conjecture. *Bioscience*, **25**, 730–7.

Bateman, R. M. (1996). Nonfloral homoplasy and evolutionary scenarios in living and fossil land plants. In *Homoplasy: The Recurrence of Similarity in Evolution*, ed. M. J. Sanderson & L. Hufford, pp. 94–130. San Diego: Academic Press.

Bateman, R. M., Crane, P. R., DiMichele, W. A., Kenrick, P. R., Rowe, N. P., Speck, T., and Stein, W. E. (1998). Early evolution of land plants: phylogeny, physiology, and ecology of the primary terrestrial radiation. *Annual Review of Ecology and Systematics*, **29**, 263–92.

Beckert, S., Steinhauser, S., Muhle, H., & Knoop, V. (1999). A molecular phylogeny of bryophytes based on nucleotide sequences of the mitochondrial nad5 gene. *Plant Systematics and Evolution*, **218**, 179–92.

Behrensmeyer, A. K., Damuth, J. D., DiMechele, W. A., Potts, R., Sues, H.-D., & Wing, S. L. (1992). *Terrestrial Ecology Through Time. Evolutionary Paleoecology of Terrestrial Plants and Animals*. Chicago: University of Chicago Press.

Bhattacharya, D. & Medlin, L. (1998). Algal phylogeny and the origin of land plants. *Plant Physiology*, **116**, 9–15.

Bhattacharya, D., Weber, K., An, S. S., & Berning-Koch, W. (1998). Actin phylogeny identifies *Mesostigma viride* as the flagellate ancestor of the land plants. *Journal of Molecular Evolution*, **47**, 544–50.

Bierhorst, D. W. (1971). *Morphology of Vascular Plants*. New York: Macmillan.

Bopp, M. & Capesius, I. (1996). New aspects of bryophyte taxonomy provided by a molecular approach. *Botanica Acta* 109, 368–72.

Bower, F. O. (1890). On antithetic as distinct from homologous alternation of generations in plants. *Annals of Botany*, **4**, 374–70.

(1908). *The Origin of a Land Flora: A Theory Based upon Facts of Alternation*. London: Macmillan.

Bremer, K. (1985). Summary of green plant phylogeny and classification. *Cladistics*, **1**, 369–85.

Bremer, K., Humphries, C. J., Mishler, B. D., & Churchill, S. P. (1987). On cladistic relationships in green plants. *Taxon*, **36**, 339–49.

Capesius, I. (1995). A molecular phylogeny of bryophytes based on the nuclear encoded 18S rRNA genes. *Journal of Plant Physiology*, **146**, 59–63.

Capesius, I. & Bopp, M. (1997). New classification of liverworts based on molecular and morphological data. *Plant Systematics and Evolution*, **207**, 87–97.

Celakovsky, J. (1874). Über die verschiedenen Formen und Bedeutung des Generationswechsels der Pflanzen. *Sitzungsbericht der königlischen böhmischen Gesellschaft der Wissenschaft Prag*, **1874**, 22–61.

Chapman, R. L., Buchheim, M. A., Delwiche, C. F., Friedl, T., Huss, V. A. R., Karol, K. G., Lewis, L. A., Manhart, J., McCourt, J. M., Olsen, J. L., & Waters, D. A. (1997). Molecular systematics of the Green Algae. In *Molecular Systematics of Plants*, vol. 2, *DNA Sequencing*, ed. D. E. Soltis, P. S. Soltis, & J. J. Doyle, pp. 509–40. Boston: Kluwer.

Cook, M. E. & Graham, L. E. (1998). Structured similarities between surface layers of selected charophycean algae and bryophytes and the cuticles of vascular plants. *International Journal of Plant Sciences*, **159**, 780–6.

Crandall-Stotler, B. (1980). Morphogenetic designs and a theory of bryophyte origins and divergence. *Bioscience*, **30**, 580–5.

(1984). Musci, Hepatics and Anthocerotes – an essay on analogues. In *New Manual of Bryology*, vol. 2, ed. R. M. Schuster, pp. 1093–129. Nichinan: Hattori Botanical Laboratory.

(1986). Morphogenesis, developmental anatomy and bryophyte phylogenetics: contraindications of monophyly. *Journal of Bryology*, **14**, 1–23.

Crandall-Stotler, B. & Bozzola, J. J. (1988). Fine structure of the meristematic cells of *Takakia lepidozioides* Hatt. et H. Inoue (Takakiophyta). *Journal of the Hattori Botanical Laboratory*, **64**, 197–218.

Crowe, C. T., Pike, L. M., Cross, K. S., & Renzaglia, K. S. (1997). The *psb*A gene sequence can be used to infer phylogenetic relationships among the major lineages of bryophytes. *American Journal of Botany*, **84** (suppl. vol. 6), 14–15.

Delwiche, C. F., Graham, L. E., & Thomson, N. (1989). Lignin-like compounds and sporopollenin in *Coleochaete*, an algal model for land plant ancestry. *Science*, **245**, 399–401.

Duff, R. J. & Nickrent, D. L. (1999). Phylogenetic relationships of land plants using mitochondrial small-subunit rDNA sequences. *American Journal of Botany*, **86**, 372–86.

Edwards, D., Fanning, U., & Richardson, J. B. (1994). Lower Devonian coalified sporangia from Shropshire: *Salopella* Edwards & Richardson and *Tortilicaulis* Edwards. *Botanical Journal of the Linnean Society*, **116**, 89–110.

Edwards, D., Duckett, J. G., & Richardson, J. B. (1995). Hepatic characters in the earliest land plant. *Nature*, **374**, 635–6.

Edwards, D., Abbott, G. D., & Raven, J. A. (1996). Cuticles of early land plants: a palaeoecophysiological evaluation. In *Plant Cuticles – An Integrated Approach*, ed. G. Kierstiens, pp. 1–31. Oxford: BIOS.

Edwards, D., Wellman, C. H., & Axe, L. (1998). The fossil record of early land plants and interrelationships between primitive embryophytes: too little and too late? In *Bryology for the the Twenty-first Century*, ed. J. W. Bates, N. W. Ashton, & J. G. Duckett, pp. 15–43. Leeds: Maney and British Bryological Society.

Frederick, S. E., Gruber, P. J., & Tolbert, N. E. (1973). The occurrence of glycolate dehydrogenase and glycolate oxidase in green plants: an evolutionary survey. *Plant Physiology*, **52**, 318–23.

Frey, W., Hilger, H. H., & Hofman, M. (1996). Water-conducting cells of extant *Symphyogyna*-type Metzgerialean taxa: ultrastructure and phylogenetic implications. *Nova Hedwigia*, **63**, 471–81.

Friedl, T. (1997). The evolution of the Green Algae. *Plant Systematics and Evolution, (suppl.)* **11**, 87–101.

Friedman, W. E. & Cook, M. E. (2000). The origin and early evolution of tracheids in vascular plants: intergration of paleobotanical and neobotanical data. *Philosophical Transactions of the Royal Society*, in press.

Garbary, D. J. & Renzaglia, K. S. (1998). Bryophyte phylogeny and the evolution of land plants: evidence from development and ultrastructure. In *Bryology for the the Twenty-first Century*, ed. J. W. Bates, N. W. Ashton, & J. G. Duckett, pp. 45–63. Leeds: Maney and British Bryological Society.

Garbary, D. J., Renzaglia, K. S., & Duckett, J. G. (1993). The phylogeny of land plants: a cladistic analysis based on male gametogenesis. *Plant Systematics and Evolution*, **188**, 237–69.

Gensel, P. G. & Andrews, H. N. (1987). The evolution of early land plants. *American Scientist*, **75**, 478–89.

Graham, L. E. (1985). The origin of the life cycle of plants. *American Scientist*, **73**, 178–86. (1993). *The Origin of Land Plants*. New York: Wiley.
(1996). Green algae to land plants: An evolutionary transition. *Journal of Plant Science*, **109**, 241–51.

Graham, L. E. & Wilcox, L. W. (1999). *Algae*. Upper Saddle River: Prentice-Hall.

Graham, L. E., Delwiche, C. F., & Mishler, B. D. (1991). Phylogenetic connections between the 'green algae' and the 'bryophytes'. *Advances in Bryology*, **4**, 213–44.

Gray, J. (1985). The microfossil record of early land plants: advances in understanding of early terrestrialization, 1970–1984. *Philosophical Transactions of the Royal Society of London B*, **309**, 167–95.

Gray, J. & Boucot, A. J. (1977). Early vascular land plants: proof and conjecture. *Lethaia*, **10**, 145–74.

Gray, J., Boucot, A. J., Grahn, Y., & Himes, G. (1992). A new record of early Silurian land plant spores from the Paraná Basin, Paraguay (Malvinokaffric Realm). *Geological Magazine*, **129**, 741–52.

Halle, T. G. (1916). A fossil sporogonium from the Lower Devonian of Röragen in Norway. *Botaniska Notiser*, 79–81.
(1936). Notes on the Devonian genus *Sporogonites*. *Svensk Botanisk Tidskrift*, **30**, 613–23.

Hébant, C. (1977). The conducting tissues of bryophytes. *Bryophytorum Bibliotheca*, **10**, 1–157 + 80 plates.

Hedderson, T. A., Chapman, R. L., & Rotes, W. L. (1996). Phylogenetic relationships of bryophytes inferred from nuclear-encoded rRNA gene sequences. *Plant Systematics and Evolution*, **200**, 213–24.

Hedderson, T. A., Chapman, R. L., & Cox, C. (1998). Bryophytes and the origins and diversification of land plants: new evidence from molecules. In *Bryology for the Twenty-first Century*, ed. J. W. Bates, N. W. Ashton, & J. G. Duckett, pp. 65–7. Leeds: Maney and British Bryological Society.

Hemsley, A. R. (1994). The origin of the land plant sporophyte: an interpolational scenario. *Biological Review*, **69**, 263–73.

Hofmeister, W. (1851). *Vergleichende Untersuchungen der Keimung, Entfaltung und Fruchtbildung höherer Kryptogamen (Moose, Farne, Equisetaceen, Rhizocarpeen und Lycopodiaceen) und der Samenbildung der Coniferen*. Leipzig: F. Hofmeister.

Holloway, J. E. (1939). The gametophyte, embryo and young rhizome of *Psilotum triquetrum* Swartz. *Annals of Botany*, **3**, 313–36.

Hori, H., Lim, B.-L., & Osawa, S. (1985). Evolution of green plants as deduced from 5S rRNA sequences. *Proceedings of the National Academy of Sciences USA*, **82**, 820–3.

Hueber, F. M. (1961). *Hepaticites devonicus*, a new fossil liverwort from the Devonian of New York. *Annals of the Missouri Botanical Garden*, **48**, 125–32.

Hyvönen, J. & Piippo, S. (1993). Cladistic analysis of the hornworts (Anthocerotophyta). *Journal of the Hattori Botanical Laboratory*, **74**, 105–19.

Johnson, N. G. (1985). Early Silurian palynomorphs from the Tuscarora formation in central Pennsylvania and their paleobotanical and geological significance. *Review of Palaeobotany and Palynology*, **45**, 307–60.

Kenrick, P. (1994). Alternation of generations in land plants: new phylogenetic and palaeobotanical evidence. *Biological Review*, **69**, 293–30.

Kenrick, P. & Crane, P. R. (1991). Water-conducting cells in early fossil land plants: implications for the early evolution of tracheophytes. *Botanical Gazette*, **152**, 335–56.

(1997a). The origin and early evolution of land plants. *Nature*, **389**, 33–9.

(1997b). *The Origin and Early Diversification of Land Plants. A Cladistic Study*. Washington: Smithsonian Institution Press.

Knoll, A. H. & Niklas, K. J. (1987). Adaptation, plant evolution, and the fossil record. *Review of Palaeobotany and Palynology*, **50**, 127–49.

Knoll, A. H., Niklas, K. J., Gensel, P. G., & Tiffney, B. H. (1984). Character diversification and pattern of evolution in early vascular plants. *Paleobiology*, **10**, 34–7.

Kobiyama, Y. & Crandall-Stotler, B. (1999). Studies of specialized pitted parenchyma cells of the liverwort *Conocephalum* Hill and their phylogenetic implications. *International Journal of Plant Science*, **160**, 351–70.

Kodner, R. & Graham, L. (1999). Hydrolyzed remains of *Polytrichum* moss resemble Silurian/Devonian microfossils. 16th International Botanical Congress, St Louis, abstracts 326.

Kranz, H. D., Miks, D., Siegler, M.-L., Capesius, I., Sensen, C. W., & Huss, V. A. R. (1995). The origin of land plants: phylogenetic relationships among charophytes, bryophytes, and vascular plants inferred from complete small subunit ribosomal RNA gene sequences. *Journal of Molecular Evolution*, **41**, 74–84.

Krassilov, V. A. & Schuster, R. M. (1984). Paleozoic and Mesozoic fossils. In *New Manual of Bryology*, vol. 2, ed. R. M. Schuster, pp. 1173–93. Nichinan: Hattori Botanical Laboratory.

Kroken, S. B., Graham, L. E., & Cook, M. E. (1996). Occurrence and evolutionary significance of resistant cell walls in Charophytes and Bryophytes. *American Journal of Botany*, **83**, 1241–54.

Lacey, W. S. (1969). Fossil bryophytes. *Biological Review*, **44**, 189–205.

Lewis, L. A., Mishler, B. D., & Vilgalys, R. (1997). Phylogenetic relationships of the liverworts (Hepaticae): a basal embryophyte lineage, inferred from nucleotide sequence data of the chloroplast gene *rbcL*. *Molecular Phylogenetics and Evolution*, **7**, 377–93.

Ligrone, R., Duckett, J. G., & Renzaglia, K. S. (1993). The gametophyte–sporophyte junction in land plants. *Advances in Botanical Research*, **19**, 231–317.

Maden, A. R., Witthier, D. P., Garbary, D. J., & Renzaglia, K. S. (1997). Ultrastructure of the spermatozoid of *Lycopodiella lateralis* (R. Br.) B. Øllgaard (Lycopodiaceae). *Canadian Journal of Botany* 75, 1728–38.

Malek, O., Lättig, K., Hiesel, R., Brennicke, A., & Knoop, V. (1996). RNA editing in bryophytes and a molecular phylogeny of land plants. *European Molecular Biology Organization Journal*, **15**, 1403–11.

Manhart, J. R. & Palmer, J. D. (1990). The gain of two chloroplast tRNA introns marks the green algal ancestors to land plants. *Nature*, **345**, 268–70.

Marchant, H. J. & Pickett-Heaps, J. D. (1973). Mitosis and cytokinesis in *Coleochaete scutata*. *Journal of Phycology*, **9**, 461–71.

Melkonian, M. (1982). Structural and evolutionary aspects of the flagellar apparatus in green algae and land plants. *Taxon*, **31**, 255–65.

Melkonian, M. & Surek, B. (1995). Phylogeny of the Chlorophyta: congruence between ultrastructural and molecular evidence. *Bulletin de la Société Zoologique de France*, **120**, 191–208.

Miller, H. A. (1982). Bryophyte evolution and geography. *Biological Journal of the Linnean Society*, **18**, 145–86.

Mishler, B. D. & Churchill, S. P. (1984). A cladistic approach to the phylogeny of the "bryophytes." *Brittonia*, **36**, 406–24.

Mishler, B. D. & Churchill, S. P. (1985). Transition to a land flora: phylogenetic relationships of the green algae and bryophytes. *Cladistics*, **1**, 305–28.

Mishler, B. D., Lewis, L. A., Buchheim, M. A., Renzaglia, K. S., Garbary, D. J., Delwiche, C. F., Zechman, F. W., Kantz, T. S., & Chapman, R. L. (1994). Phylogenetic relationships of the "green algae" and "bryophytes." *Annals of the Missouri Botanical Garden*, **81**, 451–83.

Murray, B. M. (1988). Systematics of the Andreaopsida (Bryophyta): Two orders with links to *Takakia*. *Beiheft zur Nova Hedwigia*, **90**, 289–336.

Neuburg, M. F. (1960). Mosses from the Permian deposits of Angaraland. *Trudy Geologicheskogo Instituta. Akademiya nauk SSSR*, **19**, 1–104. [in Russian]

Niklas, K. J. (1979). Simulations of apical developmental sequences in bryophytes. *Annals of Botany*, **44**, 339–52.

(1997). *The Evolutionary Biology of Plants*. Chicago: University of Chicago Press.

Nickrent, D. L., Parkinson, C. L., Palmer, J. D., & Duff, J. R. (2000). Multigene phylogeny of plants: hornworts are basal and mosses are sister to liverworts. (submitted)

Nishiyama, T. & Kato, M. (1999). Molecular phylogenetic analysis among bryophytes and tracheophytes based on combined data of plastid coded genes and the 18S rRNA gene. *Molecular Biology and Evolution*, **16**, 1027–36.

Oostendorp, C. (1987). The bryophytes from the Palaeozoic and Mezosozoic. *Bryophytorum Bibliotheca*, **34**, 1–112 + 49 plates and 7 tables.

Paolillo, D. J. & Bazzaz, F. A. (1968). Photosynthesis in sporophytes of *Polytrichum* and *Funaria*. *Bryologist*, **71**, 61–9.

Pringsheim, N. (1876). Über vegetative Sprossung der Moosfruchte. *Monatsbericht der kaiserlichen Akademie für Wissenschaften Berlin 1876*, 425–9.

Pringsheim, N. (1878). Über die Sprossung der Moosfruchte. *Jahrbuch für wissenschaftliche Botanik*, **11**, 1–46.

Qiu, Y.-L. & Palmer, J. D. (1999). Phylogeny of early land plants: insights from genes and genomes. *Trends in Plant Sciences*, **4**, 26–30.

Qiu, Y.-L., Cho, Y., Cox, J. C., & Palmer, J. D. (1998). The gain of three mitochondrial introns identifies the liverworts as the earliest land plants. *Nature*, **394**, 671–4.

Remy W., Gensel, P. G., & Haas, H. (1993). The gametophyte generation of some early Devonian land plants. *International Journal of Plant Science*, **154**, 35–58.

Renzaglia, K. S., Brown, R. C., Lemmon, B. E., Duckett, J. G., & Ligrone, R. (1993). Occurrence and phylogenetic significance of monoplastidic meiosis in liverworts. *Canadian Journal of Botany*, **72**, 65–72.

Renzaglia, K. S., McFarland, K. D., & Smith, D. K. (1997). Anatomy and ultrastruture of the sporophyte of *Takakia ceratophylla* (Bryophyta). *American Journal of Botany*, **84**, 1337–50.

Renzaglia, K. S., Duff, R. J. & Garbary, D. J. (2000). Vegetative and reproductive innovations of early land plants: implications for a unified phylogeny. *Philosophical Transactions of the Royal Society* (in press).

Richardson, J. B. (1985). Lower Palaeozoic sporomorphs: their stratigraphical distribution and possible affinities. *Philosophical Transaction of the Royal Society of London, Botany*, **309**, 201–5.

Rothwell, G. W. (1995). The fossil history of branching: implications for the phylogeny of land plants. In *Experimental and Molecular Approaches to Plant Biosystematics*, ed. P. C. Hoch & A. G. Stephenson, pp. 71–6. St Louis: Missouri Botanical Garden.

Samigullin, T. H., Valiejo-Roman, K. M., Troitsky, A. V., Bobrova, V. K., Filin, V. R., Martin, W., & Antonov, A. S. (1998). Sequences of rDNA internal transcribed spacers from the chloroplast DNA of 26 bryophytes: properties and phylogenetic utility. *Federation of European Biochemical Societies Letters*, **422**, 47–51.

Sattler, R. (1998). On the origin of symmetry, branching and phyllotaxis in land plants. In *Symmetry in Plants*, ed. R. V. Jean & D. Barabé, pp. 775–93. Singapore: World Scientific.

Scheirer, D. C. (1980). Differentiation of bryophyte conducting tissues: structure and histochemistry. *Bulletin of the Torrey Botanical Club*, **107**, 298–307.

Schofield, W. B. (1985). *Introduction to Bryology*. New York: Macmillan.

Schuster, R. M. (1966). *The Hepaticae and Anthocerotae of North America*, vol. 1. New York: Columbia University Press.

(1977). The evolution and early diversification of the Hepaticae and Anthocerotae. In *Beitrage zur Biologie der niederen Pflanzen*, ed. W. Frey, H. Hurka, & F. Oberwinkler, pp. 107–15. Stuttgart: Fischer Verlag.

(1979). The phylogeny of the Hepaticae. In *Bryophyte Systematics*, ed. G. C. S. Clarke & J. G. Duckett, pp. 41–82. London: Academic Press.

(1984). Evolution, phylogeny and classification of the Hepaticae. In *New Manual of Bryology*, vol. 2., ed. R. M. Schuster, pp. 892–1070. Nichinan: Hattori Botanical Laboratory.

(1997). On *Takakia* and the phylogenetic relatioships of the Takakiales. *Nova Hedwigia*, **64**, 281–310.

Schweitzer, H. J. (1983). Die Unterdevonflora des Rheinlandes, 1. Teil. *Palaeontographica*, **189**B, 1–138.

Sluiman, H. J. (1985). A cladistic evaluation of the lower and higher green plants (Viridiplantae). *Plant Systematics and Evolution*, **149**, 217–32.

Smith, A. J. E. (1986). Bryophyte phylogeny: fact or fiction. *Journal of Bryology*, **14**, 83–9.

Smith, D. K. & Davison, P. G. (1993). Antheridia and sporophytes in *Takakia ceratophylla* (Mitt.) Grolle: evidence for reclassification among the mosses. *Journal of the Hattori Botanical Laboratory*, **73**, 263–71.

Springer, E. (1935). Über Apogame (vegetativ enstandene) Sporogone an der bivalente Rasse des Laubmooses *Phascum cuspidatum*. *Zeitschrift für induktive Abstammungs- und Vererbungslehre*, **69**, 249–62.

Steemans, P. (1999). Paléodiversification des spores et des cryptospores de l'Ordovicien au Dévonien inférieur. *Geobios*, **32**, 341–52.

Steere, W. C. (1969). A new look at evolution and phylogeny in Bryophytes. In *Current Topics in Plant Sciences*, ed. J. E. Gunkel, pp. 134–42. New York: Academic Press.

Steinhauser, S., Beckert, S., Capesius, I., Malek, O., and Knoop, V. (1999). Plant mitochondrial RNA editing. *Journal of Molecular Evolution*, **48**, 303–12.

Stewart, W. N. (1983). *Paleobotany and Evolution of Plants*. Cambridge: Cambridge University Press.

Stewart, W. N. & Rothwell, G.W. (1993). *Paleobotany and the Evolution of Plants*. New York: Cambridge University Press.

Strother, P. K., Al-Hajri, S., & Traverse, A. (1996). New evidence of land plants from the lower Middle Ordovician of Saudi Arabia. *Geology*, **24**, 55–8.

Sztein, A. E., Cohen, J. D., Slovin, J. P., & Cooke, T. J. (1995). Auxin metabolism in representative land plants. *American Journal of Botany*, **82**, 1514–21.

Taylor, T. N. (1982). The origin of land plants: A paleobotanical perspective. *Taxon*, **31**, 155–77.

Taylor, T. N. & Taylor, E. L. (1993). *The Biology and Evolution of Fossil Plants*. Englewood Cliffs: Prentice-Hall.

Taylor, W. A. (1996). Ultrastructure of lower Palaeozoic dyads from southern Ohio. *Review of Palaeobotany and Palynology*, **92**, 269–79.

Thomas, B. A. (1972). A probable moss from the Lower Carboniferous of the Forest of Dean, Gloucestershire. *Annals of Botany*, **36**, 155–61.

Wettstein, F. von (1942). Über einige Beobachtungen und experimentelle Befunde bei Laubmoosen. II. Über die vegetativ enstehenden Sporogone von *Phascum cuspidatum* und die willkurliche Anderung des Gestaltwechsels bei diesem Laubmoos. *Bericht der deutschen botanischen Gesellschaft*, **40**, 399–405.

Wiencke, C. & Schulz, D. (1978). The development of transfer cells in the haustorium of the *Funaria hygrometrica* sporophyte. *Bryophytorum Bibliothecum*, **13**, 147–67.

Zimmerman, W. (1932). Phylogenie. In *Manual of Bryology*, ed. F. Verdoorn, pp. 433–64. The Hague: Martinus Nijhoff.

5

Chemical constituents and biochemistry

Only 30 years ago the chemistry of bryophytes was virtually unknown. Recent research on the biology of bryophytes and progress in analytical techniques has resulted in a deeper knowledge about the chemical constituents of bryophytes, although our understanding of their biochemical processes, especially biosynthetic pathways, compared to vascular plants, is still rather poor. In the first part of this chapter the present state of the art regarding the chemistry of bryophytes will be presented, the second part deals with some aspects of chemosystematics, and in the third part knowledge of the biochemistry of bryophytes is summarized. A recent thorough and comprehensive review on the topic has been published by Asakawa (1995). A further review, in 1997, on "heterocyclic compounds in bryophytes" was published by the same author. He deals mainly with secondary metabolites and does not mention inorganic compounds. Primary metabolites are partly discussed, e.g., lipids, carbohydrates, and phaeophytins. Biochemistry was not a topic of his review; this field of research was recently surveyed in Chopra and Bhatla (1990) and Rudolph (1990).

In the section 'Chemistry of bryophytes', present knowledge about inorganic compounds and primary metabolites, their structural analogues, and other ubiquitous compounds in bryophytes will be discussed, followed by the main classes of secondary products. The examples mentioned here are selected from recent original papers **not** cited in Asakawa (1995, 1997); they are more or less confined to those compounds typical of bryophytes in general. The following section on 'Chemosystematics' will discuss marker components for at least the bryophyte **classes**. Examples mentioned in the section on 'Biochemistry' will also focus on recent original papers. Because of limited space in this chapter, only the most important examples of each compound class and biochemical processes can be discussed. Structures are drawn only for those compounds not illustrated in Asakawa (1995, 1997) or in Huneck (1983).

5.1 Chemistry of bryophytes

5.1.1 Inorganic compounds

Among bryophytes there does not appear to be a general accumulation of special inorganic compounds as known for example in lycopods with aluminium oxide, or in horsetails with silica (Hegnauer 1962). Inorganic ions found in most bryophytes do not differ qualitatively and quantitatively from those found in vascular plants (Huneck 1983). A few recent papers deal with this topic and in an increased number of studies bryophytes are used to monitor dispersion patterns of atmospheric pollutants (Muhle 1984, Steinnes 1995). It is frequently observed that certain bryophytes accumulate potentially toxic levels of metal ions without apparent damage. Some species show specific tolerance to metals present in their natural environment (Wells & Brown 1995). There is one liverwort species with an unusually high dry-weight content of 1.3% mercury: *Jungermannia vulcanicola* Steph. The mercury is deposited in the cell walls in a mercury–sulfur component (Satake & Miyasaka 1984). Satake *et al.* (1984) found in some water mosses in New Caledonia rather unusual high contents of inorganics. In the moss *Ectropothecium subobscurum* Thér. they detected high contents of nickel (690 μg g^{-1} dry weight), manganese (470 μg g^{-1}), chromium (254 μg g^{-1}), and bromine (210 μg g^{-1}) and in the liverwort *Lopholejeunea* sp. an incredibly high amount of manganese with 15300 μg g^{-1}. In general, according to those authors, the amounts of nickel, chromium, and bromine in plants do not exceed 100 μg g^{-1} dry weight.

5.1.2 Primary metabolites, their structural analogues, and other ubiquitous compounds

Primary metabolites are regarded as essential for almost all living organisms and are derived from reactions of primary metabolism. Among them should be noted carbohydrates, nucleotides, proteins, intermediates of the tricarboxylic acid cycle, lipids, common pigments for photosynthetic processes, and lignin. For a detailed introduction to "primary" and "secondary" plant products see Mothes (1980). As bryophytes definitely belong to the "green line" of land plants developing from early chlorophytes (Graham *et al.* 1991, Hedderson *et al.* 1998), their primary metabolism is very similar to that of vascular plants (Hegnauer 1962, 1986). This holds true for such essential compounds as cellulose in the cell walls, chlorophylls a and b, main carotenoids, starch, nucleic acids, sugars, and certain lipids. So far no real **lignin** has been found in the cell walls of bryophytes, only "lignin-like" aromatic compounds (Wilson *et al.* 1989, Edelmann *et al.* 1998). Some of the few recent papers on **carbohydrates** in bryophytes are discussed in Asakawa (1995). Konno *et al.* (1987) analyzed cell cultures of the liverwort *Marchantia polymorpha* L. for its pectic polysaccharides in cell walls and classified them as three types. According to these authors, carbohydrates correspond in general to those of vascular plants, even in their absolute configuration. The protein **amino acids** of bryophytes do not seem to be significantly different from those of higher plants (Huneck 1983). In a survey of the occurrence of amino acids, Dutt (1996) found common protein amino acids in 23 species of mosses and liverworts, and in the liverwort *Pellia epiphylla* (L.) Corda also the two non-protein amino acids: γ-aminobutyric acid (GABA) and ornithine. Salm *et al.* (1998) isolated the amino acid N-(4-hydroxy-3,5-dimethoxy-benzoyl)aspartic acid, called fontinalin (1), from *Fontinalis squamosa* Hedw. Trennheuser *et al.* (1994)

(1)

detected six glutamic acid amides in *Anthoceros agrestis* Paton. Hegewald and Kneifel (1987) investigated three liverwort and nine moss species for their **polyamine** content. They found nine polyamines, spermidine being the dominant one. Kraut *et al.* (1997a) isolated and synthesized the oligopeptide rufulamide (2) from *Metzgeria rufula* Spruce. Adam and Becker (1993) screened 13 liverwort, one hornwort and seven moss species for the occurrence of lectins. From *Marchantia polymorpha* they purified the first bryophyte lectin and showed it to be a monomeric protein of around 16 kDa. Among the **lipids** one may distinguish between n-alkanes and their derivatives, and fatty acids, either free or bound as triglycerides. The occurrence of aliphatic compounds in liverworts and mosses was summarized by Asakawa (1995), whereas no report of compounds of this class seems to have been published for hornworts. A few additional aliphatic compounds have since been reported. According to Asakawa *et al.* (1996)

(2)

the liverwort *Chiloscyphus pallidus* Mitt. in Hook. f. emits a "very strong stink bug smell" from crushed fresh material. The authors found as compounds responsible for the odor simple aliphatic aldehydes, of which (E)-dec-2-enal (3) is the major one. Nissinen and Sewón (1994) studied

(3)

hydrocarbon patterns in different tissues of *Polytrichum commune* Hedw. Neinhuis and Jetter (1995) found n-alkanes, n-alkyl esters, n-alkanals, primary n-alkanols, and n-alkanoic acids, closely resembling those of moss gametophytes and higher plants, in waxes of the sporophytes of *Pogonatum belangeri* (C. Muell.) Jaeg. and *P. urnigerum* (Hedw.) P. Beauv. The dominant compound was nonacosan-10-ol. In addition to the common fatty acids ubiquitous in green land plants, mosses and liverworts contain typical unsaturated ones, most of them mentioned by Asakawa (1995). Examples for other recent papers on lipids in bryophytes are those of Kohn *et al.* (1994), Dembitsky and Rezanka (1995), and Toyota *et al.* (1997*b*). Further primary compounds, including enzymes, will be discussed under *Biochemistry*.

5.1.3 Secondary products

Table 5.1 shows the classes of secondary compounds and their distribution in bryophytes.

Table 5.1. *Secondary compounds from bryophytes*

Compound class	Liverworts	Hornworts	Mosses
Terpenoids	++	(+)	+
Aromatic compounds	++	+	++
Nitrogen-containing compounds	+	(+)	(+)
Sulfur-containing compounds	(+)	–	–
Chlorine-containing compounds	+	–	–

Notes: ++ >100 compounds; + 10–100 compounds; (+) <10 compounds.

Table 5.2. *Terpenoids from bryophytes*

Terpenoid-class	Liverworts	Hornworts	Mosses
Monoterpenes	60	–	3
Sesquiterpenes	~600	5	9
Diterpenes	~300	–	4
Triterpenes	18	2	25
Steroids	17	4	10
Tetraterpenes	26	?	28
Polyterpenes	+	?	+

Terpenoids

The distribution of terpenoids in bryophytes and the approximate numbers of single compounds of the different terpenoid classes known from bryophytes are shown in Table 5.2.

Monoterpenes and related compounds Besides aliphatic compounds (see above) and certain sesquiterpenes (see below), monoterpenes are mainly responsible for the characteristic odour of many liverworts. They are typically stored in the oil bodies and are therefore almost limited to liverworts. Examples of monoterpenes responsible for this odour are given by Asakawa (1995). Only a few additional species have been investigated for the occurrence of monoterpenes and a few new compounds have been isolated. Toyota *et al.* (1997*d*) reinvestigated the fragrant liverwort *Conocephalum conicum* (L.) Underw., one of the chemically best-known liverwort species. Nevertheless, they succeeded in detecting three new monoterpene esters, the main one being (+)-bornyl cis-ferulate (4). In hornworts monoterpenes have so far not been detected and in mosses they have up to now been found only in the two Splachnaceae species, *Splachnum luteum* Hedw. and *S. rubrum* Hedw. (Asakawa 1995). For most monoterpenes the

(4)

absolute configuration has not been determined, since almost all of them were detected by GC-MS (gas chromatography – mass spectroscopy) analysis (Asakawa 1995).

Sesquiterpenes and related compounds Sesquiterpenes are by far the largest class of secondary metabolites in bryophytes, although they mainly occur in liverworts (Table 5.2). Liverworts synthesize both the same sesquiterpenes as higher plants and others with characteristic liverwort skeletons. In some cases the compounds correspond to the same enantiomers as in higher plants, in others they represent the optical antipodes. So far acyclic, mono-, di-, tri-, and tetracyclic sesquiterpenes have been detected. Asakawa (1995, 1997) gave a detailed discussion of the sesquiterpene types found up to the middle of the 1990s. He listed no fewer than 61 types of sesquiterpenes from liverworts, eudesmane types being most frequent. Many of them show remarkable biological activities (Asakawa 1995, 1997). Among the sesquiterpene skeletons found to be unique to liverworts are those shown in Fig. 5.1. Since 1994 about 70 papers have reported on the isolation and identification of sesquiterpenes from liverworts. Only a few of those papers can be discussed here.

Since their detection in liverworts, the barbatane- and bazzanane-type sesquiterpene skeletons had been regarded as liverwort-specific. König *et al.* (1996) identified the sesquiterpene hydrocarbons β-bazzanene, α- and β-barbatene, and isobazzanene for the first time as constituents of the roots of *Meum athamanticum* (L.) Jacq. (Umbelliferae). For α- and β-barbatene the authors determined the opposite configurations to those in liverworts. In other cases both enantiomers of one sesquiterpene may occur in one liverwort species (e.g., Saritas *et al.* 1998). In a first study on the chemistry of the liverwort family Cephaloziellaceae, Wu *et al.* (1996) detected in *Cephaloziella recurvifolia* Hatt. the aromadendrane-type sesquiterpene alcohol (-)-ledol and other oxygenated sesquiterpenes as major constituents. Toyota *et al.* (1997*a*) isolated two sesquiterpene alcohols belonging to the new "dumortane-type" sesquiterpene skeleton from

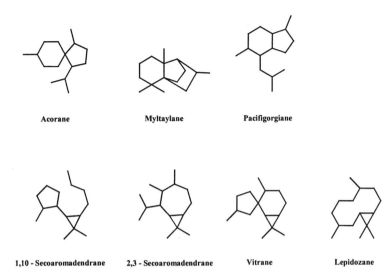

Fig. 5.1. Sesquiterpene skeletons unique to liverworts.

Argentinian *Dumortiera hirsuta* (Sw.) Nees and determined the structure of one "dumortenol" as 3,4,11-trimethyl-7-methylenebicyclo[6.3.0]undec-2-en-11α-ol **(5)**. Other interesting new sesquiterpenes have been reported by

(5)

Liu *et al.* (1996) and Hashimoto *et al.* (1998a). Toyota *et al.* (1996a) discovered (-)-ent-spathulenol isolated from liverworts to be an artefact. Only three sesquiterpenes are so far known from hornworts (Asakawa 1995). Sesquiterpenes had not been recorded in mosses until *Takakia* was recognized to be a moss rather than a liverwort (Smith & Davison 1993, Renzaglia *et al.* 1997). From both *Takakia* species, sesquiterpenes were identified by GC-MS analysis (Asakawa 1995).

Diterpenes Diterpenes detected in bryophytes are the second largest terpenoid class in bryophytes. As with sesquiterpenes, almost all diterpenes have been detected in liverworts. They are unknown for hornworts and except for phytol, which is a ubiquitous part of chlorophylls a and b in all green bryophytes, only four are known from mosses (Table 5.2).

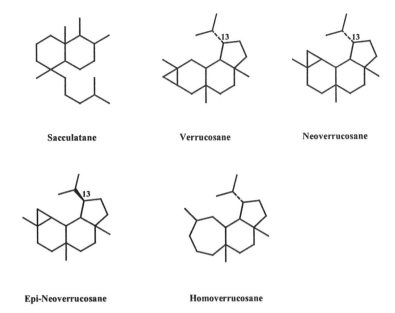

Fig. 5.2. Diterpene skeletons unique to liverworts.

Regarding the absolute configuration of diterpenes, as for sesquiterpenes, both enantiomers may occur, either the same as or opposite to
those found in higher plants. So far, acyclic, di-, tri-, tetra-, and pentacyclic
diterpenes have been detected. Asakawa (1995, 1997) mentioned at least 18
diterpene types found in liverworts, labdanes, clerodanes, and kauranes
being the most common ones. Typical diterpene skeletons for liverworts
are shown in Fig. 5.2. Since 1994, more than 40 additional papers have
reported diterpenes in liverworts, some of them exhibiting remarkable
biological activities. Thus Perry *et al.* (1999) isolated four new 8,9-secokau-
ranes and three new kauren-15-ones from *Lepidolaena taylorii* (Gott.) Trev.,
which is endemic to New Zealand. Some of these compounds showed differential cytotoxic activity against human tumor cell lines with 8,9-seco-
kaurane (**6**) being the main cytotoxin. Cullmann and Becker (1998*a*)

(**6**)

(7)

reported the isolation of pellialactone (7), a new nor-diterpenoid with an unusual skeleton from *Pellia epiphylla*. Other new diterpenes have been reported by Nagashima *et al.* (1997), Yoshida *et al.* (1997), and Hashimoto *et al.* (1998*b*). The four diterpenes so far known from mosses are cited in Asakawa (1995).

Triterpenes Triterpenes are found in species of all three bryophyte classes, but only two are known from hornworts (Table 5.2). So far no bryophyte-specific triterpenoid skeleton has been detected. About 40 different triterpenes are known from bryophytes, with around 20 from mosses only. Four have been found in both liverworts and mosses. Asakawa (1995) reported on GC-MS investigations of about 700 liverwort species. These studies documented the presence of squalene in almost all species, whereas it was only once found in a moss: *Racomitrium japonicum* Dozy & Molk. (Asakawa 1995). The most common triterpenoid in mosses is diploptene (= hop-22(29)-ene) (Asakawa 1995). Grammes *et al.* (1994) isolated 22-hydroxy-29-methylhopanoate, 20-hydroxy-22(29)-hopene, 22(30)-hopene-29-oic acid, adianton (= 30-nor-21β-hopane-22-one), and tetrahymenol (= 3β-hydroxy-gammaceran) from *in vitro* cultures of *Fossombronia alaskana* Steere & Inoue and *F. pusilla* (L.) Nees. Cullmann and Becker (1998*a*) and Sanders (1996) detected betulin (= 20(29)-lupen-3β,28-diol) in *Pellia epiphylla* and *Lepidozia reptans* (L.) Dum. respectively. Toyota *et al.* (1998) isolated the new dammarane-type triterpenoid dammara-17Z,21-diene (8) from the moss *Floribundaria aurea* subsp. *nipponica* (Nog.)

(8)

Nog. This was the first record of this triterpene type for bryophytes. From the hornwort *Phaeoceros laevis* (L.) Prosk. Trennheuser (1992) isolated the triterpenoid acids betulic and ursolic acid, both common in higher plants. Ursolic acid is also known from mosses (Asakawa 1995) and liverworts (Schmidt 1996).

Steroids The main phytosteroids, campesterol, sitosterol, and stigmasterol, are found in almost all bryophytes investigated (Table 5.2). In some mosses and liverworts, Patterson *et al.* (1990) found only the 24α-ethyl epimer (sitosterol), whereas in other moss and liverwort species mixtures with the 24β epimer (clionasterol) occur. In higher plants only the 24α epimers (sitosterol and campesterol) have been detected, in algae always the 24β ones (clionasterol and dihydrobrassicasterol). Regarding the epimers of phytosteroids, bryophytes take an intermediate position between algae and tracheophytes. Besides the steroids listed in Asakawa (1995) a range of others, new for bryophytes, was later detected in several liverwort species and in *Hypnum cupressiforme* Hedw. (e.g., Sievers 1992, Martini 1996, Sanders 1996, Schmidt 1996, Cullmann & Becker 1998a, Hashimoto *et al.* 1998b). One new phytosteroid, stigmast-5,28-dien-3β-ol (**9**), was isolated by Martini (1996) from *Riella helicophylla* (Bory & Mont.) Mont.

(**9**)

Tetraterpenes (carotenoids) For the distribution of carotenoids in bryophytes see Table 5.2. Besides the 12 carotenoids mentioned by Asakawa (1995) for mosses and liverworts, Czeczuga (1980, 1985a, b) and Czeczuga *et al.* (1982) found an additional 22 carotenoids in 10 liverwort and 32 moss species. Among them seven are found only in liverworts and nine only in mosses. There seem to be no reports of carotenoids in hornworts.

Polyterpenes Such polyterpenes as rubber and gutta-percha from certain higher plants are unknown from bryophytes, but according to

Table 5.3. *Aromatic compounds from bryophytes*

Compound class	Liverworts	Hornworts	Mosses
Benzoic and cinnamic acid derivatives	+	(+)	+
Phenolethers, alkylphenols, phenylglycosides	+	–	–
Bibenzyls, bisbibenzyls, bisbibenzyl dimers, stilbenes, and related compounds	+ +	–	(+)
Phenanthrenes	+	–	–
Naphthalenes	(+)	–	–
Acetophenones	(+)	–	–
Lignans	+	(+)	–
Flavonoids	+ +	–	+ +
Coumarins, isocoumarins, coumestans	(+)	–	+
Benzonaphthoxanthenones	–	–	(+)

Notes: + + >100 compounds; + 10–100 compounds; (+) <10 compounds.

Rezanka and Dembitsky (1993), polyisoprenoid alcohols appear to be universally present in bacteria, plants, and animals. They investigated eight mosses and *Marchantia polymorpha* for the occurrence of prenols in bryophytes. The mosses contain mainly betulaprenols with an average chain length of C_{60}, *M. polymorpha* had two maxima of chain lengths, the first at C_{70} and the second at C_{115}, and contained predominantly ficaprenols with up to 5% betulaprenols.

Aromatic compounds

Besides the terpenoids, aromatic compounds constitute the most important secondary products of bryophytes. They occur as a wide range of different types, but rather unequally distributed among the bryophyte classes (Table 5.3). Only one type has been found in species of each class; most types are known only from liverworts, others only from mosses. No type is found exclusively in hornworts. Most of these types are also known from vascular plants. Only the sphagnorubins from peat mosses, the benzonaphthoxanthenones from *Polytrichum*, and with very few exceptions also the bisbibenzyls and bisbibenzyl dimers of liverworts are regarded as bryophyte-specific.

Benzoic and cinnamic acid derivatives Asakawa (1995) recorded about 30 compounds of this class. Only a few more papers have reported such compounds in bryophytes. Perry *et al.* (1996) found the new cytotoxic monoterpene phenyl ether (**10**) in *Trichocolea mollissima* (Hook. f. & Tayl.) Gott. They also corrected the structures of benzoates earlier reported from *Tri-*

(10)

chocolea tomentella (Ehrh.) Dum. Kraut *et al.* (1993) isolated from two *Frullania* species (one of them at that time still known as *F. muscicola* Steph., now confirmed to be a new species and renamed *F. azorica* Sim-Sim, Sérgio, Mues & Kraut [Sim-Sim *et al.* 1995]) three glycerol glucosides acylated with caffeic acid, the main compound being (**11**). Schoeneborn (1996) detected the isocitric acid esters of caffeic (**12**) and ferulic acid, respectively, in *Mar-*

(11)

(12)

chesinia bongardiana (Lehm. & Lindenb.) Trev. Further benzoic and cinnamic acid derivatives have been reported by Trennheuser (1992), Brinkmeier (1996), and Kraut and Mues (1999).

Phenylethers, alkylphenols, phenylglycosides Asakawa (1995) mentioned a few examples of this type of aromatic compounds in liverworts. According to Toyota *et al.* (1997c), the intensely fragrant odour of crushed *Leptolejeunea elliptica* (Lehm. & Lindenb.) Schiffn. is due to 1-ethyl-4-hydroxy-, -4-methoxy- and -4-acetoxybenzene. The first allopyranosyl aromatic glycoside for bryophytes was isolated by Toyota *et al.* (1996*b*) from *Conocephalum conicum*: 2-(3,4-dihydroxyphenyl)-ethyl-O-β-allopyranoside.

Bibenzyls, bisbibenzyls, bisbibenzyl dimers, stilbenes, and related compounds
After flavonoids this class of aromatic compounds is the second most important one in liverworts, still unknown from hornworts and mosses with the exception of pallidisetins A [13, 2-(E)] and B [13, 2-(Z)], two novel

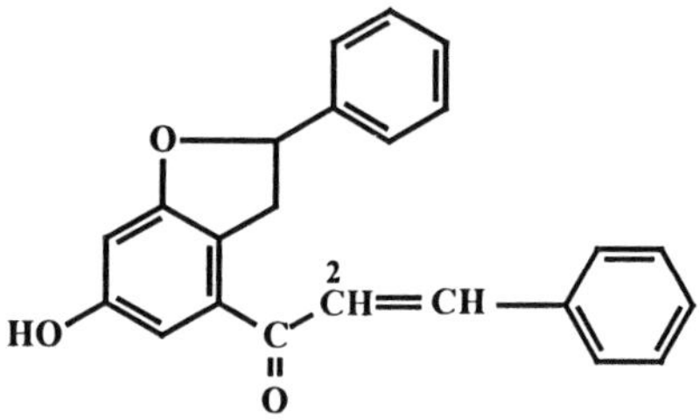

(13)

cinnamoyl bibenzyls isolated by Zheng *et al.* (1994) from *Polytrichum pallidisetum* Funck. Both have cytotoxic activities against the human tumor cell lines RPMI-7951 melanoma and U-251 glioblastoma multiforme. Bibenzyls also occur regularly in species of certain angiosperm families, e.g., Cannabaceae, Dioscoreaceae, and Orchidaceae. Until recently bisbibenzyls and dimeric bisbibenzyls had been regarded as liverwort-specific, but Bai *et al.* (1996) isolated a bibenzyl-dihydrophenanthrene dimer, biogenetically related to the bisbibenzyls, from tubers of the orchid *Pleione bulbocodioides* (Franch.) Rolfe. Asakawa (1995, 1997) recorded more than 90 bibenzyls and their derivatives, including about 50 bisbibenzyls and seven bisbibenzyl dimers. Since then about 30 papers describing this class of compounds from liverworts have appeared. Kunz and Becker (1994) isolated the new 2,5,4′-trihydroxybibenzyl from *Ricciocarpos natans* (L.) Corda and Cullmann *et al.* (1997) found the new 7-hydroxypellepiphyllin (14) in

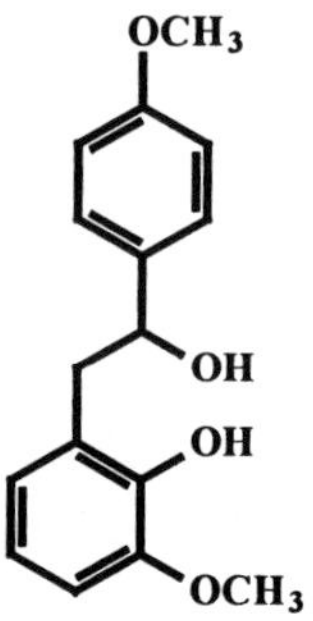

(14)

Pellia epiphylla. Other new bibenzyls have been reported by Kraut *et al.* (1997*b*), Rycroft *et al.* (1998*a*), and Schoeneborn (1996). A new bisbibenzyl was detected by Kraut *et al.* (1997*b*) in *Lethocolea glossophylla* (Spruce) Grolle, the bisprenylated glossophyllin (**15**) with a chromene and a chromane

(15)

moiety. Anton *et al.* (1997) found the two new bisbibenzyls isoplagiochins E (**16**) and F (**17**). Nabeta *et al.* (1998) isolated from cell cultures of *Heteroscy-*

R = H :   16
R = OH :   17   (16/17)

phus planus (Mitt.) Schiffn. two bisbibenzyls, the known isoplagiochin A and the structurally related planusin A (**18**). Cullmann *et al.* (1997) revised the structures of some perrottetin E-type bisbibenzyls from *Pellia epiphylla*

(18)

and isolated the new bisperrottetin, 13′, 13‴-bis-(10′-hydroxy-perrottetin E) **(19)**. The first and still the only monomeric stilbene from bryophytes,

(19)

3,4-dihydroxy-3′-methoxystilbene, was isolated by Speicher and Schoeneborn (1997) from *Marchesinia bongardiana*, its structure being confirmed by independent synthesis.

Phenanthrenes Although biogenetically related to the bibenzyls via the dihydro derivatives, phenanthrenes are treated separately here. Asakawa (1995) recorded only three dihydrophenanthrenes and three phenanthrenes. Subsequently, a few papers have reported this compound type: Adam and Becker (1994), Anton *et al.* (1997), and Connolly (1997).

Naphthalenes Besides naphthalene itself, only 2,4,7-trimethoxy-naphthalene and the methylene-dioxy substituted naphthalenes wettstein A–C were discussed in Asakawa (1995, 1997). Later, Rycroft *et al.* (1998*b*) isolated the new naphthalenes 1,2,4-trimethoxy- and 1,2,3,4-tetramethoxynaphthalene from *Adelanthus decipiens* (Hook.) Mitt.

Acetophenones Acetophenones are reported by Asakawa (1995) only from *Trocholejeunea sandvicensis* (Gott.) Mizut. Lorimer and Perry (1994) isolated two antifungal acetophenones (**20, 21**) from *Plagiochila fasciculata*

20 $R_1 = R_3 = H$; $R_2 = OCH_3$
21 $R_1 = R_2 = R_3 = H$
22 $R_1 = H$; $R_2 = R_3 = OCH_3$
23 $R_1 = CH_3$; $R_2 = OCH_3$; $R_3 = H$
24 $R_1 = CH_3$; $R_2 = R_3 = OCH_3$ (20–24)

Lindenb. These were also detected by Rycroft *et al.* (1998*b*) in *Adelanthus decipiens*, which also contained additionally the new acetophenones **22–24**. Cullmann and Becker (1999) detected the first acetophenone C-glycoside from bryophytes, the 2,4,6-trihydroxyacetophenone-3,5-di-C-β-D-glucopyranoside, in *Lepicolea ochroleuca* (Spreng.) Spruce.

Lignans The lignans are a class of phenolics of which at the time of Asakawa's reviews (1995, 1997) only a few compounds were known from bryophytes. Among them the cyclolignan 2,3-dicarboxy-6,7-dihydroxy-1-(3,4-dihydroxy)-phenyl-1,2-dihydronaphthalene (**25**), first isolated by

(25)

Cullmann *et al.* (1993) from *Pellia epiphylla*, seems to be a common basic element of this class. This compound was later detected in other liverwort species. Lignans of this type are still limited to liverworts, whereas other lignan types are typical phenolics of hornworts. They are yet unknown in mosses (Asakawa 1995). The 1,2-dehydro derivative of **25** was first detected by Tazaki *et al.* (1995) in *Jamesoniella autumnalis* (DC.) Steph. The 1,2-dehydro-2-decarboxy derivative of **25** was the first compound of this type found as a natural product, first isolated from *Pellia epiphylla* by Rischmann *et al.* (1989), and later from other liverworts, e.g., *Bazzania trilobata* (L.) S. Gray (Martini *et al.* 1998a), *Lepidozia incurvata* Lindenb., *Chiloscyphus polyanthos* (L.) Corda, and *Jungermannia exsertifolia* Steph. subsp. *cordifolia* (Dum.) Vána (Cullmann *et al.* 1999). Other new lignans have been reported by Martini *et al.* (1998a). Cullmann and Becker (1999) isolated the first glycosides and further esters of **25** from *Lepicolea ochroleuca*. Because of its wide occurrence, they proposed for **25** the trivial name "epiphyllic acid."

Flavonoids Flavonoids are the most widespread phenolics in bryophytes, although still unknown from hornworts. Since Asakawa's review (1995), about 30 papers on the occurrence of flavonoids in mosses and liverworts have appeared, among them some reports of other flavonoid monomeric aglycones, e.g., Morais and Becker (1991), Cullmann and Becker (1998b), and Hashimoto *et al.* (1998b). Sievers *et al.* (1994) found traces of kaempferol in *Hypnum cupressiforme*, the first time for a moss as a free aglycone. They detected the new dihydroflavonol derivatives, hypnum acid (**26**) and its methyl ester. About 15 new flavone glycosides

(26)

have since been isolated, e.g., by Anhut *et al.* (1992), Kraut *et al.* (1993,1995,1996), and Brinkmeier *et al.* (1998). From *in vitro* cultures Kunz *et al.* (1994) isolated the red cell-wall pigments of *Ricciocarpos natans*. The main compound, riccionidin A (**27**), is the first natural anthocyanidin derivative with an additional connection betweeen the B- and C-ring and with a hydroxylation pattern, unusual for flavonoids. The second pigment, riccionidin B, is a dimer of A. As cell-wall pigments, both

HO

HO   O

HO   OH

HO   O

(27)

compounds are not glycosylated like the common anthocyanins of vascular plants. They were also detected as red pigments of *Marchantia polymorpha*, *Riccia duplex* Lorbeer, and *Scapania undulata* (L.) Dum. The typical red pigments reported from certain *Sphagnum* species, the sphagnorubins A–C, may have a chemical relation to 3-deoxyanthocyanidins. They were discussed by Asakawa (1995) and no other compounds of this type have since been isolated. Much progress has been made in the field of biflavonoid research in bryophytes since Asakawa's review (1995). Almost 20 papers reported biflavonoids in mosses and the first paper on a biflavone in a liverwort was published by Kraut and Mues (1999). They found dicranolomin in *Chandonanthus hirtellus* (Web.) Mitt. subsp. *giganteus* (Steph.) Vanden Berghen. Geiger *et al.* (1997) published a review on the occurrence of flavonoids in arthrodontous mosses. They recorded eight biflavonoids and five triflavonoids not listed in Asakawa (1995), provided new references, and drew the structures of all bi- and triflavonoids from mosses known up to the publication of their paper. They also presented a table with the known flavonoid glycosides reported from mosses. Later Brinkmeier (1996) and Brinkmeier *et al.* (1999) reported six new biflavonoids from *Mnium hornum* Hedw. and *Pilotrichella flexilis* (Hedw.) Angstr. The actual flavonoid types and their numbers in bryophytes are presented in Table 5.4. Altogether 356 different flavonoids have been reported from bryophytes.

Coumarins, isocoumarins, coumestans The first coumarins isolated and identified from bryophytes were reported by Jung *et al.* (1994, 1995) from *Atrichum undulatum* (Hedw.) P. Beauv., *Polytrichum formosum* Hedw., and *Tetraphis pellucida* Hedw. These authors characterized 11 highly oxygenated new coumarins found to be unique to nematodontous mosses; the structures of most of these compounds are shown in Asakawa (1997). Besides these new coumarins, Jung *et al.* (1994) detected daphnin, the 7,8-dihydroxy-7-O-β-D-glucopyranosylcoumarin, in *P. formosum*, also known from higher plants. This is the only coumarin so far to be found also in some liverworts; e.g., *Lepidozia reptans* (Sanders 1996) and *Bazzania trilobata* (Martini 1996). Raubuch (1998) investigated other nematodontous mosses

Table 5.4. *Flavonoids from bryophytes*

Flavonoid type	Liverworts	Mosses
Flavone aglycones and glycosides	218	77
Flavonol aglycones and glycosides	10	9
Anthocyanins and derivatives	2	5
Aurones	1	3
Biflavonoids	1	36
Flavanones	4	–
Dihydrochalcones	2	–
Dihydroflavonols	–	3
Isoflavones	–	7
Triflavones	–	5

and *Encalypta streptocarpa* Hedw. for the occurrence of coumarins and iso-lated 13 different coumarin glycosides, among them five new ones, but of a similar type as those reported from Jung *et al.* (1994, 1995). Thus, until now 16 different coumarins have been reported from mosses. The few iso-coumarins known from bryophytes are listed by Asakawa (1995, 1997). Brinkmeier (1996) found the first coumestans as derivatives of 4,2′-epoxy-3-phenylcoumarins in a bryophyte: she isolated six different hydroxy-lated and methoxylated compounds of this type from *Mnium hornum*, the main one being 4,2′-epoxy-4′-hydroxy-5,7,5′-trimethoxy-3-phenyl-coumarin (**28**).

(**28**)

Benzonaphthoxanthenones The only known compounds of this type from bryophytes mentioned by Asakawa (1995, 1997) are the ohioensins A–E from *Polytrichum ohioense* Ren. & Card., based on the novel benzo-naphthoxanthenone skeleton. They exhibited cytotoxicity against 9 PS murine leukemia cells and certain human tumor cells in culture. Zheng *et al.* (1994) isolated three other novel benzonaphthoxanthenones from *Poly-trichum pallidisetum*, of which 1-O-methyl-ohioensin B (**29**) was the main compound. They also showed cytotoxic activity against several human tumor cell lines.

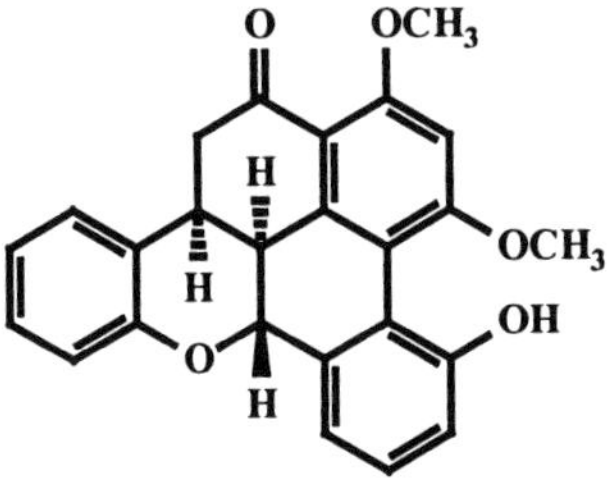

(29)

Nitrogen-containing compounds

Bryophytes do not regularly produce nitrogen-containing secondary products, e.g. alkaloids. Thus, Asakawa (1995, 1997) recorded only two alkylindols, skatol and indoleacetic acid (IAA) from liverworts, the new alkaloid anthocerodiazonin from *Anthoceros agrestis*, four maytansinoids from *Isothecium subdiversiforme* Broth. and *Thamnobryum sandei* (Besch.) Iwats., an alkylamide from *Thuidium kanedae* Sak., and indole and pyrrolidine from *Splachnum rubrum*. Only one other paper on nitrogen-containing compounds has since been reported for bryophytes. Besides the known harmane alkaloid, 7-hydroxyharmane (= harmol), Salm *et al.* (1998) found for the first time in a bryophyte, the new harmol propionic acid ester (30) in *Fontinalis squamosa*.

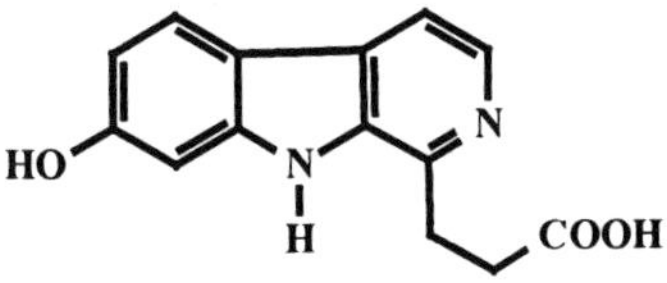

(30)

Sulfur-containing compounds

The only sulfur-containing compounds reported from bryophytes are dimethyl sulfide and the thioacrylates isotachins A–C from *Isotachis* and *Balantiopsis* species (Asakawa 1995).

Chlorine-containing compounds

Asakawa (1995, 1997) recorded as the only chlorine-containing compound from bryophytes the drimane-type sesquiterpene 7α-chloro-6β-hydroxyconfertifolin detected in *Makinoa crispata* (Steph.) Miyake. Two other papers reported chlorine-containing compounds in liverworts. Anton *et al.* (1997) isolated 12-chloro-isoplagiochin D from a *Plagiochila* species. Martini *et al.* (1998*b*) detected 10 cyclic bisbibenzyls substituted with one to six chlorine atoms and bazzanin K (31), a new dichlorinated

(31)

macrocycle consisting of a phenanthrene and a bibenzyl moiety, in *Bazzania trilobata*.

5.2 Chemosystematics

As mentioned above, most compounds of primary and secondary metabolism synthesized by bryophytes are comparable to those produced by vascular plants. Liverworts are clearly the chemically best investigated bryophyte class. Asakawa (1995, 1997) did not give a precise number of chemically investigated liverwort species, but estimated that about "6% of all liverwort species and less than 2% of all mosses have been studied chemically." This does not mean that each "chemically examined" species has been investigated for all classes of natural products. In some cases, only the lipophilic extract and its main compound(s) has been characterized. Therefore, although it is difficult to give a precise number of chemically investigated species, at least one compound has been detected in approximately 700–900 liverworts and about 400 mosses. For the occurrence of flavonoids only, Geiger *et al.* (1997) counted nearly 300 moss species from 59 different families. From hornworts fewer than 10 species in five genera have been studied; therefore knowledge about the chemistry of hornworts is only fragmentary. Significant differences in the chemistry of the bryophyte classes are summarized in Tables 5.2 and 5.3. Aside from mono- and diterpenes, which are definitely unknown for hornworts, all other terpenoid types occur in each bryophyte class. The marker components for liverworts are the sesqui- and diterpenes. Certain liverwort-typical sesqui- and diterpene skeletons have been detected, but the majority exhibit similar structural types to those found in vascular plants. For aromatic compounds (Table 5.3) only the benzoic and cinnamic acid derivatives are known from all three groups; all others are either limited to liverworts or have been found only in liverworts and mosses. Only the coumestans and benzonaphthoxanthenones appear to be restricted in bryophytes to mosses. The typical aromatic compounds of liverworts are the bisbibenzyls and bisbibenzyl dimers. The liverwort and hornwort lignans are characteristic for each class, absent from mosses and different from the lignans of vascular plants. Flavonoids are common compounds in the green land plants. Thus, their absence in hornworts is somewhat surprising and might be due to the unsufficient chemical investigation of this class. Among the flavonoids, only the tricetin 6,8-di-C-glycosides, especially tricetin 6,8-di-C-β-D-glucopyranoside (32), are

(32)

unique for liverworts in the plant kingdom. Most other flavone mono-meric aglycones and glycosides are found in higher plants as well. All bi- and triflavonoids of mosses (included dicranolomin from *Chandonanthus*) are unique and differ from those of pteridophytes and seed plants. Other flavonoid types unique to bryophytes are the riccionidins from liverworts and the sphagnorubins from *Sphagnum*. The highly oxygenated coumar-ins of mosses are aromatics reported only for nematodontous groups and for *Encalypta streptocarpa*. Primary and secondary metabolites of bryophy-tes are most variable in the liverworts. This is true of both structure types and of single compounds. The lowest diversity is found in hornworts, probably due to poor knowledge of their chemistry in general.

Differences in chemistry between the bryophyte classes, and between bryophytes and vascular plants, are by far less substantial than the differ-ences between the different groups of algae (excluding the green algae) on one hand, and bryophytes and tracheophytes on the other hand. There-fore, chemistry clearly supports the evolution of bryophytes and tracheo-phytes as being monophyletic from the green line of land plants.

5.3 Biochemistry

The biochemistry of bryophytes has been recently reviewed, e.g., by Chopra and Bhatla (1990) and Rudolph (1990).

5.3.1 Enzyme activities in primary metabolism

Most of the papers on enzyme activities in the primary metabolism of bryophytes deal with nitrogen metabolism. Thus the nitrogen metabolism of *Sphagnum* was discussed in detail by Rudolph *et al.* (1993). Studies on the *in vivo* activity of nitrate reductase, an enzyme catalyzing the conversion of nitrate to nitrite, were carried out by Marschall (1998). The enzyme aspartate aminotransferase (AspAT) plays a key role in nitrogen and carbon metabolism of plants, animals, and microorganisms, catalyzing the reversible transfer of the amino group from aspartate to glutamate. Heeschen *et al.* (1996) separated four AspAT isoforms from extracts of *Sphagnum fallax* (Klinggr.) Klinggr., each exhibiting identical molecular weights. Heeschen *et al.* (1997) purified the glutamate dehydrogenase of *S. fallax*, the enzyme catalyzing the reversible reductive amination of 2-oxoglutarate to L-glutamate. Kahl *et al.* (1997, 1998) worked on the purification and characterization of L-glutamine synthetase from *S. fallax*. Only a few papers have reported on other aspects of primary metabolism in bryophytes. Lipoxygenase catalyzes the oxidation of linoleic acid and certain other polyunsaturated fatty acids to form conjugated diene hydroperoxides. Fatty acid hydroperoxides (HPOs) thus formed are further converted to various lipid breakdown compounds in plant tissues. HPOlyase cleaves the HPOs to form C_6-aldehydes and C_{12}-oxo acids. Matsui *et al.* (1996) found HPOlyase activity in cultured cells of *Marchantia polymorpha* and they succeeded in isolating and characterizing the HPOlyase partially, the first time for a bryophyte. They also isolated and identified n-hexanal and 12-oxo-(9Z)-dodeceonoic acid, products of HPOlyase activity. Matlok *et al.* (1989) described a method for the isolation of cell-wall-bound peroxidase isozymes of *Sphagnum magellanicum* Brid.

5.3.2 Biosynthesis and metabolism of secondary products

Substantial progress has been made during the last three years in the research on the biosynthesis of terpenoids in liverworts. In the past, the mevalonic acid pathway leading to the biosynthesis of isopentenyldiph-

osphate (IPP) had been accepted as the only biosynthetic route for all terpenoid producing organisms. Only recently, an alternative nonmevalonoid
pathway of IPP biosynthesis, the "glyceraldehyde–pyruvate" pathway,
has been discovered in bacteria and plants (Rohmer 1998). Adam *et al.*
(1998) studied the incorporation of ^{13}C-labeled glucose into the monoterpenes borneol and bornyl acetate, the sesquiterpenes cubebanol and ric-
ciocarpin A, the diterpene phytol, and the phytosteroid stigmasterol in
axenic cultures of *Ricciocarpos natans* and *Conocephalum conicum*. They
showed that the isoprene building blocks of sesquiterpenes and stigmasterol are made via the mevalonic acid pathway, whereas isoprene units of
the monoterpenes and phytol are exclusively derived from the glyceraldehyde–pyruvate pathway. Other papers on terpenoid biosynthesis in liverworts were published by Adam *et al.* (1996), Adam and Croteau (1998), and
Nabeta *et al.* (1994).

Some progress has also been made in research on biosynthesis and
metabolism of phenolics in bryophytes. Rasmussen *et al.* (1995) worked on
biosynthesis of the *Sphagnum*-specific trans-sphagnum acid, and Rasmussen *et al.* (1996) on the biosynthesis and accumulation of 4'-O-β-D-
glucosyl-cis-p-coumaric acid in axenic cultures of *S. fallax*. Later
Rasmussen and Rudolph (1997) isolated a stereospecific glucosyltransfer-
ase from *S. fallax*, that catalyzes the transfer of glucose from UDP-glucose
to the 4'-hydroxy group of cis-p-coumaric acid. Fischer *et al.* (1995) clearly
showed activity of the key enzyme in flavonoid biosynthesis, chalcone
synthase, in *Marchantia polymorpha* and identified naringenin as the
typical cyclization product of the corresponding chalcone. The first paper
dealing with the biosynthesis of cyclic bisbibenzyls of liverworts was published by Friederich *et al.* (1999), using sterile thallus tissue of *Marchantia
polymorpha*. Feeding experiments with ^{13}C-labeled precursors showed
that the rings A and C of the marchantin A molecule (**33**) are derived from

(**33**)

the benzene ring of L-phenylalanine via trans-cinnamic acid and p-coumaric acid. A phenylpropane–polymalonate pathway using dihydro-p-coumaric acid and acetate/malonate was proposed for the biosynthesis of the bibenzyl monomers which were confirmed to be the building blocks of the marchantin molecule. The bibenzyls in turn are coupled to form the bisbibenzyl structure, but the mechanism of this final coupling step is unknown.

Closing remarks

Substantial progress has been made on the chemistry of bryophytes in the last two decades, but still fewer than 10% of all bryophyte species have been thoroughly studied. Almost all papers deal with the chemistry of the gametophyte, while only a few compare the chemistry of the gameto-phyte and the sporophyte of a given species, one of the most impressive examples being the dissertation of Cullmann (1996) and the papers result-ing from that dissertation mentioned in this review. The biological activ-ity of bryophyte compounds has been documented in several papers, e.g., Asakawa (1995, 1998). Numerous bryophyte species, especially liverworts, are now cultivated under axenic conditions (e.g., Becker 1994) and a substantial number of bryophyte compounds have been synthesized (Asakawa 1995, Eicher *et al.* 1996, 1998). Bryophytes are chemically rich and their chemistry and biochemistry are active areas of research.

Acknowledgments

My sincere thanks are given to my colleague, Professor Dr H. Rudolph, Kiel, for valuable help with references to the section on *Biochemistry*, and to my co-workers, Mrs C. Zehren and Dr U. Lion, for drawing the struc-tures presented in this review.

REFERENCES

Adam, K.-P. & Becker, H. (1993). A lectin from the liverwort *Marchantia polymorpha* L. *Experientia,* **49,** 1098–100.
    (1994). Phenanthrenes and other phenolics from in vitro cultures of *Marchantia polymorpha. Phytochemistry,* **35,** 139–43.
Adam, K.-P. & Croteau, R. (1998). Monoterpene biosynthesis in the liverwort *Conocephalum conicum*: demonstration of sabinene synthase and bornyl diphosphate synthase. *Phytochemistry,* **49,** 475–80.
Adam, K.-P., Crock, J., & Croteau, R. (1996). Partial purification and characterization of a monoterpene cyclase, limonene synthase, from the liverwort *Ricciocarpos natans. Archives of Biochemistry and Biophysics,* **332,** 352–6.

Adam, K.-P., Thiel, R., Zapp, J., & Becker, H. (1998). Involvement of the mevalonic acid pathway and the glyceraldehyde–pyruvate pathway in terpenoid biosynthesis of the liverworts *Ricciocarpos natans* and *Conocephalum conicum*. *Archives of Biochemistry and Biophysics*, **354**, 181–7.

Anhut, S., Biehl, J., Seeger, T., Mues, R., & Zinsmeister, H. D. (1992). Flavone-C-glycosides from the mosses *Plagiomnium elatum* and *Plagiomnium cuspidatum*. *Zeitschrift für Naturforschung*, **47c**, 654–60.

Anton, H., Kraut, L., Mues, R., & Morales, Z. M. I. (1997). Phenanthrenes and bibenzyls from a *Plagiochila* species. *Phytochemistry*, **46**, 1069–75.

Asakawa, Y. (1995). Chemical constituents of the bryophytes. In *Progress in the Chemistry of Organic Natural Products*, vol. 65, ed. W. Herz, G. W. Kirby, R. E. Moore, W. Steglich, & Ch. Tamm, pp. 1–618. New York: Springer.

(1997). Heterocyclic compounds found in bryophytes. *Heterocycles*, **46**, 795–848.

(1998). Biologically active compounds from bryophytes. *Journal of the Hattori Botanical Laboratory*, **84**, 91–104.

Asakawa, Y., Toyota, M., Nakaishi, E., & Tada, Y. (1996). Distribution of terpenoids and aromatic compounds in New Zealand liverworts. *Journal of the Hattori Botanical Laboratory*, **80**, 271–95.

Bai, L., Yamaki, M., & Takagi, S. (1996). Stilbenoids from *Pleione bulbocodioides*. *Phytochemistry*, **42**, 853–6.

Becker, H. (1994). Secondary metabolites from bryophytes *in vitro* cultures. *Journal of the Hattori Botanical Laboratory*, **76**, 283–91.

Brinkmeier, E. (1996). Phenolische Verbindungen aus *Mnium hornum*, *Leptostomum macrocarpon* und *Pilotrichella flexilis*. PhD dissertation, Saarbrücken University.

Brinkmeier, E., Geiger, H., & Zinsmeister, H. D. (1998). Flavone-7-O-sophorotriosides and biflavonoids from the moss *Leptostomum macrocarpon* (Leptostomataceae). *Zeitschrift für Naturforschung*, **53c**, 1–3.

(1999). Biflavonoids and 4,2′-epoxy-3-phenylcoumarins from the moss *Mnium hornum*. *Phytochemistry*, in press.

Chopra, R. N. & Bhatla, S. C. (eds) (1990). *Bryophyte Development: Physiology and Biochemistry*. Boca Raton: CRC Press.

Connolly, J. D. (1997). Natural products from around the world. *Revista Latinoamericana de Química*, **25**, 77–85.

Cullmann, F. (1996). Phytochemische Untersuchungen an Gametophyten, Sporophyten und Sporen des Lebermooses *Pellia epiphylla* (L.) Corda. PhD dissertation, Saarbrücken University.

Cullmann, F. & Becker, H. (1998a). Terpenoid constituents of *Pellia epiphylla*. *Phytochemistry*, **47**, 237–45.

(1998b). Terpenoid and phenolic constituents of sporophytes and spores from the liverwort *Pellia epiphylla*. *Journal of the Hattori Botanical Laboratory*, **84**, 285–95.

(1999). Lignans from the liverwort *Lepicolea ochroleuca*. *Phytochemistry*, **52**, 1651–6.

Cullmann, F., Adam, K.-P., & Becker, H. (1993). Bisbibenzyls and lignans from *Pellia epiphylla*. *Phytochemistry*, **34**, 831–4.

Cullmann, F., Becker, H., Pandolfi, E., Roeckner, E., & Eicher, Th. (1997). Bibenzyl derivatives from *Pellia epiphylla*. *Phytochemistry*, **45**, 1235–47.

Cullmann, F., Schmidt, A., Schuld, F., Trennheuser, M. L., & Becker, H. (1999). Lignans from the liverworts *Lepidozia incurvata*, *Chiloscyphus polyanthos* and *Jungermannia exsertifolia* ssp. *cordifolia*. *Phytochemistry*, **52**, 1647–50.

Czeczuga, B. (1980). Investigations on carotenoids in Embryophyta. I. Bryophyta. *Bryologist*, **83**, 21–8.

(1985*a*). Investigations on carotenoids in Embryophyta. III. Representatives of the Hepaticae. *Phyton*, **25**, 113–21.

(1985*b*). Investigations on carotenoids in Embryophyta. IV. The presence of apocarotenals in peatmosses. *Acta Societatis Botanicorum Poloniae*, **54**, 77–83.

Czeczuga, B., Gutkowski, R., & Czerpak, R. (1982). Investigations of carotenoids in Embryophyta. II. Musci from the Antarctic. *Nova Hedwigia*, **36**, 695–701.

Dembitsky, V. M. & Rezanka, T. (1995). Distribution of acetylenic acids and polar lipids in some aquatic bryophytes. *Phytochemistry*, **40**, 93–7.

Dutt, B. (1996). Analytik und Isolierung von Aminosäuren unter besonderer Berücksichtigung nichtproteinogener Aminosäuren aus ausgewählten Laub- und Lebermoosen. PhD dissertation, Saarbrücken University.

Edelmann, H. G., Neinhuis, C., Jarvis, M., Evans, B., Fischer, E., & Barthlott, W. (1998). Ultrastructure and chemistry of the cell wall of the moss *Rhacocarpus purpurascens* (Rhacocarpaceae): a puzzling architecture among plants. *Planta*, **206**, 315–21.

Eicher, Th., Servet, F., & Speicher, A. (1996). Bryophyte constituents 6. Synthesis of herbertene-derived sesquiterpenes from *Herberta adunca*. *Synthesis*, 963–70.

Eicher, Th., Fey, S., Puhl, W., Büchel, E., & Speicher, A. (1998). Bryophyte constituents 9. Syntheses of cyclic bisbibenzylsystems. *European Journal of Organic Chemistry*, **108**, 877–88.

Fischer, S., Böttcher, U., Reuber, S., Anhalt, S., & Weissenböck, G. (1995). Chalcone synthase in the liverwort *Marchantia polymorpha*. *Phytochemistry*, **39**, 1007–12.

Friederich, S., Maier, U. H., Deus-Neumann, B., Asakawa, Y., & Zenk, M. H. (1999). Biosynthesis of cyclic bis(bibenzyls) in *Marchantia polymorpha*. *Phytochemistry*, **50**, 589–98.

Geiger, H., Seeger, T., Zinsmeister, H. D. & Frahm, J.-P. (1997). The occurrence of flavonoids in arthrodontous mosses – an account of the present knowledge. *Journal of the Hattori Botanical Laboratory*, **83**, 273–308.

Graham, L. E., Delwiche, C. F., & Mishler, B. D. (1991). Phylogenetic connections between the "Green Algae" and the "Bryophytes." *Advances in Bryology*, **4**, 213–44.

Grammes, C., Burkhardt, G., & Becker, H. (1994). Triterpenes from *Fossombronia* liverworts. *Phytochemistry*, **35**, 1293–6.

Hashimoto, T., Irita, H., Tanaka, M., Takaoka, S., & Asakawa, Y. (1998*a*). Two novel Diels-Alder reaction-type dimeric pinguisane sesquiterpenoids and related compounds from the liverwort *Porella acutifolia* subsp. *tosana*. *Tetrahedron Letters*, **39**, 2977–80.

Hashimoto, T., Irita, H., Yoshida, M., Kikkawa, A., Toyota, M., Koyama, H., Motoike, Y., & Asakawa, Y. (1998*b*). Chemical constituents of the Japanese liverworts *Odontoschisma denudatum*, *Porella japonica*, *P. acutifolia* subsp. *tosana* and *Frullania hamatiloba*. *Journal of the Hattori Botanical Laboratory*, **84**, 309–14.

Hedderson, T. A., Chapman, R., & Cox, C. J. (1998). Bryophytes and the origins and diversification of land plants: new evidence from molecules. In *Bryology for the Twenty-first Century*, ed. J. W. Bates, N. W. Ashton, & J. G. Duckett, pp. 65–77. Leeds: Maney and British Bryological Society.

Heeschen, V., Jacubowski, S., & Rudolph, H. (1996). Purification, characterization, and compartmentalization of aspartate aminotransferase isoforms from *Sphagnum*. *Journal of Plant Physiology*, **149**, 267–76.

Heeschen, V., Gerendás, J., Richter, C. P., & Rudolph, H. (1997). Glutamate dehydrogenase of *Sphagnum*. *Phytochemistry,* **45**, 881–7.

Hegewald, E. & Kneifel, H. (1987). Das Vorkommen von Polyaminen in Moosen. *Journal of the Hattori Botanial Laboratory,* **62**, 201–3.

Hegnauer, R. (1962). *Chemotaxonomie der Pflanzen,* vol. 1. Basel: Birkhäuser.

(1986). *Chemotaxonomie der Pflanzen,* vol. 7. Basel: Birkhäuser.

Huneck, S. (1983). Chemistry and biochemistry of bryophytes. In *New Manual of Bryology,* vol. 1, ed. R. M. Schuster, pp. 1–116. Nichinan: Hattori Botanical Laboratory.

Jung, M., Zinsmeister, H. D., & Geiger, H. (1994). New tri- and tetraoxygenated coumarin glucosides from the mosses *Atrichum undulatum* and *Polytrichum formosum*. *Zeitschrift für Naturforschung,* **49c**, 697–702.

Jung, M., Geiger, H., & Zinsmeister, H. D. (1995). Tri- and tetrahydroxycoumarin derivatives from *Tetraphis pellucida*. *Phytochemistry,* **39**, 379–81.

Kahl, S., Gerendás, J., Heeschen, V., Ratcliffe, R. G., & Rudolph, H. (1997). Ammonium assimilation in bryophytes. L-glutamine synthetase from *Sphagnum fallax*. *Physiologia Plantarum,* **101**, 86–92.

Kahl, S., Berswordt-Wallrabe, P. von, Heeschen, V., Schmidt, H., & Rudolph, H. (1998). Plastidic L-glutamine synthetase: a membrane-bound enzyme in *Sphagnum fallax*. *Journal of Plant Physiology,* **153**, 270–5.

Kohn, G., Hartmann, E., Stymne, S., & Beutelmann, P. (1994). Biosynthesis of acetylenic fatty acids in the moss *Ceratodon purpureus* (Hedw.) Brid. *Journal of Plant Physiology,* **144**, 265–71.

König, W. A., Rieck, A., Saritas, Y., Hardt, I. H., & Kubeczka, K.-H. (1996). Sesquiterpene hydrocarbons in the essential oil of *Meum athamanticum*. *Phytochemistry,* **42**, 461–6.

Konno, H., Yamasaki, Y., & Katoh, K. (1987). Fractionation and partial characterization of pectic polysaccharides in cell walls from liverwort (*Marchantia polymorpha*) cell cultures. *Journal of Experimental Botany,* **38**, 711–22.

Kraut, L. & Mues, R. (1999). The first biflavone found in liverworts and other phenolics and terpenoids from *Chandonanthus hirtellus* ssp. *giganteus* and *Plagiochila asplenioides*. *Zeitschrift für Naturforschung,* **54c**, 6–10.

Kraut, L., Mues, R., & Sim-Sim, M. (1993). Acylated flavone and glycerol glucosides from two *Frullania* species. *Phytochemistry,* **34**, 211–18.

Kraut, L., Scherer, B., Mues, R., & Sim-Sim, M. (1995). Flavonoids from some *Frullania* species (Hepaticae). *Zeitschrift für Naturforschung,* **50c**, 345–52.

Kraut, L., Mues, R., Speicher, A., Wagmann, M., & Eicher, Th. (1996). Carboxylated α-pyrone derivatives and flavonoids from the liverwort *Dumortiera hirsuta*. *Phytochemistry,* **42**, 1693–8.

Kraut, L., Klaus, T., Mues, R., Eicher, Th., & Zinsmeister, H. D. (1997a). Isolation and synthesis of rufulamide, an oligopeptide analogue from *Metzgeria rufula*. *Phytochemistry,* **45**, 1621–6.

Kraut, L., Mues, R., & Zinsmeister, H. D. (1997b). Prenylated bibenzyl derivatives from *Lethocolea glossophylla* and *Radula voluta*. *Phytochemistry,* **45**, 1249–55.

Kunz, S. & Becker, H. (1994). Bibenzyl derivatives from the liverwort *Ricciocarpos natans*. *Phytochemistry,* **36**, 675–7.

Kunz, S., Burkhardt, G., & Becker, H. (1994). Riccionidins A and B, anthocyanidins from the cell walls of the liverwort *Ricciocarpos natans*. *Phytochemistry,* **35**, 233–5.

Liu, H.-J., Wu, C.-L., Hashimoto, T., & Asakawa, Y. (1996). Nudenoic acid: a novel tricyclic sesquiterpenoid from the Taiwanese liverwort *Mylia nuda*. *Tetrahedron Letters*, **37**, 9307–8.

Lorimer, S. D. & Perry, N. B. (1994). Antifungal hydroxyacetophenones from the New Zealand liverwort, *Plagiochila fasciculata*. *Planta Medica*, **60**, 386–7.

Marschall, M. (1998). Nitrate reductase activity during desiccation and rehydration of the desiccation tolerant moss *Tortula ruralis* and the leafy liverwort *Porella platyphylla*. *Journal of Bryology*, **20**, 273–85.

Martini, U. (1996). In vitro Kultur und phytochemische Untersuchung des Lebermooses *Riella helicophylla* (Bory & Montagne) Montagne sowie Isolierung phenolischer Inhaltsstoffe aus Freilandmaterial des Lebermooses *Bazzania trilobata* (L.) S. F. Gray. PhD dissertation, Saarbrücken University.

Martini, U., Zapp, J., & Becker, H. (1998*a*). Lignans from the liverwort *Bazzania trilobata*. *Phytochemistry*, **49**, 1139–46.

(1998*b*). Chlorinated macrocyclic bisbibenzyls from the liverwort *Bazzania trilobata*. *Phytochemistry*, **47**, 89–96.

Matlok, J., Krzakowa, M., & Rudolph, H. (1989). Peroxidase patterns in bryophytes: a critical evaluation. *Journal of the Hattori Botanical Laboratory*, **67**, 407–14.

Matsui, K., Kaji, Y., Kajiwara, T., & Hatanaka, A. (1996). Developmental changes of lipoxygenase and fatty acid hydroperoxide lyase activities in cultured cells of *Marchantia polymorpha*. *Phytochemistry*, **41**, 177–82.

Morais, R. M. S. C. & Becker, H. (1991). Growth and secondary product formation of in vitro cultures from the liverwort *Reboulia hemisphaerica*. *Zeitschrift für Naturforschung*, **46c**, 28–32.

Mothes, K. (1980). Historical Introduction. In *Encyclopedia of Plant Physiology, New Series*, ed. A. Pirson & M. H. Zimmermann, vol. 8, *Secondary Plant Products*, ed. E. A. Bell & B. V. Charlwood, pp. 1–21. Berlin: Springer.

Muhle, H. (1984). Moose als Bioindikatoren. *Advances in Bryology*, **2**, 65–89.

Nabeta, K., Mototani, Y., Tazaki, H., & Okuyama, H. (1994). Biosynthesis of sesquiterpenes of cadinane type in cultured cells of *Heteroscyphus planus*. *Phytochemistry*, **35**, 915–20.

Nabeta, K., Ohkubo, S., Hozumi, R., & Katoh, K. (1998). Macrocyclic bisbibenzyls in cultured cells of the liverwort, *Heteroscyphus planus*. *Phytochemistry*, **49**, 1941–3.

Nagashima, F., Tamada, A., Fujii, N., & Asakawa, Y. (1997). Terpenoids from the Japanese liverworts *Jackiella javanica* and *Jungermannia infusca*. *Phytochemistry*, **46**, 1203–8.

Neinhuis, C. & Jetter, R. (1995). Ultrastructure and chemistry of epicuticular wax crystals in Polytrichales sporophytes. *Journal of Bryology*, **18**, 399–406.

Nissinen, R. & Sewón, P. (1994). Hydrocarbons of *Polytrichum commune*. *Phytochemistry*, **37**, 179–82.

Patterson, G. W., Wolfe, G. R., Salt, T. A., & Chiu, P.-L. (1990). Sterols of bryophytes with emphasis on the configuration at C-24. In *Bryophytes: Their Chemistry and Chemical Taxonomy*, Proceedings of the Phytochemical Soceity of Europe vol. 29, ed. H. D. Zinsmeister & R. Mues, pp. 103–7. Oxford: Clarendon Press.

Perry, N. B., Foster, L. M., Lorimer, S. D., May, B. C. H., Weavers, R. T., Toyota, M., Nakaishi, E., & Asakawa, Y. (1996). Isoprenyl phenylethers from liverworts of the genus *Trichocolea*: cytotoxic activity, structural corrections, and synthesis. *Journal of Natural Products*, **59**, 729–33.

Perry, N. B., Burgess, E. J., Baek, S.-H., Weavers, R. T., Geis, W., & Mauger, A. B. (1999). 11–Oxygenated cytotoxic 8,9-secokauranes from a New Zealand liverwort, *Lepidolaena taylorii*. *Phytochemistry,* **50,** 423–33.

Rasmussen, S. & Rudolph, H. (1997). Isolation, purification and characterization of UDP-glucose:cis-p-coumaric acid-β-D-glucosyltransferase from *Sphagnum fallax*. *Phytochemistry,* **46,** 449–53.

Rasmussen, S., Peters, G., & Rudolph, H. (1995). Regulation of phenylpropanoid metabolism by exogenous precursors in axenic cultures of *Sphagnum fallax*. *Physiologia Plantarum,* **95,** 83–90.

Rasmussen, S., Wolff, C., & Rudolph, H. (1996). 4′-O-β-D-glucosyl-cis-p-coumaric acid – a natural constituent of *Sphagnum fallax* cultivated in bioreactors. *Phytochemistry,* **42,** 81–7.

Raubuch, J. (1998). Neue Cumarinderivate aus einigen Vertretern der Laubmoosklasse Bryopsida. PhD dissertation, Saarbrücken University.

Renzaglia, K. S., McFarland, K. D., & Smith, D. K. (1997). Anatomy and ultrastructure of the sporophyte of *Takakia ceratophylla* (Bryophyta). *American Journal of Botany,* **84,** 1337–50.

Rezanka, T. & Dembitsky, V. M. (1993). Occurrence of C_{40}–C_{130} polyisoprenoid alcohols in lower plants. *Phytochemistry,* **34,** 1335–9.

Rischmann, M., Mues, R., Geiger, H., Laas, H. J., & Eicher, Th. (1989). Isolation and synthesis of 6,7-dihydroxy-4–(3,4-dihydroxyphenyl)naphthalene-2-carboxylic acid from *Pellia epiphylla*. *Phytochemistry,* **28,** 867–9.

Rohmer, M. (1998). Isoprenoid biosynthesis via the mevalonate-independent route, a novel target for antibacterial drugs? *Progress in Drug Research,* **50,** 135–54.

Rudolph, H. (1990). Biochemical and physiological aspects of bryophytes. In *Bryophytes: Their Chemistry and Chemical Taxonomy*, Proceedings of the Phytochemical Society of Europe vol. 29, ed. H. D. Zinsmeister & R. Mues, pp. 227–52. Oxford: Clarendon Press.

Rudolph, H., Hohlfeld, J., Jacubowski, S., von der Lage, P., Matlok, H., & Schmidt, H. (1993). Nitrogen metabolism of *Sphagnum*. *Advances in Bryology,* **5,** 79–105.

Rycroft, D. S., Cole, W. J., & Aslam, N. (1998a). 3,4–Dihydroxy-3′-methoxybibenzyl from the liverwort *Plagiochila exigua* from Scotland. *Phytochemistry,* **49,** 145–8.

Rycroft, D. S., Cole, W. J., & Rong, S. (1998b). Highly oxygenated naphthalenes and acetophenones from the liverwort *Adelanthus decipiens* from the British Isles and South America. *Phytochemistry,* **48,** 1351–6.

Salm, R. F., Zinsmeister, H. D., & Eicher, Th. (1998). Nitrogen-containing compounds from the moss *Fontinalis squamosa*. *Phytochemistry,* **49,** 887–92.

Sanders, S. (1996). Phytochemische Untersuchungen der Lebermoose *Jamesoniella rubricaulis* (Nees) Grolle und *Lepidozia reptans* (L.) Dum. PhD dissertation, Saarbrücken University.

Saritas, Y., Bülow, N., Fricke, C., König, W., & Muhle, H. (1998). Sesquiterpene hydrocarbons in the liverwort *Dumortiera hirsuta*. *Phytochemistry,* **48,** 1019–23.

Satake, K. & Miyasaka, K. (1984). Evidence of high mercury accumulation in the cell wall of the liverwort *Jungermannia vulcanicola* Steph. to form particles of a mercury–sulphur compound. *Journal of Bryology,* **13,** 101–5.

Satake, K., Iwatsuki, Z., & Nishikawa, M. (1984). Inorganic elements in some aquatic bryophytes from streams in New Caledonia. *Journal of the Hattori Botanical Laboratory,* **57,** 71–82.

Schmidt, A. (1996). Phytochemische Untersuchung des Lebermooses *Lepidozia incurvata* Lindenberg. PhD dissertation, Saarbrücken University.

Schoeneborn, R. (1996). Die sekundären Inhaltsstoffe des neotropischen Lebermooses *Marchesinia bongardiana* (Lehm. & Lindenb.) Trev. und Untersuchungen zur Chemotaxonomie der Gattung *Marchesinia* S. Gray. PhD dissertation, Saarbrücken University.

Sievers, H. (1992). Inhaltsstoffe und in vitro-Kultur des Laubmooses *Hypnum cupressiforme* Hedw. PhD dissertation, Saarbrücken University.

Sievers, H., Burkhardt, G., Becker, H., & Zinsmeister, H. D. (1994). Further biflavonoids and 3'-phenylflavonoids from *Hypnum cupressiforme*. *Phytochemistry,* **35**, 795–8.

Sim-Sim, M., Sérgio, C., Mues, R., & Kraut, L. (1995). A new *Frullania* species (Trachycolea) from Portugal and Macaronesia, *Frullania azorica* sp. nov. *Cryptogamie, Bryologie et Lichénologie,* **16**, 111–23.

Smith, D. K. & Davison, P. G. (1993). Antheridia and sporophytes in *Takakia ceratophylla* (Mitt.) Grolle: evidence for reclassification among the mosses. *Journal of the Hattori Botanical Laboratory,* **73**, 263–71.

Speicher, A. & Schoeneborn, R. (1997). 3,4–Dihydroxy-3'-methoxystilbene, the first monomeric stilbene derivative from bryophytes. *Phytochemistry,* **45**, 1613–15.

Steinnes, E. (1995). A critical evaluation of the use of naturally growing moss to monitor the deposition of atmospheric metals. *Science of the Total Environment,* **160/161**, 243–9.

Tazaki, H., Adam, K.-P., & Becker, H. (1995). Five lignan derivatives from in vitro cultures of the liverwort *Jamesoniella autumnalis*. *Phytochemistry,* **40**, 1671–5.

Toyota, M., Koyama, H., Mizutani, M., & Asakawa, Y. (1996*a*). (-) ent-Spathulenol isolated from liverworts is an artefact. *Phytochemistry,* **41**, 1347–50.

Toyota, M., Saito, T., & Asakawa, Y. (1996*b*). A phenethyl glycoside from *Conocephalum conicum*. *Phytochemistry,* **43**, 1087–8.

Toyota, M., Bardón, A., Kamiya, N., Takaoka, S., & Asakawa, Y. (1997*a*). Dumortenols, novel skeletal sesquiterpenoids from the Argentinian liverwort *Dumortiera hirsuta*. *Chemical and Pharmaceutical Bulletin,* **45**, 2119–21.

Toyota, M., Konoshima, M., Nagashima, F., Hirata, S., & Asakawa, Y. (1997*b*). Butenolides from *Marchantia paleacea* subspecies *diptera*. *Phytochemistry,* **46**, 293–6.

Toyota, M., Koyama, H., & Asakawa, Y. (1997*c*). Volatile components of the liverworts *Archilejeunea olivacea*, *Cheilolejeunea imbricata* and *Leptolejeunea elliptica*. *Phytochemistry,* **44**, 1261–4.

Toyota, M., Saito, T., Matsunami, J., & Asakawa, Y. (1997*d*). A comparative study on three chemo-types of the liverwort *Conocephalum conicum* using volatile constituents. *Phytochemistry,* **44**, 1265–70.

Toyota, M., Masuda, K., & Asakawa, Y. (1998). Triterpenoid constituents of the moss *Floribundaria aurea* subsp. *nipponica*. *Phytochemistry,* **48**, 297–9.

Trennheuser, F. (1992). Phytochemische Untersuchung und in vitro Kultur ausgewählter Vertreter der Anthocerotopsida. PhD dissertation, Saarbrücken University.

Trennheuser, F., Burkhard, G., & Becker, H. (1994). Anthocerodiazonin, an alkaloid from *Anthoceros agrestis*. *Phytochemistry,* **37**, 899–903.

Wells, J. M. & Brown, D. H. (1995). Cadmium tolerance in a metal-contaminated population of the grassland moss *Rhytidiadelphus squarrosus*. *Annals of Botany,* **75**, 21–9.

Wilson, M. A., Sawyer, J., Hatcher, P. G., & Lerch, H. E., III (1989). 1,3,5-Hydroxybenzene structures in mosses. *Phytochemistry*, **28**, 1395–400.

Wu, C.-L., Huang, Y.-M., & Chen, J.-R. (1996). (-)-Ledol from the liverwort *Cephaloziella recurvifolia* and the clarification of its identity. *Phytochemistry*, **42**, 677–9.

Yoshida, T., Toyota, M., & Asakawa, Y. (1997). Scapaundulins A and B, two novel dimeric labdane diterpenoids, and related compounds from the Japanese liverwort, *Scapania undulata* (L.) Dum. *Tetrahedron Letters*, **38**, 1975–8.

Zheng, G.-Q., Ho, D. K., Elder, P. J., Stephens, R. E., Cottrell, C. E., & Cassady, J. M. (1994). Ohioensins and pallidisetins: novel cytotoxic agents from the moss *Polytrichum pallidisetum*. *Journal of Natural Products*, **57**, 32–41.

6

Molecular genetic studies of moss species

6.1 Introduction

Research into the classical genetics of mosses is of long standing (see Cove 1983, Reski 1998) but molecular genetic studies of mosses have been slow to gather momentum and are much less extensive than such studies of flowering plants. Nevertheless, mosses present unique opportunities for the study of plant genetics and development, combining simplicity with technical convenience (see reviews by Cove *et al.* 1997, Reski 1998, Knight 2000). As a result, molecular genetic studies are gaining impetus. DNA, RNA, and proteins are easy to extract particularly from protonemal tissue and genetic transformation is routine for a number of species. The recent establishment that, in *Physcomitrella patens*, recombination occurs between a homologous sequence in transforming DNA and the corresponding sequence in the genome, allows not only gene knockout but also the prospect of more sophisticated gene manipulation. Furthermore, the characterization of the homologous recombination system has focused attention on mosses and has already resulted in a considerable increase in the numbers of researchers using moss species for molecular genetic studies.

The most extensively studied moss species at the molecular level is *Physcomitrella patens*. It was Harold Whitehouse, famous for his genetic research on fungi, but a lifelong bryologist, who suggested that *P. patens* would be suitable for modern genetic research, and the wild-type strain isolated by him from near Cambridge, UK, remains the stock wild type used in most studies of this species. Engel (1968) was the first to isolate mutants in *P. patens* and to begin to realize its potential. This species was thus the first in which genetic transformation was achieved (Schaefer *et al.*

1991), rather than such species as *Funaria hygrometrica*, *Ceratodon purpureus*, and *Tortula ruralis*, where physiological studies are more extensive.

6.2 Genome size

With the development of large-scale DNA sequencing technology, genomic research is in its ascendancy (see section 6.6.1 below). This impetus alone is sufficient to focus attention on genome size and karyotype analysis in moss species, although there is also considerable intrinsic interest in these topics for evolutionary studies. An extensive analysis of genome size has recently been carried out by Johann Greilhuber and his co-workers at the University of Vienna (Temsch *et al.* 1998, Voglmayr 1998, 1999, Voglmayr & Greilhuber 1999). These studies used both Feulgen densitometry and propidium iodide flow cytometry and have included over 130 taxa. They confirmed that within species or genera where polyploid series are thought to exist, there is a correlation between chromosome number and genome size. A few other studies have established a similar correlation (Abderrahman 1998), and a doubling of nuclear DNA content has also been observed following the production of a somatic diploid by way of protoplast fusion (Lamparter *et al.* 1998). Voglmayr's study shows that there is much less variability for haploid genome size between different moss species than is found in angiosperms. In the 273 accessions examined that comprised 132 different taxa, there was only an about-12-fold variation in the DNA content of haploid genomes (range 0.17 to 2.05 pg), compared with the almost 1000-fold range observed in angiosperms. More than 80% of the moss species examined have haploid genomes between 0.25 and 0.6 pg, and thus fall within a 2.4-fold range of variation. Voglmayr relates his genome size determinations to chromosome numbers according to Fritsch (1991) (see Fig. 6.1). Although it has been thought that $n = 7$ represents the basic haploid number for mosses and species with chromosome numbers in the range 10 to 14 were derived from diploids, and 20 to 28 from tetraploids, genome size determinations do not support this conclusion. For example, *Rhytidiadelphus triquetrus* has a genome size of 0.52 pg and a haploid chromosome number of 6, whereas for *R. subpinnatus*, the genome size is 0.42 pg while the chromosome number is 10. The largest genome in this survey for which a chromosome number had been published was for *Mnium marginatum* (2.05 pg, $n = 12$) and the smallest was *Aulacomnium androgynum* (0.26 pg, $n = 10$–12). Since parallel genome size measurements and chromosome counts have not

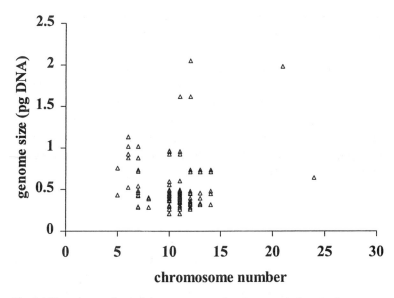

Fig. 6.1. Moss genome size and chromosome number. Data matrix from Voglmayr (1998) revised by Voglmayr (unpublished data). The chromosome counts have been taken from Fritsch (1991). Chromosome counts made before *ca.* 1950 and also counts made of material from a continent different from the material used for the genome size determination have been omitted. Where several close but differing chromosome counts are given for a species (e.g., 10, 11, 12), that species has multiple entries in the matrix, one for each chromosome number. For genome sizes in species with several accessions, the overall mean for propidium iodide flow cytometry determinations was taken (Voglmayr 1998, 1999).

been made on the same collections, the conclusions must remain some-what guarded, but it does indeed appear that different evolutionary forces may be at work in the selection of genome size in bryophytes and angiosperms. It may be that homologous recombination occurs not only following transformation, but also between duplicated sequences in somatic cells, which would slow down the accumulation of disperse repetitive DNA sequences (Voglmayr 1998, 1999). Another hypothesis has been offered by Renzaglia *et al.* (1995) who propose that the motility of moss sperm may be an important selective force. The mass of a sperm is important in determining its motility and is due mainly to its nucleus. Thus increase in genome size would lead to reduced sperm motility, and would therefore be counter-selective.

Not all of the moss species most commonly used for molecular genetic procedures were included in Voglmayr's survey, although he determined the genome size of a *Tortula ruralis* accession as 0.39 pg, and of two *Ceratodon purpureus* accessions both as 0.39 pg. The figure for *C. purpureus* agrees with the figure, published by Lamparter *et al.* (1998), of between 2.4 and 2.7 times that of *Arabidopsis thaliana* (i.e., 0.36–0.40 pg). Voglmayr has recently determined the genome size of *Funaria hygrometrica* collected in the Vienna Botanic Garden as 0.40 pg (Voglmayr, unpublished data). The genome size of *Physcomitrella patens* has not yet been determined using the same range of procedures, but using DAPI (4′,6-diamidino-2-phenylindole) staining and flow cytometry, Reski (1998) estimates a size of about three times larger than *A. thaliana*, i.e., 0.45 pg.

6.3 Transformation

6.3.1 Transformation procedures

The first successful transformations of a moss species were achieved in *Physcomitrella patens* using PEG(polyethylene glycol)-mediated uptake of DNA by protoplasts (Schaefer *et al.* 1991). Large numbers of protoplasts can be isolated in this species with ease, and protoplasts regenerate at high frequency directly to form protonemal filaments. Transformation using protoplasts is therefore straightforward. Genes conferring resistance to antibiotics are used routinely to select transformants and include *npt*II (conferring resistance to kanamycin and geneticin), *hpt* (resistance to hygromycin), and *sul* (resistance to sulphonamide). Transformation of *P. patens* can also be achieved using a biolistic procedure in which DNA is delivered by particle bombardment (Sawahel *et al.* 1992).

After transformation with DNA of plasmids containing antibiotic-resistance-coding genes, both techniques yield three classes of resistant regenerants (Knight 1994, Schaefer *et al.* 1994). Some regenerants sub-cultured from the initial screen for resistance, fail to grow further on selective medium and are presumed to express resistance only transiently. Many regenerants show only weak resistance and lose this phenotype when selection pressure is removed, after which plasmid DNA can no longer be detected. Such transgenics are described as unstable. Although antibiotic resistance is lost rapidly in the absence of selection, the phenotype can be maintained for many serial subcultures provided selection is maintained. Free plasmid can be recovered from unstable transgenics and this, coupled with their phenotypic instability, has led to the proposal that plasmid DNA is maintained extra-chromosomally in this class of transgenic (Knight 1994, Machuka *et al.* 1999*b*). In addition to transients and unstables, it is also possible to isolate stable transformants. Such transgenics show strong antibiotic resistance, growing in the presence of selection at nearly the same rate that untransformed strains grow on non-selective media. Stable transformants do not lose their transformed phenotype (nor their plasmid DNA content) after culture in the absence of selection. The plasmid DNA is transmitted through crosses in a Mendelian manner (Schaefer *et al.* 1991), and consequently it has been concluded that stable transformation results from the integration of transforming DNA at a chromosomal locus, a conclusion confirmed by molecular analyses. The relative frequency at which these three classes are recovered following transformation depends on how long selection is maintained before the first subculture. If regenerants are subcultured 14 days after transformation about half fail to grow further on selective medium and are therefore transients, most of the remainder are unstables, and only a few percent are stable transformants.

Schaefer and co-workers (Schaefer *et al.* 1994) propose that a further class of transgenic exists, subdividing unstable transformants into two classes. The description of unstable transformants given above corresponds to their "unstable replicative transformant" class. In addition, they recognized a further class of unstable transformants that occurred at a low frequency, similar to that of stable transformants, which they term "fast-growing unstable" transformants. This class grows as well on selective medium as stable transformants but loses its resistant phenotype in the absence of selection. Meiotic transmission of transformed phenotypic characters is occasionally observed with this class but does not occur in the regular manner observed for stable transformants. The same paper (Schaefer *et al.* 1994) reports that both classes of unstable transformant can spontaneously convert to a stable phenotype suggesting that integration into the genome need not occur coincidentally with transformation. Knight (1994) suggests that the two classes of unstable transformants recognized by Schaefer and co-workers represent extremes of a continuum, but recently it has been shown that free plasmid can be recovered from transgenics which have the growth rate on selective medium which is characteristic of a stable transformant (Machuka *et al.* 1999b). It is possible that these transgenics are fast-growing unstable transformants rather than stable transformants, and further investigation will be required to examine this.

The frequency of transformation rates can be expressed in number of ways, of which rate per microgram of transforming DNA can be used to compare the protoplast and biolistic transformation procedures. The biolistic procedure yields from one to five stable transformants per μg of DNA, while the protoplast procedure yields stable transformants at about one-tenth of this rate. However, more practically, both procedures yield about the same number of regenerants using the standard procedures. The effect of the physical state of DNA used for transformation by way of protoplasts has received some attention, but no significant differences in rates of stable transformant generation appear to be associated with linearization of circular plasmids. Recently, it has been reported that denaturing transforming DNA, before use in the biolistic transformation procedure, increases the frequency of transformation 5-fold (Cho *et al.* 1999).

Transformation procedures for other moss species are now being developed. Transformation of *Ceratodon purpureus* has been achieved using a procedure based on the protoplast delivery system used for *P. patens* (Zeidler *et al.* 1999) and is at least as efficient. In this species, as with *P. patens*, both stable and unstable transformants are recovered. Zeidler *et al.* (1999) report that a 2- to 3-fold higher frequency of stable transformation is obtained with constructs using the actin 1 promoter to drive the selected gene, compared with those attained when the weaker *35S* promoter is used (see section *Promoters* below). They also investigated the efficiency of cotransformation in *C. purpureus*. Protoplasts were transformed with an equal mixture of a plasmid containing the *hpt* gene, and a plasmid coding for the *gfp* reporter gene (coding for green fluorescent protein; see section 6.4.1 below). Seven days after selection with hygromycin, 40% of regenerants were shown to express the reporter gene coded for by the unselected plasmid, showing that co-transformation occurs at a high frequency (Zeidler *et al.* 1999).

6.3.2 Consequences of transformation

The consequences of transformation have, as yet, only been studied in *Physcomitrella patens*. In this species, when transformation is carried out with plasmids containing no DNA homologous to moss genomic sequences, insertion of the plasmid sequence into the genome appears to occur at random. However, where sequences homologous to genomic sequences are contained within plasmid DNA, then plasmid insertion is targeted to the homologous sequence. Stable transformants usually have more than one copy of the transforming DNA integrated at the insertion site. This could arise by the initial insertion of a single copy, followed by homologous recombination between the inserted sequence and further free plasmid genomes. Alternatively it is possible that concatemers are formed between plasmid DNA copies before insertion. There is some evidence that linear DNA molecules may be circularized and concatemerized following transformation (see below) and that recombination may occur between plasmids without genomic integration. Co-transformation with two constructs which contain non-overlapping deletions of the *npt*II gene generate unstable antibiotic resistant regenerants (Kammerer & Cove, unpublished data).

The occurrence of homologous recombination between sequences in transforming DNA and the *P. patens* genome was identified initially by parallel molecular studies using single-copy genomic sequences (Schaefer & Zrÿd, 1997) and genetic studies in which the outcome of the retransformation of a stable transgenic strain with a related plasmid was analyzed (Kammerer & Cove 1996). Schaefer and Zrÿd obtained transformation rates, using the protoplast delivery system, that were 10 times higher than those obtained with a construct containing no homologous DNA sequences and over 90% of stable transgenic strains obtained after transformation with constructs containing homologous DNA were shown to be targeted to the homologous genomic locus, thus establishing that targeting occurs with very high efficiency. A number of studies have since exploited this finding to achieve gene targeting.

A cDNA homologue of the bacterial *fts*Z, a gene that is required for bacterial cell division, has been used to disrupt the homologous *P. patens* genomic locus (Strepp *et al.* 1998). Using a PCR-based method, a 1775-bp sequence was identified from a *P. patens* cDNA library. This was used to make a construct that lacked part of 5' sequence of the open reading frame and that had an *npt*II expression cassette inserted into the open reading frame, such that there was 247 bp of 5' cDNA sequence to one side of the selective marker, and 658 bp of cDNA sequence to the other side. The linearized construct was used for transformation using the protoplast procedure. Of 51 stable transgenics obtained, 44 were phenotypically normal, but seven had cells containing a single very large chloroplast. PCR and Northern blotting was used to establish the molecular basis of the transgenics and demonstrate that the phenotypically abnormal transgenics had the genomic *fts*Z locus disrupted, but in those with a normal phenotype, the locus remained intact. Disruption of the moss *fts*Z gene therefore blocks plastid division. This study demonstrates dramatically the power of homologous recombination as a tool for the study of gene function and establishes that some steps in plastid division and prokaryote cell division must show remarkable evolutionary conservation. The expression of the *npt*II gene varied considerably between transgenics and it is proposed that this reflects variation in the number of copies of the transforming DNA that had been inserted. Since linear DNA was used for transformation this may provide further evidence that transforming DNA is circularized after uptake.

Using a similar approach, a *P. patens* cDNA has been isolated that has the potential to code for Δ-6-acyl-lipid desaturase, and this has been used to clone the corresponding moss genomic sequence (Girke *et al.* 1998). A construct was made with an *npt*II expression cassette substituted for *ca.* 250 bp of genomic sequence in the central portion of the putative gene. Transformation was carried out using the protoplast procedure and a linear fragment having *ca.* 900 bp of 5′ genomic sequence flanking one side of the selection cassette and by *ca.* 1000 bp on the other, such that both the 5′ and 3′ ends of the gene were deleted. Five stable transgenics were chosen at random and all were shown to have the genomic locus disrupted. These transgenics also showed alterations in their pattern of fatty acid production consistent with them lacking a novel Δ-6-acyl-lipid desaturase, again demonstrating the power of this homologous recombination system for establishing gene function.

The specificity of homologous recombination between transforming and genomic DNA has been investigated using a member of the *P. patens cab* gene family that codes for chlorophyll-binding proteins (Hofmann *et al.* 1999). Transformation was carried out using the protoplast procedure and a construct in which DNA corresponding to an internal fragment of the genomic sequence of one member of the 14-member multigene family was cloned adjacent to an *npt*II expression cassette. Of nine stable transgenics analyzed, three were found to have the plasmid inserted into the homologous *cab* gene. Whether the other six transgenics had the plasmid inserted at unrelated sequences or into other members of the *cab* family has not yet been established; however since members of the family show considerable sequence homology and since three out of nine transgenics were accurately targeted, the sequence homology requirements for targeting appear to be stringent, although further studies will be required before this can be established more precisely. It is also not yet clear whether the sequence itself can lead to differences in recombination rates. It appears that most sequences so far tried have yielded at least some homologous recombinants but negative results may not have been communicated, and it would anyhow be predicted that targeting an essential gene, so that it was inactivated, would lead to lethality and a failure to recover homologous recombinants. The length of sequence homology affects recombination frequencies and high frequencies of targeting appear to require about 1 kbp or more of homologous sequence.

Most of these results, together with others in progress that have not yet been published, involve transformation with intact plasmid. Here a single cross-over between the homologous sequence in the plasmid and the corresponding genomic sequence would be sufficient to achieve targeting (see Fig. 6.2a). In those cases where a linear fragment containing a selective marker inserted within the homologous sequence is used, integration can be achieved by a cross-over on either side of the selective marker (see Fig. 6.2b) or by a single cross-over if the transforming DNA can be circularized after delivery, with or without concatemerization (see Fig. 6.2c). Further molecular analyses will be necessary to establish whether both pathways are possible and what are their relative frequencies. All successful targeting experiments performed so far have employed the protoplast transformation procedure and there is some evidence that the biolistic procedure does not result in homologous recombination (Knight, unpublished data).

6.4 Expression of heterologous genes in mosses

6.4.1 Coding sequences

As well as the genes coding for antibiotic resistance described above, the expression of a number of other heterologous genes has been achieved in mosses. In *Physcomitrella patens*, the *gus* reporter gene, coding for β-glucuronidase (Jefferson *et al.* 1987) has been used to investigate ABA-mediated regulation of gene expression (Knight *et al.* 1995). Zeidler and co-workers have used not only *gus* but also *gfp* (coding for green fluorescent protein, Sheen *et al.* 1995) and *luc* (coding for luciferase) as reporter genes for their studies of promoter function (see section 6.4.2) in *P. patens* and *Ceratodon purpureus* (Zeidler *et al.* 1999).

The gene coding for apoaequorin, the protein component of the calcium-sensitive luminescent photoprotein from the jellyfish *Aequoria victoria*, has been transformed in to *P. patens*, and shown to code for a functional protein that can be used to detect changes in calcium concentration within moss cells (Russell *et al.* 1996). Transient increases in cytosolic calcium concentration were detected following cold shock, mechanical

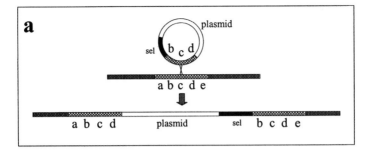

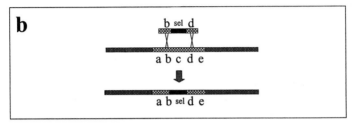

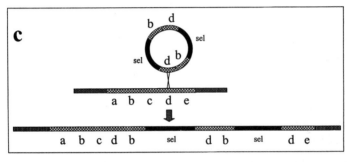

Fig. 6.2. Pathways for gene inactivation following homologous recombination between transforming DNA and genomic DNA. For convenience, the chromosomal locus has been labeled arbitrarily to consist of five regions, **a, b, c, d, e**. The coding cassette for a selective marker is labeled **sel**. (a) Gene inactivation by a single cross-over between a circular plasmid and the genomic locus. An internal fragment of the targeted gene (**bcd**) is cloned adjacent to the selective cassette. A single cross-over yields two copies of the gene separated by plasmid DNA. Neither copy is functional since one is has a 3′ deletion and the other a 5′ deletion. (b) Gene inactivation by a double cross-over between a transformed genomic fragment and the genomic locus. A selective cassette has been substituted for part of the centre of the gene to be targeted. A cross-over either side of the selective cassette results in a stable transgenic with an inactive gene. (c) An alternative pathway of gene activation using a linear construct as in (b). Circularization of the transforming DNA has occurred and in the case illustrated here a concatemer between two copies of the sequence has been generated. A single cross-over results in the inactivation of the chromosomal locus. This illustrates one of a number of different possibilities involving circularization and concatemerization.

perturbation, or lowered pH (Russell *et al.* 1996), and after brief treatment of dark-grown protonemal filaments with blue light (Russell *et al.* 1998).

The *ipt* gene from the bacterium *Agrobacterium tumefaciens*, that encodes an enzyme involved in cytokinin biosynthesis, has been expressed in mutant strains of *P. patens* (Reutter *et al.* 1998). Following transformation with a plasmid containing an *ipt* expression cassette, the ability to produce buds was restored in a mutant defective in gametophore production. This is consistent with the mutant either having reduced levels of cytokinins or having a reduced sensitivity to cytokinin, the latter explanation being favoured by Reutter and co-workers, since their direct measurements of cytokinin levels failed to detect differences between the mutant and wild-type strains. Transformation with the same construct, of another mutant defective in both plastid division and gametophore development, restored plastid division but had no effect on the abnormal gametophore development, a result consistent with the proposed requirement of cytokinin for plastid division, but suggesting that the defect in gametophore development was not due to a cytokinin deficiency.

Aphototropic mutants of *C. purpureus* are unable to orient the direction of growth of protonemal apical cells to a unilateral light source. Such mutants fall into two classes depending on whether or not phototropic activity can be restored by feeding with precursors of the phytochrome chromophore (Lamparter *et al.* 1996). Repairable mutants are therefore thought to be deficient in chromophore synthesis. Recently, phototropic competence has been restored to this latter class of mutant, by injecting into protonemal apical cells a plasmid containing an expression cassette for a rat heme oxygenase cDNA (Brücker *et al.* 1999). Heme oxygenase is thought to catalyze a step in phytochrome chromophore biosynthesis, and so this result supports the conclusion that this class of aphototropic mutant is indeed deficient in chromophore biosynthesis. The success of this technique opens exciting possibilities for the direct study of gene function particularly in the area of cellular morphogenesis.

6.4.2 Promoters

A number of promoters have been used in molecular genetic studies of mosses, but most studies involving *Physcomitrella patens* have employed the cauliflower mosaic virus *35S* promoter. Although the *35S* promoter works well enough in *P. patens* (Schaefer *et al.* 1991, Kammerer & Cove 1996) and *C. purpureus* (Cove, unpublished data) to drive sufficient expression of the *npt*II and *hpt* genes to generate an antibiotic-resistant phenotype, it does not lead to high levels of gene expression. For example, in *P. patens*, the level of *gus* gene expression from the *35S* promoter is only one-third of that from the wheat *Em* promoter in the absence of abscisic acid, and one-hundredth of that from the *Em* promoter when abscisic acid is present (Knight *et al.* 1995, Cove *et al.* 1997). Zeidler *et al.* (1999) have recently carried out a comparison of various promoters in *P. patens* and *C. purpureus*, using a number of different reporter genes. They confirm the weak expression associated with the *35S* promoter. The rice *Actin1* promoter (Zhang *et al.* 1991) gave strong *gus* expression in both *P. patens* and *C. purpureus*, but neither a heat shock (Schoffl *et al.* 1989) nor a *Chlamydomonas RbcS* (Kozminski *et al.* 1993) promoter gave stronger expression than the *35S* promoter.

The identification of promoters that can be used experimentally to regulate gene expression is important for molecular genetic studies. In the study described above (Knight *et al.* 1995), the wheat *Em* promoter was shown to drive approximately 30 times as much reporter gene activity in the presence of added abscisic acid as in its absence. The finding may be of limited utility for general use, since levels of expression from this promoter even in the absence of added abscisic acid are higher than those using the *35S* promoter, and furthermore the use of a phytohormone to control gene expression may be undesirable particularly in developmental genetic studies. The tetracycline-regulatable promoter system may have more utility and has been used successfully in *P. patens* (Zeidler *et al.* 1996) and *C. purpureus* (Zeidler *et al.* 1999). Using strains transgenic for the tetracycline repressor, the *Top10* promoter was shown to be repressed by concentrations of tetracycline as low as $1 \mu g \, ml^{-1}$, at which concentration no morphological effect was observed for either species. In the absence of tetracycline, repression was relieved and expression levels were at least 100 times higher in *P. patens*. The derepressed expression levels in *C. purpureus* were comparable to those in *P. patens*, but the repressed levels were higher, derepression leading to only a 5- to 10-fold increase in expression; the constructs used were, however, not identical.

6.5 Expression of moss genes in heterologous systems

The expression of genes in heterologous systems is a powerful tool for the isolation of genes coding for functionally homologous products. This technique has not yet been used extensively to identify moss genes. The

Physcomitrella patens gene coding for adenosine kinase (*adk*), an enzyme that is also involved in the conversion of cytokinins to their corresponding nucleotides, has been identified by functional complementation of an *Escherichia coli* strain deficient in purine synthesis (von Schwartzenberg *et al.* 1998).

Heterologous expression also allows the possibility of the production of a gene product in larger quantities that may be achievable in the moss itself. As part of studies on the mode of action of phytochrome (Zeidler *et al.* 1998), the *Ceratodon purpureus* gene which codes for the phytochrome apoprotein (*Cerpu; PHY2*) has been expressed in *Saccharomyces cerevisiae*. The apoprotein synthesized in yeast could be reconstituted using either phytochromobilin or phycocyanobilin to give holoproteins that showed photoreversibility, demonstrating the utility of this heterologous expression system for further molecular studies on the mode of action of phytochrome.

6.6 Prospect

6.6.1 Genomics

The molecular genetics of mosses has now reached the stage that a more detailed knowledge of moss genomes at the sequence level would be invaluable. A starting point for such genomic studies is the acquisition of a database of sequences of cDNAs (also known as expressed sequence tags, or ESTs). This database is an invaluable research tool both for evolutionary studies and for the identification of sequences homologous to those from other species. It is particularly valuable to have such a resource for mosses, since homologous recombination allows targeted knockout and the possibility of attributing a function to the moss homologue of a gene, the function of which is as yet unknown. Three recent papers (Reski *et al.* 1998, Machuka *et al.* 1999*a*, Wood *et al.* 1999) report a start on acquiring such a resource but only together amount to 251 ESTs from *Physcomitrella patens* and 152 from *Tortula ruralis*. Many of these ESTs show no homology to sequences from other plant species, but this may be due not only to the evolutionary distance between mosses and flowering plants but also to the relatively incomplete plant EST database. A target of at least 30 000 moss ESTs needs to be aimed at and it is likely that work to achieve this will be undertaken in the near future. Such studies need to be complemented by detailed genomic mapping but this appears to still be some time distant.

6.6.2 Tagging

Although mosses promise to be excellent material for the exploitation of molecular genetic techniques to study a wide range of developmental, physiological, and biochemical problems, a technique that is needed to complete their utility is gene tagging. Tagging achieves mutation by the insertion of DNA into a gene, thereby not only inactivating the gene but also providing a molecular label to allow the disrupted gene to be isolated. No successful tagging technique has so far been reported for mosses and a number of studies have explored the possibility of using heterologous tagging systems. The Ac/Ds transposon system from *Zea mays* does not appear to function in *P. patens* (Kammerer & Cove, unpublished data), nor does *Agrobacterium tumefaciens* T DNA (Knight & Hohn, unpublished data). New *Agrobacterium* strains with wider host ranges have recently become available, and these may allow T DNA transfer to moss. However, the moss homologous recombination system itself allows the possibility of gene knockout and tagging. It is probably necessary to increase the frequency of stable transformation before this pathway for gene tagging becomes useful, but the reward is such that it worth putting considerable effort in achieving this goal.

Acknowledgments

I wish to thank all those who made material available to me and especially Celia Knight and Hermann Voglmayr for their critical reading of the manuscript.

REFERENCES

Abderrahman, S. (1998). DNA content of two cytotypes of *Funaria hygrometrica*. *Korean Journal of Genetics*, **20**, 103–8.

Brücker, G., Zeidler, M., Kohchi, T., Hartmann, E., & Lamparter, T. (1999). Microinjection of heme oxygenase genes rescues phytochrome-chromophore-deficient mutants of the moss *Ceratodon purpureus*. *Planta*, in press.

Cho, S.-H., Chung, Y.-S., Cho, S.-K., Rim, Y.-W., & Shin, J.-P. (1999). Particle bombardment mediated transformation and *gfp* expression in the moss *Physcomitrella patens*. *Molecules and Cells*, **9**, 14–19.

Cove, D. J. (1983). Genetics of Bryophyta. In *New Manual of Bryology*, ed. R. M. Schuster, pp. 222–31. Nichinan: Hattori Botanical Laboratory.

Cove, D. J., Knight, C. D., & Lamparter, T. (1997). Mosses as model systems. *Trends in Plant Sciences*, **2**, 99–105.

Engel, P. P. (1968). The induction of biochemical and morphological mutants in the moss *Physcomitrella patens*. *American Journal of Botany*, **55**, 438–46.

Fritsch, R. (1991). Index to bryophyte chromosome counts. *Bryophytorum Bibliotheca*, **40**.

Girke, T., Schmidt, H., Zahringer, U., Reski, R., & Heinz, E. (1998). Identification of a novel Δ-6-acyl-group desaturase by targeted gene disruption in *Physcomitrella patens*. *Plant Journal*, **15**, 39–48.

Hofmann, A. H., Codon, A. C., Ivascu, C., Russo, V. E. A., Knight, C. D., Cove, D. J., Schaefer, D. G., Chakhparonian, M., & Zrÿd, J. P. (1999). A specific member of the *Cab* multigene family can be efficiently targeted and disrupted in the moss *Physcomitrella patens*. *Molecular and General Genetics*, **261**, 92–9.

Jefferson, R. A., Kavanagh, T. A., & Bevan, M. W. (1987). GUS-fusions: β-glucuronidase as a sensitive and versatile gene fusion marker in higher plants. *European Molecular Biology Organization Journal*, **6**, 3901–7.

Kammerer, W. & Cove, D. J. (1996). Genetic analysis of the effects of re-transformation of transgenic lines of the moss *Physcomitrella patens*. *Molecular and General Genetics*, **250**, 380–2

Knight, C. D. (1994). Studying plant development in mosses: the transgenic route. *Plant, Cell and Environment*, **17**, 669–74.

(2000). Molecular tools for phenotypic analysis in moss. In *Encyclopaedia of Cell Technology*, ed. R. Spier. New York: Wiley.

Knight, C. D., Sehgal, A., Atwal, K., Wallace, J. C., Cove, D. J., Coates, D., Quatrano, R. S., Bahadur, S., Stockley, P. G., and Cuming, A. C. (1995). Molecular responses to abscisic acid and stress are conserved between moss and cereals. *Plant Cell*, **7**, 499–506.

Kozminski, K. G., Diener, D. R., and Rosenbaum, J. L. (1993). High level expression of non-acetylatable α-tubulin in *Chlamydomonas reinhardii*. *Cell Motility and the Cytoskeleton*, **25**, 158–70.

Lamparter, T., Esch, H., Cove, D. J., Hughes, J., & Hartmann, E. (1996). A phototropic mutants of the moss *Ceratodon purpureus* with spectrally normal and with spectrally dysfunctional phytochrome. *Plant, Cell and Environment*, **19**, 560–8.

Lamparter, T., Brücker, D., Esch, H., Hughes, J., Meister, A., & Hartmann, E. (1998). Somatic hybridisation with aphototropic mutants of the moss *Ceratodon purpureus*: genome size, phytochrome photoreversibility, tip-cell phototropism and chlorophyll regulation. *Journal of Plant Physiology*, **153**, 394–400.

Machuka, J., Bashiardes, S., Ruben, E., Spooner, K., Cuming, A., Knight, C. D., & Cove, D. J. (1999a). Sequence analysis of expressed sequence tags from an ABA-treated cDNA library identifies stress response genes in the moss *Physcomitrella patens*. *Plant Cell Physiology*, **40**, 378–87.

Machuka, J., Birgul, N., Reski, R., & Knight, C. D. (1999b). Functional plasmids are rescued from extrachromosomal DNA of transgenic *Physcomitrella patens*. *Molecular and General Genetics* (submitted).

Renzaglia, K. S., Rasch, E. M., & Pike, L. M. (1995). Estimates of nuclear DNA content in bryophyte sperm cells: phylogenetic considerations. *American Journal of Botany*, **82**, 18–25.

Reski, R. (1998). Development, genetics and molecular biology of mosses. *Botanica Acta*, **111**, 1–15.

Reski, R., Reynolds, S., Wehe, M., KleberJanke, T., & Kruse, S. (1998). Moss (*Physcomitrella patens*) expressed sequence tags include several sequences which are novel for plants. *Botanica Acta*, **111**, 143–9.

Reutter, K., Atzorn, R., Hadeler, B., Schmulling, T., & Reski, R. (1998). Expression of the bacterial *ipt* gene in *Physcomitrella* rescues mutations in budding and in plastid division. *Planta*, **206**, 196–203.

Russell, A. J., Knight, M. R., Cove, D. J., Knight, C. D., Trewavas, A. J., & Wang, T. L. (1996). The moss, *Physcomitrella patens*, transformed with apoaequorin cDNA responds to cold shock, mechanical perturbation and pH with transient increases in cytoplasmic calcium. *Transgenic Research*, **5**, 167–70.

Russell, A. J., Cove, D. J., Trewavas, A. J., & Wang, T. L. (1998). Blue light but not red light induces a calcium transient in the moss *Physcomitrella patens* (Hedw.) B., S. & G. *Planta*, **206**, 278–83.

Sawahel, W., Onde, S., Knight, C. D., & Cove, D. J. (1992). Transfer of foreign DNA into *Physcomitrella patens* protonemal tissue using the gene gun. *Plant Molecular Biology Reporter*, **10**, 315–16.

Schaefer, D. G. & Zrÿd, J.-P. (1997). Efficient gene targeting in the moss *Physcomitrella patens*. *Plant Journal*, **11**, 1195–206.

Schaefer, D. G., Zrÿd, J.-P., Knight, C. D., & Cove, D. J. (1991). Stable transformation of the moss, *Physcomitrella patens*. *Molecular and General Genetics*, **226**, 418–24.

Schaefer, D. G., Bisztray, G., & Zrÿd, J. P. (1994). Genetic transformation of the moss *Physcomitrella patens*. In *Biotechnology in Agriculture and Forestry*, vol. 29, *Plant Protoplasts and Genetic Engineering V*, ed. Y. P. S. Bajaj, pp. 349–64. Berlin: Springer Verlag, Heidelberg.

Schöffl, F., Rieping, M., Baumann, G., Bavan, M. W., & Angermüller, S. (1989). The function of plant heat shock promoter elements in the regulated expression of chimaeric genes in transgenic tobacco. *Molecular and General Genetics*, **217**, 246–53.

Schwartzenberg, K. von, Kruse, S., Reski, R., Moffatt, B., & Laloue, M. (1998). Cloning and characterization of an adenosine kinase from *Physcomitrella* involved in cytokinin metabolism. *Plant Journal*, **13**, 249–57.

Sheen, J., Hwang, S., Niwa, Y., Kobayashi, H., & Galbraith, D. W. (1995). Green fluorescent protein as a new vital marker in plant cells. *Plant Journal*, **8**, 777–84.

Strepp, R., Scholz, S., Kruse, S., Speth, V., & Reski, R. (1998). Plant nuclear gene knockout reveals a role in plastid division for the homolog of the bacterial cell division protein FtsZ, an ancestral tubulin. *Proceedings of the National Academy of Sciences USA*, **95**, 4368–73.

Temsch, E. M., Greilhuber, J., & Krisai, R. (1998). Genome size in *Sphagnum* (peat moss). *Botanica Acta*, **111**, 325–30.

Voglmayr, H. (1998). Genome size analysis in mosses (Musci) and downy mildews (Peranosporales). PhD dissertation, University of Vienna.

(1999). Genome size in mosses (Musci). *Annals of Botany* (submitted).

Voglmayr, H. & Greilhuber, J. (1999). Genomgrössanalyse bei Laubmoosen. *Abhandlungen der Zoologischen-Botanischen Gesellschaft in Österreich*, **30**, 169–77.

Wood, A. J., Duff, R. J., & Oliver, M. J. (1999). Expressed sequence tags (ESTs) from desiccated *Tortula ruralis* identify a large number of novel plant genes. *Plant Cell Physiology*, **40**, 361–8.

Zeidler, M., Gatz, C., Hartmann, E., & Hughes, J. (1996). Tetracyclin-regulated reporter gene expression in the moss *Physcomitrella patens*. *Plant Molecular Biology*, **30**, 199–205.

Zeidler, M., Lamparter, T., Hughes, J., Hartmann, E., Remberg, A., Braslavsky, S. E., Schaffner, K., & Gärtner, W. (1998). Recombinant phytochrome of the moss *Ceratodon purpureus*: heterologous expression and kinetic analysis of Pr > Pfr conversion. *Photochemistry and Photobiology*, **68**, 857–63.

Zeidler, M., Hartmann, E., & Hughes, J. (1999). Transgene expression in the moss *Ceratodon purpureus*. *Journal of Plant Physiology*, **5/6**, 641–50.

Zhang, W., McElroy, D., & Wu, R. (1991). Analysis of the rice 5′ region activity in transgenic rice. *Plant Cell*, **3**, 1155–65.

7

Control of morphogenesis in bryophytes

Experimental studies of morphogenesis in the bryophytes certainly date to the work of von Wettstein early in the 20th century and has its beginnings even earlier. In part because of this extensive history, in part because the mosses, liverworts, and hornworts represent a paraphyletic assemblage that includes three-fourths of the systematic range in land plant lineages, and in part because of the taxonomic diversity within any particular type of bryophyte, the literature in this area is particularly rich and voluminous (to adapt a phrase used by Katherine Esau [1965]). There are many still-excellent reviews in this area, in particular, reviews that concentrate on hormone physiology, as well as book-length treatments of bryophyte development and physiology, including the exhaustive treatments by Chopra and Kumar (1988) and Bhatla (1994).

This chapter will briefly sketch out what is known about hormonal and other external triggers to morphogenesis in the bryophytes, including factors described recently in higher plants but not yet examined in bryophytes. That overview will be followed by case studies, selected for their importance to our understanding of bryophyte biology, their potential for elucidating general features of plant biology, or for their use of novel and generally useful approaches.

7.1 Plant growth regulators and bryophyte development

The concept of a plant "hormone" has beguiled botanists for some period of time, and this concept is particularly problematic when botanists restrict their view of plants to just the angiosperms. In animals, hormones are easily described as molecules produced in one part of the body that move throughout the body and act on another part of the body. This simple definition has broad applicability, particularly when problematic substances can be categorized as "growth factors" rather than hormones. As pointed out by many workers over the years, while it is possible to demonstrate the transport of plant hormones and possible to document preferential biosynthesis of particular hormones in particular tissues (e.g., the classic example: cytokinin in roots, auxin in young leaves), the local action of hormones as well as the multiple and startlingly distinct effects that can result from application of any one plant hormone make it difficult to stretch the animal definition around the untidy bundle of phenomena in plants.

There is another, and perhaps more fundamental, difference in animal and plant hormones, and this difference captures one of the essential features of plant biology. While animal hormones were once thought to have only one function (insulin and sugar metabolism, for example), it is now clear that they often have effects on more than one tissue (e.g., follicle-stimulating hormone, FSH, stimulates oocyte development, but also acts on Sertoli cells in the testes). While animal hormones are now known to affect more than one tissue, their repertoire is limited. There are many animal hormones and each has specific developmental effects.

Early work on plant hormones searched for molecules with specific morphogenetic effects and called these partially characterized activities by eponymous names (e.g., rhizocauline). When these activities became defined chemically, these partially characterized factors turned out to be the same molecules (e.g., auxin) again and again, even though the developmental events they controlled were widely different.

The paradigm for plant hormones was changed forever by the insightful experiments of Skoog and Miller (1957). As all undergraduate botanists know, by evaluating the outcome from exposing tobacco pith to combinations of auxin and cytokinin, Skoog and Miller discovered that it was the ratio of these two known plant hormones that controlled whether the tissue would grow shoots, grow roots, or grow into an undifferentiated mass of callus tissue. There was no specific root-hormone, no specific shoot-hormone, no specific callus-hormone. With these experiments, the central mystery of plant hormones and their essential difference from animal hormones was described. Animals had a series of chemically distinct molecules, each with a single target, the pancreas, the thyroid, and so on. Plants use a more limited repertoire of chemicals, but these compounds can be combined to generate a large variety of outcomes. An appropriate analogy might be to a written language: in Chinese, there is a distinct symbol for each word, while in Hebrew, words are built from combinations of a small number of symbols.

The combinatorial mechanics of plant hormones, both the fact that plant hormones work in concert and at particular ratios, as well as the fact that treatment with one hormone often causes the production of another hormone, makes work in this area especially challenging. Experiments must distinguish an effect caused by auxin from an effect caused by the ethylene produced in auxin-treated tissues. Moreover, determining the intracellular concentrations that result from exogenous application of hormone pushes biochemical techniques to their limits. These problems are most extreme in the vascular land plants. The presence of a thick cuticle and the large amounts of cell or tissue differentiation that accompany morphological specialization in the relatively (or actually) massive plant body of the vascular plants confound biochemical analysis.

Most bryophytes live tightly coupled to their moist habitats, and their biology seems to make use of the substrate to coordinate overall growth. While it is certain that bryophytes can transport and mobilize hormones directly from cell to cell (Chopra & Kumar 1988), a variety of experiments observing the effects of conditioned medium or substrate show that bryophytes excrete and then use exogenous hormone to regulate development (Beutelmann & Bauer 1977). The fact that bryophytes use exogenous sources of hormones to regulate and coordinate their development simplifies experimental design and analysis. While it may be convenient for scientists to view this source of hormones as exogenous, that distinction may be both artificial and misleading when applied to bryophytes. Bryophytes are known to use the space outside their cells walls for other important physiological processes. For example, while it is possible to show that the conducting elements in the leafy shoots of mosses can and do transport water (especially in species with very tall shoots), for many mosses, most of the water is transported over the surface of the leafy shoot, and not through the stem.

The apparent lack of morphological complexity in bryophytes is a trap for the unwary. Recent molecular studies have confirmed what careful thought and observation established long ago: the first land plants were bryophytes (Crandall-Stotler 1980, Mishler *et al.* 1992). Contemporary liverworts are not primitive, but represent some 400 million years of successful evolutionary experimentation on a body plan. This can be contrasted to the 130–145 million year history for the angiosperms, the johnnies-come-lately of the plant kingdom (Sun *et al.* 1998). The relatively small diversity of cell types in any one tissue or organ in bryophytes and the interestingly large assemblage of developmental and morphological responses of bryophytes make bryophytes especially useful for experimental work. Both these features reflect a central principle of bryophyte biology. Most bryophytes are intimately connected to their environment and are completely dependent on local conditions. They have no roots to bring in water from a distance; they have no massive plant body to buffer environmental assault.

Vascular plants capitalize on their extended and multicellular body when they respond to stress. Whether the response is the local closing of stomata in response to low leaf water potential, or the global signaling mediated by abscisic acid synthesis in roots as the soil volume dries, the response of these plants is a response of tissues or organs. In contrast, the responses of bryophytes tend to be responses at the cellular level. In mosses, for example, whether the response is the transformation of protonematal filaments into brood cells, into tmema, or into various kinds of gemmae, or the special biochemistry that allows leafy shoots to simply dry down and wait to rehydrate under favorable conditions, these responses are the responses of individual cells, and will occur in small pieces or even single isolated cells.

7.1.1 Plant growth regulators: the five "classical" plant hormones

Since the bryophytes and the vascular plants shared an ancient ancestor (Mishler *et al.* 1992), it comes as no surprise to discover that plant hormones first described, isolated, and characterized from vascular plants are also found in bryophytes and that these hormones can be shown to control physiological and morphological processes in the bryophytes. Hormone metabolism has been studied in great detail in the vascular plants, and this knowledge base can serve to orient what we do, and do not, know about these hormones in bryophytes.

Auxin

While a number of auxins have been isolated from higher plants, the primary auxin is indole-3-acetic acid (IAA). Classic feeding experiments indicated that tryptophan was the natural auxin precursor, and this has now been confirmed by stable isotope labeling studies (Kende & Zeevaart 1997). Although tryptophan is the precursor for the biosynthesis of IAA, this biosynthesis seems to occur in at least three ways, through indole-3-

pyruvate, tryptamine, or indole-3-acetonitrile, and certain plant families rely more on one than any other of these paths. There are indications that IAA can also be made from tryptophan precursors (Bartel 1997). While this evidence is strong, the recent demonstration in *Arabidopsis* of alternate, low activity, pathways for the biosynthesis of tryptophan (Radwanski & Last 1995) shows how difficult it will be to rule out any involvement of tryptophan.

Vascular plants usually contain relatively little free IAA; most of the IAA is present as conjugates (Bandurski *et al.* 1995). A range of bryophytes has now been surveyed for IAA and IAA conjugates by one of the few laboratories capable of such a feat (Sztein *et al.* 1995). The bryophytes do contain IAA, and Sztein *et al.* (1995) recognize a distinct liverwort pattern as well as a moss–hornwort pattern. Each pattern of IAA metabolites is as different from every other as they all are from the tracheophyte pattern. The liverworts contain relatively large amounts of IAA (40–400 μg-equivalents g^{-1} freshweight) and almost all this IAA is unconjugated. Small amounts of IAA metabolites are present (to 10% of the total IAA) but these appear to be intermediates in the natural turnover and degradation of auxin.

The mosses and hornworts contain similar amounts of auxin (and these amounts range as widely as they do in the liverworts: 20–100 μg-equivalents g^{-1} fresh weight). In the mosses and hornworts, IAA is conjugated, and more than one conjugated form can be found within and between species. While no free IAA was recovered from *Sphagnum*, the other mosses and the single hornwort examined contained a relatively large proportion of their auxin as unconjugated IAA ($\sim$50–85%).

Cytokinin

The natural cytokinins are substituted adenines, and are made not only by plants but by some microbes as well (Kende & Zeevaart 1997). In higher plants and in the plant pathogen *Agrobacterium*, the committed biosynthetic step seems to be the addition of an isopentenyl group to adenosine monophosphate by AMP-isopentenyl transferase (IPT). Isopentenyl adenine itself is active as a cytokinin, but in many angiosperms it is converted via hydroxylation or reduction of the side chain into zeatin or dihydrozeatin. The enzyme IPT has been cloned and characterized from *Agrobacterium*. While a similar enzymatic activity has been observed in plant extracts, searches using the cloned IPT gene from *Agrobacterium* on (vascular) plant DNA has failed to find a homologous gene, suggesting that the plant enzyme is quite different from the bacterial gene.

This search for an *Agrobacterium*-like IPT may well be successful in mosses, however. Adding virulent Agrobacteria to moss results in gametophore production, indicating the production of cytokinin (Speiss *et al.* 1971). This effect requires physical contact between the bacteria and the moss cells (Speiss *et al.* 1976), and can be competed by the addition of vascular plant cell wall preparations (Speiss *et al.* 1984). Both these observations indicate that the cytokinin production is the result of transformation of cells by the *Agrobacterium*. Moss shows no evidence of tumor formation from *Agrobacterium* infection, in marked contrast to the response of angiosperms to infection with wild-type strains of *Agrobacterium*. If surveys of moss DNA find *Agrobacterium* IPT genes, evidence of natural infections by *Agrobacterium* in the past, bryologists will still need to determine if this IPT gene has been co-opted for use by moss, or whether the analysis of cytokinin biosynthesis in moss will be complicated by introduced genes supplementing an endogenous pathway.

Like auxins, the free forms of the cytokinins are the active versions of the molecules. Cytokinins can be glycosylated reversibly or inactivated irreversibly (by cytokinin oxidase or by N-glycosylation) to regulate the level of active cytokinin in the cell.

Cytokinins have been isolated from a few rather closely related mosses. Reports include isopentenyl adenine alone (to about 1 μmol, Beutelmann & Baur 1977) as well as isopentenyl adenine with small amounts of zeatin (Wang *et al.* 1981). Some caution might be applied in interpreting this report of zeatin, since it was not detected in wild-type moss, but only in some, but not all, cytokinin-overproducing mutants. Other mutants have been identified with decreased response to exogenous cytokinin. Complementation mapping of these mutants finds one group of mutants with a block in cytokinin biosynthesis and three distinct groups of mutants that result in overproduction of cytokinin (Cove & Ashton 1984). Those results may indicate that biosynthesis of isopentenyl adenine proceeds by a single step (IPT) and that degradation happens by three distinct mechanisms.

Gibberellins

An enormous number of gibberellins have been isolated from the vascular plants (112 by 1997, Kende & Zeevaart 1997). The biosynthesis of the central ring structure that characterizes all gibberellins has been worked out by a number of investigators, starting with the cyclization of geranyl-

geranyl phosphate, culminating in oxidation by series of membrane bound P450 monooxygenases to produce GA_{12}. A variety of subsequent oxidations generates additional structural changes, producing other gibberellic acids (GAs). Only a few of these GAs are active (those with a 3β-hydroxyl group). Bioactive GAs are inactivated by 2β-hydroxylation; GAs can be stored conjugated to glycosyl esters or as glycosides.

In contrast, gibberellin metabolism in bryophytes has attracted very little attention. GA-like substances have been extracted from moss and liverworts (Chopra & Kumar 1988). There are some indications that GA plays a role in the normal physiology of bryophytes, but none of these effects is as dramatic as the internode elongation seen in vascular plants and may be doubtful. Application of the GA antagonists AMO 1618 or CCC has been shown to retard growth and elongation in liverworts. Exposure to exogenous GAs has been shown to promote (or retard) spore germination, protonemal growth, and gametophore production in mosses (Mitra & Allsopp 1959, Vaarama & Taren 1963).

Abscisic acid

Experiments using carotenoid-deficient mutants or inhibitors of carotenoid biosynthesis linked the biosynthesis of abscisic acid (ABA) to the carotenoid pathway in vascular plants (Kende & Zeevaart 1997). The first committed step in ABA biosynthesis is the epoxidation of zeaxanthin, and control of the overall rate of synthesis is suspected to reside at the next step, the cleavage to xanthoxin. ABA can be inactivated by oxidation to phaseic acid or by conversion to a glucose ester (Zeevaart & Creelman 1988).

While the biosynthetic pathway for ABA has not been confirmed in the bryophytes, bryophytes are known to make ABA. Slow drying of the moss *Funaria hygrometrica*, for example, increases the ABA content to 10 nmol g^{-1} dry weight (Werner *et al.* 1991). While it has been traditional to say that liverworts do not make or use ABA (e.g., Chopra & Kumar 1988), more recent experiments have shown that at least some liverworts (members of the Marchantiales as well as *Riccia*) contain significant amounts of ABA (Bopp & Werner 1993). Similarly, examination of hornworts under dry conditions recovers ABA in amounts to 10–600 nmol kg^{-1} fresh weight. Support for the presence of ABA in liverworts and hornworts is strengthened by demonstrations of the responses of these plants to exogenous application of ABA: changed growth form of *Riccia*, the closing of stomata in hornwort and moss sporophytes.

Ethylene

Unraveling the cyclic biosynthetic pathway for ethylene (Yang & Hoffman 1984) was a triumph of modern plant physiology. In vascular plants, the immediate precursor of ethylene is amino-cyclopropane-1-carboxylic acid (ACC) and ethylene is produced by the action of ACC oxidase. Ethylene production is governed by the enzyme ACC synthase (with some exceptions: Osborne *et al.* 1996). Not only is the synthesis of ACC the result of a cyclic pathway, many if not most ethylene-generating tissues in vascular plants have a positive feedback loop: ethylene stimulates the production of additional ethylene from the tissue. While various microbial organisms can make ethylene, this synthesis can proceed in the absence of methionine and seems to make use of citric cycle intermediates (Kende & Zeevaart 1997).

Bryophytes do make ethylene. Ethylene production has been measured in both liverworts (*Marchantia*: Fredericq *et al.* 1977) and mosses (*Funaria*: Rohwer & Bopp 1985). The evidence for the biosynthetic pathway for ethylene in bryophytes is convincing, although scanty and indirect. Rohwer and Bopp (1985) have shown that ACC is absent from young cultures of moss, and that ACC appears at exactly the time that ethylene begins to be produced. As the cultures age further, the amount of ACC rises (content per mg protonema), and the rate of ethylene synthesis rises to a maximum level. Rohwer and Bopp (1985) tried to evaluate potential precursors for ethylene biosynthesis with feeding experiments, but did not see the kinds of results such experiments give with vascular plants. In retrospect, since *Funaria* accumulates ACC, such experiments would be much better done measuring increases in ACC rather than searching for increased rates of ethylene evolution. Indeed, feeding ACC resulted only in the slightest of increases in ethylene evolution (4.62×10^{-2} nl h^{-1} vs. 3.84×10^{-2}), and resulted in only a net of 33 pmol of ethylene over a 3-day period from feeding 1 μmol of ACC. Such situations are known in vascular plants: tobacco cells convert ACC to ethylene in stoichiometric amounts while cells of *Convolvulus*, where ACC can be a major proportion of the free amino acids in the cell, do not show increased rates of ethylene production when fed ACC (Rhodes *et al.* 1982).

While the precursor for the ACC in mosses remains uncharacterized, Rohwer and Bopp (1985) show convincingly that auxin exposure leads quickly to increased amounts of ACC and to increased rates of ethylene production. Such ethylene production from exposure to auxin is a well-known phenomemon in vascular plants. We do not know if bryophytes show the classic positive feedback loop of ethylene-stimulated ethylene production. Rohwer and Bopp's (1985) observation of scant increases in ethylene evolution from feeding ACC to moss confined in sealed vials for three days might be advanced as evidence against the existence of such positive feedback in mosses.

7.1.2 The "neo-classic" plant hormones

In the last decade, the number of molecules that are know to function as hormones in vascular plants has increased. To the classic five hormones, we must now add oligosaccharins, brassinolides, jasmonates, salicylates, and, it seems, even some peptide hormones (Creelman & Mullet 1997). Unfortunately, none of this work has included even a cursory survey of the bryophytes for the presence of such molecules or for some morpho-logical or physiological result after exogenous application of these new vascular plant growth substances. Hopefully some bryologist looking for a research opportunity will remedy this oversight in the near future. A brief survey in my own laboratory found no obvious effects on the growth and development of *Funaria* from exposure to salicylates (to μmol levels), but it may well be that more sophisticated bioassays, looking for inhibi-tion of bud formation, of brood cell formation, of tmema formation, or interference with other changes stimulated by one of the classic five hor-mones, could document biological activities in bryophytes for these neo-classical vascular plant growth substances.

7.1.3 Plant growth regulators: novel bryophytic factors

Novel growth regulators: lunularic acid

The earlier and erroneous belief that liverworts did not contain ABA stemmed in part from the isolation of the structurally distinct hormone, lunularic acid (LA), that seemed to play the same role as ABA in liverworts. LA is a stilbene (3,4'-dihydroxybibenzyl-2-carboxylic acid) that restricts thallus growth. Dormant and desiccation-resisitant thalli contain the largest amounts of LA; reports from various species range from 1–66 μg g^{-1} fresh weight or 4–250 μmol (Chopra & Kumar 1988). Vascular plants do make a large variety of stilbenes including LA itself (Hashimoto *et al.* 1988) but there is no evidence that LA acts a natural growth regulator in vascular plants. For example, LA does not compete with ABA for uptake (Milborrow & Rubery 1985). LA does compete for sites on maize and zucchini membranes that bind an inhibitor of auxin transport, naphthyl-phthalmic acid, and while this competitor has an ED_{50} (the dose that gives a 50% reduction in response) that could be physiologically relevant, 10 μmol (Katekar *et al.* 1993), there is no suggestion that LA functions as an inhibitor of auxin transport in liverworts.

The biosynthesis of LA in liverworts was thought to involve hydrangeic acid, but more recent work argues for prelunularic acid as the LA precursor from the phenylpropanoid pathway (Ohta *et al.* 1984). Liverworts contain free LA as well as a variety of conjugates, including glycosides (Kunz & Becker 1992). It seems likely that the free acid is the active form of the molecule and that conjugation targets LA for degradation.

7.1.4 Other morphogens and factors regulating plant growth

Light-regulation

Bryophytes, like all other plants, receive and respond to developmental cues from their environment. Many of these cues are conveyed by light. Plants can perceive and respond to the direction, quality, and intensity of incident light as well as the duration of the photoperiod, and a thorough and comprehensive review of this subject would take many volumes. In the vascular plants, these responses are mediated by red and by blue wavelengths of light and the pigments phytochrome and cryptochrome, respectively. The bryophytes also have distinct red- and blue-light reponses, and have long been assumed to contain and use the pigments phytochrome and cryptochrome (Wada & Kadota 1989). Since bryophytes are not only phototropic but also show responses to polarized light, they should contain two cryptochrome molecules, a soluble molecule responsible for phototropism and a molecule immobilized on a membrane responsible for the polarotropic response. If bryophytes do not contain two distinct cryptochromes, they must have mechanisms for distributing the single cryptochrome to these two subcellular locations.

Cryptochrome mediates the so-called blue-light responses in plants. Much work with the vascular plants has shown that there are several blue-light receptors (Briggs & Liscum 1997, Cashmore 1997) and this may well be the case in the bryophytes. Phytochrome mediates the so-called red-light responses in plants. In the vascular plants, phytochrome is a multigene family, with five genes that code for phytochrome (Mathews & Sharrock 1997). The N-terminal portion of these proteins contains the site for chromophore attachment, and is responsible for the photoreception by the molecule, and for the differences in photosensitivity and photo-lability that characterize the various classes of phytochrome (Quail 1997). The carboxy-terminal portions of phytochrome seem completely interchangeable in the vascular plants, each containing an homologous domain necessary for signal transfer from phytochrome to the signal transduction pathway. While this domain is similar to a domain in pro-karyotic sensor histidine kinases, experiments have not yet shown that vascular plant phytochromes are light-activated histidine kinases.

In the bryophytes (mosses), molecular cloning and Southern hybridiza-tion found a single phytochrome gene in the moss *Physcomitrella* similar to the phytochrome genes in vascular plants (Schneider-Poetsch *et al.* 1994). In contrast, molecular cloning with the moss *Ceratodon* recovered a novel phytochrome gene (Thummler *et al.* 1992). The cloning strategy used the conserved amino acid sequences required for binding of the chromophore to the phytochrome molecules (Thummler *et al.* 1990), and, as expected, the N-terminal portion of the novel moss phytochrome is similar to vascu-lar plant phytochomes. The carboxy-terminal portion of the molecule shows homology to eukaryotic tyrosine kinases as well as serine/threonine kinases; expression of this gene in a human embryonic kidney fibroblast cell line shows that the protein product does not have tyrosine kinase activity, and that it may represent a light-activated serine/threonine kinase (Thummler *et al.* 1995). The moss *Ceratodon* has now been shown to contain a conventional phytochrome as well as the novel phytochrome (Pasentsis *et al.* 1998). Phylogenetic analysis of the N-terminal portions of phytochrome molecules shows that moss phytochromes form a clade dis-tinct from the clade formed by the diverse angiosperm phytochromes (Schneider-Poetsch *et al.* 1994). This result confirms speculation by others that the common ancestor of the mosses and the vascular plants had a single phytochrome gene and that diversification of the gene families in these two clades occurred independently.

There are some intriguing differences between the responses to light by vascular plants and bryophytes. For example, while the vascular plants use cryptochrome for phototropic responses (Briggs & Liscum 1997), it is thought that most bryophytes use phytochrome for some of their photo-tropic responses (Wada & Kadota 1989). Bryophytes can also store a photo-tropic stimulus (Hartmann & Weber 1988). Protonemata of the moss *Ceratodon* are positively phototropic; when exposed to unilateral light in the presence of monesin, a drug that inhibits growth, the protonemata acquire a memory of the photostimulus. If such protonemata are moved to the dark and held for lengths of time sufficent for dark reversion of phytochrome to the Pfr form, they will still grow in the direction of the previous light stimulus when the monesin is washed out. The phototropic memory is qualitative (direction of stimulus) as well as quantitative (the degree of bending).

While this example may seem rather extreme, requiring special manipulation of the protonemata to demonstrate the effect, the storage of light signal may be a more general phenomenon. Chloronema of the moss *Funaria* forms brood cells when exposed to ABA, and this response has been shown to be calcium-dependent and to require light (Yu & Christianson 1995). The experiments demonstrating the requirement for light exposure show only moderate reductions in numbers of brood cells when the chloronema is treated with ABA immediately after transfer to the dark. ABA treatment of chloronema held in the dark for 24 hours, more than sufficient time for dark reversion of phytochrome, still results in some brood cells; the memory of light exposure, assayed by the ability to form brood cells after ABA treatment, takes more than 48 hours to fade.

Temperature

Vernalization, the requirement for a period of cool temperature as a precondition for the switch from vegetative growth to flowering of sporophytes, is a traditional topic in textbooks of vascular plant physiology. Moss gametophytes show similar requirements for cold treatments in order to switch from vegetative growth to the production of gametangia (Monroe 1965, Dietert 1980, Collier & Hughes 1982). These studies all use mosses from temperate climatic zones and are hardly a surprise to field bryologists who can correlate collection date and the presence of gametangia or young sporophytes on herbarium specimens. It seems likely that the sexual stage of tropical mosses will be triggered by factors such as alternation of dry and wet periods that have been shown to control growth and flowering in tropical vascular plants.

Other factors

The bryological literature contains reports of a variety of chemical factors produced by bryophytes that affect their growth. For the mosses, these include an uncharacterized diffusible substance produced by leafy gametophores that inhibits the germination of spores (Mishler & Newton 1988),

and diffusible substances that coordinate the growth of adjacent protonemata (Bopp 1963, Watson 1981, Christianson 1998b). These substances
also include factors partially characterized by work in the 19th century
(Pfeffer 1904) that provide chemotactic signals which guide sperm to the
archegonium. Each of these interactions might be studied further using
analysis by means of mutants and the techniques of molecular biology.

7.2 Cases studies in bryophyte development

We recognize bryophytes as a group based on a similar, bryophytic, grade
of morphological organization; DNA-based phylogeny tells us the three
kinds of bryophytes are as distinct from each other as they are to all the
vascular plants combined (Mishler *et al.* 1992). While bryophytes encompass an enormous range of developmental complexity, from the presumably simple strap-like plants of *Riccia* with their embedded sporophytes,
to those mosses with elaborate peristomes, this complexity often has
simple geometry and involves responses at the level of the cell rather than
the organ. Coupled with the ability of bryophytes to regenerate from
small fragments, these properties make bryophytes admirably suited for
studies of basic questions of plant biology.

7.2.1 Protonemal switching

While it is a convenient didactic tool to draw parallels between the filamentous protonemata of mosses and the filamentous algae, the filamentous protonema characteristic of so many mosses is actually a derived
condition. Nevertheless, the change in protonemal physiology as the
protonemal mat ages is a central feature of the developmental biology of
most mosses, and a wonderful example of that most elusive of features in
plant biology, competence to respond to hormones. While not all mosses
show the clear morphological distinctions seen in *Funaria*, the basic
pattern of development in *Funaria* is true for all mosses (Table 7.1). Spores
germinate and the protonemal filaments during this first stage of the life
cycle are chloronema, chloroplast-rich cells with perpendicular end-
walls. Cells in these filaments respond to exogenous auxin, making caulonemal side branches, respond to ABA by dividing and becoming brood
cells, and respond to the withdrawal of calcium by an unequal division
and maturation into tmema. While low concentrations of cytokinin stimulate the formation of chloronemal side branches, chloronemal filaments
do not respond to cytokinin exposures by forming shoot buds.

Table 7.1. *Chloronema and caulonema compared (expanded from list of Bopp 1990)*

	Chloronema	Caulonema
Morphological	branching irregular transverse septa	branching regular oblique septa
Cytological	walls colorless round chloroplasts	walls pigmented spindle-shaped chloroplasts
Physiological	positive phototropism auxins give caulonema no buds after cytokinin brood cells with ABA tmema in low Ca^{2+}	negative phototropism anti-auxins give chloronema buds from cytokinin treatment no brood cells after ABA does not become tmema
Molecular	other *myb*-sequences –	certain *myb*-like genes novel extracellular proteins

In contrast, caulonemal filaments, as their name suggests, do make shoot buds when exposed to adequate levels of cytokinin. Recent molecular work has shown that chloronema and caulonema differ in expression of *myb*-like sequences, and that the formation of shoot buds involves the secretion of characteristic proteins (Leech *et al.* 1993, Neuenschwander *et al.* 1994). In species where this change in responsiveness to cytokinin correlates with changes in morphology and cytology, caulonemal filaments can be recognized by the presence of wall pigments, spindle-shaped chloroplasts, and oblique end-walls. The ability to respond to cytokinin is not a simple addition to the progressive developmental program of the protonema, but reflects a profound rearrangement of the physiology and hormone-responsiveness of the protonema. Caulonema does not respond to ABA by making brood cells, and it does not respond to low calcium by making tmema. Caulonema and chloronema even differ in their phototropic responses (Table 7.1).

The transition from chloronema to caulonema involves changes in the tip cells of the filaments in a protonemal mat (or in the way the daughter cell matures), but does not usually include changes in the existing cells of the filaments. It is possible, for example, to grow *Funaria* protonema on filters in the presence of anti-auxins, with light provided from only one side, to produce an extended mass of parallel chloronemal filaments. Transferred to auxin and cytokinin, cells are added to the filaments by the division of the tip cell. These newly produced cells will form buds. Caulonemal side branches will form in the chloronemal portion of the filaments after the transfer to auxin and cytokinin, and the cells in these side branches will form buds. The chloronemal cells in the original filaments, however, do not form buds, showing that the physiology and responsiveness of chloronema and caulonema is a property of individual cells, not filaments, and that this physiology is not communicated from cell to cell.

Similarly, observations of the normal development of protonema reveal that buds form only from caulonemal cells or from the presumably chloronemal cell at the base of a filament of secondary chloronema (a chloronemal side branch from a caulonemal filament) (Saunders & Hepler 1982). While the strict cell autonomy of caulonemal and chloronemal physiology characterizes *Funaria* and *Funaria*-type mosses, other named types of protonemal development show evidence for a more global conversion of the protonemal mat (Bhatla 1994). The moss *Physcomitrella*, for example, shows distinct chloronemal and caulonemal morphologies and the ability of protonemal mats to form shoot buds does depend on the presence of caulonemal filaments. Although controlled growth conditions may restrict the bud-forming response to caulonema (Cove & Ashton 1984), *Physcomitrella* will form buds from chloronema as well as caulonema under many culture conditions (Reski & Abel 1985, Bhatla 1994).

A variety of mutants blocked in the transition from chloronema to caulonema have been reported (Cove & Ashton 1984, Bhatla *et al.* 1996, Marsh & Christianson 1997). Genetic and biochemical analysis has found both predicted classes of mutant, those with defects in auxin biosynthesis, as well as those with increased rates of auxin turnover (Cove & Ashton 1984). Analysis of mutants blocked as chloronema has also recovered defects in adenylate cyclase, suggesting the involvement of cAMP in this transition (Cove & Ashton 1984). It should be noted, however, that while a biochemical assay confirmed the absence of adenylate cyclase activity in the mutant, assays using other nucleotides were not done, raising the possibility of the involvement of other cyclic nucleotides (cGMP for example) in the transduction of the auxin signal. G-proteins have been linked to auxin responses in angiosperms, and it seems likely that the transduction mechanisms in mosses would also involve G-proteins.

As might be predicted from the two main mechanisms, defects in auxin biosynthesis and increased rates of auxin catabolism, some chloronemal mutants can make the transition to caulonema if exogenous auxin is supplied (those with defects in auxin biosynthesis), while others will not, even when very high concentrations of auxin are supplied (Bhatla *et al.* 1996). Experiments with antagonists of intracellular calcium (TMB-8) and inhibitors of calmodulin (TFP) demonstrate that the auxin-mediated transition from chloronema to caulonema in the wild-type requires calcium and involves calmodulin (Bhatla *et al.* 1996). Chloronemal mutants that can make the transition when supplied with exogenous auxin can also be triggered by high levels of calcium, either as elevated amounts of calcium in the medium or by treatment with calcium iono- phore. In contrast, those mutants that cannot have their ability to make caulonema restored by treatment with exogenous auxin do not have that ability restored by high external levels of calcium. While treatment of such mutants with calcium ionophore does result in the formation of oblique septa, a cytological detail characteristic of caulonema, treated cul- tures of such mutants do not acquire the ability to make shoot buds.

Given the recent isolation of receptors for cytokinin (Kakimoto 1996) and key genes in the cytokinin and auxin response pathways in vascular plants (Estelle 1998, Gray & Estelle 1998), bryologists have a wonderful opportunity to use the power of molecular biology to understand how responsiveness to one hormone arises while responsiveness to others dis- appears. With proper culture conditions, the conversion of chloronema into brood cells or tmema is both rapid and extensive. Chloronema of *Funaria* grown on anti-auxins can have more than 99% of the cells con- verted into brood cells within 72 hours, and a similarly high percentage of cells converted to tmema over 96 hours (Christianson, Squier, & Cart- wright, unpublished data).

7.2.2 Gametophore formation on moss protonema
The formation of shoot buds along protonemal filaments may be one of the most studied events in the life cycle of mosses and has been the subject of many reviews and book-length treatments (Bopp 1990, Bhatla 1994). The major features are well known to most bryologists and, as an example of the dramatic response of plants to cytokinin, to plant biologists in general. Once protonema contain caulonemal filaments (see above), expo- sure to exogenous cytokinin leads within minutes to an influx of calcium, and within hours to nuclear migration and division in target cells, result- ing, by some 72 hours, in a five- to six-celled young shoot bud, containing a distinct apical cell (Saunders & Hepler 1982).

Much of the interest in this process has focused on the initial percep- tion of cytokinin and the subsequent rapid influx of calcium. Electro- physiological measurements and the use of calcium-sensitive dyes document that influx of calcium begins within minutes and that external currents return to basal levels perhaps within 1–2, and certainly within

24, hours (Saunders & Hepler 1981, Saunders 1986). Other experiments show a mobilization of calcium (Saunders 1986) as well as cytokinin (Brandes & Kende 1968) to target cells; calcium shows further localization within the target cells themselves (Saunders 1986).

This influx of calcium is now known to involve channels controlled by membrane potential. Dihydropyridines (DHP) interact with these channels, and experiments using DHP agonists result in enormous numbers of buds forming along the caulonema of *Funaria*, while experiments using DHP antagonists in the presence of cytokinin block bud formation almost completely (Conrad & Hepler 1988). Other experiments characterized these channels in chloronemal protoplasts of *Physcomitrella*, demonstrating that calcium uptake is cytokinin-dependent and that the degree to which various cytokinins stimulate calcium uptake correlates with the degree to which they stimulate bud formation (Schumaker & Gizinski 1993). There seems to be a single DHP-binding protein in *Physcomitrella* protoplasts, localized to the plasma membrane; DHP binding is calcium-dependent, and is stimulated by cytokinin (Schumaker & Gizinski 1995). The binding of DHP can be stimulated by a nonhydrolyzable GTP analog, and inhibited by GDP analogs, but only in partially permeabilized vesicles (Schumaker & Gizinski 1996). Such results suggest that interactions with the channel on the outer face of the plasma membrane are coupled to G-proteins on the inner face, and point toward the involvement of G-proteins in the signal transduction pathway leading to bud formation.

While cytokinin exposure triggers calcium uptake and is the first step in the signal transduction toward bud formation, a brief exposure to cytokinin is not sufficient to result in bud formation. Two kinds of experiments showed that cytokinin was also required to mediate some event well after the initial perception of the hormone. Brandes and Kende (1968) exposed protonemata to cytokinin, and observed the effects of washing out the cytokinin after various periods of time. If cytokinin was washed out before 72 hours had elapsed, some of the buds that were forming reverted to filamentous growth. Similarly, Saunders and Hepler (1982) observed that exposure to calcium ionophore A23187 would substitute for initial exposure to cytokinin, producing an influx of calcium, inducing nuclear migration, and resulting in the first divisions characteristic of bud formation. None of these nascent buds would complete development, however, and would revert to filamentous growth unless cytokinin was added to the medium.

This second interaction with cytokinin is responsible for the stable

commitment of nascent buds, the biological end-point of the signal transduction that began with the cytokinin-stimulated influx of calcium.
Recent experiments have shown that the well-known log-linear relationship between concentration of cytokinin and the number of buds
produced from protonematal colonies is controlled by this second interaction (Christianson 1998a). While bud formation requires calcium, this
second interaction has not been shown explicitly to require calcium
(although it may), and it is not necessary that both cytokinin-mediated
events use the same receptor. Indeed, there is some evidence that suggests
the two cytokinin receptors are chemically distinct.

A survey for the ability of phenylurea cytokinins to induce bud formation in *Funaria* finds that one phenylurea cytokinin, diphenylurea (DPU),
although active in other cytokinin bioassays, is almost completely unable
to stimulate bud formation in *Funaria* (Christianson & Hornbuckle 1999).
However, caulonema exposed to DPU and then transferred to the adenine
cytokinin benzyladenine (under conditions where additional buds
cannot begin to form) do develop buds. Since DPU can initiate bud formation, DPU must interact with the first cytokinin receptor. The failure of
DPU to stimulate complete bud formation must be due to an inability to
interact with the cytokinin receptor used during the second cytokinin-
mediated event. A simple experiment confirms the absence of an interaction with this second receptor by showing that DPU does not compete
with adenine cytokinins for this second receptor. Adding 20-fold molar
excess DPU to small concentrations of the adenine cytokinin benzyladenine does not result in reduced numbers of buds as it would if DPU and
benzyladenine competed for binding sites on this second receptor (Christianson & Hornbuckle 1999).

While it is true that cytokinin triggers the formation of buds in moss
caulonema, the details uncovered over the past 30 years make it clear that
the trigger is not simple. Fortunately, this increasingly complicated scenario (Fig. 7.1) may still be simple enough to be completely understood.
There are still more details to be added, of course. The cytokinin-stimulated formation of shoot buds in moss is known to be inhibited by
simultaneous treatment with other hormones. Abscisic acid (ABA), for
example, inhibits bud formation and this inhibition is so quantitative
that it can be used as a bioassay to determine ABA concentrations
(Valadon & Mummery 1971). Experiments moving caulonema between
media with cytokinin alone and cytokinin plus ABA find that ABA does
not interfere with the initial response to cytokinin; since transfers to

a. *First View*

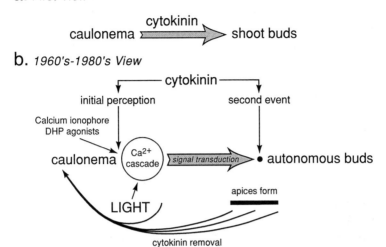

b. *1960's-1980's View*

c. *Current View*

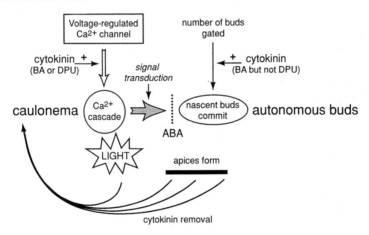

Fig. 7.1. The process of bud formation in moss. The observation that conditioned medium contained a factor that facilitated bud formation and the subsequent identification of the factor as cytokinin was a major conceptual advance. Later experiments showed that cytokinin was required at two distinct times in the process. Recent work characterizing both the initial perception of hormone and the second involvement of cytokinin is resulting in less emphasis on the hormone itself and more emphasis on the developmental events triggered by the hormone.

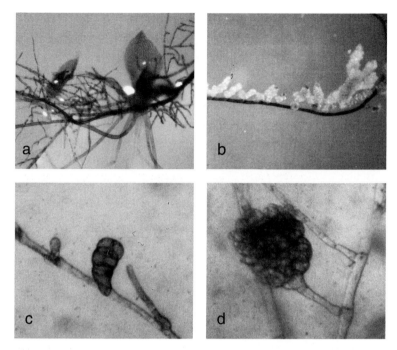

Fig. 7.2. Buds formed by wild-type and mutant strains of *Funaria hygrometrica*.
Exposure of protonemal filaments to mutagens followed by mass culture and
visual selection for areas showing atypical bud formation has recovered a
number of mutant phenotypes. UV light, the chemical mutagen N-nitroso-N-
methylurea, and the ICR acridine frameshift mutagens have all been shown to
be effective. (a) Wild-type strain, showing formation of leafy buds along
caulonemal filaments. (b) Mutant strain #9 produces aberrant buds, and
determinant axile structures with lateral branching but without any leaf
formation. (c) Mutant strain #5mtr1 produces buds that arrest without
establishing an apical cell. (d) Mutant strain #5mtr3 produces buds arrested
after the formation of the first merophytes.

cytokinin plus ABA result in the same numbers of buds as transfers to
medium without any cytokinin, ABA seems to prevent the second
cytokinin-mediated event and the final commitment of nascent buds
(Christianson & Marsh 1997).

It is possible to recover mutants impaired in their ability to form buds
(Fig. 7.2). Now that transformation of mosses is routine in several laboratories (Schaefer *et al.* 1991; Christianson & Archibald, unpublished data), it
is even possible to envisage recovering such mutants via insertional
mutagenesis, a technique used to isolate genes in *Arabidopsis* (Feldmann *et
al.* 1989). The isolation and characterization of the genes used to make
shoot buds in moss will fill in the details of the physiology of bud forma-

tion even as what we already know about the physiology will facilitate and inform the characterization and testing of the mutant strains.

7.3 Towards the future

The bryophytes are regaining the serious attention from plant scientists that these fascinating plants so deserve. Reviews of aspects of bryophyte biology are appearing in journals that usually feature articles on *Arabidopsis* (Christianson 1996, Cove *et al.* 1997, Schumaker & Dietrich 1997). The very first articles reporting the conservation of signaling mechanisms between the mosses and the angiosperms have now appeared, and molecular biologists are about to realize (in both meanings of the word) the potential of the bryophytes for the fine dissection of phenomena originally identitifed in model angiosperms (Knight *et al.* 1995).

Scientists interested in the evolution of gene families and the diversification of gene function have begun to include the bryophytes as part of their surveys of the plant kingdom. The small heat-shock genes, for example, have been characterized in the moss *Funaria* and this work shows that these genes diversified prior to the divergence of the mosses and the vascular plants (Waters & Vierling 1999). Members of the small heat-shock genes are differentially expressed in seeds, flowers, and other angiosperms organs, and that observation had led to speculation that the emergence of such specialized organs favored the duplication and divergence of family members. Discovering that the duplication and divergence had already occurred in the common ancestor of the mosses and angiosperms certainly eliminates flowers and seeds from speculations about causes of the diversification.

The future seems likely to include cases where a body of knowledge accumulated by bryologists will now be extended to encompass the angiosperms. I certainly hope that work on shoot formation in mosses will give us new insights into the processes used by angiosperm embryos as they form shoot apices, for example. Desiccation-tolerant mosses have been known for a long time and the biology, both cellular and molecular, of this response is being unraveled by Oliver and his collaborators (Oliver 1996). The strategy used by the moss *Tortula* is quite different from the strategies used for desiccation tolerance in the angiosperms (the response in *Tortula* is not mediated by the hormone ABA, for example). A large number of desiccation-responsive genes have now been identified in *Tortula*, and this work is now being included in reviews on plant stress that previously focused only on angiosperms (compare Ingram & Bartels

1996 to previous *Annual Reviews* in this area). While it may seem far-fetched to imagine improved stress tolerance in crop plants tracing back to research on mosses, such applied work is certainly possible to envisage. Indeed, using transgenic technology to introduce a novel stress metabolite from one angiosperm into another angiosperm did result in transgenic plants with increased stress tolerance (Bohnert & Sheveleva 1998). Similar work introducing moss metabolism into angiosperms may well be successful.

This chapter has touched on aspects of an enormous subject. The key features, however, are simple. Plant biology is conserved, and the three types of bryophytes, as the basal clades of the four clades of green plants, preserve three-fourths of what it is to be a land plant. Bryophytic organization means that responses to environmental and developmental signals are majorly the responses of cells, and not the responses of organs or tissues. The open nature of the bryophyte plant, from the ability to regenerate from fragments to the fact that these plants use the local environment as a natural part of their developmental and metabolic signaling system, simplifies the design and analysis of many experiments. The future is full of research opportunity. Bryologists should expect to be surprised by all the attention and by the influx of naïve scientists who may look at familiar organisms in unsettling but productive ways.

Acknowledgment

My work on the developmental biology of mosses has been supported by grants from the Kansas University General Research Fund, NSF OSR-9550487 and matching support from the State of Kansas. I thank Sharon Hagen for final preparation of the figures.

REFERENCES

Bartel, B. (1997). Auxin biosynthesis. *Annual Review of Plant Physiology and Plant Molecular Biology*, **48**, 51–66.

Bandurski, R. S., Cohen, J. D., Slovin, J. P., & Reinecke, D. M. (1995). Auxin biosynthesis and metabolisim. In *Plant Hormones: Physiology, Biochemistry and Molecular Biology*, ed. P. J. Davies, pp. 39–65. Dordrecht: Kluwer.

Beutelmann, P. & Bauer, L. (1977). Purification and identification of a cytokinin from moss callus cells. *Planta*, **133**, 215–17.

Bhatla, S. C. (1994). *Moss Protonema Differentiation*. New York: Wiley.

Bhatla, S. C., Kapoor, S., & Khurana, J. P. (1996). Involvement of calcium in auxin-induced cell differentiation in the protonema of the wild strain and auxin mutants of the moss *Funaria hygrometrica*. *Journal of Plant Physiology*, **147**, 547–52.

Bohnert, H. J. & Sheveleva, E. (1998). Plant stress adaptations—making metabolism move. *Current Opinion in Plant Biology*, 1, 267–74.

Bopp, M. (1963). Development of the protonema and bud formation in mosses. *Journal of the Linnean Society (Botany)*, 58, 305–9.

(1990). Hormones of the moss protonema. In *Bryophyte Development: Physiology and Biochemistry*, ed. R. N. Chopra & S. C. Bhatla, pp. 55–77. Boca Raton: CRC Press.

Bopp, M. & Werner, O. (1993). Abscisic acid and desiccation tolerance in mosses. *Botanica Acta*, 106, 103–6.

Brandes, H. & Kende, H. (1968). Studies on cytokinin-controlled bud formation in moss protonema. *Plant Physiology*, 43, 827–37.

Briggs, W. R. & Liscum, E. (1997). The role of mutants in the search for the photoreceptor for phototropism in higher plants. *Plant Cell and Environment*, 20, 768–72.

Cashmore, A. R. (1997). The cryptochrome family of photoreceptors. *Plant Cell and Environment*, 20, 764–767.

Chopra, R. N. & Kumar, P. K. (1988). *Biology of Bryophytes*. New York: Wiley.

Christianson, M. L. (1996). Morphogenesis and the coordination of cell division in the bryophytes. *Seminars in Cell and Developmental Biology*, 7, 881–9.

(1998a). The quantitative response to cytokinin in the moss *Funaria hygrometrica* does not reflect differential sensitivity of initial target cells. *American Journal of Botany*, 85, 144–8.

(1998b). The mosses *Funaria hygrometrica* and *Ceratodon purpureus* use different molecules to regulate growth of adjacent protonema. *American Journal of Botany*, 85, S7.

Christianson, M. L. & Hornbuckle, J. S. (1999). Phenylurea cytokinins assayed for induction of shoot buds in the moss *Funaria hygrometrica*. *American Journal of Botany*, 86, 1645–8.

Christianson, M. L. & Marsh, I. W. (1997). Studies on the abscisic acid inhibition of bud formation in the moss *Funaria hygrometrica*. *American Journal of Botany*, 84, S39–40.

Collier, P. A. & Hughes, K. W. (1982). Life cycle of the moss, *Physcomitrella patens*, in culture. *Journal of Tissue Culture Methods*, 7, 19–22.

Conrad, P. A. & Hepler, P. K. (1988). The effect of 1,4-dihydropyridines on the initiation and development of gametophore buds in the moss *Funaria*. *Plant Physiology*, 86, 684–7.

Cove, D. J. & Ashton, N. W. (1984). The hormonal regulation of gametophytic development in bryophytes. In *The Experimental Biology of Bryophytes*, ed. A. F. Dyer & J. G. Duckett, pp. 177–201. London: Academic Press.

Cove, D. J., Knight, C. D., & Lamparter, T. (1997). Mosses as model systems. *Trends in Plant Science*, 2, 99–105.

Crandall-Stotler, B. (1980). Morphogenetic designs and a theory of bryophyte origins. *BioScience*, 30, 580–5.

Creelman, R. A. & Mullet, J. E. (1997). Oligosaccharins, brassinolides, and jasmonates: nontraditional regulators of plant growth, development and gene expression. *Plant Cell*, 9, 1211–23.

Dietert, M. F. (1980). The effect of temperature and photoperiod on the development of geographically isolated populations of *Funaria hygrometria* and *Weissia controversa*. *American Journal of Botany*, 67, 369–80.

Esau, K. (1965). *Plant Anatomy*, 2nd edn. New York: Wiley.

Estelle, M. (1998). Cytokinin action: two receptors better than one? *Current Biology*, **16**, 539–41.

Feldmann, K. A., Marks, M. D., Christianson, M. L., & Quatrano, R. S. (1989). A dwarf mutant of *Arabidopsis* generated by T-DNA insertion mutagenesis. *Science*, **243**, 1351–4.

Fredericq, H., Veroustraete, F., DeGreef, J., & Rethy, R. (1977). Light enhanced ethylene production in *Marchantia polymorpha* L. *Archives Internationales de Physiologie et de Biochimie*, **85**, 977–8.

Gray, W. M. & Estelle, M. (1998). Biochemical genetics of plant growth. *Current Opinion in Biotechnology*, **9**, 196–201.

Hartmann, E. & Weber, M. (1988). Storage of the phytochrome-mediated phototropic stimulus of moss protonemal tip cells. *Planta*, **175**, 39–49.

Hashimoto, T., Tori, M., & Asakawa, Y. (1988). A highly efficient preparation of lunularic acid and some biological activities of stilbene and dihydrostilbene derivatives. *Phytochemistry*, **27**, 109–13.

Ingram, J. & Bartels, D. (1996). The molecular basis of dehydration tolerance in plants. *Annual Review of Plant Physiology and Plant Molecular Biology*, **47**, 377–403.

Kakimoto, T. (1996). CKI1, a histidine kinase homologue implicated in cytokinin signal transduction. *Science*, **274**, 982–5.

Katekar, G. F., Venis, M. A., & Geissler, A. E. (1993). Binding of lunularic acid, hydrangeic acid and related compounds to the receptor for 1-N-naphthyl-phthalmic acid. *Phytochemistry*, **32**, 527–31.

Kende, H. & Zeevaart, J. A. D. (1997). The five "classical" plant hormones. *Plant Cell*, **9**, 1197–210.

Knight, C. D., Sehgal, A., Atwal, K., Wallace, J. C., Cove, D. J., Coates, D., Quatrano, R. S., Bahdur, S., Stockley, P. G., & Cumming, A. C. (1995). Molecular responses to abscisic acid are conserved between moss and cereals. *Plant Cell*, **7**, 499–506.

Kunz, S. & Becker, H. (1992). Bibenzyl glycosides from the liverwort *Ricciocarpos natans*. *Phytochemistry*, **31**, 3981–3.

Leech, M. J., Kammerer, W., Cove, D. J., Martin, C., & Wang, T. L. (1993). Expression of *myb*-related genes in the moss, *Physcomitrella patens*. *Plant Journal*, **3**, 51–61.

Marsh, I. W. & Christianson, M. L. (1997). Growth rates and cell sizes in wild-type and auxin-insensitive chloronemal cell lines of the moss *Funaria hygrometrica*. *American Journal of Botany*, **84**, s25–6.

Mathews, S. & Sharrock, R. A. (1997). Phytochrome gene diversity. *Plant Cell and Environment*, **20**, 666–71.

Milborrow, B. V. & Rubery, P. H. (1985). The specificity of the carrier-mediated uptake of ABA by root segments of *Phaseolus coccineus* L. *Journal of Experimental Botany*, **36**, 807–22.

Mishler, B. D. & Newton, A. E. (1988). Influence of mature plants and desiccation on germination of spores and gametophytic fragments of *Tortula*. *Journal of Bryology*, **15**, 327–42.

Mishler, B. D., Thrall, P. H., Hopple, J. S., Jr., De Luna, E., & Vilgalys, R. (1992). A molecular approach to the phylogeny of bryophytes: cladistic analysis of chloroplast-encoded 16S and 23S ribosomal genes. *Bryologist*, **95**, 172–80.

Mitra, G. C. & Allsopp, A. (1959). The effects of various physiologically active substances on the development of the protonemata and bud formation in *Pohlia nutans* (Hedw.) Linb. *Phytomorphology*, **9**, 64–71.

Monroe, J. H. (1965). Some factors evoking formation of sex organs in *Funaria*. *Bryologist*, **68**, 337–9.

Neuenschwander, U., Fleming, A. J., & Kuhlemeier, C. (1994). Cytokinin induces the developmentally restricted synthesis of an extracellular protein in *Physcomitrella patens*. *Plant Journal*, **5**, 21–31.

Ohta, Y., Abe, S., Komura, H., & Kobayashi, M. (1984). Prelunularic acid in liverworts. *Phytochemistry*, **23**, 1607–9.

Oliver, M. J. (1996). Desiccation tolerance in vegetative plant cells. *Physiologia Plantarum*, **97**, 779–87.

Osborne, D. J., Walters, J., Milborrow, B. V., Norville, A., & Stange, L. M. C. (1995). Evidence for a non-ACC ethylene biosynthesis pathway in lower plants. *Phytochemistry*, **42**, 51–60.

Pasentsis, K., Paulo, N., Algarra, P., Dittrich, P., & Thummler, F. (1998). Characterization and expression of the phytochrome gene family in the moss *Ceratodon purpureus*. *Plant Journal*, **13**, 51–61.

Pfeffer, W. (1904). *Planzenphysiologie*. vol. 2. Leipzig: Wilhelm Englemann.

Quail, P. R. (1997). An emerging molecular map of the phytochromes. *Plant Cell and Environment*, **20**, 657–65.

Radwanski, E. R. & Last, R. L. (1995). Tryptophan biosynthesis and metabolism: biochemical and molecular genetics. *Plant Cell*, **7**, 921–34.

Reski, R. & Abel, W. O. (1985). Induction of budding on chloronema and caulonema of the moss, *Physcomitrella patens*, using isopentenyladenine. *Planta*, **165**, 354–8.

Rhodes, D., Jamieson, G., & Christianson, M. L. (1982). Analysis of (^{15}N)H$_4^+$ assimilation in suspension cultures of the field bindweed, *Convolvulus arvensis* L., using combined gas chromatography and computer simulation techniques. *In Vitro*, **18**, 296.

Rohwer, F. & Bopp, M. (1985). Ethylene synthesis in moss protonema. *Journal of Plant Physiology*, **117**, 331–8.

Saunders, M. J. (1986). Cytokinin activation and redistribution of plasma-membrane ion channels in *Funaria*. *Planta*, **167**, 402–9.

Saunders, M. J. & Hepler, P. K. (1981). Localization of membrane-associated calcium following cytokinin treatment in *Funaria* using chlorotetracycline. *Planta*, **152**, 272–81.

(1982). Calcium ionophore A23187 stimulates cytokinin-like mitosis in *Funaria*. *Science*, **217**, 943–5.

Schaefer, D., Zrÿd, J.-P., Knight, C. D., & Cove, D. J. (1991). Stable transformation of the moss *Physcomitrella patens*. *Molecular and General Genetics*, **226**, 418–24.

Schneider-Poetsch, H. A. W., Marx, S., Kolukisaogu, U., Hanelt, S., & Braun, B. (1994). Phytochrome evolution: phytochrome genes in ferns and mosses. *Physiologia Plantarum*, **91**, 241–50.

Schumaker, K. S. & Dietrich, M. A. (1997). Programmed changes in form during moss development. *Plant Cell*, **9**, 1099–107.

Schumaker, K. S. & Gizinski, M. J. (1993). Cytokinin stimulates dihydropyridine-sensitive calcium uptake in moss protoplasts. *Proceedings of the National Academy of Science USA*, **90**, 10937–41.

(1995). 1,4–Dihydropyridine binding sites in moss plasma membranes. *Journal of Biological Chemistry*, **270**, 23461–7.

(1996). G proteins regulate dihydropyridine binding to moss plasma membranes. *Journal of Biological Chemistry*, **272**, 21292–6.

Skoog, F. & Miller, C. O. (1957). Chemical regulation of growth and organ formation in plant tissues cultured *in vitro*. *Symposium of the Society for Experimental Biology*, **11**, 118–31.

Speiss, L. D., Lippincott, B. B., & Lippincott, J. A. (1971). Development and gametophore initiation in the moss *Pylaisiella selwynii* as influenced by *Agrobacterium tumefaciens*. *American Journal of Botany*, **58**, 726–31.

(1976). The requirement of physical contact for moss gametophore induction by *Agrobacterium tumefaciens*. *American Journal of Botany*, **63**, 324–8.

(1984). Role of the moss cell wall in gametophore formation induced by *Agrobacterium tumefaciens*. *Botanical Gazettte*, **145**, 302–7.

Sun, G., Dilcher, D. L., Zheng, S., & Zhou, Z. (1998). In search of the first flower: a Jurassic angiosperm, *Archaefructus*, from northeast China. *Science*, **282**, 1692–5.

Sztein, A. E., Cohen, J. D., Slovin, J. P., & Cooke, T. J. (1995). Auxin metabolism in representative land plants. *American Journal of Botany*, **82**, 1514–21.

Thummler, F., Beetz, A., & Rudiger, W. (1990). Phytochrome in lower plants. Detection and partial sequence of a phytochrome gene in the moss *Ceratodon purpureus* using the polymerase chain reaction. *FEBS Letters*, **275**, 125–129.

Thummler, F., Dufner, M., Kreisl, P., & Dittrich, P. (1992). Molecular cloning of a novel phytochrome gene of the moss *Ceratodon purpureus* which encodes a putative light-regulated protein kinase. *Plant Molecular Biology*, **20**, 1003–17.

Thummler, F., Herbst, R., Algarra, P., & Ullrich, A. (1995). Analysis of the protein kinase activity of moss phytochrome expressed in fibroblast cell culture. *Planta*, **197**, 592–6.

Vaarama, A. & Taren, N. (1963). On the separate and combined effects of calcium, kinetin and gibberellic acid on the development of moss protonemata. *Journal of the Linnean Society (Botany)*, **58**, 297–304.

Valadon, L. R. G. & Mummery, R. S. (1971). Quantitative relationship between various growth substances and bud production in *Funaria hygrometrica*. A bioassay for abscisic acid. *Physiologia Plantarum*, **24**, 232–4.

Wada, M. & Kadota, A. (1989). Photomorphogenesis in lower green plants. *Annual Review of Plant Physiology and Plant Molecular Biology*, **40**, 169–91.

Wang, T. L., Horgan, R., & Cove, D. J. (1981). Cytokinins from the moss *Physcomitrella patens*. *Plant Physiology*, **68**, 735–8.

Waters, E. R. & Vierling, E. (1999). The diversification of plant cytosolic heat shock proteins preceded the divergence of mosses. *Molecular Biology and Evolution*, **16**, 127–39.

Watson, M. A. (1981). Chemically mediated interactions among juvenile mosses as possible determinants of their community structure. *Journal of Chemical Ecology*, **7**, 367–76.

Werner, O., Espin, R. M. R., Bopp, M., & Atzorn, R. (1991). Abscisic-acid-induced drought tolerance in *Funaria hygrometrica* Hedw. *Planta*, **186**, 99–103.

Yang, S. F. & Hoffman, N. E. (1984). Ethylene biosynthesis and its regulation in higher plants. *Annual Review of Plant Physiology*, **35**, 155–89.

Yu., L. P. & Christianson, M. L. (1995). The ABA-triggered formation of brood cells in *Funaria hygrometrica* requires light and calcium. *American Journal of Botany*, **82**, s28–9.

Zeevaart, J. A. D. & Creelman, R. A. (1988). Metabolism and physiology of abscisic acid. *Annual Review of Plant Physiology and Plant Molecular Biology*, **39**, 439–73.

8

Physiological ecology

8.1 Introduction

Bryophytes share much of their physiology with other green land plants, but there are also important differences; the similarities and differences do not necessarily fall in line with simple expectations. Because most bryophytes have "stems" and "leaves," and tradition has regarded them as "lower" plants, it is too easy to think of them as underdeveloped miniatures of vascular plants – as organisms that have evolutionarily not yet "made the grade." Raven (1977, 1984) has emphasized the importance of supracellular transport systems in the evolution of land plants, and the physiological correlates that we must read alongside the anatomical structures of fossil plants. But the highly differentiated supracellular conducting systems exemplified by xylem and phloem are really only a prerequisite for *large* land plants. For simple physical reasons of scale, conduction of water and metabolites in bryophytes can be much more diffuse. Similarly, bryophyte and vascular-plant leaves and leafy canopies must be thought through and compared as photosynthetic systems from first principles, not by simple analogies between structures operating at radically different scales, which can be seriously misleading.

In adapting to the erratic subaerial supply of water, vascular land plants evolved xylem, bringing water from the soil to meet the needs of the above-ground shoots and leaves. Bryophytes in general adopted the alternative strategy of evolving desiccation tolerance, photosynthesizing and growing during moist periods and suspending metabolism during times of drought. These two patterns of adaptation are in many ways complementary. On the one hand, bryophytes are limited by their lack of roots, but their poikilohydric habit means that they can colonize hard and impermeable surfaces like tree trunks and rock outcrops, impenetrable to roots, from which vascular plants are excluded. Bryophytes typically take up water and nutrients over the whole surface of the shoots. They efficiently intercept and absorb solutes in rainwater, cloud and mist droplets, and airborne dust. This ability underlies both their conspicuous success in many nutrient-limited habitats and the vulnerability of many species to atmospheric pollution.

Physiologically, bryophytes are neither simple nor primitive. The divergence of bryophytes and the various vascular-plant groups goes back to the early history of plant life on land – certainly 400 million years, and probably longer (Edwards 1998). Mosses, Hepaticae, and Anthocerotae may well have been evolutionarily independent for equally long. Thus bryophytes should be seen not as primitive precursors of vascular plants, but as the diverse and highly evolved representatives of an alternative strategy of adaptation, making up a prominent part of the vegetation in such habitats as oceanic temperate forests, tropical cloud forests, bogs and fens, and polar and alpine fellfields and tundras.

The physiological ecology of bryophytes has been the subject of a number of reviews in recent years (Longton 1981, 1988, Proctor 1981a, 1982, 1990); mineral nutrition and pollution responses are reviewed by Brown (1982, 1984), Brown and Bates (1990), Bates and Farmer (1992) and Bates in chapter 9 of this volume. Bryophyte production, and its responses to major environmental factors, has been reviewed by Russell (1990), Frahm (1990), Vitt (1990) and Sveinbjörnsson and Oechel (1992). The present chapter concentrates on some of the ecophysiological features more particularly characteristic of bryophytes.

8.2 Bryophyte water relations

Vascular plants have internal water conduction, and the surface of the leaves and young stems is typically covered with a more or less waterproof and water-repellent cuticle. In bryophytes, much of the movement of water about the shoots takes place externally, in capillary spaces around the leaf bases, within a felt of rhizoid tomentum or paraphyllia on the stems, or in the channels between papillae on the leaves. To be sure, some bryophytes have well-developed internal conducting structures (Hébant 1977), but it is only in a limited number of large acrocarpous mosses that anything approaching a vascular-plant transpiration stream is seen.

Buch (1945, 1947) distinguished *endohydric, mixohydric,* and *ectohydric* bryophytes. In endohydric species, exemplified by the tall, robust *Polytri-chum* and *Dawsonia* species and the large Mniaceae, the stem possesses a well-developed central strand of hydroids, and much or most water conduction is internal. But even in these large endohydric mosses conduction in the open lumina of the hydroids in the central strand of the stem is only part of the story. Significant conduction takes place externally in the capillary spaces of sheathing leaf bases and rhizoid tomentum, and as in peripheral vascular-plant tissues at a comparable scale, much internal water movement must be relatively diffuse, within the cell walls, through the cells themselves, or some combination of the two; most water movement must be of this kind in the large marchantialean liverworts. Bryophytes are likely to be scarcely less complex in respect of tissue water movement than vascular plants (Proctor 1979*a*, Steudle & Petersen 1998). Many bryophytes, especially those of intermittently dry habitats, are ectohydric. Water conduction is predominantly external, in an interconnecting network of capillary spaces on the outer surface of the plant. These include the spaces within sheathing leaf bases, in the concavities of overlapping imbricate leaves as in *Scleropodium* or *Pilotrichella*, within felts of rhizoids or paraphyllia that cover the stem in such genera as *Philonotis* and *Thuidium*, in the interstices between the papillae that cover the leaf surfaces in, e.g., *Encalypta*, *Syntrichia*, and *Anomodon*, and between tightly-packed shoots or between shoots and the substrate. The external water in ectohydric bryophytes is as much an essential part of the plant's physiological functioning as the water in the xylem of vascular plants. In some bryophytes, especially smaller terrestrial species, neither internal nor external conduction is overwhelmingly predominant. These mixohydric species rely on a (probably always variable) balance between the two. More detailed reviews of water conduction in bryophytes are given by Proctor (1979*a*, 1981*a*, 1982).

The cell water relations of bryophytes are essentially the same as those of other plant cells and are illustrated by the "Höfler diagram" of Fig. 8.1(a). In a fully turgid cell the osmotic potential Ψ_π is exactly balanced by the turgor pressure Ψ_P of the cell wall; the cell is externally in equilibrium with pure liquid water, and its water potential Ψ_w (by definition) is zero. If the external water potential becomes negative, the cell must lose water. The reduction in cell volume causes turgor pressure to fall and osmotic potential to become more negative (numerically greater). When the turgor pressure falls to zero, the water potential of the cell is equal to the osmotic potential of its contents. At any lower water content, osmotic potential

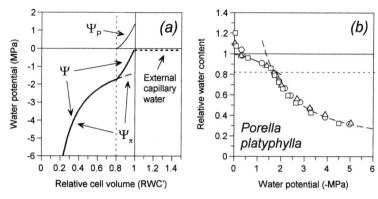

Fig. 8.1. (a) Höfler diagram for a bryophyte illustrating the relationship of cell water
potential (Ψ) and its components osmotic potential (Ψ_π) and turgor pressure
(Ψ_P) to cell water content and external capillary water. Based on the data of Fig.
8.1.(b). (b) The relation of relative water content to water potential for the leafy
liverwort *Porella platyphylla*, from thermocouple psychrometer measurements.
Water content was originally plotted as % dry weight, and the full-turgor point
estimated by inspection from the graph, as described by Proctor *et al.* (1998).
The horizontal dotted line indicates the turgor-loss point. A rectangular
hyperbola has been fitted to the data points below this, and a polynomial
regression to the points between full turgor and turgor loss. This graph is in
effect a Höfler diagram with water potential taken as the *x* axis, and matches
the presentation used by Proctor *et al.* (1998) and Proctor (1999). Compare Fig.
8.1(a) and the "pressure–volume" curve of Fig. 8.2(a).

and cell water potential are equal and inversely proportional to the
volume of water in the cell. The relation between osmotic potential and
cell volume plots onto the Höfler diagram as a rectangular hyperbola. The
relation of cell water potential to cell water content follows this hyperbola
up to the turgor-loss point. It then breaks away to follow a line, generally
slightly concave to the water-potential axis, to the full-turgor point, where
the relative water content (RWC) = 1.0 (by definition) and Ψ_w = 0. Practical
measurements are generally of tissues rather than individual cells, but if
the cells all have similar properties the same principles apply. Bryophyte
shoots generally carry some external water, held at small negative water
potentials determined by the dimensions of the capillary spaces in which
it is held. The effect of this water in a Höfler diagram for a bryophyte shoot
is illustrated by the dotted line in Fig. 8.1(a).

If one of the axes of the graph relating water potential to water content
is plotted on a reciprocal scale, the hyperbola of Fig. 8.1 becomes a straight
line. The graph of 1/Ψ against (1 – RWC) (Fig. 8.2(a)) is referred to as a pressure–volume (P–V) curve (Jones 1992). Turgor loss is marked by the point

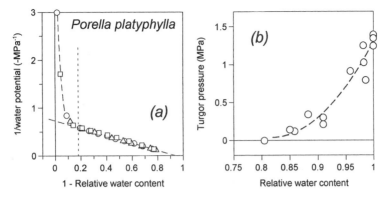

Fig. 8.2. (a) Pressure–volume graph from the same data as Fig. 8.1(b). Water content is plotted as 1 − RWC and decreases from left to right; the y axis is the reciprocal of water potential. Turgor loss is indicated by the vertical dotted line. A linear regression has been fitted to the points to the right of this. It intersects the y axis at the reciprocal of the full-turgor osmotic potential, the turgor-loss line at the reciprocal of the osmotic potential at turgor loss, and the x axis at a point which gives a measure of the effective osmotic volume of the cells. (b) The relation of turgor pressure to relative water content for *Porella platyphylla*, from thermocouple psychrometer measurements. The curve leaves the x axis at the turgor-loss point and cuts the y axis at the full-turgor osmotic potential. The slope of the curve gives a rough measure of the bulk modulus of elasticity ε_B of the tissues.

at which the relation of $1/\Psi$ to $(1 - RWC)$ breaks away from linearity, and the reciprocal of the osmotic potential at this point can be read from the graph. The intercept of the straight line on the $1/\Psi$ axis gives the reciprocal of the osmotic potential at full turgor; the intercept on the RWC axis is commonly taken as a measure of non-osmotic (or "apoplast") water but its exact significance is debatable (Proctor *et al.* 1998). From the data in the P–V curve, turgor pressure can be calculated for water contents between full turgor and the turgor-loss point (Fig. 8.2(b)). The steepness of slope of this curve (and the difference in water content between full turgor and turgor loss) depends on cell-wall extensibility, measured by the bulk elastic modulus ε_B (which itself generally varies continuously between turgor loss and full turgor, in a manner depending on the exact physical properties of the cell walls).

Some representative water-relations data for bryophytes are summarized in Table 8.1. Osmotic potentials at full turgor mostly lie between −1.0 and −2.0 MPa, but are generally less negative (numerically around half these values) in thalloid liverworts. *Metzgeria furcata*, matching leafy

Table 8.1. *Water-relations parameters of bryophytes*

Figures are in general mean ± s.d. from three or four replicates.

Species	Osmotic potential at full turgor (− MPa)	x-intercept of P–V curve (RWC)	RWC at turgor loss	Bulk elastic modulus ε_B at RWC 1.0 (MPa)	Water content at full turgor (% d.w.)	Water content blotted (% d.w.)
Targionia hypophylla	0.74 ± 0.03	−0.069 ± 0.003	0.70	n.d.	1003 ± 45	940 ± 37
Conocephalum conicum	0.54 ± 0.08	−0.002 ± 0.032	0.45	2.2 ± 0.8	1400 ± 132	1277 ± 108
Marchantia polymorpha	0.38 ± 0.02	0.052 ± 0.027	0.60	1.5*	1025 ± 35	956 ± 65
Dumortiera hirsuta	0.49 ± 0.05	−0.014 ± 0.023	0.90	7.6 ± 1.2	1636 ± 118	1628 ± 109
Metzgeria furcata	1.11 ± 0.03	0.043 ± 0.017	0.75	11.3 ± 0.7	300‡	363 ± 22
Pellia epiphylla	0.72 ± 0.08	−0.031 ± 0.032	0.80	4.8*	1020‡	1046 ± 157
Bazzania trilobata	1.41 ± 0.07	0.081 ± 0.025	0.80	17.3 ± 3.5	253 ± 6	300 ± 11
Porella platyphylla	1.37 ± 0.03	0.053 ± 0.013	0.80	13.3 ± 1.2	273 ± 5	312 ± 8
Frullania tamarisci	1.78 ± 0.20	0.189 ± 0.017	0.60	7.6 ± 0.5	134 ± 3	216 ± 7
Jubula hutchinsiae	1.02 ± 0.04	0.097 ± 0.010	0.70	6.3 ± 2.8	353 ± 21	353 ± 17
Andreaea alpina	1.59 ± 0.03	0.265 ± 0.006	0.70	6.8 ± 0.4	110 ± 4	141 ± 9
Polytrichum commune	2.09 ± 0.09	0.116 ± 0.023	0.75	19.2 ± 0.4	179 ± 6	186 ± 11
Dicranum majus	1.27 ± 0.04	0.126 ± 0.025	0.80	12.2 ± 1.2	185 ± 15	193 ± 7
Tortula ruralis	1.36 ± 0.18	0.266 ± 0.093	0.75	5.8 ± 1.5	108 ± 11	n.d.
Racomitrium lanuginosum	1.29 ± 0.08	0.224 ± 0.030	0.65	5.3 ± 1.9	121 ± 4	135 ± 5
Mnium hornum	1.21 ± 0.07	0.099 ± 0.049	0.70	6.1 ± 1.7	215 ± 7	175 ± 6
Antitrichia curtipendula	1.47 ± 0.28	0.175 ± 0.033	0.65	5.9 ± 0.6	152 ± 11	174 ± 16
Neckera crispa	1.27 ± 0.09	0.271 ± 0.092	0.65	7.7 ± 1.6	140 ± 5	150 ± 13
Hookeria lucens	0.95 ± 0.03	0.021 ± 0.004	0.70	6.2 ± 1.5	571 ± 42	n.d.
Anomodon viticulosus	1.65 ± 0.07	0.230 ± 0.009	0.65	8.5 ± 2.3	133 ± 3	176 ± 10
Homalothecium lutescens	2.08 ± 0.08	0.086 ± 0.054	0.70	18.8 ± 2.9	193 ± 15	218 ± 27
Rhytidiadelphus triquetrus	1.44 ± 0.16	0.136 ± 0.028	0.75	9.6 ± 1.3	182‡	n.d.
Rhytidiadelphus loreus	1.34 ± 0.02	0.237 ± 0.049	0.70	5.9 ± 1.2	142 ± 10	180 ± 13

Notes: The sign * indicates a single value from the combined data of all replicates, ‡ that the values from individual replicates were not distinguishable.
Sources: Data from Proctor *et al.* (1998) and Proctor (1999).

liverworts and mosses in ecological adaptation, is an interesting exception. There is no clear indication of more negative values in species of dry habitats; many of the more extreme published figures based on plasmolysis are certainly wrong. The intercept of the P–V curve on the water-content axis correlates with cell-wall thickness relative to the cell lumen; it is high in such species as *Andreaea alpina*, *Racomitrium lanuginosum*, and *Neckera crispa*, and low in, e.g., *Hookeria lucens* and the big thalloid liverworts. Water content at full turgor as a percentage of dry weight is also related to the amount of cell-wall material, and varies widely from about 100% dry weight in small desiccation-tolerant species of sun-baked rocks to 2000% or more in thalloid liverworts of moist habitats. Both these measures change as the shoots develop and mature, and are sensitive to the inclusion of moribund older material, so they vary with the seasons and can never be very precise. Relative water content at turgor loss and ε_B are also correlated, but somewhat loosely. By vascular-plant standards, bryophyte cell walls are typically rather readily extensible (low ε_B), especially in the large thalloid liverworts and some ectohydric mosses, but some mosses (e.g. *Polytrichum commune*, *Dicranum majus*, *Homalothecium lutescens*) and leafy liverworts (e.g. *Bazzania trilobata*, *Porella platyphylla*) show ε_B values which would pass unnoticed among herbaceous vascular plants (Zimmermann & Steudle 1978). Cell-wall extensibility also varies with time, ε_B increasing as the shoots mature.

The division between apoplast water in the cell walls, symplast water within the cells, and external capillary water, and especially the latter two, is important for several reasons (Dilks & Proctor 1979, Beckett 1996, Proctor *et al.* 1998). First (from the point of view of the physiological investigator), it is essential to know the full-turgor water content in order to calculate RWC values physiologically comparable with those for vascular plants; "RWC" values based on "saturated" water contents can be wholly misleading. As Table 8.1 shows, acceptable approximate estimates of full-turgor water content can often be obtained by carefully blotting samples of saturated shoots; underestimates may arise through thumb pressure expressing symplast water from large-celled species, and overestimates through incomplete removal of external water from species with intricate external capillary spaces, or the presence of large amounts of apoplast water. Second (from the point of view of the bryophyte), the external capillary water can be exceedingly important physiologically. Its significance in relation to external water movement has already been alluded to. External water is also of prime importance in relation to water storage, which in turn is a major determinant of the length of time the shoots remain turgid and able to photosynthesize and grow. It is often the major component of water associated with the plant, and it can vary widely without affecting the water status of the cells. It is common to find that external capillary water exceeds symplast water by a factor of five or more; a not especially wet-looking sample of the pendulous African forest moss *Pilotrichella ampullacea* that I took to make measurements for a P–V curve turned out to have a total water content corresponding to a RWC of more than 12! Most of this water would have been held in the concavities of the closely-overlapping "ampulla-like" leaves. The effect of external storage of large amounts of water is that for most of the time the shoots are either functioning at full turgor, or they are too dry to support metabolism, with only brief interludes at intermediate water potentials between these states. From the bryophyte's point of view, *any* habitat is "wet" during and following rain, and "dry" during drought. The only difference is in the relative times spent wet and dry; drought stress and drought tolerance as they affect vascular plants hardly enter the picture, and the drought metabolites of vascular plants are conspicuously absent in bryophytes. We should remember that desert ephemeral flowering plants are mesophytes, which flourish following occasional periods of rain and escape drought by means of their desiccation-tolerant seeds. Bryophytes escape drought by means of their desiccation-tolerant vegetative shoots. Desiccation-tolerant bryophytes and vascular desert ephemerals may equally be seen as "drought-escaping" plants. It is a paradox that "poikilohydric" bryophytes may spend less time metabolizing at sub-optimal water content than many "homoiohydric" vascular plants!

8.3 Bryophyte shoots as photosynthetic systems; light and water-stress responses

The first essential in considering bryophytes as photosynthetic systems is to put out of mind vascular-plant leaves with their complex ventilated mesophyll, epidermis, and stomata, and to go back to physical and cell-biological first principles. Vascular-plant leaves are typically deployed in the turbulent air well above the ground. The diffusion resistance of the relatively thin laminar boundary layer of the individual leaves is quite small. The epidermis with its cuticle and stomata effectively marks the boundary between (relatively slow) diffusive mass transfer within the leaf and (much faster) turbulent mixing in the surrounding air, and it is easy to show by experiment in the laboratory that the rate of water loss is largely determined by stomatal aperture. However, this leaves out of consideration two important factors in the field situation, one general, and one particularly applicable to bryophytes. First, the latent heat of evaporation must come from the surroundings – by convective exchange with the air, by conduction from the substrate, or by radiative exchange with the wider environment. In a laboratory experiment with an isolated plant the amount of heat involved is small and easily left out of consideration; in the vegetation cover of a landscape it becomes a major factor in determining water loss (Jarvis & McNaughton 1986). Second, boundary-layer conditions for bryophytes are often largely determined by the extensive substrates on which they grow; further, many bryophytes grow in the shelter of trees or smaller vascular plants which reduce the ambient windspeed to varying degrees. Thus, various environmentally-determined parameters are major controls of water loss from bryophytes, and laboratory experiments on isolated bryophyte shoots or cushions that do not take this into account may have little relevance to what goes on in the field.

The small leaves of many bryophytes lie largely or wholly within the laminar boundary layer of the bryophyte carpet or cushion, or of the substrate on which it grows. The thickness of the boundary layer is in the region of a few hundred μm at a windspeed of $1\ \mathrm{m\ s^{-1}}$. Wind-tunnel measurements show that at very low windspeeds a moss cushion behaves as a smooth simple object; water loss increases approximately as the square root of the windspeed, reflecting the corresponding decrease in boundary-layer thickness. Hair-points on the leaves can have the effect of separating the sites of momentum and water-vapour transfer, in effect trapping an additional thickness of stagnant air between the moist leaf surfaces and the airstream. (Hair-points can have other effects, increasing albedo for one.) Beyond a certain point, evaporation rises more rapidly with windspeed; the rougher the cushion surface, the lower the windspeed at which this occurs (Proctor 1981*b*). At low windspeeds, the bryophyte colony functions, in effect, as a single "leaf," and gas exchange in the spaces between the leaves proceeds mainly by the comparatively slow process of molecular diffusion. Increasing evaporation at higher windspeeds reflects both the increasing tendency of the moss surface to generate turbulence in the airstream, and the fractally increasing area of the evaporating surface of the cushion as measured by a boundary layer of progressively decreasing thickness. Moss or leafy-liverwort canopies operate at a scale intermediate between vascular-plant leafy canopies on the one hand, and the cells of a vascular-plant mesophyll on the other, and analogies may be sought in both directions. Bryophytes commonly show very high leaf-area index (LAI) values. A few estimates of my own gave figures of ~6 in *Tortula intermedia*, 18 in *Mnium hornum*, and 20–5 in *Pseudoscleropodium purum* (Proctor 1979a), in the same range as the few other (unpublished) figures I have encountered. They are nearer the range of vascular-plant ratios of mesophyll area to leaf area (~14–40; Nobel 1974) than to LAIs for vascular-plant canopies which are usually less than 10 and commonly ~5.

The diffusive path for water loss is from the *leaf surface* to the atmosphere; that for CO_2 uptake is from the atmosphere to the *chloroplasts*. In other words, CO_2 uptake encounters additional liquid-phase diffusive resistance in the cell walls and cytoplasm. As molecular diffusion is slower in water than in air by a factor of about 10^4, this addition resistance is large, even if the cell-wall thickness is only a few μm, and underlies the selection pressure for evolution of high LAI values in bryophytes and high mesophyll/leaf-area ratios in vascular-plant leaves. In addition to these diffusive resistances, the photosynthetic system of the chloroplasts may be regarded as imposing a "carboxylation resistance" to CO_2 uptake. The generally similar values of $\delta^{13}C$ for bryophytes (averaging around $-27‰$) and C3 vascular plants (Rundel *et al.* 1979, Teeri 1981, Proctor *et al.* 1992) suggests that the two show a similar balance of diffusion and biochemical limitations on CO_2 uptake, despite the generally thicker cell walls of bryophyte leaves compared with vascular-plant mesophyll cells. This probably reflects convergence on an adaptive optimum in the deployment of Rubisco relative to supporting tissues (Raven 1984: appendix 3) in the two groups. Substantially more negative $\delta^{13}C$ values are seen in aquatic bryophytes utilizing a substantial proportion of respired CO_2 (e.g., *Fontinalis antipyretica* [Rundel *et al.* 1979, Raven *et al.* 1987], *Sphagnum cuspidatum* [Proctor *et al.* 1992]). Less negative $\delta^{13}C$ values can be the consequence of high diffusive limitation by excess superincumbent water (Williams & Flanagan 1996). Anthocerotae such as *Anthoceros* and *Phaeoceros* show consistently low discrimination against $^{13}CO_2$, giving $\delta^{13}C$ values of -15 to $-20‰$, because uniquely among bryophytes they have a carbon-concentrating mechanism associated with the pyrenoid (Smith & Griffiths 1996a, b). C4 vascular plants typically have $\delta^{13}C$ values around -10 to $-12‰$.

Morphological adaptation in bryophytes must reconcile the potentially conflicting requirements of water conduction and storage, and free gas exchange for photosynthesis. This is achieved in various ways. Many, and probably most, bryophyte leaf surfaces carry at least a thin layer of water-repellent cuticular material, and some bear conspicuous granular or crystalline epicuticular wax (Proctor 1979b). This is most striking in some glaucous-looking endohydric species of moist places, such as *Pohlia cruda*, *P. wahlenbergii*, *Saelania glaucescens*, many Bartramiaceae, and liverworts such as *Douinia ovata* and *Gymnomitrion obtusum*. Many mosses (and some leafy liverworts) have shoot systems with closely overlapping concave leaves, the inner faces functioning for water storage, and the outer surfaces, kept free of superincumbent water by surface tension, serving for gas exchange. Striking instances of shoots of this kind are seen in, e.g., *Anomobryum filiforme*, *Scleropodium* spp., *Myurium hochstetteri*, *Pleurozium schreberi*, *Pilotrichella*, *Weymouthia*, and *Nowellia curvifolia* but there are many less extreme variations on the same theme. Densely papilla-covered or mammillate leaf surfaces are also common, and in many cases these too appear to provide a division between water conduction and gas exchange, the papilla (or mammilla) apices remaining dry while the interstices between them provide a continuous network of water-conducting channels (Buch 1945, 1947, Proctor 1979a).

The leaves of Polytrichales and thalli of Marchantiales have complex ventilated photosynthetic tissues paralleling leaves of vascular plants. In elementary accounts these tend to be seen in terms of restriction of water loss, but we probably ought to see them rather as increasing the area for CO_2 uptake relative to the area for evaporation during periods when the plant *is* adequately supplied with water. Conspicuous surface wax is notably a feature of the lamella margins of the leaves of *Dawsonia* and the larger Polytrichaceae (Clayton-Greene *et al.* 1985, Proctor 1992) where it serves the same function of preventing entry of water into the interlamellar spaces. The water-repellent pore margins of Marchantiales similarly prevent flooding of the ventilated photosynthetic tissues of the thallus, in the same way that the sharp water-repellent edges of the stomata prevent waterlogging of the mesophyll in vascular plants (Schönherr & Ziegler 1975).

When compared in terms of true relative water content (i.e., cell water content relative to that at full turgor), photosynthesis in bryophytes of widely differing adaptive types, and higher-plant cells, responds similarly to water deficit (Fig. 8.3). Bryophytes have often been said to show "shade-plant-like" features in their photosynthetic physiology (Valanne 1984). Some of these characteristics may have more to do with the long

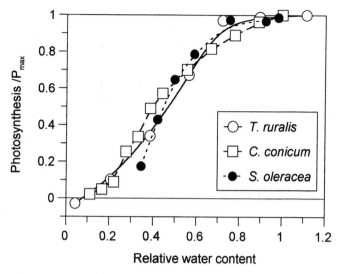

Fig. 8.3. Response of net photosynthesis to cell water deficit in two contrasting bryophytes, from gas-exchange measurements. The data for the desiccation-tolerant moss *Tortula ruralis* are recalculated from Tuba *et al.* (1996), taking as full turgor a value of 165% dry weight estimated from measurements at their field site in July 1998, and assuming 10% of the full-turgor water content to be apoplast water. The data for the thalloid liverwort *Conocephalum conicum* are recalculated from Slavik (1965), assuming that full-turgor water content coincides with the maximum value for net photosynthesis (900% dry weight). Measurements for spinach (*Spinacia oleracea*), a mesophytic vascular plant, are included for comparison (Kaiser 1987).

evolutionary independence of bryophytes and vascular plants of the two groups than with real adaptive differences. Bryophytes typically have rather low chlorophyll a/b quotients, in the range of shade-adapted vascular plants (Egle 1960). Martin and Churchill (1982) found an overall mean value of 2.69 (SD 0.27) from 14 species of exposed habitats in Kansas, and 2.38 (SD 0.20) for 20 forest species after canopy closure. Kershaw & Webber (1986) found a progressive change in the chlorophyll a/b quotient in an old-orchard population of *Brachythecium rutabulum* from 2.9 in young shoots before tree-canopy expansion to ~2.0 in deep shade in autumn. However, bryophytes show wide variation in their light responses, with big differences between species, and substantial seasonal and plastic variation. Photosynthesis of shade-loving forest species (e.g., *Rhytidiadelphus loreus, Thamnobryum alopecurum*) is commonly saturated at a photosynthetic photon flux (PPFD) of 100–300 μmol m^{-2} s^{-1} (corresponding to 5–10% of full sunlight), whereas some species of open, sun-

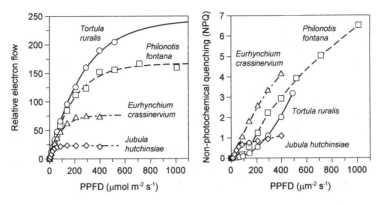

Fig. 8.4. Light responses of bryophytes, from chlorophyll-fluorescence measurements. (a) The relation of relative photosynthetic electron flow (calculated as ΦPSII $\times$ PPFD) to irradiance (measured as μmol photons $m^{-2}s^{-1}$ in the photosynthetically active range, $\lambda = 400 - 700$ nm). The curves are of the form $y = A(1 - e^{-kx})$, which empirically give a good fit to the data; the following "95% saturation" values should be taken as indicative for purposes of comparison, not as precise estimates: *Tortula ruralis* 830, *Philonotis fontana* 623, *Anomodon viticulosus* 216, *Jubula hutchinsiae* 76 (all μmol $m^{-2}s^{-1}$). (b) The relation of non-photochemical quenching (NPQ = $[F_m - F_m']/F_m'$) to irradiance. NPQ gives a measure of zeaxanthin-dependent photoprotection, and is largely suppressed by the inhibitor dithiothreitol (DTT); it reaches very high levels at high irradiance in many bryophytes of sun-exposed habitats. Note that NPQ in *Jubula hutchinsiae* is approaching saturation at a relatively modest irradiance. NPQ in *Philonotis fontana* and *Anomodon viticulosus* rises near-linearly with irradiance over the range of the measurements. In *Tortula ruralis*, NPQ is still climbing steeply at the highest irradiance measured.

exposed habitats (e.g., *Grimmia pulvinata*, *Tortula ruralis*, *Aulacomnium palustre*) may not reach saturation until 1000 μmol $m^{-2}s^{-1}$ PPFD or more (Fig.8.4(a)). Chlorophyll-fluorescence measurements (Schreiber *et al.* 1995) on bryophytes generally show high levels of non-photochemical quenching at high irradiances (Fig. 8.4(b)). This suggests correspondingly high levels of xanthophyll-cycle-mediated photoprotection (Björkman & Demmig-Adams 1995, Horton *et al.* 1996, Gilmore 1997), dissipating harmlessly as heat potentially damaging excess energy absorbed by the chlorophyll.

8.4 Desiccation tolerance

Desiccation tolerance is a common and characteristic but not universal feature of bryophytes (Proctor 1981*a*, Bewley & Krochko 1982, Stewart

1989). Some species of constantly moist or shady habitats are very sensitive to drying out, and there is every gradation between these and species
of sunbaked bare soil or rock surfaces, which not only survive but flourish
in habitats where they spend a large part of their time in a state of intense
desiccation. Two points about desiccation tolerance may be made at the
outset. The first is that it is a very widespread phenomenon among living
organisms, occurring among microorganisms, fungi, algae, lichens, bryophytes, vascular plants (where it is uncommon in vegetative tissues but
the norm in spores and seeds), as well as in such animal groups as ciliates,
rotifers, tardigrades, nematodes, and the eggs of crustacea of imperma-
nent water bodies. The second point is that desiccation tolerance has certainly evolved (or re-evolved) independently a number of times in the
plant kingdom. While we might expect some common features throughout, there is no reason to suppose that the details will be the same in every
case.

The ecological context of desiccation tolerance is intermittent availability of water to the plant. The limiting minimum is set by the duration
of periods of precipitation sufficient to bring the bryophyte to full turgor.
In practice all bryophytes store sufficient water to extend the moist
periods substantially beyond this – how much beyond, depends on envi-
ronmentally determined rates of evaporation (Proctor 1990, Proctor &
Smith 1994). Generally, in the open under the high-radiation conditions
of late spring and summer, moist periods for bryophytes tend to be closely
tied to precipitation events. With declining radiation income and the prevailing leafy canopies of late summer and autumn, water storage in bryophyte mats and cushions can bridge progressively longer gaps between
spells of rain. This will work out differently for different species. Heavy
dewfall may provide sufficient water for significant early-morning photosynthesis by such species as the steppe-grassland and sand-dune moss
Tortula (Syntrichia) ruralis, and the pendulous species of tropical cloud
forests and temperate rainforests (e.g. *Pilotrichella, Floribundaria, Papillaria*)
depend on limited storage of water from frequent rainfall or interception
of cloudwater droplets. There clearly must be a lower limit to the time for
which a bryophyte can usefully be moist. "Moist-break" experiments
showed that *Hylocomium splendens* desiccated for 11 days at 32% relative
humidity recovered to substantially its pre-desiccation state after 24 h,
but not after 6 h, moist. *Rhytidiadelphus squarrosus* showed not only substantially complete recovery within 6 h, but a marked degree of drought
hardening (Dilks & Proctor 1976). Species such as *Tortula ruralis* and the

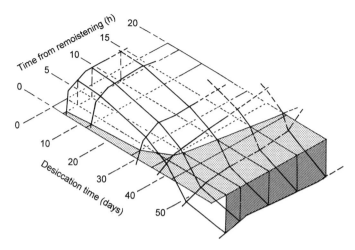

Fig. 8.5. Desiccation tolerance: the relation of net photosynthesis to desiccation time (at 50% relative humidity, ∼20 °C), and time after subsequent remoistening, in the moss *Anomodon viticulosus*. Shading shows the area over which net assimilation is negative. Redrawn (simplified) from the manometric gas-exchange data of Hinshiri and Proctor (1971). See text for further explanation.

pendant forest mosses certainly recover quicker than this, but just how quickly has yet to be established. We can envisage a spectrum of adaptation from "high-inertia" desiccation-tolerant species (e.g., on the forest floor) experiencing relatively long periods wet and dry, and in which drying and recovery may be relatively slow, and "low-inertia" species (on exposed rocks and twigs, or short turfs in open dry grasslands) able to make a living in rapidly alternating wet and dry conditions (Tuba *et al.* 1996, 1998, Marschall & Proctor 1999).

Ecophysiological responses of bryophytes to desiccation are complex. Some basic general features are illustrated in Fig. 8.5, which shows the results of a series of experiments with *Anomodon viticulosus*. After short periods of desiccation (up to about 2 weeks in this case), the moss recovers rapidly and completely on remoistening and normal rates of photosynthesis are re-established with a few hours. With longer desiccation, the recovery process becomes progressively more prolonged, and final recovery less complete. In the example here, recovery is markedly slowed, but still substantially complete within 10 h after 22 days' desiccation. After 35 days' desiccation recovery on remoistening is very slow, and net photosynthesis has reached less than half its pre-desiccation value after 20 h remoistening. Prolonged desiccation (beyond ∼40 days in this instance) leads to prolonged net carbon loss on remoistening, with only limited ultimate recovery.

Results with other species differ from the pattern seen here in the time scales on the two axes. Responses also depend on the intensity of desiccation to which the plant has been exposed. Dilks and Proctor (1974) found that *Racomitrium lanuginosum* survived much better at low than at high humidity (like most seeds), barely attaining the compensation point within 24 h following 6 months at 76% relative humidity, but showing substantially complete recovery after 8 months at 32% relative humidity *Plagiothecium undulatum*, a characteristic moss of moist woods, showed the opposite response (like "recalcitrant" seeds), surviving best at the highest humidity, and worse damaged the more intense the desiccation.

Bryophytes are much more tolerant of high (or very low) temperatures dry than wet. Lethal temperatures for moist, metabolically active bryophytes are in the same range as for C3 vascular plants, about 40–50 °C. Survival of desiccation is related to temperature. For desiccation-tolerant species including *Anomodon viticulosus*, *Tortula intermedia*, and *Frullania tamarisci*, the relationship follows Arrhenius's equation; the logarithm of survival time is proportional to the reciprocal of the absolute temperature (Hearnshaw & Proctor 1982); in a few, such as *Racomitrium lanuginosum* and *R. aquaticum*, the relationship is curvilinear. In either case, survival times are measured in minutes at 100 °C and in months or years at 0 °C.

Two sets of questions may be asked about desiccation tolerance. First, what are the factors that enable a plant to be tolerant of desiccation? Second, given that a plant is desiccation tolerant, what is the effect of drying–rehydration cycles upon the various processes of the plant's metabolism? These are probably not always clearly distinguished, and answers to questions of the second kind may give few clues to answers about the first. In very general terms, desiccation tolerance clearly requires a cell structure that can lose most of its water without disruption, and membranes that retain the essentials of their structure in the dry state or are readily and quickly reconstituted on remoistening. Obviously *all* the essential metabolic systems of the plant must remain intact or be readily reconstituted, and it is a reasonable prima facie supposition that there is unlikely to be any one critically-sensitive system because selection pressure will bear heavily on any weak link. One might expect to find "molecular packaging" materials, perhaps most likely carbohydrates (Seel *et al.* 1992, Smirnoff 1992) or proteins (Oliver 1996), protecting macromolecules; another reasonable expectation would be well-developed

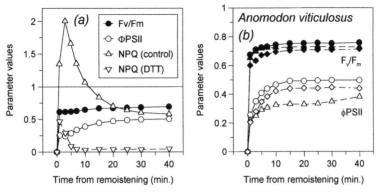

Fig. 8.6. Recovery of the chlorophyll-fluorescence parameter F_v/F_m (measuring the maximum quantum efficiency of PSII) in the dark, and of ΦPSII in the light ($50\ \mu$mol m^{-2} s^{-1}), following remoistening of the moss *Anomodon viticulosus*, after 21 days' desiccation, with non-photochemical quenching (NPQ) estimated from the mean values of F_m and F_m' of the two sets of matched samples ($n = 3$). See also Csintalan *et al.* (1999). (a) The peak of NPQ is almost wholly suppressed by remoistening with 5 mmol l^{-1} dithiothreitol, which had little effect on recovery of F_v/F_m or ΦPSII. (b) Recovery of F_v/F_m (filled symbols) and ΦPSII (open symbols) in *Anomodon viticulosus* desiccated for 8–10 days and remoistened with water (circles), 3 mmol l^{-1} chloramphenicol (triangles) or 0.3 mmol l^{-1} cycloheximide (diamonds). The protein-synthesis inhibitors have little effect in the dark. The data show a statistically significant effect of chloramphenicol (which inhibits translation of chloroplast-encoded proteins) in the light, but this is relatively small. Some of the differences seen here may be due to unspecific side-effects of the inhibitors.

antioxidant systems minimizing damage by active oxygen species during drying and rehydration (Smirnoff 1993, Alscher *et al.* 1997). Oliver (1996) and Oliver *et al.* (1998) hypothesize that survival in "modified" desiccation-tolerant plants (primarily vascular "resurrection plants") is a function of inducible protective mechanisms, while "true" desiccation-tolerant plants (including bryophytes) rely more heavily on "repair" processes.

Protein synthesis is quickly re-established on remoistening (Gwóźdz *et al.* 1974, Oliver 1991), and is clearly a necessary part of full recovery. However, protein synthesis seems to play rather little part in recovery of major metabolic systems of desiccation-tolerant bryophytes from moderate periods of desiccation. This recovery can be very rapid. The gas-exchange data of Tuba *et al.* (1996) show re-establishment of normal rates of net photosynthesis in *Tortula ruralis* within 30 minutes of remoistening. Modulated chlorophyll-fluorescence techniques have underlined how fast recovery of the photosystems can be (Fig. 8.6(a)). It is virtually

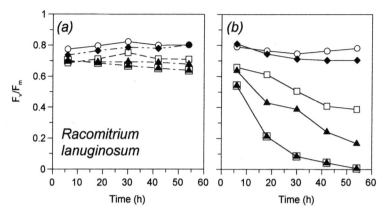

Fig. 8.7. A longer-term experiment on the effect of metabolic inhibitors on recovery of F_v/F_m in the moss *Racomitrium lanuginosum* after ~10 days' desiccation. Open circles, control (water); open squares, 3 mmol l^{-1} dithiothreitol; solid triangles, 3 mmol l^{-1} chloramphenicol; solid diamonds, 0.3 mmol l^{-1}, 0.3 mmol l^{-1} cycloheximide; superimposed squares and triangles, dithiothreitol + chloramphenicol. (a) In dark; the cycloheximide treatment is scarcely distinguishable from the control, and dithiothreitol and chloramphenicol have only modest effects alone or in combination. (b) In light (~125 μmol m^{-2} s^{-1}), cycloheximide (inhibiting cytoplasmic protein synthesis) produces a slow but progressive fall in F_v/F_m. With dithiothreitol, F_v/F_m drops rather quickly to values around 0.4, probably reflecting suppression of the photoprotection associated with NPQ. Chloramphenicol produces a rapid and progressive drop in F_v/F_m, especially in the presence of dithiothreitol. This is likely to reflect the rapid turnover of the D1 protein of photosystem II in the light (Anderson *et al.* 1997), accentuated when photoprotection is suppressed by dithiothreitol, and inhibition of synthesis of this chloroplast-encoded protein by chloramphenicol.

unaffected by inhibitors of either chloroplast or cytoplasmic protein synthesis (Figs. 8.6(b), 8.7). The factors that enable these plants to be so highly desiccation-tolerant must be mainly constitutive. Some "repair" of membranes and integrated systems must be necessary, but it appears that this must be largely a matter of reassembly of pre-existing components, rather than synthesis of new ones. However, there is undoubtedly a spectrum of possibilities between extreme constitutive desiccation-tolerant bryophytes, and species in which tolerance depends heavily on "hardening" processes or is under abscisic-acid control as in *Funaria hygrometrica* (Werner *et al.* 1991, Bopp & Werner 1993), or seasonally switched as in *Lunularia cruciata* (Schwabe & Nachmony-Bascomb 1963). That *some* degree of hardening/dehardening can occur in even the most tolerant species is suggested by the results of Dilks and Proctor (1976) and Schonbeck and

Bewley (1981). The problems of desiccation tolerance are clearly many-faceted and complex, and there may well be significant differences in mechanisms between different groups of species. In particular, when dealing with bryophytes we may need to get away from the assumption that conditions in vascular plants in any way represent "normality."

8.5 Overview

How can the special ecophysiological characteristics of bryophytes be summarized in a few words? First, bryophytes grow and function largely within the laminar boundary layer next to the ground or other substrate; their immediate surroundings are in effect an integral part of their physiology. Second, most bryophytes carry substantial amounts of external water, which can vary widely without affecting the water status of the cells, so bryophytes spend most of their time *either* fully turgid, *or* dry and metabolically inactive. It is *essential* to know (at least approximately) the true full-turgor water content for research on effects of water stress on bryophyte cells and tissues. Third, intermittent availability of water is the norm for many bryophytes; their desiccation tolerance may be more usefully thought of as a means of evading drought, than as an extreme form of drought tolerance. Initial recovery from normal desiccation is very rapid, and seems to depend little on protein synthesis; but we are still far from understanding the fundamental basis or the consequences of desiccation tolerance. Fourth, all bryophytes show some shade-plant characteristics in their photosynthetic physiology, although they span a wide range in habitat and light responses. Species of dry sunny habitats have moderately high light-saturation levels, and very high levels of NPQ (and thus presumably photoprotection) at high irradiances. All bryophytes are C3 plants; anthocerotes are unique in having a carbon-concentrating mechanism. Finally, this last point is a reminder that bryophytes are phylogenetically diverse; the major groups – mosses, hepatics, anthocerotes – have been evolutionarily independent from one another and from vascular plants through most of the history of plant life on land.

REFERENCES

Alscher, R. G., Donahue, J. L., & Cramer, C. A. (1997). Reactive oxygen species and antioxidants: relationships in green cells. *Physiologia Plantarum*, **100**, 224–33.
Anderson, J. M., Park, Y.-I., & Chow, W. S. (1997). Photoinactivation and photoprotection of photosystem II in nature. *Physiologia Plantarum*, **100**, 214–23.

Bates, J. W. & Farmer, A. M. (1992). *Bryophytes and Lichens in a Changing Environment*. Oxford: Clarendon Press.

Beckett, R. P. (1996). Pressure volume analysis of a range of poikilohydric plants implies the existence of negative turgor in vegetative cells. *Annals of Botany*, **79**, 145–52.

Bewley, J. D. & Krochko, J. E. (1982). Desiccation-tolerance. In *Encyclopaedia of Plant Physiology, New Series*, vol. 12B, ed. O. L. Lange, P. S. Nobel, C. B. Osmond, & H. Ziegler, pp. 325–78. Berlin: Springer-Verlag.

Björkman, O. & Demmig-Adams, B. (1955). Regulation of photosynthetic light energy capture, conversion, and dissipation in leaves of higher plants. In *Ecophysiology of Photosynthesis*, ed. E.-D. Schulze & M. M. Caldwell, pp. 17–47. Berlin: Springer-Verlag.

Bopp, M. & Werner, O. (1993). Abscisic acid and desiccation tolerance in mosses. *Botanica Acta*, **106**, 103–6.

Brown, D. H. (1982). Mineral nutrition. In *Bryophyte Ecology*, ed. A. J. E. Smith, pp. 383–444. London: Chapman & Hall.

(1984). Uptake of mineral elements and their use in pollution monitoring. In *The Experimental Biology of Bryophytes*, ed. A. F. Dyer & J. G. Duckett, pp. 229–55. London: Academic Press.

Brown, D. H. & Bates, J. W. (1990). Bryophytes and nutrient cycling. *Botanical Journal of the Linnean Society*, **104**, 129–47.

Buch, H. (1945). Über die Wasser- und Mineralstoffversorgung der Moose. Part 1. *Commentationes Biologici Societas Scientiarum Fennicae* **9(16)**, 1–44.

(1947). Über die Wasser- und Mineralstoffversorgung der Moose. Part 2. *Commentationes Biologici Societas Scientiarum Fennicae* **9(20)**, 1–61.

Clayton-Greene, K. A., Collins, N. J., Green, T. G. A., & Proctor, M. C. F. (1985). Surface wax, structure and function in leaves of Polytrichaceae. *Journal of Bryology*, **13**, 549–62.

Csintalan, Zs., Proctor, M. C. F., & Tuba, Z. (1999). Chlorophyll fluorescence during drying and rehydration in the mosses *Rhytidiadelphus loreus* (Hedw.) Warnst., *Anomodon viticulosus* (Hedw.) Hook. & Tayl. and *Grimmia pulvinata* (Hedw.) Sm. *Annals of Botany*, **84**, 235–44.

Dilks, T. J. K. & Proctor, M. C. F. (1974). The pattern of recovery of bryophytes after desiccation. *Journal of Bryology*, **8**, 97–115.

(1976). Effects of intermittent desiccation on bryophytes. *Journal of Bryology*, **9**, 249–64.

(1979). Photosynthesis, respiration, and water content in bryophytes. *New Phytologist*, **82**, 97–114.

Edwards, D. (1998). Climatic signals in Palaeozoic land plants. *Philosophical Transactions of the Royal Society of London B*, **353**, 141–57.

Egle, K. (1960). Menge und Verhältnis der Pigmente. In *Encyclopaedia of Plant Physiology*, vol. 5, ed. W. Ruhland, pp. 444–96. Berlin: Springer-Verlag.

Frahm, J.-P. (1990). Bryophyte phytomass in tropical ecosystems. *Botanical Journal of the Linnean Society*, **104**, 23–33.

Gilmore, A. M. (1997). Mechanistic aspects of xanthophyll cycle-dependent photoprotection in higher plant chloroplasts and leaves. *Physiologia Plantarum*, **99**, 197–209.

Gwózdz, E. A., Bewley, J. D., & Tucker, E. B. (1974). Studies on protein synthesis in *Tortula ruralis*: polyribosome reformation following desiccation. *Journal of Experimental Botany*, **25**, 599–608.

Hearnshaw, G. F. & Proctor, M. C. F. (1982). The effect of temperature on the survival of dry bryophytes. *New Phytologist*, **90**, 221–8.

Hébant, C. (1977). *The Conducting Tissues of Bryophytes*. Vaduz: J. Cramer.

Hinshiri, H. N. & Proctor, M. C. F. (1971). The effect of desiccation on subsequent assimilation and respiration of the bryophytes *Anomodon viticulosus* and *Porella platyphylla*. *New Phytologist*, **70**, 527–38.

Horton, P., Ruban, A. V., & Walters, R. G. (1996). Regulation of light harvesting in green plants. *Annual Review of Plant Physiology and Plant Molecular Biology*, **47**, 655–84.

Jarvis, P. G. & McNaughton, K. G. (1986). Stomatal control of transpiration: scaling up from leaf to region. *Advances in Ecological Research*, **15**, 1–49.

Jones, H. G. (1992). *Plants and Microclimate*, 2nd edn. Cambridge: Cambridge University Press.

Kershaw, K. A. & Webber, M. R. (1986). Seasonal changes in the chlorophyll content and quantum efficiency of the moss *Brachythecium rutabulum*. *Journal of Bryology*, **14**, 151–8.

Longton, R. E. (1981). Physiological ecology of mosses. In *The Mosses of North America*, ed. R. J. Taylor & S. E. Leviton, pp. 77–113. Washington: Pacific Division, American Academy of Science.

(1988). *The Biology of Polar Bryophytes and Lichens*. Cambridge: Cambridge University Press

Martin, C. E. & Churchill, S. P. (1982). Chlorophyll concentrations and *a/b* ratios in mosses collected from exposed and shaded habitats in Kansas. *Journal of Bryology*, **12**, 297–304.

Marschall, M. & Proctor, M. C. F. (1999). Desiccation tolerance and recovery of the leafy liverwort *Porella platyphylla* (L.) Pfeiff.: chlorophyll-fluorescence measurements. *Journal of Bryology*, **21**, 257–62.

Nobel, P. S. (1974). *Biophysical Plant Physiology*. San Francisco: Freeman.

Oliver, M. J. (1991). Influence of protoplasmic water loss on the control of protein synthesis in the desiccation-tolerant moss *Tortula ruralis*. Ramifications for a repair-based mechanism of desiccation tolerance. *Plant Physiology*, **97**, 1501–11.

(1996). Desiccation tolerance in vegetative plant cells. *Physiologia Plantarum*, **97**, 779–87.

Oliver, M. J., Wood, A. J., & O'Mahony, P. (1998). "To dryness and beyond" – preparation for the dried state and rehydration in desiccation-tolerant plants. *Plant Growth Regulation*, **24**, 193–201.

Proctor, M. C. F. (1979a). Structure and eco-physiological adaptation in bryophytes. In *Bryophyte Systematics*, ed. G. C. S. Clarke & J. G. Duckett, pp. 479–509. London: Academic Press.

(1979b). Surface wax on the leaves of some mosses. *Journal of Bryology*, **10**, 531–8.

(1981a). Physiological ecology of bryophytes. *Advances in Bryology*, **1**, 79–166.

(1981b). Diffusion resistances in bryophytes. In *Plants and their Atmospheric Environment*, ed. J. Grace, E. D. Ford, & P. G. Jarvis, pp. 219–29. Oxford: Blackwell.

(1982). Physiological ecology: water relations, light and temperature responses, carbon balance. In *Bryophyte Ecology*, ed. A. J. E. Smith, pp. 333–81. London: Chapman & Hall.

(1990). The physiological basis of bryophyte production. *Botanical Journal of the Linnean Society*, **104**, 61–77.

(1992). Scanning electron microscopy of lamella-margin characters and the phytogeography of the genus *Polytrichadelphus*. *Journal of Bryology*, **17**, 317–33.

(1999). Water-relations parameters of some bryophytes evaluated by thermocouple psychrometry. *Journal of Bryology*, **21**, 263–70.

Proctor, M. C. F. & Smith, A. J. E. (1994). Ecological and systematic implications of branching patterns in bryophytes. In *Experimental and Molecular Approaches to Plant Biosystematics*, ed. P. C. Hoch & A. G. Stephenson, pp. 87–110. St Louis: Missouri Botanical Garden.

Proctor, M. C. F., Raven, J. A., & Rice, S. K. (1992). Stable carbon isotope discrimination measurements in *Sphagnum* and other bryophytes: physiological and ecological implications. *Journal of Bryology*, **17**, 193–202.

Proctor, M. C. F., Nagy, Z., Csintalan, Zs., & Takács, Z. (1998). Water-content components in bryophytes: analysis of pressure–volume curves. *Journal of Experimental Botany*, **49**, 1845–54.

Raven, J. A. (1977). The evolution of land plants in relation to supracellular transport processes. *Advances in Botanical Research*, **5**, 153–219.

(1984). Physiological correlates of the morphology of early vascular plants. *Botanical Journal of the Linnean Society*, **88**, 105–26.

Raven, J. A., Macfarlane, J. J., & Griffiths, H. (1987). The application of carbon isotope discrimination techniques. In *Plant Life in Aquatic and Amphibious Habitats*, ed. R. M. M. Crawford, pp. 129–49. Oxford: Blackwell.

Rundel, P. W., Stichler, W., Zander, R. H., & Ziegler, H. (1979). Carbon and hydrogen isotope ratios of bryophytes from arid and humid regions. *Oecologia*, **4**, 91–4.

Russell, S. (1990). Bryophyte production and decomposition in tundra ecosystems. *Botanical Journal of the Linnean Society*, **104**, 3–22.

Schonbeck, M. W. & Bewley, J. D. (1981). Responses of the moss *Tortula ruralis* to desiccation treatments. II Variations in desiccation tolerance. *Canadian Journal of Botany*, **59**, 2707–12.

Schönherr, J. & Ziegler, H. (1975). Hydrophobic cuticular ledges prevent water entering the air pores of liverwort thalli. *Planta*, **124**, 51–60.

Schreiber, U., Bilger, W., & Neubauer, C. (1995). Chlorophyll fluorescence as a nonintrusive indicator for rapid assessment of *in vivo* photosynthesis. In *Ecophysiology of Photosynthesis*, ed. E.-D. Schulze & M. M. Caldwell, pp. 49–70. Berlin: Springer-Verlag.

Schwabe, W. & Nachmony-Bascomb, S. (1963). Growth and dormancy in *Lunularia cruciata* (L.) Dum. II. The response to daylength and temperature. *Journal of Experimental Botany*, **14**, 353–78.

Seel, W. E., Hendry, G. A. F., & Lee, J. A. (1992). Effects of desiccation on some activated oxygen processing enzymes and anti-oxidants in mosses. *Journal of Experimental Botany*, **43**, 1031–7.

Slavik, B. (1965). The influence of decreasing hydration level on photosynthetic rate in the thalli of the hepatic *Conocephallum conicum*. In *Water Stress in Plants. Proceedings of a Symposium held in Prague, September 30–October 4, 1963*, ed. B. Slavik, pp. 195–201. The Hague: W. Junk.

Smirnoff, N. (1992). The carbohydrates of bryophytes in relation to desiccation tolerance. *Journal of Bryology*, **17**, 185–91.

(1993). The role of active oxygen in the response of plants to water deficit and desiccation. *New Phytologist*, **125**, 27–58.

Smith, E. C. & Griffiths, H. (1996a). The occurrence of the chloroplast pyrenoid is correlated with the activity of a CO_2-concentrating mechanism and carbon isotope discrimination in lichens and bryophytes. *Planta*, **198**, 6–16.

Smith, E. C. & Griffiths, H. (1996b). A pyrenoid-based carbon-concentrating mechanism is present in terrestrial bryophytes of the class Anthocerotae. *Planta*, **200**, 203–12.

Steudle, E. & Peterson, C. A. (1998). How does water get through roots? *Journal of Experimental Botany*, **49**, 775–88.

Stewart, G. R. (1989). Desiccation injury, anhydrobiosis and survival. In *Plants under Stress*, ed. H. G. Jones, T. J. Flowers, & M. B. Jones, pp. 115–30. Cambridge: Cambridge University Press.

Sveinbjörnsson, B. & Oechel, W. C. (1992).Controls on growth and productivity of bryophytes: environmental limitations under current and anticipated conditions. In *Bryophytes and Lichens in a Changing Environment*, ed. J. W. Bates & A. M. Farmer, pp. 77–102. Oxford: Clarendon Press.

Teeri, J. A. (1981). Stable carbon isotope analysis of mosses and lichens growing in xeric and moist habitats. *Bryologist*, **84**, 82–4.

Tuba, Z., Csintalan, Zs., & Proctor, M. C. F. (1996). Photosynthetic responses of a moss, *Tortula ruralis* ssp. *ruralis*, and the lichens *Cladonia convoluta* and *C. furcata* to water deficit and short periods of desiccation, and their ecophysiological significance: a baseline study at present-day CO_2 concentration. *New Phytologist*, **133**, 353–61.

Tuba, Z., Proctor, M. C. F., & Csintalan, Zs. (1998). Ecophysiological responses of homochlorophyllous and poikilochlorophyllous desiccation tolerant plants: a comparison and an ecological perspective. *Plant Growth Regulation*, **24**, 211–17.

Valanne, N. (1984). Photosynthesis and photosynthetic products in mosses. In *The Experimental Biology of Bryophytes*, ed. A. F Dyer & J. G. Duckett, pp. 257–73. London: Academic Press.

Vitt, D. H. (1990). Growth and production dynamics of boreal mosses over climatic, chemical and topographic gradients. *Botanical Journal of the Linnean Society*, **104**, 35–59.

Werner, O., Espin, R. M. R., Bopp, M., & Atzorn, R. (1991) Abscisic-acid induced drought tolerance in *Funaria hygrometrica* Hedw. *Planta*, **186**, 99–103.

Williams, T. G. & Flanagan, L. B. (1996). Effect of changes in water content on photosynthesis, transpiration and discrimination against $^{13}CO_2$ and $C^{18}O^{16}O$ in *Pleurozium* and *Sphagnum*. *Oecologia*, **108**, 38–46.

Zimmermann, U. & Steudle, E. (1978). Physical aspects of water relations of plant cells. *Advances in Botanical Research*, **6**, 46–117.

9

Mineral nutrition, substratum ecology, and pollution

9.1 Introduction

Bryophytes are familiar and attractive ingredients in many types of natural landscape. Their shaggy coverings on branches and boughs, crags and boulders, in waterfalls and on woodland banks, add distinction to the larger scene. Less appealingly, they grow occasionally on bizarre materials, like the leather of a discarded boot, or a rusty iron pipe. Even in modern cities where air pollution and the built environment may seem unrelenting, there are bryophytes able to colonize crevices in masonry, soil accumulations in gutters, and to soften the otherwise geometrical wildernesses of roof tiles with their rounded cushions. To the scientist all these situations provide taxing problems concerning the supply of necessary resources, the impact of the bryophytes on their habitat, and their responses to undesirable chemicals in the environment. This chapter describes the special problems that bryophytes encounter in obtaining essential mineral nutrients, and in dealing with non-essential elements and compounds. The substratum† on which a bryophyte grows can be a source of nutrients and of other chemicals that may cause stresses. Also, the periods of time for which different types of substrata are available for colonization by bryophytes vary enormously. Therefore, both the chemical properties and the wider ecological characteristics of different substrata are considered from the point of view of their suitability for bryophyte growth. Many examples of bryophytes behaving as specific indicators of particular chemical environments are given. Their uses in biomonitoring provide practical instances of this.

† Frequently also termed the "substrate" in the literature, although that word is better restricted to the substance on which an enzyme acts.

[248]

9.2 Mineral nutrition

9.2.1 Mineral nutrient requirements

Evidence about the elemental requirements of bryophytes has been derived from culture experiments employing defined nutrient solutions, from chemical analyses of tissues and by studies of ion uptake and cell electrophysiology.

Growth on defined media

An example is provided by Hoffman's (1966) study of the nutrient relations of the cosmopolitan moss *Funaria hygrometrica*. The effects of nutrient deficiencies were investigated in protonemata grown on agar containing Hoagland's solution with individual elements lacking. The results showed that *F. hygrometrica* has closely similar macronutrient and micronutrient requirements to vascular plants. Hoffman also performed "nutrient triangle" experiments to define the optimal ratios of the major anions (N:P:S) and cations (K:Ca:Mg) for growth. These overly complex factorial experiments are difficult to analyze effectively. Nevertheless, an important conclusion was that the growth of protonemata and the development of numerous gametophores were favored by quite different cation combinations. By contrast, in the anion experiments, the absence of any one of the elements resulted in poor protonemal growth and no gametophores. Hoffman and many subsequent investigators of nutrient and pollutant effects in bryophytes have used mature shoots (and protonemata) or thalli as the experimental material. Compared to experiments starting with spores, these have the potential problem that elements initially present in the plants may be in sufficient quantity to mask or confound the effects of the applied nutrient treatments.

Bryophytes need relatively dilute nutrient media in contrast to those required for optimal growth of crop plants. Working with the thalloid liverwort *Marchantia polymorpha*, Voth (1943) varied the dilution of a basic nutrient solution to alter osmotic pressure but not the ratios of the elements. When a concentrated solution was used (solution 1), many of the thallus tips and wings were killed, the thallus dry weight and area were small, and production of gemmae cups was low. Over the intermediate concentration range (solutions 3–5), the plants increased in size, were darker, had more ascending tips, and developed more rhizoids in response to greater dilution of the nutrients. At the lowest concentrations (solutions 6–10) a greater intensity of red-purple coloration developed in the rhizoids, scales, and lower epidermis, and rhizoids were especially numerous whereas gemmae cups became fewer. Cell walls were extremely thin in the strongest solutions with many collapsed cells seen, but a maximum thickness of cell walls was seen in the most dilute solutions with most cells appearing healthy. Whilst survival of M. *polymorpha* is possible over a wide range of extreme concentrations, the species clearly grows best in dilute media. Only a handful of bryophytes has been subjected to scrutiny in solution culture experiments (Brown 1982) and further careful work is desirable.

Chemical analysis of tissues
Chemical analyses of bryophytes can provide useful clues about mineral requirements, about tolerance of non-essential elements and offer a means of biomonitoring the deposition of elements such as heavy metals. Numerous studies have investigated total element concentrations of bryophyte tissues, usually employing dry-ashing or wet-ashing techniques to solubilize the minerals for analysis by spectrophotometric methods. Many authors have discussed the protocols for preparing materials for analysis and particularly the need to remove surface contamination by soil particles and rock fragments (e.g., Shacklette 1965, Woollon 1975, Brown 1982). However, washing of previously dried material is not recommended because of the risk of leakage of cell solutes during rehydration (Brown & Buck 1979). Washing with tapwater, a potentially mineral-rich solution, is also likely to alter element levels through cation exchange (see below). Therefore, before embarking upon a program of chemical analyses, it is important to consider the possible cellular locations of the elements in question and to design the sampling and extraction method to provide the maximum information for the effort involved.

9.2.2 Mineral uptake mechanisms

Cellular ion uptake and membrane function
The few studies of the kinetics of absorption of ions by bryophytes have concentrated upon heavy metal pollutants and radionuclides (Brown 1984). For metals, zinc absorption by the aquatic moss *Fontinalis antipyretica* (Pickering & Puia 1969) is typical in showing a phase of rapid (30 minutes) uptake of 50% of the absorbed zinc. The remainder is absorbed slowly over several days and is sensitive to light, temperature, and metabolic inhibitors. The rapid uptake represents passive sorption of zinc ions onto the extracellular cation exchange sites of the moss tissue (see below), whereas the slower phase is believed to represent true uptake into the

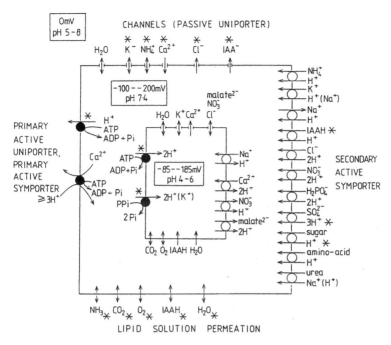

Fig. 9.1. Transport processes at the plasmalemma and tonoplast of an embryophyte cell, with those characterized for bryophytes indicated by an asterisk. (Reproduced from Raven *et al.* (1998) with permission of the British Bryological Society.)

cells. Wells and Richardson (1985) reported that a range of physiological anions and non-essential analogs displayed similar saturation kinetics in *Hylocomium splendens*. Cadmium uptake by *Rhytidiadelphus squarrosus* exhibited Michaelis–Menten kinetic constants (Km, V_{max}) that differed quite markedly between field populations exhibiting slightly different morphologies (Brown & Beckett 1985). This is particularly relevant to the use of bryophytes in monitoring heavy metal deposition.

Knowledge about the specific mechanisms of cellular ion uptake in bryophytes is summarized by Raven *et al.* (1998). Fig. 9.1 shows the various transport processes occurring at the plasmalemma and tonoplast of an embryophyte cell with pathways specifically demonstrated in bryophytes indicated by asterisks. The electro-negativity of the cell interior is in part due to active efflux of H^+, probably catalyzed by a "P" type ATPase which expends one mole of ATP per mole of H^+ pumped out. This pump also regulates cytoplasmic pH in bryophytes at around 7.3–7.6. This activity also provides the driving force for a number of specific symporter proteins allowing passage through the plasmalemma and against a concentration gradient for NH_4^+, sugars, and amino acids. Related mechanisms probably operate for active entry of K^+, NO_3^-, SO_4^{2-} and $H_2PO_4^-$ and for efflux of Ca^{2+} but have not yet been unequivocally demonstrated in bryophytes. Much less is known about the specific details of tonoplast transport in bryophytes and still less about any physiological differences that may exist between species with vacuoles and those lacking them (Oliver & Bewley 1982).

Action potentials are losses (depolarization) of the normal transmembrane potential difference that last in plant cells typically for a few seconds. They have been observed in the liverwort *Conocephalum conicum* and the hornwort *Anthoceros*. The depolarization in *C. conicum* is caused by Ca^{2+} influx and Cl^- efflux. The following repolarization is connected with entry of K^+ and efflux of H^+ (Trebacz *et al.* 1994). Virtually nothing is known about the occurrence of action potentials in the major groups of bryophytes or whether they have any involvement in membrane repair as implicated in higher plants.

Cation exchange

Clymo (1963) emphasized the importance of *cation exchange* in accumulation of cations by *Sphagnum* in mires. In fact, the cell walls of most plants possess a net negative charge owing to the ionization of weak acid moieties built into their fibrillar structure. The *cation exchange capacity* (CEC) can be determined by saturating these sites with a cation (e.g., Ca^{2+}) and then displacing this with another cation (e.g., Mg^{2+}) and measuring the quantity of Ca^{2+} released. The plasmalemma is probably not exposed directly to the ionic composition of the exterior solution as the negative charges tend to repel anions and alter the ratio of cations entering the wall environment. Detailed study (Richter & Dainty 1989*a*) of the cation-exchanger in *Sphagnum russowii* suggests that polymeric uronic acids account for over half the CEC, whilst phenolic compounds are responsible for about 25%, and amino acids, silicates, and sulfate esters deposited in the wall all make lesser contributions. Dependent on the pH of the external solution, all or a fraction of the acid moieties may ionize, e.g., for carboxyls in uronic acids and amino acids,

$$R.COOH = R.COO^- + H^+.$$

Under strongly acid conditions the reaction is driven to the left and CEC falls as ionization is suppressed, but progressively through less acidic, neutral, and alkaline conditions the net negative charge increases. The extent of ionization of a given weak acid group is indicated by its pK, i.e., the pH at which 50% has ionized and 50% remains un-ionized. By varying pH stepwise in the presence of metals with contrasted valencies (Na^+, Ca^{2+}, La^{3+}), Richter and Dainty (1989a) showed that *S. russowii* possesses two classes of cation-binding sites. One, with a low pK (2–4), appears to be principally due to uronic acids and amino acids, while the other, with a high pK (> 5), is almost certainly due mainly to weak phenolic acids.

Metals and other cations permeating the cell wall easily displace the protons from the ionized weak acids and may become relatively firmly held by the negative charges. At the same time the external medium receives the displaced protons and, in some circumstances, this may lead to its acidification. Exchangeably-bound cations are readily displaced by other cations in the external medium, particularly if the latter: (a) are present at higher concentration, (b) have larger hydrated atomic radii, or (c) possess a higher valency. The data presented in Fig. 9.2 were obtained by incubating the moss *Hylocomium splendens* in a mixture of cations (Rühling & Tyler 1970). They reveal an order of binding affinity for several heavy metal cations (Cu, Pb > Ni > Co > Zn, Mn) that appears to be widespread. The heavy ions Cu and Pb were adsorbed preferentially onto the exchange sites even when supplied in the presence of much higher concentrations of the lighter cations Ca, K, Mg, and Na. The behavior of the exchange sites varies with the species of cation employed to determine CEC. This is probably because the larger polyvalent cations combine strongly with and "condense" the fixed anions to varying extents (Sentenac & Grignon 1981, Richter & Dainty 1989b). Wide variations occur in CEC between bryophyte taxa and some of this variation appears to have ecological significance.

In *Sphagnum*-dominated mires the cation-exchanger of the *Sphagnum* plants is believed to be the principal mechanism by which acidic conditions are preserved (Clymo 1963, 1967, Brehm 1971, Clymo & Hayward 1982). Incoming cations are adsorbed and the released protons are added to those already present in the mire water, the production of fresh exchange sites by new growth keeping pace with cation inputs. Clymo (1963) observed strong correlations between the CEC of *Sphagnum* spp., their optimal height above the water table, and the hydrogen ion concentration of the interstitial water. Thus hummock species had the highest CECs and hummock water had the lowest pHs. It is probable that the cation-exchanger also has a role in nutrient absorption, the higher values

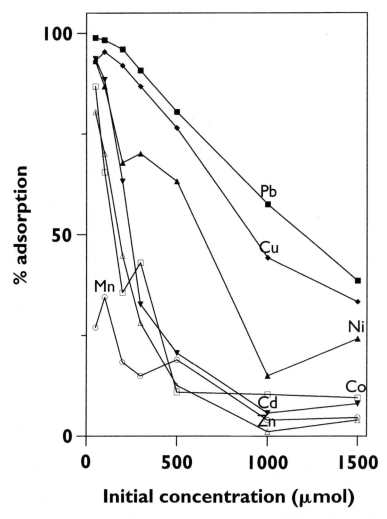

Fig. 9.2. Percentage absorption of metal ions onto the cation-exchanger of *Hylocomium splendens* from solutions containing equimolar concentrations of seven metal ions. The solutions were supplied (5 g of air-dried moss to 500 ml) at a range of initial concentrations and their final metal levels were determined after incubation for 2 h. (Redrawn from Rühling & Tyler 1970.)

of hummock species perhaps compensating for the shorter periods of hydration in this position. Among plants *Sphagnum* has unusually high CEC under acid conditions, a factor which coincidentally favours heavy metal accumulation; however elevated CEC is also characteristic of calcicole bryophytes (see pp. 277–80).

Element location within the tissues

Much of the natural variability in total cation levels of bryophytes appears to reflect extracellular accumulations by the cation-exchanger rather than wide variations in the living cells. In many situations a clearer picture can be obtained if the intracellular and cation-exchanger compartments are analyzed separately. This can be achieved by employing a sequential elution technique as described by Brown and Wells (1988) and Bates (1992*a*).

Clear patterns emerge for the major cations when bryophyte taxa from different habitats are compared. Those with a clear metabolic function are present within the cells in consistently high concentrations: a relatively high potassium concentration is believed to be essential for the normal folding of cytoplasmic enzymes; magnesium is present in chlorophyll and is an activator of several enzymes; calcium is believed to act primarily as a "messenger" in plant cells and is largely absent from the cytoplasm but often the predominant cation on the cation-exchanger, reflecting its abundance in many natural situations. Up to half of the total magnesium may also be exchangeable. Many other metals and some other cations, including the ammonium ion and cationic pesticides, may also enter the cation-exchanger.

Element concentrations alter as tissues age. Tamm (1953) neatly demonstrated this in *Hylocomium splendens*. The shoots or "fronds" of *H. splendens* consist of chains of annual "segments." Each segment is normally clearly demarcated from its forbears and offspring owing to a predominantly sympodial pattern of growth which makes dating of the tissues comparatively simple. Nitrogen, phosphorus, and potassium reached their highest concentrations in the young shoot apices and declined in older segments (Fig. 9.3). Calcium, however, increased in the older segments on a dry-weight basis. According to Bates (1979) this is partly an artefact arising through an increase in the cell wall/protoplasm ratio owing to slow degradation of the cell walls. Eckstein and Karlsson (1999) have provided a modern analysis of nitrogen dynamics in the segment chains of *H. splendens* (see below).

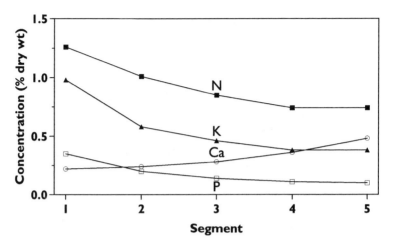

Fig. 9.3. Concentrations of some major nutrient elements in the annual stem segments of *Hylocomium splendens* on 7 August 1948 in boreal forest at Grenholmen, Uppland, Sweden. Segment 1 was initiated in 1948, segment 2 in 1947, and so on. (After Tamm 1953.)

Mineral supply to the sporophyte

Mineral nutrients appear to reach the developing sporophyte from the gametophyte via the conducting tissues of the seta. When Chevalier *et al.* (1977) supplied radioactively-labeled orthophosphate to gametophores of *Funaria hygrometrica*, a proportion of the ^{32}P was eventually detected in the capsule and its spores. The proportion translocated was highest (18% of total absorbed) when the capsule was green without recognizable spores, but fell to zero in plants with mature brown capsules. Uptake of ^{32}P also occurred when the solution was applied directly to the capsule indicating that absorption of nutrients from wet deposition by young sporophytes may occur in nature. Brown and Buck (1978) used an analytical approach to infer a similar pattern of nutrient cation movements from the leafy gametophores of *F. hygrometrica* to the developing sporophyte. Rydin (1997) suggested that the production of sporophytes may be an important sink for resources in bryophyte populations but this needs verification.

9.2.3 Nutrient inputs in nature

Fig. 9.4 shows the three likely sources of nutrients for terrestrial bryophyte gametophores in nature: 1) the substratum 2) wet deposition, i.e., precipitation including leachates from any plant or other surfaces over which it flows, 3) dry deposition, i.e., dust and gases (e.g., NH$_3$, SO$_2$,

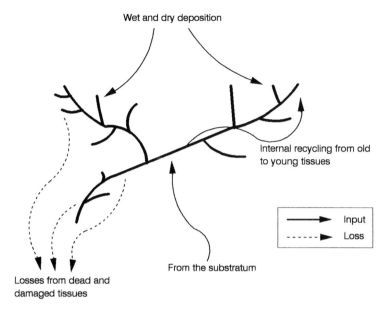

Fig. 9.4. A dynamic model of the potential inputs and losses of nutrients and non-essential elements to a bryophyte. (Reproduced from Bates (1992*a*) with permission of the British Bryological Society.)

NO₂). Bryophytes may utilize several sources for the different essential elements. Techniques that have been used to study nutrient supply include: 1) analysis of tissues and of precipitation (including canopy throughfall and tree stemflow) before and after passing through a bryophyte layer, and 2) nutrient application experiments.

Analytical studies

Tamm's (1953) study of growth and nutrition of the boreal forest moss *Hylocomium splendens* is widely regarded as a "classic" in bryology having provided the foundation for many later investigations. Uptake of water from the soil by the ectohydric "fronds" of *H. splendens* is poor and he considered it unlikely that mineral nutrients were input by this pathway. Growth rates of *H. splendens* were higher under the forest canopy than in clearings, and particularly rapid in the zone under the boundaries of tree canopies. This was also the region where Tamm demonstrated the greatest nutrient enrichment of throughfall by leachates from the tree canopy. Thus, a major conclusion was that *H. splendens* received mineral nutrients predominantly as wet deposition. Canopy leachates appeared to be important as a source of P which is present at very low concentrations in precipitation. Tamm also deduced that, despite a strong dependency of growth on moisture supply, nutrient limitation was the most important obstacle to the productivity of H. *splendens* in Norwegian forests.

The importance of wet deposition in supplying mineral elements to *Sphagnum* species in Scandinavian ombrotrophic ("rain fed") mires was also inferred by Malmer (1988). Total concentrations of several elements and especially nitrogen and sulfur showed, in some cases, correlations with known wet depositions of the elements at the mires. Press and Lee (1983) demonstrated significant acid phosphatase activity in 11 species of *Sphagnum* surveyed in Britain and Sweden. Acid phosphatase activity was negatively correlated with the total phosphorus concentration of the plants and, in experiments, increased under conditions of phosphate starvation. The study was presented against the background of increased phosphorus supply due to atmospheric pollution, but the results imply that *Sphagnum* may often utilize simple organic forms of phosphorus in its peaty environment, the supply of inorganic phosphorus in precipitation being poor (also see p. 298). With regard to nitrogen utilization by ombrotrophic peat-mosses, Woodin *et al.* (1985) demonstrated a remarkably close coupling between the atmospheric supply of nitrate ions in wet deposition and the induction and utilization of this nitrogen source by the nitrate reductase enzyme. During dry periods the nitrate reductase activity in *S. fuscum* at an unpolluted site in northern Sweden remained low, but activity rapidly increased during natural precipitation containing dilute NO_3^- (Fig. 9.5), or during experimental treatments of the moss carpet with 1 mmol NO_3^-. Woodin *et al.* (1985) noted that by efficient capture of the NO_3^- in rainwater the *Sphagnum* plants deprive higher plants rooted in the peat of this nutrient supply.

An experimental approach to the problem of determining the sources of mineral nutrients to *Calliergonella cuspidata* in Dutch chalk grassland was adopted by van Tooren *et al.* (1990). They determined nutrient concentrations in rainfall, in the water dripping from *C. cuspidata*, and in the shoots of the moss. NH_4^+ was the only ion that appeared to be absorbed by the moss in significant amounts from the natural wet and dry deposition received. Interestingly, nitrogen and phosphorus levels were significantly higher at the end of the experiment in plants on soil compared to those on acid-washed sand, and uptake of nitrogen, phosphorus, and potassium were all significantly higher in shoots on soil, and these also had a higher growth rate than the sand plants, when they were main-

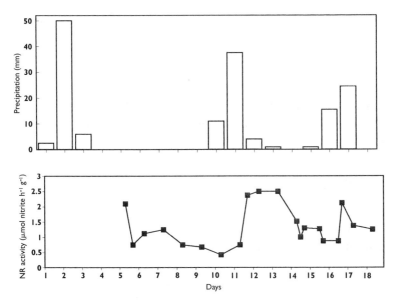

Fig. 9.5. Nitrate reductase (NR) activity in *Sphagnum fuscum* in relation to natural
precipitation (upper graph) on an ombrotrophic, subarctic mire at Abisko,
Sweden. Day 1 corresponds to 18 July 1983. NR activities are means of four
replicates. (Redrawn from Woodin *et al.* 1985.)

tained in a humid garden frame. These results, although achieved under
artificial conditions, show that nutrient uptake from soil cannot be
ignored in ectohydric bryophytes.

Nutrient application experiments
The application of nutrients to natural swards of bryophytes has been
increasingly used to examine the effectiveness of different supply pathways and to assess the likely impacts of anthropogenic increases in nutrient (mainly nitrogen) supply.

Major shifts in vegetation composition, especially the ousting of bryophytes by higher plants, invariably accompany the application of macronutrients to natural bryophyte-rich ground communities (Mickiewicz
1976, Brown 1982, Jäppinen & Hotanen 1990, Kellner & Mårshagen 1991).
However, much depends on the nature of the initial community, its
degree of isolation from sources of potential invaders, and the type and
intensity of fertilization. One of the first effects of the quickly dissolving
solid fertilizers used to improve timber yield in northern European
forests is to cause "burning" of the bryophyte tissues contacted (Jäppinen

& Hotanen 1990). Most of the common species (*Pleurozium schreberi, Hylocomium splendens, Dicranum* spp., *Sphagnum* spp.) decreased markedly in these studies, but *Polytrichum commune* appeared more resistant. The decline of *Rhytidiadelphus squarrosus* in acidic and calcareous grassland plots observed by Morecroft *et al.* (1994) in response to ammonium nitrate or ammonium sulphate additions was not accompanied by a detectable increase in higher plant cover and appears to have resulted from direct disturbance of the moss's nitrogen metabolism.

Elements present at elevated concentrations in the substratum are often found in high concentrations in bryophytes indicating direct uptake. Hébrard *et al.* (1974) provided a unique demonstration of this by fashioning artificial boulders from a concrete mixture into which a solution of the radionuclide ^{90}Sr had been mixed. The isotope readily entered shoots of *Grimmia orbicularis* and *Leucodon sciuroides* later implanted into cracks in the boulders. Maximal concentrations in the shoots were attained during prolonged wet periods, these providing the most suitable conditions for solubilization and uptake of the ^{90}Sr. Although pleurocarpous mosses in forests generally have a poorer contact with their underlying soil, Bates and Farmer (1990) eventually found elevated calcium concentrations in the young apices of *Pleurozium schreberi* plants growing over a layer of calcium carbonate powder. It was concluded that Ca^{2+} ions had moved to the apices through the cell wall (apoplast) system under the influence of an evaporative moisture flow. Nutrient flow from underlying litter is also implied in a study of grassland bryophytes by Rincón (1988).

The importance of wet deposition in supplying macronutrients to ectohydric mosses was investigated by Bates (1987, 1989*a, b*) in *Pseudoscleropodium purum* in nutrient application experiments performed under field conditions. This pleurocarpous species, like *Hylocomium splendens* and *Pleurozium schreberi*, forms monospecific carpets that are separated from the underlying soil by a layer of accumulated litter in grassland, scrub, and open forest habitats. Addition of potassium and calcium caused immediate increases of these metals in the cation-exchanger, moreover the addition of calcium displaced natural exchangeable magnesium, but levels of all three cations gradually equilibrated with their ambient availabilities as revealed by the control (Fig. 9.6). Interestingly, the calcium-treated shoots, in which exchangeable magnesium had been displaced, experienced a period of significantly lowered magnesium concentration in the intracellular fraction (Fig. 9.7) providing a first suggestion that exchangeably-held cations may be available for absorption into the cells.

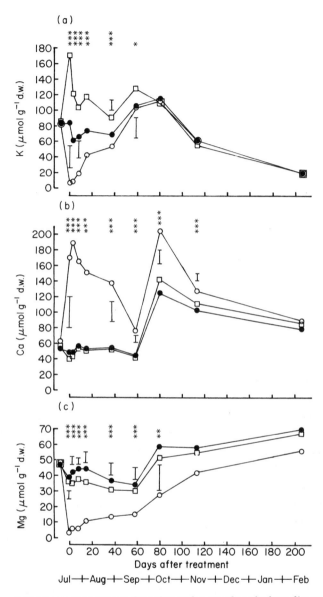

Fig. 9.6. Metal concentrations on the cation-exchanger of *Pseudoscleropodium purum* immediately before nutrient application, and at intervals afterwards, in Windsor Forest, Berkshire, UK. ● = untreated, □ = KH₂PO₄ treated, ○ = CaCl₂ treated. (a) Potassium, (b) calcium, (c) magnesium. Significance of treatment effect at each harvest: *** P < 0.001, ** P < 0.01, * P < 0.05. Vertical bars represent least significant difference (P = 0.05). (Reproduced from Bates (1989a) with permission of the British Bryological Society.)

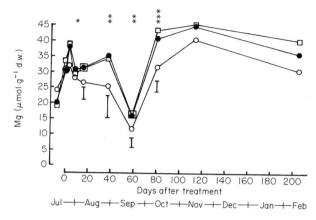

Fig. 9.7. Intracellular magnesium concentration of *Pseudoscleropodium purum* before nutrient application, and at intervals afterwards. See Fig. 9.6 for details. (Reproduced from Bates (1989*a*) with permission of the British Bryological Society.)

In these experiments, a marked and protracted rise in the level of intracellular phosphorus was observed, but cellular absorption of potassium occurred only under moist conditions.

The ecological importance of the nutrient-retaining capacity of *Pseudoscleropodium purum* has been emphasized in a comparative study with *Brachythecium* (Bates 1994). In field conditions *B. rutabulum* maintains a higher productivity than *P. purum* (Rincón 1988, Rincón & Grime 1989) by exploiting nutrients in plant litter (see p. 000). However, when nutrients were supplied in a short "pulse" and the plants cultivated in a nutrient-free environment, the relative growth rate of *P. purum* was higher than that of *B. rutabulum*. It was concluded that *P. purum* utilizes the unpredictable nutrient supply in wet deposition in an efficient and opportunistic manner, whereas *B. rutabulum* relies upon a more or less continuous input of nutrients from its litter substratum.

9.2.4 Desiccation effects on nutrient retention

Bryophytes leak cell solutes during rehydration following a period of desiccation (e.g., Gupta 1977, Brown & Buck 1979). Until recently this process had been investigated only under laboratory conditions, but Coxson (1991) reported major losses of electrolytes from tropical forest epiphytes during rehydration. The extent to which these losses may be disadvantageous, or benefit other organisms, merits further investigation.

The effects of desiccation–rehydration cycles on ion uptake and nutrient utilization have been little studied but results from several field studies (Bates 1987, 1989*a*, *b*, van Tooren *et al.* 1990, Bakken 1994) suggest that desiccation may often impair the capacities of bryophytes to benefit from favorable nutrient regimes. Bates (1997) compared the growth and nutrient accumulation from regular applications of nitrogen, phosphorus, and potassium in the mosses *Brachythecium rutabulum* and *Pseudoscleropodium purum* growing continuously hydrated or subject to intermittent drying. *P. purum* proved to be significantly more tolerant of desiccation than *B. rutabulum* and was able to absorb nutrients (e.g., phosphorus) almost as well under intermittent desiccation as under continuous hydration. Growth and nutrient uptake were both strongly suppressed by intermittent desiccation in *B. rutabulum* whose high productivity depends on long periods of continuous hydration. These observations have been extended by Bates and Bakken (1998). These studies, and those of Brown and Buck (1979), show that the cell-wall exchange sites probably sequester and recycle cations like K^+ and Mg^{2+} leaked from the protoplasts during rehydration episodes.

9.2.5 Evidence for internal recycling of nutrients

Several writers (e.g., Malmer 1988, Brown & Bates 1990, Bates 1992*a*) have speculated that internal recycling of essential elements (i.e., from old to young tissues) may occur in bryophytes, thus removing the need for continued ion absorption. What evidence is there for such recycling?

Older work (e.g., Collins & Oechel 1974, Callaghan *et al.* 1978) had suggested that translocation of resources such as photosynthates did not occur in bryophytes, except in taxa like *Polytrichum* with obvious conducting tissues. However, Alpert (1989) demonstrated movement of photoassimilate (but not mineral nutrients) from leaves to stem bases and underground stems of the ectohydric moss *Grimmia laevigata*. Rydin and Clymo (1989) also obtained evidence of movement, from old to young tissues of *Sphagnum recurvum*, of both carbon and phosphorus compounds, and demonstrated the presence of numerous plasmodesmata linking stem cells and thus affording a possible symplast pathway. Neither of these species possesses recognizable conducting tissues in the sense of Hébant (1977). Ligrone and Duckett (1994, 1996) have described "food conducting" cells in many mosses that lack conventional conducting tissues but it is unknown whether these are involved in translocation of inorganic nutrients.

Wells and Brown (1996) devised an ingenious method for testing the recycling hypothesis employing shoots of *Rhytidiadelphus squarrosus* cut to different initial sizes (4 and 8 cm) and cultivated in nutrient-free conditions. Nutrient contents were determined in the new growth (nitrogen) and in the existing 2-cm segment. New growth was being supported entirely by the nutrient content of the existing growth, with elements moving from the latter for this purpose. When the shorter (4-cm) segments were used, the withdrawal of nutrients from the parent segments was proportionately greater in response to the smaller overall pool size. Brümelis and Brown (1997) working with segment chains on *Hylocomium splendens* adjusted the nutrient pool available to the developing juvenile segment by removing branches from the parent segment. Branch removal led to a lowered potassium, magnesium, calcium, and zinc content of the juveniles. A similar approach was employed by Bates and Bakken (1998) except that nutrient pools were altered by killing (steaming) sections of stem and the two ecologically contrasted mosses *Brachythecium rutabulum* and *Pseudoscleropodium purum* were compared. Internal relocation of nutrients was important in the "low productivity" *P. purum* but not in the "high productivity" *B. rutabulum*.

An important contribution has been made by Eckstein and Karlsson (1999) who compared recycling of ^{15}N among segments of *H. splendens* with that occurring in ramets of *Polytrichum commune* in arctic Sweden. Young growth of both species was an important "sink" for nitrogen. In late summer in *P. commune* all older segments showed a net loss as the element was moved to subterranean stems. In *H. splendens* the dynamics can be summarized as follows: current-year's segments are totally dependent upon older segments for nitrogen and received a disproportionate supply of ^{15}N (i.e., the most recently absorbed nitrogen); one-year-old segments still import ^{15}N from older segments but also absorb external nitrogen and so act largely as conduits for nitrogen supply to the juveniles; two-year-old segments act as storage sites for resources; three-year-old segments are degenerating and act only as sources of nitrogen.

Collectively, these studies indicate that the relocation of elements within growing bryophytes is probably a widespread and important facet of nutrition.

9.2.6 Role of bryophytes in ecosystem nutrient dynamics
Bryophytes assume long-term dominance only in peatlands and some tundra environments where competition from higher plants is absent

(Bates 1998). Nevertheless, they can also form conspicuous components of ecosystems dominated by higher plants, notably in moist forests, or become dominant for short periods in successional or ephemeral communities. In these situations they may have an importance in the overall nutrient economy of the ecosystem that is disproportionate in relation to their often modest biomass (see reviews by Longton 1988, 1992, Slack 1988, Brown & Bates 1990, Bates 1992a).

In a successional community on glacial sands dominated by the mosses *Polytrichum juniperinum* and *Polytrichum piliferum* in New Hampshire, USA, Bowden (1991) concluded that measurements of nitrogen in bulk precipitation accounted for 58%, and nitrogen fixation and coarse organic nitrogen for 7% of the total nitrogen input. The remaining 35% of nitrogen was input as wet deposited organic nitrogen, dry deposition, and dew. Nitrogen was retained, with only small losses, by both the mosses and the accumulating organic matter in the soil. This moss-ecosystem was extremely efficient at removing nitrogen from precipitation and when the moss was removed experimentally, nitrogen losses from the ecosystem temporarily exceeded inputs.

Bryophytes may be important in more complex ecosystems by absorbing nutrients in precipitation, dust and litter before they can be taken up by the roots of higher plants (Oechel & Van Cleve 1986). This has already been mentioned with respect to utilization of wet-deposited NO_3^- by *Sphagnum* in ombrotrophic mires (see above). Detailed estimates of nutrient inputs by throughfall and litterfall to the bryophyte layer in two forests are summarized in Table 9.1. Nutrient accumulation by bryophytes, which formed about 90% of the ground flora in the Welsh oakwood, was comfortably exceeded by the inputs, but there is little excess potassium for tree and other higher-plant growth. In the nutrient-poor Alaskan black spruce forest (Table 9.1b) the bryophyte layer appears to have accumulated more of every element except calcium than was input as throughfall and litter. Moreover, the moss layer retained considerable further potential for nutrient sequestering on its cation-exchange complex. Additional nutrients may have been obtained by the bryophytes from the underlying soil (Oechel & Van Cleve 1986). Despite this efficiency in nutrient capture, two species (*Hylocomium splendens*, *Sphagnum nemoreum*) responded with higher photosynthesis when fertilized with Hoagland's solution suggesting that they had been nutrient-limited. Bryophytes may also have significance in the nutrient economies of other communities where they are less conspicuous than in forest. In Dutch

Table 9.1. *Mineral inputs and accumulation in two forests with luxuriant bryophyte ground covers*

(a) Coed Cymerau oakwood, Wales (after Rieley *et al.* 1979)	kg ha^{-1} yr^{-1}			
	Ca	Mg	K	Na
Throughfall	10.0	13.9	19.0	103.8
Litterfall	21.0	4.2	10.2	3.1
Combined throughfall and litterfall	31.0	18.1	29.2	106.9
Bryophyte accumulation	4.1	3.9	14.3	1.6
(b) Washington Creek black spruce forest, Alaska (after Oechel & Van Cleve 1986)	meq m^{-2} per season				
	N	P	Ca	Mg	K
Combined throughfall and litterfall	24.0	0.6	29.0	5.0	4.0
Bryophyte accmulation	92.0	5.0	14.0	12.0	16.0

chalk grassland they grow and absorb nutrients during autumn and winter when higher plants are inactive, and they release nutrients by decomposition in spring and summer which are utilized by the higher plants (van Tooren *et al.* 1988).

They may also influence ecosystems by retaining nutrients for long periods in undecomposed dead matter (Longton 1988, van Tooren 1988, Brown & Bates 1990). Bryophyte tissues decompose at much slower rates than those of higher plants, major factors being low temperatures, water-logging, acidity, high cation exchange capacity, presence of high contents of lignin-like compounds, accumulation of lipids, and high carbon:nitrogen ratios. Insufficient data on bryophyte decomposition rates exist for many key habitats (Brown & Bates 1990).

Whether fungi are involved in nutrient cycling between bryophytes and other ecosystem components remains uncertain. Chapin *et al.* (1987) presented data suggesting that mycorrhizal fungi of the dominant tree *Picea mariana* (black spruce) in Alaskan forest stimulated the release of phosphorus from the overlying bryophyte carpet to the tree roots. Quite a different picture was obtained by Wells and Boddy (1995) who observed translocation of ^{32}P from pieces of inoculated wood (buried in the leaf litter) to the living apices of the moss *Hypnum cupressiforme*. This occurred via the saprotrophic basidiomycete *Phanerochaete velutina* which was observed to connect to the older parts of *H. cupressiforme*. Although true mycorrhizas involving mosses are unknown, this type of association might account for the uptake of scarce elements like phosphorus from the underlying soil. Liverwort–fungus symbioses, in contrast, are relatively common (e.g., Duckett *et al.* 1991, Duckett & Read 1995) but little is known about their importance in mineral nutrition. Further important discoveries can be anticipated in this field.

9.3 Substratum ecology

9.3.1 Range of substrata occupied

Bryophytes grow on a wide range of natural substrata: soil, rock, bark, rotting wood, dung, animal carcasses, and leaf cuticles (Smith 1982*a*). From an ecological viewpoint (During, 1979, 1992, Bates 1998) the main properties that determine whether a substratum can be colonized by a particular bryophyte species are: 1) the life-span of the surface, 2) its chemical properties, and 3) its water-holding capacity. It will be appreciated that if each of these properties offered just a few distinct habitat classes, collectively they would yield a range of contrasted ecological niches. In fact many bryophytes are faithful indicators for particular sets of substratum-related conditions.

9.3.2 Longevity of substrata

During (1979) emphasized the importance of the life-span of the substratum (or its surface) in determining the kinds of bryophytes that might colonize it and successfully reproduce. It had already been established in higher plants that certain integrated sets of morphological and physiological characteristics favored particular *life-strategies* (e.g., Grime 1974). During (1979, 1992) argued that the most important habitat properties shaping the evolution of bryophytes are 1) longevity of the substratum, determining the effort to be put into rapid reproduction, and 2) the need for long-range dispersal to colonize new substratum patches, influencing the number and size of spores produced (many, small spores give the greatest chance of successful long-range dispersal). Where there is little need for long-range dispersal, a third factor, the evolutionary option to avoid any unfavorable season as a dormant spore rather than a more tender gametophyte, presents itself and favors large spore size.

The main life-strategies recognized by During (1992) are shown in Table 9.2. The first column shows types that, before the end of their lives, must colonize new substratum patches at some distance from the existing patch. Thus they produce many light spores to increase the chance of

Table 9.2. *Bryophyte life-strategies based on the revised system of During (1992)*

Potential life-span	Spores		Reproductive effort
	Numerous, small ($<$20 μm)	Few, large ($>$20 μm)	
$<$1 year	*Fugitives*	*Annual shuttle*	High
A few years	*Colonists*[a]	*Medium shuttle*[c]	
Many years	*Perennial stayers*[b]	*Dominants*	Low

Notes:
[a] Consisting of *Ephemeral colonists, Colonists sensu stricto* and *Pioneers.*
[b] Consisting of *Competitive perennials* and *Stress-tolerant perennials.*
[c] Consisting of *Short-lived shuttle* and *Long-lived shuttle.*

success. *Funaria hygrometrica* is the best-known example of a *fugitive*, a mobile species that colonizes briefly available habitat patches (e.g., gaps in turf), then spreads to other, often distant sites. *Colonists* occupy similar unpredictably-appearing habitats that persist for longer. Local population multiplication may be brought about by gemmae and rhizoid tubers (e.g., *Bryum bicolor*). Bryophytes that form long-lasting carpets on relatively stable forest floors (e.g., *Hylocomium splendens*) are *perennial stayers*. Their annual spore output is small but occurs over many successive years. The right-hand column includes species with larger spores where there is less need for long-range dispersal. *Annual shuttle* species are ephemeral plants that recolonize almost the same place (suitable "microsites") year after year from large immobile spores left by the previous generation(s). Their wider dispersal is often positively hindered by production of cleistocarpic capsules (e.g., *Ephemerum, Riccia*). Some of these species possess long-lived spores that remain dormant and enter a *diaspore bank* in the soil from which they may germinate in any of several successive years (During 1997). *Short-lived shuttle* and *long-lived shuttle* species occupy longer-lived microsites. Examples are provided by *Splachnum* spp. on dung patches and many epiphytes on twigs and branches. The longer-lived types also commonly have asexual propagules. Lastly, *dominants* refers to large-spored bryophytes that dominate certain ecosystems. *Sphagnum* species in peatlands are the only clear example (During 1992).

During's classification rests upon a subjective assessment of the correlations between bryophyte habitats and their morphological and physiological attributes. Hedderson and Longton (1995) employed multivariate analysis to study these plant and habitat concordances more objectively in three large orders of mosses (also see Longton 1997). They concluded that many of the life-strategies did exist in fact, but that these should be regarded as "noda" in a continuous network of life history variation.

9.3.3 Communities, competition, and succession

Colonization of a particular substratum by a bryophyte clearly depends on certain fundamental physiological requirements being met of which three (duration of surface, chemistry, moisture supply) can be directly determined by the substratum. Although communities composed entirely of bryophytes are frequently found on most substrata, many others in which bryophytes are ecologically important also include algae, lichens, non-lichenized fungi, filmy ferns, and other higher plants. A large and disparate literature exists describing "bryophyte" communities with many of the more recent studies using multivariate methods and other objective numerical techniques to delimit communities and explore vegetation–environment relationships (Bates 1982a, Birks et al. 1998). Much of the earlier work is summarized in Smith (1982a), but no modern synthesis exists.

An important question is whether interspecific competition plays a part in shaping the composition of bryophyte-dominated communities. Earlier reviewers emphasized the abilities of bryophytes in avoiding competition, largely by tolerating stressful habitats (e.g., Slack 1977, Grime et al. 1990). More recently several studies have indicated that both negative and positive interactions may occur between bryophyte species and influence the compositions of communities (van der Hoeven & During 1997, Rydin 1997, Bates, 1998).

It should be evident from the definitions of life-strategies employed by During (1979, 1992) that bryophytes often appear transiently within communities that represent stages in succession or cyclical processes. They are often important pioneers in the colonization of bare areas, where their longevity, tolerances of desiccation and extreme temperatures, and modest nutrient requirements are critical attributes (Longton 1992). Their presence promotes soil formation by encouraging weathering, trapping wind-blown inorganic particles and by contributing organic matter together with plant nutrients like nitrogen. They may also both facilitate and inhibit the regeneration of vascular species by altering microclimatic features and shading seedlings (Longton 1992, Rydin 1997). The dynamic aspect of bryophyte communities on most substrata requires much further investigation.

9.3.4 Substratum and chemical specialists

Numerous bryophytes exhibit a strong association with particular substrata. Sometimes these associations have a chemical basis, although in other cases there appears to be some other eco-physiological explanation. The following sections describe some of the more important substratum specialisms found amongst bryophytes.

Epiphytes

Plants that grow upon the stems of other plants without deriving sustenance from their living tissues are called epiphytes. The "host" is termed a *phorophyte*. The bark of trees in many parts of the world supports a diverse flora of epiphytic bryophytes, although they become scarce in very deep forest shade, on very acid surfaces (e.g., some conifers), under atmospheric pollution, and where the bark is abraded by winter ice or rubbed by livestock. The epiphytic flora reaches greatest luxuriance under continuously moist conditions, notably in high altitude *cloud forest* (Pócs 1982). Numerous earlier studies of epiphytic communities have been reviewed by Smith (1982b). He separated epiphytes into "obligate" and "facultative" kinds, the latter also occurring in other habitats. Trees at maturity frequently support distinct vertical zones of communities (e.g., Trynoski & Glime 1982, Cornelissen & ter Steege 1989). Young twigs often support open communities with desiccation-tolerant taxa of Orthotrichaceae, *Frullania*, and small Lejeuneaceae, whereas the lower trunk may become completely swathed in a carpet of more desiccation-sensitive Brachytheciaceae and Hypnaceae. Phorophyte axes probably support a successional progression of communities as they age, but most studies have inferred this from measurements made at only one time (Tewari *et al.* 1985, Stone 1989, Lara & Mazimpaka 1998). Direct studies of epiphyte successional dynamics are badly needed. Obligate epiphytes mostly appear to be early successional species, whereas the luxuriant climax communities of the trunk base are usually dominated by facultative epiphytes (Smith 1982b, Bates *et al.* 1997).

Although a degree of "host specificity" is encountered among epiphyte communities, it is now clear that individual epiphytes respond to the nature of the environment rather than "recognize" a particular phorophyte species (Palmer 1986, Schmitt & Slack 1990). Much ecological work has been directed at discovering the main environmental factors affecting epiphytic communities. Numerous earlier data are brought together in the monumental *Phytosociology and Ecology of Cryptogamic Epiphytes* (Barkman 1958), which may still be consulted with profit. Sampling problems are also considered by Bates (1982*a*) and John and Dale (1995). Major substratum factors influencing community composition include longevity of the tree, rate of renewal of the bark surface, water-holding capacity of the bark, its acidity and nutrient content. Life-span of the tree becomes important if one likens it to an island that is progressively acquiring a flora. This concept has mostly been discussed in the context of conservation of "old forest" taxa by sympathetic forest management (Rose 1992). Trees with rapidly flaking or peeling bark like many conifers, *Eucalyptus*, and *Betula* will clearly only be able to support rapidly establishing bryophytes with colonist and short-lived shuttle life-strategies. In dry climates a high water-holding capacity of the bark (e.g., as in *Sambucus nigra*) may be critical in allowing some species to survive as epiphytes, but this aspect has been little studied. Among trees with similar physical properties, bark acidity assumes major importance in determining community composition (Studlar 1982, Bates 1992*b*). Bark pH ranges from neutrality (e.g., *Ulmus*) to markedly acid (pH 3.5 or less in many conifers), but there is much intraspecific variation. Acidity and nutrient content of bark appear to be at least partially influenced by soil conditions (Bates 1992*b*, Gustafsson & Eriksson 1995) as well as by acid atmospheric pollutants which can easily overwhelm the limited buffering capacity (Farmer *et al.* 1991). Nutrients in precipitation, canopy leachate, and dust are likely to be the main sources for epiphytes, together with any from decomposing bark (Bengstrom & Tweedie 1998). Trees and their epiphytic coverings may also be highly effective in scavenging aerosol droplets from mists, especially in cloudy upland situations. Luxuriant epiphytes must frequently have a significant effect on forest nutrient dynamics (Rieley *et al.* 1979).

Epiphylls

In certain types of constantly humid forest bryophytes can colonize almost any relatively stable surface. In these conditions some species (called *epiphylls*) are capable of growing on the leaves of higher plants (Fig. 9.8). The phenomenon is most noticeable in the tropics and subtropics where large-leaved evergreens are prominent (e.g., Sjögren 1975, Pócs 1982). The most frequent epiphylls are tiny leafy liverworts of the Lejeuneaceae but larger taxa of *Frullania*, *Plagiochila*, *Radula*, and mosses frequently occur facultatively. Most of the specialist epiphylls are short-lived shuttle species (Table 9.2). Surprisingly, the most long-lived leaves support the lightest epiphyllous coverings and possibly these have adaptations that inhibit epiphyll growth (Coley *et al.* 1993). Following a long

Fig. 9.8. Leaves bearing epiphyllous liverworts and lichens from tropical–montane forest, near Ruhija, Bwindi Impenetrable Forest National Park, Uganda.

debate, it is generally agreed that epiphylls do not significantly reduce the photosynthetic output of host leaves. Coverings of epiphylls may actually deter herbivores but colonization by fungal pathogens is probably increased. Little is known about their nutrient relationships; however, Berrie and Eze (1975) showed movement of water and phosphate from host leaves to the epiphyllous liverwort *Radula flaccida*. Some of the rhizoids of *R. flaccida* were observed to penetrate the host's cuticle and contact the walls of epidermal and mesophyll cells. This does not appear to be an instance of outright parasitism, however, as no transfer of photosynthate was detected (Eze & Berrie 1977). The epiphyll–host relationship evidently merits fuller investigation.

Epiliths
Bryophytes inhabiting rocks have received less attention than epiphytes (Smith 1982*b*). Once again it is convenient to recognize "obligate" and "facultative" types, the genera *Gymnomitrion*, *Marsupella*, *Andreaea*, *Grimmia*, and *Racomitrium* containing many obligate epiliths. Such species possibly require a considerably more permanent and less water-retentive substratum than is provided by bark, but the reasons for their substratum selection are not well understood. Their competitive exclusion from more benign habitats on other substrata by faster-growing species is

probably a factor. Bates *et al.* (1997) speculated that reduced competition, following atmospheric pollution, accounted for occurrences as epiphytes of normally epilithic bryophytes in parts of southern Britain.

Most work on epiliths has aimed at delimiting the niches of species. Alpert (1985, 1988) studied the ability of *Grimmia laevigata* to colonize xeric microsites on rock surfaces. These were not colonized naturally, but adult plants transplanted to xeric sites survived without impairment suggesting that a greater desiccation sensitivity of the establishment phase limits a wider distribution. Jonsgard and Birks (1993) and Heegaard (1997) investigated the physical and chemical niches of *Racomitrium* and *Andreaea* species, respectively, in western Norway, using multivariate analyses and generalized linear modelling methods.

Litter species

Litter, meaning undecomposed dead plant parts, constitutes substrata ranging from the ephemeral (leaves) to the reasonably long-lasting (fallen tree trunks). All are potentially rich in plant nutrients although this may often be unavailable to bryophytes.

It is now evident that several bryophytes originally thought of as normal inhabitants of the soil surface are, at least seasonally, exploiters of litter deposited by dominant vascular plants. Rincón (1988) investigated the effects of a range of plant litter types on growth of some common grassland bryophytes. Nutrient-rich litter of the stinging nettle (*Urtica dioica*) stimulated the growth of all species and notably that of *Brachythecium rutabulum*, a moss that is frequently associated with the dense stands of *U. dioica* and other tall herbs. Rhizoidal attachments to the litter are important in nutrient exploitation (Rincón 1990).

A number of bryophytes, sometimes called *epixylic* species, occur more often on rotting logs than on other types of substratum. Most studies have been concerned with description of the succession of their bryophyte communities as fallen logs and cut stumps decay (Muhle & LeBlanc 1975). These go through a number of physical changes, loss of bark, softening of the wood, and break-up, and they may present a moister environment in later stages than initially. In a Swedish spruce forest Söderström (1988) recognized four stages in the succession: 1) facultative epiphytes that had fallen with the log (mostly lichens and *Ptilidium pulcherrimum*); 2) early epixylics (*Anastrophyllum hellerianum*, *Lophozia* spp., *Drepanocladus unciniatus*, *Cladonia* lichens) which colonize soon after the log falls; 3) late epixylics (e.g., *Lepidozia reptans*, *Brachythecium starkei*, *Dicranum scoparium*,

Plagiothecium denticulatum) which do not colonize until decay is advanced; and 4) ground flora species (e.g., *Hylocomium splendens, Pleurozium schreberi, Ptilidium crista-castrensis*) which colonize as the log becomes indistinguishable. No evidence appears to have been obtained of mineral nutrient transfers from rotting logs to bryophytes. As some of the commonest species (e.g., *Brachythecium rutabulum, Eurhynchium praelongum*) also exploit leaf litter, however, this seems highly likely. Saprotrophic fungi could be involved in cryptic nutrient transfers (cf. Wells & Boddy 1995).

Decaying logs provide a classic example of a patchily distributed habitat of limited duration and their specialist bryophytes present opportunities to test several hypotheses in population biology (e.g., Herben & Söderström 1992). Kimmerer (1994) compared the dissemination of *Dicranum flagellare*, an asexually reproducing species, with *Tetraphis pellucida*, which produces both spores and gemmae. *T. pellucida* was highly successful at rapidly colonizing new logs and stumps, *D. flagellare* persisted mainly by rapidly colonizing local gaps appearing through disturbance rather than by finding wholly new surfaces. Slugs appeared to be a major dispersal vector for the detachable branches of *D. flagellare* (Kimmerer & Young 1995).

Fire mosses

Fire is a natural event in many forest and grassland ecosystems but today man is also responsible for a very large number of accidental and deliberately started fires. These range from small bonfires, through rejuvenating "burns" (e.g., of moor and scrub), to conflagrations that engulf wildlife over vast areas of countryside. Bryophytes are usually readily destroyed in fires, but they are often a conspicuous element in the early succession on burned land. Southorn (1976) pointed out that *Funaria hygrometrica* is associated with old fire sites throughout the world. At seven experimental bonfire sites in Surrey (UK) she first observed bryophyte protonema 9 weeks after burning in spring-burnt sites but only after 25 weeks when burning was carried out in winter. Eventually a community composed of *F. hygrometrica, Ceratodon purpureus, Bryum argenteum*, and tuberous *Bryum* spp. became established in the first year after burning. This pioneer community was progressively deposed by recolonizing angiosperms in the second year, and bryophytes had vanished by the third year. In burnt *Picea mariana* forests in Labrador, Foster (1985) also recorded *F. hygrometrica* and *Ceratodon purpureus* as colonists of charred unstable surfaces, together with *Polytrichum juniperinum*. These were gradually replaced by normal forest mosses like *Pleurozium schreberi*, *Ptilium crista-castrensis* and *Hylocomium splendens* when the tree cover had re-established enabling light levels to decrease and atmospheric humidity to increase. Southorn (1976) demonstrated that under hot bonfires and slow forest fires the soil is greatly changed. Organic matter in the soils is burnt off and much soluble inorganic matter is deposited as ash including the plant nutrients potassium, magnesium, calcium, and phosphorus. The bases cause a dramatic rise in pH (up to 10.1 units in the Surrey bonfires). From culture experiments Southorn (1977) concluded that a requirement of *F. hygrometrica* for relatively high concentrations of nitrate-nitrogen and phosphorus was a major reason for its success on bonfire sites. Brown (1982), however, deduced that the raised pH may also be critical. The lag in colonization mentioned above was probably due to the presence in the ash of large quantities of ammonium-nitrogen which was found in culture work (Southorn 1977, Dietert 1979) to be detrimental to growth. Detoxification of this by leaching and the action of nitrifying microorganisms appears to be a prerequisite for colonization by *F. hygrometrica*. Southorn (1977) speculated that soluble organic toxins in ash might be responsible for delaying the colonization by vascular plants that enables *F. hygrometrica* to flourish in the first year. Brasell *et al.* (1986) reported high rates of nitrogen fixation from bryophyte/soil cores taken from *Eucalyptus* forest fire sites in Tasmania that are presumably a result of microbial activity.

Fast litter fires of grassland and heathland cause much less damage to the original vegetation and alter soil conditions comparatively little, and regrowth may occur from surviving underground parts. Following heathland fires in Scotland, Hobbs and Gimingham (1984) found that much variation in the pattern of recovery reflected the varying diversities of the pre-burn communities. The latter was determined partly by compositional differences during the heather (*Calluna vulgaris*) growth cycle. A bryophyte-dominated early recovery phase (*Campylopus paradoxus, Ceratodon purpureus, Polytrichum juniperinum, P. piliferum*) was obtained when heather stands in the "pioneer" and "building" stages were burnt. However, the pleurocarpous mosses *Hylocomium splendens, Hypnum jutlandicum*, and *Pleurozium schreberi* also regrew quickly from partly combusted mats in older stands in some cases.

Under conditions of acute drought heath fires may ignite the underlying peat with much more serious consequences for the pre-burn vegetation (Clément & Touffet 1990, Gloaguen 1990, Maltby *et al.* 1990). Here a bonfire type of succession is initiated, but the poor conditions and lack of propagules can retard higher plant recolonization for 10 years or more. On the North York Moors (UK) this was mainly due to droughting of the *Calluna* seedlings (Legg *et al.* 1992) but after some fires colonization of the invasive southern hemisphere moss *Campylopus introflexus* prevented *Calluna* regeneration (Equihua & Usher 1993).

Dung and cadaver mosses

Dung and the decomposing corpses of animals are sometimes colonized by a distinctive group of *coprophilous* bryophytes. Obligate coprophiles are restricted to the moss family Splachnaceae. Coprophiles are most frequent in otherwise nutrient-poor environments such as ombrotrophic mires, moors, and alpine and polar tundras. Among the commoner genera *Splachnum* is favored by wetter ground conditions than *Tetraplodon*, and *S. ampullaceum* has become rare in lowland Britain through widespread land drainage (Crundwell 1994). In North Wales *Tetraplodon mnioides* is most abundant beneath dangerous mountain crags where there is a steady supply of the carcasses of unfortunate sheep (Hill 1988). Under ideal conditions these mosses are highly successful so that in central Alberta uncolonized droppings are rare (Marino 1997). Useful comments on substratum preferences of individual Splachnaceae are given in Smith (1982a).

It is likely that these mosses are exploiting a rich nutrient source but surprisingly little appears to have been published on the specific nutrient requirements of the Splachnaceae. Some *Tayloria* species like *T. lingulata* are not coprophilous but grow in basic flushes. This may indicate a general high base requirement in members of the family. Those that grow on the pellets of birds of prey and carcasses may endure when only the bones remain, probably indicating a high phosphorus requirement. However, in a laboratory experiment no difference was found in the abilities of dung- and bone-inhabiting species to grow on moose (herbivore) and wolf (carnivore) dung (Marino 1991a). Marino (1997) suggested that moose droppings may lose nutrients quicker than wolf dung under field conditions by leaching. The latter author also marshaled evidence suggesting that high pH is not always characteristic of dung and carcasses, thus weakening an earlier hypothesis of Cameron and Wyatt (1989).

One of the essential traits of coprophilous mosses is their *entomophilous* behavior – a dependence on insects for spore dispersal (Koponen & Koponen 1978). The sporophyte of coprophiles is highly adapted with some flower-like properties that appear to lure flies (Koponen 1978,

Cameron & Troili 1982). The seta is long and often unusually thick, and the capsule is usually brightly coloured: red in *Splachnum rubrum*, yellow in *S. luteum*, purplish in *S. ampullaceum* and *Tetraplodon mnioides*. Its apophysis region is swollen and in some cases (e.g., *S. luteum*) it is drawn out radially into a disk-like structure (Fig. 9.9). Flies are also believed to be attracted by the emission of volatile attractants by the capsule. A range of volatile octane derivatives, organic acids, aldehydes, ketones, and alcohols has been identified but it is not known which are active (Pyysalo *et al.* 1983). The spores are sticky and readily dispersed by the flies to fresh dung or corpses.

The substrata colonized by Splachnaceae represent spectacularly small and short-lived targets for spore dispersal, but these mosses possess a highly focused dispersal mechanism. Commonly several Splachnaceae will colonize a single dung patch and a competitive struggle may ensue (Marino 1991*a*). Differences in spore maturation dates of species and the types of fly vector, however, probably enable several coprophiles to coexist in an area (Marino 1991*b*). A successional sequence may be observed with members of the Splachnaceae colonizing first, followed by less specialized colonists like *Ceratodon purpureus*, *Bryum* spp., and *Pohlia* spp., and finishing with common pleurocarpous mosses of the surrounding community (e.g., Webster & Sharp 1973, Lloret 1991). Marino (1997) has summarized what is currently known about the competitive hierarchies and population dynamics of these highly specialized and fascinating mosses.

Calcicoles and calcifuges

The distinction between *calcicole* (calcium-loving) and *calcifuge* (calcium-hating) species can be the principal dichotomy in a regional bryophyte flora (e.g., Bates 1995). Calcicoles are restricted to rocks and soils containing calcium carbonate, or inhabit waters that have flowed over or percolated through these substrata. Some calcicoles also occur on the least acid types of tree bark. Calcifuges live on substrata with an acid reaction, or in soft waters. Some bryophytes are apparently indifferent to the acidity of their substratum (e.g., the common grassland mosses *Pseudoscleropodium purum* and *Rhytidiadelphus squarrosus*) whereas others may need near neutral conditions (*neutrocline* taxa). Much less is known about the specific adaptations of calcicole and calcifuge bryophytes than their vascular plant equivalents. We may suspect that aluminium and iron, if present in the substratum, will be relatively mobile under acid conditions, but

Fig. 9.9. *Splachnum luteum* Hedw., one of the so-called "dung mosses" that are habitat specialists on decaying animal remains.

Table 9.3. *The cation exchange capacity in a range of epilithic calcicole and calcifuge mosses from western England and South Wales determined using unbuffered CaCl$_2$ solution (25 mmol) to saturate the exchange sites and SrCl$_2$ solution (25 mmol) to elute the adsorbed Ca^{2+} ions*

Values are means of three replicates and the estimated 95% confidence interval.

	Rock	Ca adsorbed (μg g^{-1} dry wt)
Calcicoles		
Ctenidium molluscum	Carboniferous limestone	15 510 ± 3497
Homalothecium sericeum	Carboniferous limestone	12 460 ± 319
Orthotrichum cupulatum	Carboniferous limestone	12 250 ± 1382
Schistidium apocarpum	Carboniferous limestone	12 940 ± 955
Tortella tortuosa	Carboniferous limestone	15 160 ± 679
Tortula ruralis	Carboniferous limestone	10 160 ± 684
	Mean	13 080
Calcifuges		
Andreaea rothii	Granite	2660 ± 124
Dicranoweisia cirrata	Old Red Sandstone	3200 ± 287
Grimmia donniana	Vitrified lead slag	2610 ± 114
Ptychomitrium polyphyllum	Old Red Sandstone	6690 ± 160
Racomitrium fasciculare	Old Red Sandstone	3330 ± 287
Racomitrium lanuginosum	Old Red Sandstone	2330 ± 287
	Mean	3470

Source: From Bates 1982*b*.

become extremely immobile under mildly alkaline conditions. This is partly supported by analyses of bryophytes growing on contrasted rock types in Scotland. A selection of calcifuges, and notably *Andreaea rothii*, contained large concentrations of iron, whereas the calcicoles contained up to 17 times more calcium (Bates 1978, 1982*b*). The highest aluminium concentrations were also found in *A. rothii* but levels of this element were not consistently higher in calcifuges than in calcicoles.

The differences in total metal concentrations correlate with marked differences in CEC of calcicole and calcifuge bryophytes (Bates 1982*b*). In epiliths CEC is 3–4 times higher in calcicoles than calcifuges (Table 9.3). Following experiments using EDTA to remove Ca^{2+} from the tissues, Bates (1982*b*) hypothesized that calcicoles had inherently leakier cell membranes than calcifuges. The elevated CEC of calcicoles was hypothesized to be necessary to ensure adequate Ca^{2+} adsorption for permeability control. Working with bryophytes of woodland soils, Büscher *et al.* (1990) found similar differentiation of CEC between calcifuge and calcicole species but reached a different conclusion. In laboratory experiments they investigated the selectivity of ion adsorption onto the cation-exchanger from mixtures of cations resembling soil solutions. They concluded that lower CEC in calcifuge mosses was an adaptation allowing avoidance of high Al^{3+} uptake, as plants with low CEC absorbed relatively less Al^{3+} from a mixed Al–Fe–Mn–Ca solution than those with high CEC. Although excessive iron has been shown to be toxic to the calcicole bryophyte *Fissidens cristatus* (Woollon 1975), we still know very little about the susceptibilities of bryophytes to elevated Al^{3+}. Indeed, the peaty ombrotrophic environments favoured by many *Sphagnum* spp. may often be acid but contain relatively little available Al^{3+}, which possibly explains the high CECs found in this genus.

By analogy with vascular plants, we may expect bryophytes to encounter iron deficiency in calcareous habitats. Structures analogous to the rhizodermal transfer cells of dicotyledonous calcicoles (e.g., Marschner 1986) have not yet been demonstrated in calcicole bryophytes but physiological adaptations to improve iron mobility from the substratum are likely to be present.

Importantly, calcium status, when isolated from variables like pH, appears to be relatively unimportant to bryophytes. *Calliergonella cuspidata*, a common moss of chalk downland and calcareous fen, shows a strong preference for pH values around neutrality and will not grow at below pH 6 even if the calcium concentration is increased (Streeter 1970). Clymo (1973) demonstrated that high calcium concentrations were relatively harmless to most *Sphagnum* spp., as were high pH values, but the combination of high pH and high calcium proved to be lethal to all taxa except those characteristic of base-rich flushes (e.g., *S. squarrosum*). Hummock species like *S. capillifolium* are the most susceptible to these conditions. Similar conclusions were drawn by Vanderpoorten and Klein (1999) based on field measurements of water chemistry and distribution patterns of some common aquatic bryophytes in waterfalls of the River Rhine. The putative calcifuges *Marsupella emarginata* and *Scapania undulata* only grew in waters with low solute content, but were relatively indifferent to pH. Conversely, the "calcicoles" *Chiloscyphus polyanthos*, *Cratoneuron filicinum*, *Rhynchostegium riparioides*, and *Thamnobryum alopecurum* usually occurred in base-rich water but they were also found when Ca^{2+} concentration was low. Although relatively indifferent to calcium status, these species are all strongly intolerant of low pH. Similar studies to these are needed over the full range of substrata occupied by bryophytes so that we can replace the inscrutable terms calcicole and calcifuge by a more explicit classification.

Halophytes

No bryophytes live permanently submerged in the oceans although the aquatic moss *Fontinalis dalecarlica* is able to grow in the northern Baltic Sea owing to its low salinity (Söderlund *et al.* 1988). On land very few species are true halophytes. Even in coastal dunes where deposition of saltwater spray is likely, Boerner and Forman (1975) found that none of the beach mosses survived a spray treatment with natural seawater. They concluded that survival in nature depended on dilution of the incoming salt before it contacted the mosses. In British saltmarshes, Adam (1976) recorded 66 bryophyte taxa living in situations where at least occasional tidal immersion would be experienced. Most of these species are widely distributed in non-saline habitats and it was suggested that some may represent halophytic ecotypes, but many probably experience relatively low salinities which are less harmful than full oceanic-strength seawater (Bates & Brown 1974). One of the best-known halophytes is *Schistidium maritimum* (syn. *Grimmia maritima*) which grows in the splash zone on non-calcareous seashore rocks (and occasionally in saltmarshes) on shores around the northern Atlantic (Fig. 9.10). It is commonly accompanied by an epilithic form of *Ulota phyllantha* and *Tortella flavovirens* which also appear to be highly salt tolerant (Bates 1975).

The physiological basis of salt tolerance in these mosses is partly understood. When *S. maritimum* is immersed in seawater the cells withstand any loss of membrane integrity, whereas glycophytic mosses like *Grimmia pulvinata* suffer a major loss of cell K^+ and influx of Na^+ from the seawater (Bates & Brown 1974). It was hypothesized that the increased permeability is caused by Na^+ ions in the seawater competing with Ca^{2+} ions that normally cross-link the polar heads of the membrane lipids. *S. maritimum* may also possess a metabolically active Na^+ efflux pump (Bates 1976). *S. maritimum* can withstand the normal ratio of calcium to sodium in seawater but when the ratio was experimentally lowered this species also suffered potassium loss and sodium entry. Its main adaptation possibly involves a high binding affinity of the membrane sites for Ca^{2+} ions. "Stress metabolites" like proline have not been reported from halophytic bryophytes; however, their high tolerance of desiccation probably means that, from an osmotic viewpoint, occasional wettings with seawater are harmless.

Fig. 9.10. Dark cushions of the halophytic moss *Schistidium maritimum* growing with lichens in the splash zone on an exposed rocky (mica-schist) seashore at Woodwick, Unst, Shetland Islands, UK.

Damage to forest bryophytes leading to severe land erosion has accompanied brine spills from oilfields in Canada. Ross *et al.* (1984) utilized calcium–potassium flushing treatments in an attempt to reverse the damage to the main mosses (*Hylocomium splendens, Pleurozium schreberi, Ptilium crista-castrensis*), but without success. They concluded that the best approach was to dilute the brine spills with freshwater as soon as possible after each accident.

9.4 Effects of pollutants

9.4.1 Pollution and bryophytes
Pollutants reach terrestrial bryophytes in the form of dry-deposited gases and particles, and in wet deposition. In the developed world the major pollutants affecting plants since the beginning of the Industrial Revolution have been smoke and sulfur dioxide (SO_2) from combustion of traditional fuels, together with various metals released into the atmosphere by smelting and other heavy industries. More recently, with socioeconomic changes, the introduction of desulfurization technology, and the switch to low-sulfur fuels, other gaseous pollutants have assumed importance. Nitrogen oxides (NO_x) have increased greatly in cities with the rise in the use of motor cars. SO_2 and NO_x may affect poikilohydric bryophytes as their solution products, but both gases are readily converted to strong acids by oxidation and solution in cloud droplets to form "acid rain." This may be deposited in precipitation and affect vegetation many miles from the point of production of the original pollutants. Ozone (O_3) is a secondary pollutant formed by a series of reactions involving hydrocarbons and NO_x ultimately derived from car exhaust gases. Some of the reactions are photochemical, and ozone is most frequently produced by day in sunny climates, although it is now known to be important over much of Europe in the summer. As ozone is also destroyed by NO_x under certain conditions it tends to reach largest concentrations in rural areas. Ammonia (NH_3) is of increasing importance as a rural pollutant being released principally by intensive animal rearing activities. Other pollutants that are of importance to bryophytes include agricultural pesticides and fertilizers (see Brown 1992), heavy metals, radionuclides, and the various forms of aquatic pollution.

Bryophytes are extremely sensitive to some of the above pollutants but less so to others. Heavy metals and radionuclides are accumulated to levels in the plant which far exceed those in the surrounding environment. These properties appear to result from: 1) the lack of a waterproof cuticle in ectohydric species and, therefore, a moist surface for deposition and free access to the cells 2) the absence of stomata in the gametophore which would otherwise allow the atmospheric environment to be excluded at night or when the plant is stressed; and 3) their high surface areas which render them highly efficient at trapping deposited molecules, particles, and droplets. Some of the most pollution-sensitive bryophytes are epiphytes, yet these have received much less attention than lichens in pollution research although of similarly high sensitivity.

9.4.2 Primary gaseous pollutants: SO_2, NO_x, and NH_3

Sulfur dioxide

SO_2 is among the most strongly phytotoxic of the atmospheric pollutants and the one that has caused most damage to bryophytes. Earlier studies (reviews: Rao 1982, Winner 1988) aimed principally at relating bryophyte distributions and performance to zones of pollution intensity using mapping and transect surveys. Following careful study of the distribution of bryophytes and lichens around the heavily polluted conurbation of Newcastle upon Tyne in north-east England, Gilbert (1968, 1970) set out the main features of bryophyte response to SO_2-polluted environments: 1) they show differential sensitivity, thus the number of species present shows a strong negative correlation with average SO_2 concentrations; 2) historically, losses of luxuriance and sporophyte production always precede extinction; 3) bryophytes in sheltered habitats like valleys are less affected than those in exposed situations; and 4) a high pH and buffer capacity of the substratum can ameliorate the effects of SO_2 on bryophytes. Certain bryophytes possess a high tolerance to SO_2 pollution and may actually benefit from components of urban pollution (e.g., *Funaria hygrometrica, Ceratodon purpureus, Leptobryum pyriforme, Bryum argenteum*), whereas the apparent tolerance of others (e.g., *Tortula muralis*) probably results from the buffering properties of their substratum (e.g., alkaline mortar). In temperate climates few bryophytes occur where the annual mean SO_2 concentration exceeds $50\,\mu g\ m^{-3}$ (0.17 ppm). Many cryptogams are so sensitive to SO_2 that recently a *critical level*[†] of $10\,\mu g\ m^{-3}$ (0.034 ppm) has been adopted to protect them by air pollution regulation in Europe (UKCLAG 1996).

A modest body of published work considers the effects of SO_2 on

[†] The concentration in the air above which adverse effects on plants and other receptors may be expected.

growth and physiology of bryophytes, and the basis of mechanisms of SO_2 resistance. Earlier studies employed fumigations with unrealistically high SO_2 concentrations and their results must be treated with caution. Bell (1973) provided direct evidence of the injurious effect of SO_2 fumigations at realistic concentrations on bryophyte growth. Significantly lower covers of *Bryum microerythrocarpum*, *Leptobryum pyriforme*, and *Bryum bicolor* were detected on soil under rye-grass fumigated for 26 weeks at 0.068 ppm SO_2 than in clean air. Working with bisulfite solutions, Ferguson *et al.* (1978) showed marked reductions in growth of some ombrotrophic *Sphagnum* species that had disappeared from the southern Pennines since the Industrial Revolution. *S. recurvum* (= *S. fallax*), which remains frequent in the southern Pennines, was least affected. Among the dissolved forms of sulfur pollution the sulfate ion (SO_4^{2-}) proved much less injurious than bisulfite (HSO_3^-).

Earlier workers demonstrated chlorophyll breakdown to phaeophytin as a major effect of SO_2 treatment but this was often caused by unrealistically high concentrations of SO_2. When *Polytrichum ohioense* was fumigated as both protonemata and mature gametophores, the protonemata were killed at concentrations around one-tenth of those needed to harm the gametophores (Nash & Nash 1974). Studies of the response of photosynthesis to SO_2 suggest that bryophytes are more sensitive to the pollutant than lichens (Inglis & Hill 1974, Ferguson & Lee 1979), but the order of sensitivity obtained by photosynthesis measurements alone often does not reflect field experience (e.g., Ferguson & Lee 1979, Goossens 1980). The effects of SO_2 on photosynthesis are immediate but little change in respiration occurs (Winner & Koch 1982, Winner & Bewley 1983). The latter authors clearly demonstrated the importance of hydration state on susceptibility to gaseous SO_2: mosses absorb less SO_2 as they dry out, but their photosynthesis is still impaired unless they are air-dry during fumigation. Studies with dissolved SO_2 have emphasized the importance of pH, solutions of low pH having the greatest toxicity. Of the three highly reactive ionic species formed, SO_3^{2-} predominates in neutral–alkaline conditions, HSO_3^- on the acid side of neutrality and "H_2SO_3" in extremely acid solutions. These readily enter the cell via ion channels or by disrupting the membrane (Wells & Richardson 1985). Inglis and Hill (1974) believed toxicity was largely due to "H_2SO_3" but Ferguson and Lee (1979) argued that the inhibitory effects could not be the result of one particular ionic component. Sulfite and bisulfite readily bind to sulfhydryl groups and deactivate enzymes, and they are known to be non-specific inhibitors of a number of membrane-based processes in plant cells (e.g., Lüttge *et al.* 1972).

Differential SO_2 sensitivity of species does not appear to be explicable by varying degrees of absorption of the pollutant (Winner & Bewley 1983). Instead various amelioration mechanisms may exist. Baxter *et al.* (1989, 1991) discovered that transition metals and particularly Fe^{3+} held on the cation-exchanger of *Sphagnum cuspidatum* and *S. recurvum* promoted the oxidation of HSO_3^- to harmless SO_4^{2-}. These metals were present in elevated concentrations in populations from the southern Pennines as a result of industrial fallout and conferred a higher SO_2 resistance that could also be produced by adding them artificially (Lee & Studholme 1992). Syratt and Wanstall (1969) concluded that tolerance in the SO_2-resistant moss *Dicranoweisia cirrata* was related to metabolic oxidation to harmless SO_4^{2-}, this being supported by the observed and continued increase in basal respiration. In *Pleurozium schreberi* and *Rhytidiadelphus triquetrus*, bisulfite suppression of photosynthesis is progressively reversed after two hours (B. Bharali, unpublished results). This is related to disappearance of HSO_3^- from the solution. Inhibitors of oxidative metabolism suppressed the disappearance of HSO_3^- from the external solution, implying that much of the oxidation is under metabolic control.

In comparison with lichens, little use has been made of bryophytes for bioindication of SO_2 levels. The *zone scale* using common urban bryophytes as indicators of mean SO_2 concentrations devised by Gilbert (1968) remains one of the best examples. The use of the popular "index of atmospheric purity" method – which derives a single number expressing to what extent the (usually epiphytic) bryophyte community is composed of pollution-resistant or pollution-sensitive taxa – has been criticized by Winner and Bewley (1978) and Winner (1988) in view of its dubious sensitivity. They concluded that a simple index of species diversity is just as effective. Transplantion of bryophytes from clean areas into polluted zones has been used in several studies (see Rao 1982 for older literature), e.g., transplants of *Tortula laevipila* exposed in polluted zones (70–125 μg SO_2 m^{-3}) of Lisbon showed modifications to the surface wax of the leaves, cell necrosis, and a failure to produce sporophytes (Sérgio 1987).

With falling levels of SO_2 in recent years recolonization of former territory by sensitive bryophytes has occurred. Greven (1992) described the reappearance and increase in the Netherlands of a number of bryophytes that had earlier been reported as lost through atmospheric pollution. Bates *et al.* (1997) obtained evidence from a transect study across southern

Britain of recolonization by several epiphytic bryophytes (e.g., *Cryphaea heteromalla, Orthotrichum affine, O. pulchellum, Ulota crispa, U. phyllantha*, and *Radula complanata*) in formerly SO_2-polluted eastern districts. Most of the recolonizers produce abundant spores or gemmae; however, recolonization by other taxa may be hindered by poor long-range dispersal. Employing transplants, Bates (1993) showed that *Rhytidiadelphus triquetrus*, a species that had become restricted to calcareous soils in polluted parts of southern England, could now survive on acid soils in formerly SO_2-polluted territory near London. Its natural spread is probably hampered by infrequent spore production. Where it is occurring, recolonization clearly offers some unique opportunities to understand the population dynamics of bryophyte communities.

Nitrogen oxides

Little work has been done on effects of realistic concentrations of nitrogen oxides on bryophytes (Farmer *et al.* 1992). Bell *et al.* (1992) subjected *Polytrichum formosum* to NO_2 at a concentration (122.4 μg m^{-3}) frequently experienced on rural roadsides. Initially growth of existing shoots was stimulated but later there was a 46% reduction of old shoots and a 36% reduction of new shoot formation. The sensitivity of the nitrate reductase enzyme (NR) to fumigations with NO and NO_2 (both at 35 ppb) was investigated in four mosses by Morgan *et al.* (1992). NO_2 stimulated NR activity probably because upon dissolution in the extracellular water it forms predominantly nitrate ions. NO fumigations caused an immediate reduction of NR activity presumably because the main solution product is the toxic nitrite ion. Continued exposure to either gas eventually led to a loss of NR activity, indicating that these pollutants may be harmful by disrupting nitrogen metabolism, but no reduction of photosynthesis was recorded.

Ammonia

Potentially ammonia (NH_3) is highly phytotoxic and in preliminary fumigation experiments Greven (1992) found that most of the species investigated were injured at concentrations above 30 μg m^{-3} (42 ppb). Leaf tip chlorosis was followed by necrosis within 3 days in *Racomitrium lanuginosum* and *Dicranum spurium*, but only after 12 days in *Campylopus flexuosus*. *Pleurozium schreberi* showed no visible injury except where NH_3 was accompanied by SO_2. These observations must be treated with caution as Lee *et al.* (1998) have pointed out that atmospheric concentrations are usually < 10 ppb and that, because of its high solubility, it is difficult to maintain steady, realistic concentrations in fumigation systems. NH_3 is probably more significant by contributing to the total nitrogen deposition, but studies in the Netherlands indicate that it has also been important in reversing acidification of tree bark caused by previous SO_2 pollution (Greven 1992).

9.4.3 Secondary pollutants: "acid rain" and ozone

"Acid rain"

"Acid rain" has two distinct effects: wet deposited acidity and wet atmospheric deposition of the nutrients nitrogen and sulfur.

Acidification has probably always accompanied dry deposition of SO_2 to some extent and their separate effects close to SO_2 sources may be difficult to distinguish. An increased acidity may damage cell membranes, solubilize potentially toxic metals like Al^{3+}, and worsen the impact of other pollutants like SO_2 (Farmer *et al.* 1992). Evidence of significant effects of acidity remote from sources of SO_2 is strong (Farmer *et al.* 1992). Farmer *et al.* (1991) presented data showing that the buffer capacity of *Quercus petraea* bark in Borrowdale (English Lake District) had been overwhelmed by acid input, to the detriment of neutrocline members of the epiphytic flora. Hallingbäck (1992) reported on the status of 10 mosses in southern Sweden showing that seven (*Antitrichia curtipendula*, *Hylocomium splendens*, *Neckera pumila*, *Orthotrichum lyellii*, *O. striatum*, and *Ulota crispa*) had declined since 1900. These declines were correlated with the low pH of rainfall, the more sensitive epiphytes remaining only on bark with high buffer capacity (e.g., *Acer*, *Fraxinus*, and *Ulmus*). Greven (1992) described changes to the bryophyte flora in several Dutch habitats that he attributed to the effects of acidification. Sjögren (1995) reported major changes in epilithic and epiphytic communities on the Swedish island of Öland in the years 1958–62 and 1988–90 that are most probably the consequence of wet acidic deposition.

With respect to nutrient levels, several studies have shown increases in tissue nitrogen levels of bryophytes that correlate with temporal deposition patterns. Farmer *et al.* (1991) reported consistently higher tissue nitrogen concentrations in epiphytic *Isothecium myosuroides* growing at Borrowdale, than at two sites with lower wet deposition. Investigating reasons for a decline in the common upland moss *Racomitrium lanuginosum* in parts of Britain, Baddeley *et al.* (1994) discovered a link between tissue nitrogen concentrations and nitrogen deposition. This was confirmed by reciprocal transplants between high and low nitrogen sites.

Table 9.4. *The increasing nitrogen content (mg g^{-1} dry wt) of* Racomitrium lanuginosum *from the summit area of Ingleborough (North Yorkshire, UK), based on analyses of herbarium material*

Year of collection	N content $\pm$ SE	Number of sites
1879	4.6 ± 0.3	6
1956	8.0 ± 0.2	4
1989	12.3 ± 0.5	12

Source: From Baddeley *et al.* 1994.

Herbarium specimens collected in the 19th century had appreciably lower tissue nitrogen than present-day plants (Table 9.4). It was concluded that high deposition of nitrogen in some areas has led to a decline of the montane *Racomitrium* heaths. Comparable results were obtained by Pitcairn *et al.* (1995). Bakken (1994) similarly reported higher nitrogen contents of *Dicranum majus* at a southern Norwegian locality where nitrogen deposition was 10-fold that experienced at a northern site. The close coupling of nitrate reductase activity in *Sphagnum* to the nitrate in wet deposition has already been mentioned (Woodin *et al.* 1985). These workers have also shown that under conditions of high nitrogen deposition, nitrate reductase inducibility is lost and the nitrogen thus becomes available to associated vascular species (Woodin & Lee 1987, Lee & Studholme 1992). Snow in upland areas is an efficient scavenger of cloud nitrogen and this can be released at high concentrations in "acid flushes" as the snow melts. Acid flushes cause loss of nitrate reductase inducibility, potassium leakage, and more serious injury to snowbed bryophytes like *Kiaeria starkei* (Woolgrove & Woodin 1996*a*, *b*).

The effects of acidity and nitrogen deposition have also been studied experimentally, usually by applying simulated acid deposition to plots of pristine ground vegetation (e.g., Raeymaekers 1987, Hutchinson & Scott 1988, Rochefort & Vitt 1988, Bakken 1993, 1995*a*) or to plants incubated or grown under artificial conditions (e.g., Farmer *et al.* 1992, Kooijman & Bakker 1994, Bakken 1995*b*, Taoda 1996). Farmer *et al.* (1992) reported that attempts to acidify bark of *Quercus* by applications of simulated acid rain were quickly reversed by the large volume of natural stemflow experienced in a high rainfall district. Most studies of ground vegetation have involved the moss layer in boreal forest or fen and there is a need to widen the range of communities and substrata investigated. Boreal forest mosses (*Hylocomium splendens, Pleurozium schreberi*) are generally harmed by solutions below pH 4.0 with losses of chlorophyll, photosynthetic capacity and nutrient content, and reductions of segment length, branch length, number of branches, and biomass. In fen, *Scorpidium scorpioides* were unharmed by pH 3.5, whereas *Tomentypnum nitens* benefited from the nutrient content of the acid rain (Rochefort & Vitt 1988). Several *Sphagnum* species also showed increased growth in the first two years but after four years the acid rain treatments produced negative effects (Rochefort *et al.* 1990). In short-term experiments, Taoda (1996) found that most epiphytes could regrow after treatment at 1.0 meq l^{-1} sulfuric–nitric acid mixture (pH $\sim$2.6) but died at 10.0 meq l^{-1}. Activity of the antioxidant enzyme superoxide dismutase was stimulated in the more acid tolerant species by dilute acid rain application but the significance of this remains unclear.

Recently, interest has switched to the fertilizing effects of wet deposition of nitrogen including ammonium inputs on nutrient-limited ecosystems like mires and acid grasslands (also see p. 258). Some investigators have added nitrogen as dilute ammonium nitrate and ammonium sulfate solutions (Aerts *et al.* 1992, Morecroft *et al.* 1994). At higher latitudes ombrotrophic mires appear to be limited by nitrogen, and its application has stimulated growth of *Sphagnum* spp. (Aerts *et al.* 1992) but in populated regions the addition of nitrogen now has little effect or is inhibitory, probably by disrupting normal nitrogen metabolism (Lee *et al.* 1998). Under conditions of nitrogen enrichment alterations in amino acid content in *Sphagnum* occur that may reduce carbon utilization for nitrogen storage (Baxter *et al.* 1992, Jauhiainen *et al.* 1998). Ultimately, phosphorus limitation appears to prevent higher productivity of *Sphagnum* in areas already experiencing high nitrogen deposition (Kooijman & Kanne 1993).

Ozone

The effects of O_3 on bryophytes have so far received little study. Only four out of 22 common British bryophytes showed impairment of photosynthesis and/or increased membrane leakage when subjected to an acute (150 ppb) ozone exposure (Lee *et al.* 1998). Gagnon and Karnosky (1992) showed differential responses of several *Sphagnum* species to chronic (3.6, 28.3, and 43.2 ppb) O_3 exposure. In a separate study (Potter *et al.* 1996) showed differences in the responses of *Sphagnum* spp. to acute ozone treatments, with *S. recurvum* (syn. *S. fallax*) – the species that is most tolerant of SO_2 – showing the greatest sensitivity. These preliminary investigations suggest that bryophytes are much less sensitive to O_3 than to other gaseous pollutants. However, Lee *et al.* (1998) point out that the effects of ozone may become much more important when other environment stresses like desiccation are experienced.

9.4.4 Heavy metals

Accumulation, toxicity, and tolerance
As a result of their efficient nutrient absorbing capabilities, bryophytes commonly accumulate metals and metalloid elements that are required only in trace quantities (e.g., Cu, Fe, Mn, Zn) or not at all (e.g., Cd, Cr, Hg, Ni, Pb, Se, Sr, Ti, V). Some of these elements may be picked up from the substratum and others from wind-blown particles or in wet deposition. The absorption appears to involve three separate processes. First, there is passive adsorption onto the bryophyte's cation-exchanger. Generally heavy metals are more effectively adsorbed than physiological cations like K^+ and Ca^{2+} and they may "condense" the exchange sites so that they are effectively immobilized (see p. 253). Second, some metals are capable of entering the cells (e.g., Brown & Beckett 1985, Brown & Sidhu 1992, Basile *et al.* 1994). Third, the numerous small leaves and intricate surfaces of bryophytes offer many possibilities for entrapment of metal-containing soil and ash particles. Collectively these processes allow bryophytes to accumulate heavy metals to concentrations far in excess of ambient concentrations or of concentrations found in vascular plants.

Rather little appears to be known about the mechanisms of metal toxicity or tolerance in bryophytes. In *Rhytidiadelphus squarrosus*, a metal-sensitive moss, Brown and Sidhu (1992) found a linear relationship between the concentration of intracellular Zn^{2+} and the inhibition of photosynthetic activity (Fig. 9.11). This suggests that enzymes and/or membranes may be poisoned when a heavy metal gains access to the cell interior. Mercury is particularly toxic: low concentrations (5–25 μmol) greatly inhibited photosynthesis, temporarily increased respiration, reduced chlorophyll levels, and caused loss of intracellular K^+ from *R. squarrosus* (Brown & Whitehead 1986). Although it had earlier been proposed that the cation-exchanger of bryophytes sequestered heavy metals, preventing their entry into the cells, there is now evidence that the more mobile metals eventually move from the cell walls into the cells (e.g.,

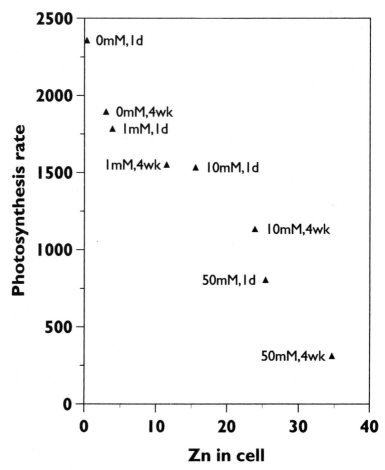

Fig. 9.11. The relationship between intracellular zinc concentration (μmol g^{-1}) and the rate of photosynthesis (ppm CO_2 min^{-1} g^{-1}) in *Rhytidiadelphus squarrosus*. Field material was supplied with zinc solutions of the stated concentration and the metal contents of 2-cm apical segments assayed after 1 day or 4 weeks. (Redrawn from Brown & Sidhu 1992.)

Brown & Sidhu 1992, Wells & Brown 1996, Brümelis & Brown 1997). With regard to metal binding within cells, Nieboer and Richardson (1980) introduced a new classification of metals intended to highlight the biological and chemical fundamentals. Metal ions were grouped into those binding electrostatically with oxygen-containing moieties (Class A), those binding covalently to S- and N-containing ligands (Class B), and those with intermediate properties (Borderline). It is fair to say that, as far as bryophytes are concerned, insufficient detailed work has been undertaken to test the usefulness of this scheme.

Some bryophytes are demonstrably "metal tolerant" being able to withstand levels of heavy metals that are toxic to other species. One famous group of species is known as the *copper mosses*. These (e.g., *Grimmia atrata, Mielichhoferia* spp., *Scopelophila cataractae*) generally occur on rocks rich in copper sulfide and may be of some use in prospecting for copper ore. It is doubtful that copper mosses have an elevated nutritional requirement for copper, but quite likely that they are extremely poor competitors with an unusually high tolerance of copper and/or its associated sulfide-generated acidity (Brown 1982). Metal-tolerant populations have also been recognized in some wide-ranging bryophytes (e.g., *Marchantia polymorpha Solenostoma crenulata, Ceratodon purpureus, Funaria hygrometrica*; see Jules & Shaw 1994) but the underlying physiological mechanisms remain obscure and require further investigation (Shaw 1994).

Monitoring heavy-metal deposition

Bryophytes have become popular organisms for *biomonitoring* – determining levels and identifying sources of – heavy-metal pollution. Two main approaches have been used: 1) surveys involving analysis of indigenous bryophytes, and 2) surveys with transplanted mosses and "moss bags." Both have the advantage that the basic materials are cheap and widely available, but there are also problems of sample reproducibility that must be carefully addressed if the results are to be meaningful. Besides revealing patterns in metal deposition, any survey of concentrations in living plant material is likely to be distorted to some extent by variations in moss growth, metal uptake, and losses imposed by habitat variations (Damman 1978).

In surveys with indigenous bryophytes, widespread pleurocarpous mosses have become the preferred subjects from a practical (and conservation) viewpoint. Careful protocols are required to standardize the samples used for analysis (reviews: Brown 1984, Burton 1986, 1990, Tyler 1990).

Country-wide surveys of metals in mosses (usually *Hylocomium splendens* or *Pleurozium schreberi*) have been repeated at intervals in Norway and Sweden. Fig. 9.12 shows some results from annual surveys in Norway. The tightly clustered nickel isopleths in the north and south are explained by emissions from copper–nickel smelters on the Kola Peninsula and by long-range transport from industrial Europe, respectively. Lead reaches

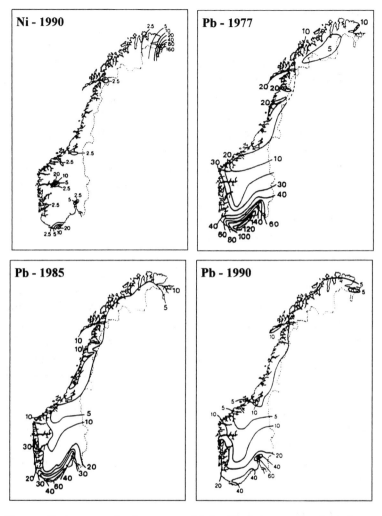

Fig. 9.12. Contour maps showing average nickel and lead concentrations (μg g^{-1}) of *Hylocomium splendens* in Norway based on samples collected at 495 sites. (Redrawn from Berg *et al.* 1995.)

Norway principally through long-range atmospheric transport and the data suggest deposition has decreased 30–40% since the first survey in 1977. Similar surveys have recently been attempted in a number of other European countries (e.g., Herpin *et al.* 1996, Markert *et al.* 1996). Multivariate analyses of these multi-element surveys enable the "signatures" from different sources to be distinguished. Berg *et al.* (1995) subjected the 1990

Norwegian data to principal component analysis and obtained the following major "axes": 1) with highest values in the south, representing long-range transport from other European countries of many elements (Bi, Pb, Sb, Mo, Cd, V, As, Zn, Tl, Hg, Ga); 2) elements associated with mineral particles in soil dust (e.g., Y, La, Al, Fe, V, Cr); 3) related to copper–nickel smelters producing Ni, Cu, Co, and As; 4) marine influence (Mg, B, Na, Sr, Ca); 5) explained by a zinc smelter in the south-west (Zn, Cd, Hg); 6) related to iron mining in the far north (Fe, Cr, Al); and 7) believed to reflect leachates from vascular plants (Cs and Rb are absorbed by roots and later transfer to mosses). Similar conclusions were reached by Kuik & Wolterbeek (1995) from a comparable analysis of heavy-metal data from *Pleurozium schreberi* samples in the Netherlands.

In areas where the natural bryophyte vegetation is poor owing to atmospheric pollution or other stresses, metal deposition can be surveyed by transplanting bryophytes or employing *moss bags*. The latter consist of small samples of moss enclosed in an inert mesh bag that can be tied to tree branches or otherwise exposed for standard periods. The moss in these bags will usually die quickly from drought stress, if it has not already been killed by acid washing to remove contamination. Therefore very long exposure times will lead to disintegration and need to be avoided. Metal accumulation depends mainly on particulate trapping and cation exchange capacity, and although several species have been used, *Sphagnum* spp. have proved most popular (Brown 1984, Burton 1986). Other types of exposure may be necessary where longer exposure periods are required. Tuba and Csintalan (1993) described the exposure of living cushions of *Tortula ruralis* within wooden boxes for three months to determine metal deposition in and around an industrial town in Hungary.

9.4.5 Radionuclides

Interest in the accumulation of radioactive isotopes by bryophytes (and lichens) was initially stimulated by the realization that fallout from nuclear weapons testing could enter foodchains, especially the tundra–caribou[†]–man foodchain in the high Arctic (Burton 1986). Bryophytes behave as almost perfect sinks for atmospheric deposition of isotopes (e.g., Svensson & Lidén 1965). Two radionuclides, ^{137}Cs and ^{90}Sr, have long half-lives and behave as analogs of the physiological elements K

† Reindeer in Europe.

and Ca. Mattsson & Lidén (1975) showed that ^{137}Cs, like K, was most abundant in the young shoot apices of *Pleurozium schreberi* and decreased with age; however, it was also retained more effectively than K in dead parts. The results indicated that ^{137}Cs was recycled from old to young tissues as the moss grew but the residence time was comparatively short (4.3 ± 0.7 years). Ultimately this is because the moss grows rapidly and leaves a proportion of the radionuclide in decaying material. Hoffman (1972) made a detailed study of ^{137}Cs transfers in a *Liriodendron* forest stand at Oak Ridge, Tennessee that is extremely revealing about sources of elements to bryophytes. He introduced the radionuclide into the tree stems through vertical slits. Eventually radioactivity was recovered in epiphytes and woodland floor bryophytes, the main pathway being via leachates from the tree canopy in the throughfall. On a dry weight basis the levels were higher in the bryophytes than in the tree foliage.

These results suggest that radionuclides could be monitored effectively utilizing bryophytes. Fig. 9.13 shows an example of moss monitoring around a point source, the Sellafield nuclear processing plant in north-west England (Sumerling 1984). Here concentrations of ^{137}Cs were determined in samples of pleurocarpous mosses. Noise in the data from uneven particulate sampling has been removed by expressing the data as *activity ratios*, in this case using the level of the uniformly distributed isotope ^{210}Pb in each sample for comparison. The data show a very steep decline in deposition within the first 4 km from the plant. Similar patterns were found for ^{144}Ce and ^{129}I.

The ban on atmospheric nuclear testing introduced in 1962 led to the cessation of many existing programs to monitor atmospheric deposition of radionuclides. However, the Chernobyl reactor accident of 26 April 1986 rekindled interest in this topic (Warner & Harrison 1993). The initial plume from this accident was carried westwards over Scandinavia, where wet deposition was greatest, and then incorporated into weather fronts moving southwestward. Some radionuclides eventually traveled northwards across Britain and locally high deposition occurred in many upland areas. Levels of radionuclides in the vegetation were so high that sales of hill sheep were suspended in affected regions. Analyses made at this time in Scotland and other European countries showed that bryophytes often had radiocesium concentrations an order of magnitude greater than in vascular plants (Livens *et al.* 1991). This is clearly a topic where further data on residence times and recycling of radionuclides within common bryophytes is much needed and may prove to have wide importance.

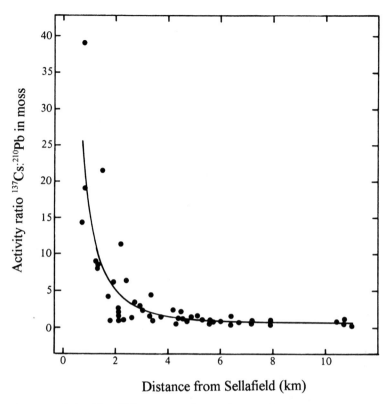

Fig. 9.13. Relationships of ^{137}Cs content of mosses (expressed as an activity ratio against ^{210}Pb content) to distance from the Sellafield nuclear processing plant, north-west England. (Redrawn from Summerling 1984.)

9.4.6 Aquatic pollution

Aquatic bryophytes may be affected by some of the pollutants already discussed (e.g., acidification, increases in nutrients, heavy metals, and radionuclides) as well as by organic pollutants, oxygen deficiency, turbidity, and temperature increases (review: Glime 1992).

Submerged aquatics accumulate heavy metals from contaminated water to a much greater extent than vascular plants, partly because their metal uptake is less seasonal, and partly because they can absorb over their entire surface. Drainage from disused metal mines is a common cause of contamination. In mid Wales the liverwort *Scapania undulata* has proved to be one of the most metal-tolerant taxa among the common species of acid upland streams. Metal concentrations in its shoots reflect those in the water (McLean & Jones 1975). In many situations levels of metals in bryophytes appear to be at equilibrium with those in the water. Kelly and Whitton (1989) established the nature of these relationships for three mosses and a liverwort based on measurements of zinc, cadmium, and lead accumulation in many European streams. Each species and element gave a different pattern, and moreover the bryophytes absorbed greater quantities of metals than three algae with which they were compared. Uptake in these cases is rapid, probably largely representing adsorption onto the cation-exchanger, and levels in bryophytes may also decrease (depuration) following a concentration spike (Mouvet *et al.* 1993, Claveri *et al.* 1994). However, uptake of mercury by *Jungermannia vulcanicola* and *Scapania undulata* from an acid stream in Japan involved formation of crystals of HgS in the cell walls (Satake *et al.* 1990). Both indigenous aquatics and transplanted samples have been widely used to monitor heavy metal pollution in rivers. Mouvet (1985) and Mouvet *et al.* (1986) give examples where pollutant releases from industrial premises have been precisely identified by monitoring metal levels in mosses at intervals along watercourses. Chlorinated hydrocarbons may also be accumulated by adsorption onto aquatic bryophytes (Mouvet *et al.* 1993).

Nutrient enrichment, particularly nitrogen and phosphorus, of waters through drainage from farmland and from sewage treatment plants is one of the most important aspects of water pollution. Rich fens, now under threat due to widespread land drainage, have been considered particularly susceptible to increased outputs of these elements from land drainage and blamed for the replacement of characteristic bryophytes like *Scorpidium* spp., *Campylium stellatum*, and *Bryum pseudotriquetrum* by *Calliergonella cuspidata* (Kooijman 1992). In fact, experiments showed little difference between the direct responses of *S. scorpioides* and *C. cuspidata* to elevated nutrient supply; however, the latter was able to benefit indirectly from improved growth of the associated vascular plants by raising itself on the framework of their litter (Kooijman & Bakker 1993). Frahm (1976) has established the tolerance levels of several aquatic mosses to high phosphate concentrations. Christmas and Whitton (1998) have described appreciable surface phosphatase activity in the aquatic mosses *Fontinalis antipyretica* and *Rhynchostegium riparioides* implying that these mosses can utilize simple organic forms of phosphorus. Activities of phosphomonoesterase (PMEase) were highest in the nutrient-poor headwaters of a stream where tissue phosphorus concentrations were low. Downstream the activity of PMEase declined progressively as water (and moss) concentrations of phosphorus increased. The authors suggest that assays of PMEase in aquatic mosses could provide a simple and reliable indication of nutrient status in streams with highly variable nutrient concentrations.

REFERENCES

Adam, P. (1976). The occurence of bryophytes on British saltmarshes. *Journal of Bryology*, **9**, 265–74.

Aerts, R., Wallén, B., & Malmer, N. (1992). Growth-limiting nutrients in *Sphagnum*-dominated bogs subject to low and high atmospheric nitrogen supply. *Journal of Ecology*, **80**, 131–40.

Alpert, P. (1985). Distribution quantified by microtopography in an assemblage of saxicolous mosses. *Vegetatio*, **64**, 131–9.

(1988). Survival of a desiccation-tolerant moss, *Grimmia laevigata*, beyond its observed microdistributional limits. *Journal of Bryology*, **15**, 219–27.

(1989). Translocation in the nonpolytrichaceous moss *Grimmia laevigata*. *American Journal of Botany*, **76**, 1524–9.

Baddeley, J. A., Thompson, D. B. A., & Lee, J. A. (1994). Regional and historical variation in the nitrogen content of *Racomitrium lanuginosum* in Britian in relation to atmospheric nitrogen deposition. *Environmental Pollution*, **84**, 189–96.

Bakken, S. (1993). Effects of simulated acid rain on the morphology, growth and chlorophyll content of *Hylocomium splendens*. *Lindbergia*, **18**, 104–10.

(1994). Growth and nitrogen dynamics of *Dicranum majus* under two contrasting nitrogen deposition regimes. *Lindbergia*, **19**, 63–72.

(1995a). Regional variation in nitrogen, protein and chlorophyll concentration in *Dicranum majus* – a reciprocal transplantation experiment. *Journal of Bryology*, **18**, 425–37.

(1995b). Effects of nitrogen supply and irradiance on growth and nitrogen status in the moss *Dicranum majus* from differently polluted areas. *Journal of Bryology*, **18**, 707–21.

Barkman, J. J. (1958). *Phytosociology and Ecology of Cryptogamic Epiphytes*. Assen: Van Gorcum.

Basile, A., Giordano, S., Cafiero, G., Spagnuolo, V., & Castaldo-Cobianchi, R. (1994). Tissue and cell localization of experimentally-supplied lead in *Funaria hygrometrica* Hedw. using X-ray SEM and TEM analysis. *Journal of Bryology*, **18**, 69–81.

Bates, J. W. (1975). A quantitative investigation of the saxicolous bryophyte and lichen vegetation of Cape Clear Island, County Cork. *Journal of Ecology*, **63**, 143–62.

(1976). Cell permeability and regulation of intracellular sodium concentration in a halophytic and a glycophytic moss. *New Phytologist*, **77**, 15–23.

(1978). The influence of metal availability on the bryophyte and macro-lichen vegetation of four types on Skye and Rhum. *Journal of Ecology*, **66**, 457–82.

(1979). The relationship between physiological vitality and age in shoot segments of *Pleurozium schreberi* (Brid.) Mitt. *Journal of Bryology*, **10**, 339–51.

(1982a). Quantitative approaches in bryophyte ecology. In *Bryophyte Ecology*, ed. A. J. E. Smith, pp. 1–44. London: Chapman & Hall.

(1982b). The role of exchangeable calcium in saxicolous calcicole and calcifuge mosses. *New Phytologist*, **90**, 239–52.

(1987). Nutrient retention by *Pseudoscleropodium purum* and its relation to growth. *Journal of Bryology*, **14**, 565–80.

(1989a). Retention of added K, Ca and P by *Pseudoscleropodium purum* growing under an oak canopy. *Journal of Bryology*, **15**, 589–605.

(1989b). Interception of nutrients in wet deposition by *Pseudoscleropodium purum*: an experimental study of uptake and retention of potassium and phosphorus. *Lindbergia*, **15**, 93–8.

(1992a). Mineral nutrient acquisition and retention by bryophytes. *Journal of Bryology*, **17**, 223–40.

(1992b). Influence of chemical and physical factors on *Quercus* and *Fraxinus* epiphytes at Loch Sunart, western Scotland: a multivariate analysis. *Journal of Ecology*, **80**, 163–79.

(1993). Regional calcicoly in the moss *Rhytidiadelphus triquetrus*: survival and chemistry of transplants at a formerly SO_2-polluted site with acid soil. *Annals of Botany*, **72**, 449–55.

(1994). Responses of the mosses *Brachythecium rutabulum* and *Pseudoscleropodium purum* to a mineral nutrient pulse. *Functional Ecology*, **8**, 686–92.

(1995). Numerical analysis of bryophyte–environment relationships in a lowland English flora. *Fragmenta Floristica et Geobotanica*, **40**, 471–90.

(1997). Effects of intermittent desiccation on nutrient economy and growth of two ecologically contrasted mosses. *Annals of Botany*, **79**, 299–309.

(1998). Is 'life-form' a useful concept in bryophyte ecology? *Oikos*, **82**, 223–37.

Bates, J. W. & Bakken, S. (1998). Nutrient retention, desiccation and redistribution in mosses. In *Bryology for the Twenty-first Century*, ed. J. W. Bates, N. W. Ashton, & J. G. Duckett, pp. 293–304. Leeds: Maney and British Bryological Society.

Bates, J. W. & Brown, D. H. (1974). The control of cation levels in seashore and inland mosses. *New Phytologist*, **73**, 483–95.

Bates, J. W. & Farmer, A. M. (1990). An experimental study of calcium acquisition and its effects on the calcifuge moss *Pleurozium schreberi*. *Annals of Botany*, **65**, 87–96.

Bates, J. W., Proctor, M. C. F., Preston, C. D., Hodgetts, N. G., & Perry, A. R. (1997). Occurrence of epiphytic bryophytes in a 'tetrad' transect across southern Britain. 1. Geographical trends in abundance and evidence of recent change. *Journal of Bryology*, **19**, 685–714.

Baxter, R., Emes, M. J., & Lee, J. A. (1989). Effects of the bisulphite ion on growth and photosynthesis in *Sphagnum cuspidatum* Hoffm. *New Phytologist*, **111**, 457–62.

(1991). Transition metals and the ability of *Sphagnum* to withstand the phytotoxic effects of the bisulphite ion. *New Phytologist*, **118**, 433–9.

(1992). Effects of an experimentally applied increase in ammonium on growth and amino-acid metabolism of *Sphagnum cuspidatum* Erhr. ex Hoffm. from differently polluted areas. *New Phytologist*, **120**, 265–74.

Bell, J. N. B. (1973). The effect of a prolonged low concentration of sulphur dioxide on the growth of two moss species. *Journal of Bryology*, **7**, 444–5.

Bell, S., Ashenden, T. W., & Rafarel, C. R. (1992). Effects of rural roadside levels of nitrogen dioxide on *Polytrichum formosum* Hedw. *Environmental Pollution, Series A*, **76**, 11–14.

Bengstrom, D. M. & Tweedie, C. E. (1998). A conceptual model for integration studies of epiphytes: nitrogen utilisation, a case study. *Australian Journal of Botany*, **46**, 273–80.

Berg, T., Røyset, O., Steinnes, E., & Vadset, M. (1995). Atmospheric trace element deposition: principal component analysis of ICP-MS data from moss samples. *Environmental Pollution*, **88**, 67–77.

Berrie, G. K. & Eze, J. M. O. (1975). The relationship between an epiphyllous liverwort and host leaves. *Annals of Botany*, **39**, 955–63.

Birks, H. J. B., Heegaard, E., Birks, H. H., & Jonsgard, B. (1998). Quantifying bryophyte–environment relationships. In *Bryology for the Twenty-first Century*, ed. J. W. Bates, N. W. Ashton, J. G. Duckett, pp. 305–19. Leeds: Maney and British Bryological Society.

Boerner, R. E. & Forman, R. T. T. (1975). Salt spray and coastal dune mosses. *Bryologist*, **78**, 57–63.

Bowden, R. D. (1991). Input, outputs, and accumulation of nitrogen in an early successional moss (*Polytrichum*) ecosystem. *Ecological Monographs*, **61**, 207–23.

Brasell, H. M., Davies, S. K., & Mattay, J. P. (1986). Nitrogen fixation associated with bryophytes colonizing burnt sites in Southern Tasmania, Australia. *Journal of Bryology*, **14**, 139–49.

Brehm, V. K. (1971). Ein *Sphagnum*-Bult als Beispiel einer natürlichen Ionenaustauschersäule. *Beiträge zur Biologie der Pflanzen*, **47**, 287–312.

Brown, D. H. (1982). Mineral nutrition. In *Bryophyte Ecology*, ed. A. J. E. Smith, pp. 383–444. London: Chapman & Hall.

(1984). Uptake of mineral elements and their use in pollution monitoring. In *The Experimental Biology of Bryophytes*, ed. A. F. Dyer & J. G. Duckett, pp. 229–55. London: Academic Press.

(1992). Impact of agriculture on bryophytes and lichens. In *Bryophytes and Lichens in a Changing Environment*, ed. J. W. Bates & A. M. Farmer, pp. 259–82. Oxford: Clarendon Press.

Brown, D. H. & Bates, J. W. (1990). Bryophytes and nutrient cycling. *Botanical Journal of the Linnean Society*, **104**, 129–47.

Brown, D. H. & Beckett, R. P. (1985). Intracellular and extracellular uptake of cadmium by the moss *Rhytidiadelphus squarrosus*. *Annals of Botany*, **55**, 179–88.

Brown, D. H. & Buck, G. W. (1978). Distribution of potassium, calcium and magnesium in the gametophyte and sporophyte generations of *Funaria hygrometrica* Hedw. *Annals of Botany*, **42**, 923–9.

(1979). Desiccation effects and cation distribution in bryophytes. *New Phytologist*, **82**, 115–25.

Brown, D. H. & Sidhu, M. (1992). Heavy metal uptake, cellular location, and inhibition of moss growth. *Cryptogamic Botany*, **3**, 82–5.

Brown, D. H. & Wells, J. M. (1988). Sequential elution technique for determining the cellular location of cations. In *Methods in Bryology*, ed. J. M. Glime, pp. 227–33. Nichinan: Hattori Botanical Laboratory.

Brown, D. H. & Whitehead, A. (1986). The effect of mercury on the physiology of *Rhytidiadelphus squarrosus*. *Journal of Bryology*, **14**, 367–74.

Brümelis, G. & Brown, D. H. (1997). Movement of metals to new growing tissue in the moss *Hylocomium splendens* (Hedw.) BSG. *Annals of Botany*, **79**, 679–86.

Burton, M. A. S. (1986). *Biological Monitoring*. London: Monitoring and Assessment Research Centre, King's College London.

(1990). Terrestrial and aquatic bryophytes as monitors of environmental contaminants in urban and industrial habitats. *Botanical Journal of the Linnean Society*, **104**, 267–80.

Büscher, P., Koedam, N., & van Spreybroeck, D. (1990). Cation-exchange properties and adaptation to soil acidity in bryophytes. *New Phytologist*, **115**, 177–86.

Callaghan, T. V., Collins, N. J., & Callaghan, C. H. (1978). Photosynthesis, growth and reproduction of *Hylocomium splendens* and *Polytrichum commune* in Swedish Lapland. *Oikos*, **31**, 73–88.

Cameron, R. G. & Troili, D. (1982). Fly-mediated spore dispersal in *Splachnum ampulaceum* (Musci). *Michigan Botanist*, **21**, 59–65.

Cameron, R. G. & Wyatt, R. (1989). Substrate restriction in entomophilous Splachnaceae. II. Effects of hydrogen ion concentration on establishment of gametophytes. *Bryologist*, **92**, 397–404.

Chapin, F. S., III, Oechel, W. C., Van Cleve, K., & Lawrence, W. (1987). The role of mosses in the phosphorus cycling of an Alaskan black spruce forest. *Oecologia*, **74**, 310–15.

Chevallier, D., Nurit, F., & Pesey, H. (1977). Orthophosphate absorption by the sporophyte of *Funaria hygrometrica* during maturation. *Annals of Botany*, **41**, 527–31.

Christmas, M. & Whitton, B. A. (1998). Phosphorus and aquatic bryophytes in the Swale–Ouse river system, north England. 1. Relationship between ambient phosphate, internal N:P ratio and surface phosphatase activity. *Science of the Total Environment*, **210**, 389–99.

Claveri, B., Morhain, E., & Mouvet, C. (1994). A methodology for the assessment of accidental copper pollution using the aquatic moss *Rhynchostegium riparioides*. *Chemosphere*. **28**, 2001–10.

Clément, B. & Touffet, J. (1990). Plant strategies and secondary succession on Brittany heathlands after severe fire. *Journal of Vegetation Science*, **1**, 195–202.

Clymo, R. S. (1963). Ion exchange in *Sphagnum* and its relation to bog ecology. *Annals of Botany*, **27**, 309–24.

(1967). Control of cation concentrations, and in particular of pH, in *Sphagnum* dominated communities. In *Chemical Environment in the Aquatic Habitat*, ed. H. L. Golterman & R. S. Clymo, pp. 273–84. Amsterdam: North Holland.

(1973). The growth of *Sphagnum*: some effects of environment. *Journal of Ecology*, **61**, 849–69.

Clymo, R. S. & Hayward, P. M. (1982). The ecology of *Sphagnum*. In *Bryophyte Ecology*, ed. A. J. E. Smith, pp. 229–89. London: Chapman & Hall.

Coley, P. D., Kursar, T. A., & Machado, J.-L. (1993). Colonization of tropical rain forest leaves by epiphylls: effects of site and host plant leaf lifetime. *Ecology*, **74**, 619–23.

Collins, N. J. & Oechel, W. C. (1974). The pattern of growth and translocation of photosynthate in a tundra moss, *Polytrichum alpinum*. *Canadian Journal of Botany*, **52**, 355–63.

Cornelissen, J. H. C. & ter Steege, H. (1989). Distribution and ecology of epiphytic bryophytes and lichens in dry evergreen forest of Guyana. *Journal of Tropical Ecology*, **5**, 131–50.

Coxson, D. S. (1991). Nutrient release from epiphytic bryophytes in tropical montane rain forest (Guadeloupe). *Canadian Journal of Botany*, **69**, 2122–9.

Crundwell, A. C. (1994). *Splachnum ampullaceum* Hedw. In *Atlas of the Bryophytes of Britain and Ireland*, vol. 3, *Mosses (Diplolepideae)*, ed. M. O. Hill, C. D. Preston, & A. J. E. Smith, p. 48. Colchester: Harley.

Damman, A. W. H. (1978). Distribution and movement of elements in ombrotrophic peat bogs. *Oikos*, **30**, 480–95.

Dietert, M. F. (1979). Studies on the gametophyte nutrition of the cosmopolitan species *Funaria hygrometrica* and *Weissia controversa*. *Bryologist*, **82**, 417–31.

Duckett, J. G. & Read, D. J. (1995). Ericoid mycorrhizas and rhizoid–ascomycete associations in liverworts share the same mycobiont: isolation of the partners and resynthesis of the associations *in vitro*. *New Phytologist*, **129**, 439–47.

Duckett, J. G., Renzaglia, K. S., & Pell, K. (1991). A light and electron microscope study of rhizoid–ascomycete associations and flagelliform axes in British hepatics. *New Phytologist*, **118**, 233–57.

During, H. J. (1979). Life strategies of bryophytes; a preliminary review. *Lindbergia*, **53**, 2–18.

(1992). Ecological classifications of bryophytes and lichens. In *Bryophytes and Lichens in a Changing Environment*, ed. J. W. Bates & A. M. Farmer, pp. 1–31. Oxford: Clarendon Press.

(1997). Bryophyte diaspore banks. In *Population Biology, Advances in Bryology*, vol. 6, ed. R. E. Longton, pp. 103–34. Berlin: J. Cramer.

Eckstein, R. L. & Karlsson, P. S. (1999). Recycling of nitrogen among segments of *Hylocomium splendens* as compared with *Polytrichum commune* – implications for clonal integration in an ectohydric bryophyte. *Oikos*, **86**, 87–96.

Equihua, M. & Usher, M. B. (1993). Impact of carpets of the invasive moss *Campylopus introflexus* on *Calluna vulgaris* regeneration. *Journal of Ecology*, **81**, 359–65.

Eze, J. M. O. & Berrie, G. K. (1977). Further investigations into the physiological relationships between an epiphyllous liverwort and its host leaves. *Annals of Botany*, **41**, 351–8.

Farmer, A. M., Bates, J. W., & Bell, J. N. B. (1991). Seasonal variations in acidic pollutant inputs and their effects on the chemistry of stemflow, bark and epiphyte tissues in three oak woodlands in N. W. Britain. *New Phytologist*, **118**, 441–51.

(1992). Ecophysiological effects of acid rain on bryophytes and lichens. In *Bryophytes and Lichens in a Changing Environment*, ed. J. W. Bates & A. M. Farmer, pp. 284–313. Oxford: Clarendon Press.

Ferguson, P. & Lee, J. A. (1979). The effects of bisulphite and sulphate upon photosynthesis in *Sphagnum*. *New Phytologist*, **82**, 703–12.

Ferguson, P., Lee, J. A., & Bell, J. N. B. (1978). Effects of sulphur pollutants on the growth of *Sphagnum* species. *Environmental Pollution, Series A*, **16**, 151–62.

Foster, D. R. (1985). Vegetation development following fire in *Picea mariana* (Black Spruce) – *Pleurozium* forests of south-eastern Labrador, Canada. *Journal of Ecology*, **73**, 517–34.

Frahm, J.-P. (1976). Weitere Toxitoleranzversuche an Wassermoosen. *Gewässer und Abwässer*, **60/61**, 113–23.

Gagnon, Z. E. & Karnosky, D. E. (1992). Physiological response of three species of *Sphagnum* to ozone exposure. *Journal of Bryology*, **17**, 81–91.

Gilbert, O. L. (1968). Bryophytes as indicators of air pollution in the Tyne Valley. *New Phytologist*, **67**, 15–30.

(1970). Further studies on the effect of sulphur dioxide on lichens and bryophytes. *New Phytologist*, **69**, 605–27.

Glime, J. M. (1992). Effects of pollutants on aquatic species. In *Bryophytes and Lichens in a Changing Environment*, ed. J. W. Bates & A. M. Farmer, pp. 333–61. Oxford: Clarendon Press.

Gloaguen, J. C. (1990). Post-burn succession on Brittany heathlands. *Journal of Vegetation Science*, **1**, 147–52.

Goossens, M. (1980). Comparison de la sensibilité de neuf espèces de bryophytes vis-à-vis du SO_2. *Bulletin du Sociéte Royale Botanique de Belgie*, **112**, 230–42.

Greven, H. C. (1992). *Changes in the Dutch Bryophyte Flora and Air Pollution*. Berlin: J. Cramer.

Grime, J. P. (1974). Vegetation classification by reference to strategies. *Nature*, **250**, 26–31.

Grime, J. P., Rincón, E. R., & Wickerson, B. E. (1990). Bryophytes and plant strategy theory. *Botanical Journal of the Linnean Society*, **104**, 175–86.

Gupta, R. K. (1977). A study of photosynthesis and leakage of solutes in relation to the desiccation effects in bryophytes. *Canadian Journal of Botany*, **55**, 1186–94.

Gustafsson, L. & Eriksson, I. (1995). Factors of importance for the epiphytic vegetation of aspen *Populus tremula* with special emphasis on bark chemistry and soil chemistry. *Journal of Applied Ecology*, **32**, 412–24.

Hällingback, T. (1992). The effect of air pollution on mosses in southern Sweden. *Biological Conservation*, **59**, 163–70.

Hébant, C. (1977). *The Conducting Tissues of Bryophytes*. Vaduz: J. Cramer.

Hébrard, J.-P., Foulquier, L., & Grauby, A. (1974). Approche expérimentale sur les possibilités de transfert du [90]Sr d'un substrat solide à une mousse terrestre: *Grimmia orbicularis* Bruch. *Bulletin de la Société Botanique de France*, **121**, 235–50.

Hedderson, T. A. & Longton, R. E. (1995). Patterns of life history variation in the Funariales, Polytrichales and Pottiales. *Journal of Bryology*, **18**, 639–75.

Heegard, E. (1997). Ecology of *Andreaea* in western Norway. *Journal of Bryology*, **19**, 527–636.

Herben, T. & Söderström, L. (1992). Which habitat parameters are most important for the persistence of a bryophyte species on patchy, temporary substrates? *Biological Conservation*, **59**, 121–6.

Herpin, U., Berlekamp, J., Markert, B., Wolterbeek, B., Grodzinska, K., Siewers, U., Lieth, H., & Weckert, V. (1996). The distribution of heavy metals in a transect of the three states the Netherlands, Germany and Poland, determined with the aid of moss monitoring. *Science of the Total Environment*, **187**, 185–98.

Hill, M. O. (1988). A bryophyte flora of North Wales. *Journal of Bryology*, **15**, 377–491.

Hobbs, R. J. & Gimingham, C. H. (1984). Studies of fire in Scottish heathland communities. II. Post-fire vegetation development. *Journal of Ecology*, **72** 585–610.

van der Hoeven, E. & During, H. J. (1997). Positive and negative interactions in bryophyte populations. In: H. de Kroon & J. van Groenendael, *The Ecology and Evolution of Clonal Plants*, ed. Leiden: Backhuys, pp. 291–310.

Hoffman, G. R. (1966). Observations on the mineral nutrition of *Funaria hygrometrica* Hedw. *Bryologist*, **69**, 182–92.

(1972). The accumulation of cesium-137 by cryptogams in a *Liriodendron tulipifera* forest. *Botanical Gazette*, **133**, 107–19.

Hutchinson, T. C. & Scott, M. (1988). The response of the feather moss *Pleurozium schreberi* (Brid.) Mitt. to five years of simulated acid precipitation in the Canadian boreal forest. *Canadian Journal of Botany*, **66**, 82–8.

Inglis, F. & Hill, D. J. (1974). The effect of sulphite and fluoride on carbon dioxide uptake by mosses in the light. *New Phytologist*, **73**, 1207–13.

Jäppinen, J.-P. & Hotanen, J.-P. (1990). Effect of fertilization on the abundance of bryophytes in two drained peatland forests in Eastern Finland. *Annales Botanici Fennici*, **27**, 93–108.

Jauhiainen, J., Silvola, J., & Vasander, H. (1998). Effects of increased carbon dioxide and nitrogen supply on mosses. In *Bryology for the Twenty-first Century*, ed. J. W. Bates, N. W. Ashton, & J. G. Duckett, pp., 343–60. Leeds: Maney and British Bryological Society.

John, E. & Dale, M. R. T. (1995). Neighbor relations within a community of epiphytic lichens and bryophytes. *Bryologist*, **98**, 29–37.

Jonsgard, B. & Birks, H. J. B. (1993). Quantitative studies on saxicolous bryophyte–environment relationships in western Norway. *Journal of Bryology*, **17**, 579–611.

Jules, E. S. & Shaw, A. J. (1994). Adaptation to metal-contaminated soils in populations of the moss *Ceratodon purpureus* – vegtative growth and reproductive expression. *American Journal of Botany*, **81**, 791–7.

Kellner, O. & Mårshagen, M. (1991). Effects of irrigation and fertilization on the ground vegetation in a 130-year-old stand of Scots pine. *Canadian Journal of Forestry Research*, **21**, 733–8.

Kelly, M. G. & Whitton, B. A. (1989). Interspecific differences in Zn, Cd and Pb accumulation by freshwater algae and bryophytes. *Hydrobiologia*, **175**, 1–11.

Kimmerer, R. W. (1994). Ecological consequences of sexual versus asexual reproduction in *Dicranum flagellare* and *Tetraphis pellucida*. *Bryologist*, **97**, 20–5.

Kimmerer, R. W. & Young, C. C. (1995). The role of slugs in dispersal of the asexual propagules of *Dicranum flagellare*. *Bryologist*, **98**, 149–53.

Kooijman, A. M. (1992). The decrease of rich-fen bryophytes in the Netherlands. *Biological Conservation*, **35**, 139–43.

Kooijman, A. M. & Bakker, C. (1993). Causes of the replacement of *Scorpidium scorpioides* by *Calliergonella cuspidata* in eutrophicated rich fens. 2. Experimental studies. *Lindbergia*, **18**, 123–30.

(1994). The acidification capacity of wetland bryophytes as influenced by simulated clean and polluted rain. *Aquatic Botany*, **48**, 133–44.

Kooijman, A. M. & Kanne, D. M. (1993). Effects of water chemistry, nutrient supply and intraspecific interactions on the replacement of *Sphagnum subnitens* by *S. fallax* in fens. *Journal of Bryology*, **17**, 431–8.

Koponen, A. M. (1978). The peristome and spores in Splachnaceae and their evolutionary and systematic significance. *Bryophytorum Bibliotheca*, **13**, 535–67.

Koponen, A. M. & Koponen, T. (1978). Evidence of entomophily in Splachnaceae (Bryophyta). *Bryophytorum Bibliotheca*, **13**, 569–77.

Kuik, P. & Wolterbeek, H. T. (1995). Factor analysis of atmospheric trace-element deposition data in the Netherlands obtained by moss monitoring. *Water, Air and Soil Pollution*, **84**, 323–46.

Lara, F. & Mazimpaka, V. (1998). Succession of epiphytic bryophytes in a *Quercus pyrenaica* forest from the Spanish Central Range (Iberian Peninsula). *Nova Hedwigia*, **67**, 125–38.

Lee, J. A. & Studholme, C. J. (1992). Responses of *Sphagnum* species to polluted environments. In *Bryophytes and Lichens in a Changing Environment*, ed. J. W. Bates & A. M. Farmer, pp. 314–32. Oxford: Clarendon Press.

Lee, J. A., Caporn, S. J. M., Carroll, J., Foot, J. P., Johnson, D., Potter, L., & Taylor, A. F. S. (1998). Effects of ozone and atmospheric nitrogen deposition on bryophytes. In *Bryology for the Twenty-first Century*, ed. J. W. Bates, N. W. Ashton, & J. G. Duckett, pp. 331–41. Leeds: Maney and British Bryological Society.

Legg, C. J., Maltby, E., & Proctor, M. C. F. (1992). The ecology of severe moorland fire on the North York Moors: seed distribution and seedling establishment of *Calluna vulgaris*. *Journal of Ecology*, **80**, 737–52.

Ligrone, R. & Duckett, J. G. (1994). Cytoplasmic polarity and endoplasmic microtubules associated with the nucleus and organelles are ubiquitous features of food conducting cells in bryoid mosses (Bryophyta). *New Phytologist*, **127**, 601–14.

Ligrone, R. & Duckett, J. G. (1996). Polarity and endoplasmic microtubules in food-conducting cells of mosses: an experimental study. *New Phytologist*, **134**, 503–16.

Livens, F. R., Horrill, A. D., & Singleton, D. L. (1991). Distribution of radiocaesium in the soil-plant systems of upland areas of Europe. *Health Physics*, **60**, 539–45.

Lloret, F. (1991). Population-dynamics of the coprophilous moss *Tayloria tenuis* in a Pyrenean forest. *Holarctic Ecology*, **14**, 1–8.

Longton, R. E. (1988). *Biology of Polar Bryophytes and Lichens*. Cambridge: Cambridge University Press.

(1992). The role of bryophytes and lichens in terrestrial ecosystems. In *Bryophytes and Lichens in a Changing Environment*, ed. J. W. Bates & A. M. Farmer, pp. 32–76. Oxford: Clarendon Press.

(1997). Reproductive biology and life-history strategies. In *Advances in Bryology*, vol. 6, *Population Studies*, ed. R. E. Longton, pp. 65–101. Berlin: J. Cramer.

Lüttge, U., Osmond, C. B., Ball, E., Brinckmann, E., & Kinze, G. (1972). Bisulfite compounds as metabolic inhibitors: nonspecific effects on membranes. *Plant and Cell Physiology*, **13**, 505–14.

McLean, R. O. & Jones, A. K. (1975). Studies of tolerance to heavy metals in the flora of the rivers Ystwyth and Clarach, Wales. *Freshwater Biology*, **5**, 431–44.

Malmer, N. (1988). Patterns in the growth and the accumulation of inorganic constituents in the *Sphagnum* cover on ombrotrophic bogs in Scandinavia. *Oikos*, **53**, 105–20.

Maltby, E., Legg, C. J., & Proctor, M. C. F. (1990). The ecology of severe moorland fire on the North York Moors: effects of the 1976 fires, and subsequent surface and vegetation development. *Journal of Ecology*, **78**, 490–518.

Marino, P. C. (1991a). Competition between mosses (Splachnaceae) in patchy habitats. *Journal of Ecology*, **79**, 1031–46.

(1991b) Dispersal and coexistence of mosses (Splachnaceae) in patchy habitats. *Journal of Ecology*, **79**, 1047–60.

(1997). Competition, dispersal and coexistence of Splachnaceae in patchy habitats. In *Advances in Bryology*, vol. 6, *Population Studies*, ed. R. E. Longton, pp. 241–63. Berlin: J. Cramer.

Markert, B., Herpin, U., Siewers, U., Berlkamp, J., & Lieth, H. (1996). The German heavy metal survey by means of mosses. *Science of the Total Environment*, **182**, 159–68.

Marschner, H. (1986). *Mineral Nutrition of Higher Plants*. London: Academic Press.

Mattsson, S. & Lidén, K. (1975). [137]Cs in carpets of the forest moss *Pleurozium schreberi*, 1961–1973. *Oikos*, **26**, 323–7.

Mickiewicz, J. (1976). Influence of mineral fertilization on the biomass of moss. *Polish Ecological Studies*, **2**, 57–62.

Morecroft, M. D., Sellers, E. K., & Lee, J. A. (1994). An experimental investigation into the effects of atmospheric nitrogen deposition on two semi-natural grasslands. *Journal of Ecology*, **82**, 475–83.

Morgan, S. M., Lee, J. A., & Ashenden, T. W. (1992). Effects of nitrogen oxides on nitrate assimilation in bryophytes. *New Phytologist*, **120**, 89–97.

Mouvet, C. (1985). The use of aquatic bryophytes to monitor heavy metals pollution of freshwaters as illustrated by case studies. *Verhein Internationale Verhein Limnologie*, **22**, 2420–5.

Mouvet, C., Pattée, E., & Cordebar, P. (1986). Utilisation des mousses aquatiques pour l'identification et la localisation précise de sources de pollution métallique multiforme. *Acta Oecologia*, **7**, 77–91.

Mouvet, C., Morhain, E., Sutter, C., & Couturieux, N. (1993). Aquatic mosses for the detection and follow-up of accidental discharges in surface waters. *Water, Air and Soil Pollution*, **66**, 333–48.

Muhle, H. & LeBlanc, F. (1975), Bryophyte and lichen succession on decaying logs. I. Analysis along an evaporational gradient in eastern Canada. *Journal of the Hattori Botanical Laboratory*, **39**, 1–33.

Nash, T. H., III, & Nash, E. H. (1974). Sensitivity of mosses to sulfur dioxide. *Oecologia*, **17**, 257–63.

Nieboer, E. & Richardson, D. H. S. (1980). The replacement of the nondescript term 'heavy metal' by a biologically and chemically significant classification of metal ions. *Environmental Pollution, Series B*, **1**, 3–26.

Oechel, W. C. & Van Cleve, K. (1986). The role of bryophytes in nutrient cycling in the taiga. In *Forest Ecosystems in the Alaskan Taiga*, ed. K. Van Cleve, F. S. Chapin III, P. W. Flanagan, L. A. Viereck, & C. T. Dyrness, pp. 121–37. New York: Springer-Verlag.

Oliver, M. J. & Bewley, J. D. (1982). Desiccation and ultrastructure in bryophytes. *Advances in Bryology*, **2**, 91–111.

Palmer, M. W. (1986). Pattern in corticolous bryophyte communities of the North Carolina Piedmont: do mosses see the forest or the trees? *Bryologist*, **89**, 59–65.

Pickering, D. C. & Puia, I. L. (1969). Mechanism for the uptake of zinc by *Fontinalis antipyretica*. *Physiologia Plantarum*, **22**, 653–61.

Pitcairn, C. E. R., Fowler, D., & Grace, J. (1995). Deposition of fixed atmospheric nitrogen and foliar nitrogen content of bryophytes and *Calluna vulgaris* (L.) Hull. *Environmental Pollution*, **88**, 193–205.

Pócs, T. (1982). Tropical forest bryophytes. In *Bryophyte Ecology*, ed. A. J. E. Smith, pp. 59–104. London: Chapman & Hall.

Potter, L., Foot, J. P., Caporn, S. J. M., & Lee, J. A. (1996). Response of four *Sphagnum* species to acute ozone fumigation. *Journal of Bryology*, **19**, 19–32.

Press, M. C. & Lee, J. A. (1983). Acid phosphatase activity in *Sphagnum* species in relation to phosphate nutrition. *New Phytologist*, **93**, 567–73.

Pyysalo, H., Koponen, A., & Koponen, T. (1983). Studies on entomophily in Splachnaceae (Musci). II. Volatile compounds in the hypophysis. *Annales Botanici Fennici*, **21**, 335–8.

Raeymaekers, G. (1987). Effects of simulated acid rain and lead on the biomass, nutrient status and heavy metal content of *Pleurozium schreberi* (Brid.) Mitt. *Journal of the Hattori Botanical Laboratory*, **63**, 219–30.

Rao, D. N. (1982). Responses of bryophytes to air pollution. In *Bryophyte Ecology*, ed. A. J. E. Smith, pp. 445–71. London: Chapman & Hall.

Raven, J. A., Griffiths, H., Smith, E. C., & Vaughn, K. C. (1998). New perspectives in the biophysics and physiology of bryophytes. In *Bryology for the Twenty-first Century*, ed. J. W. Bates, N. W. Ashton, & J. G. Duckett, pp. 261–75. Leeds: Maney and British Bryological Society.

Richter, C. & Dainty, J. (1989a). Ion behavior in plant cell walls. I. Characterization of the *Sphagnum russowii* cell wall ion exchanger. *Canadian Journal of Botany*, **67**, 451–59.

(1989b). Ion behavior in plant cell walls. II. Measurement of the Donnan free space, anion-exclusion space, anion-exchange capacity, and cation-exchange capacity in delignified *Sphagnum russowii* cell walls. *Canadian Journal of Botany*, **67**, 460–5.

Rieley, J. O., Richards, P. W., & Bebbington, A. D. L. (1979). The ecological role of bryophytes in a North Wales woodland. *Journal of Ecology*, **67**, 497–527.

Rincón, E. (1988). The effect of herbaceous litter on bryophyte growth. *Journal of Bryology*, **15**, 209–17.

(1990). Growth responses of *Brachythecium rutabulum* to different litter arrangements. *Journal of Bryology*, **16**, 120–2.

Rincón, E. & Grime, J. P. (1989). Plasticity and light interception by six bryophytes of contrasted ecology. *Journal of Ecology*, **77**, 439–46.

Rochefort, L. & Vitt, D. H. (1988). Effects of simulated acid rain on *Tomenthypnum nitens* and *Scorpidium scorpioides* in a rich fen. *Bryologist*, **91**, 121–9.

Rochefort, L, Vitt, D. H., & Bayley, S. E. (1990). Growth, production, and decomposition dynamics of *Sphagnum* under natural and experimentally acidified conditions. *Ecology*, **71**, 1986–2000.

Rose, F. (1992). Temperate forest management: its effects on bryophyte and lichen floras and habitats. In *Bryophytes and Lichens in a Changing Environment*, ed. J. W. Bates & A. M. Farmer, pp. 211–33. Oxford: Clarendon Press.

Ross, B. A., Webster, G. R., & Vitt, D. H. (1984). The role of mosses in reclamation of brine spills in forested areas. *Journal of Canadian Petroleum Technology*, **23**, 1–5.

Rühling, Å. & Tyler, G. (1970). Sorption and retention of heavy metals in the woodland moss *Hylocomium splendens* (Hedw.) Br. et Sch. *Oikos*, **21**, 92–7.

Rydin, H. (1997). Competition among bryophytes. In *Advances in Bryology*, vol. 6, *Population Studies*, ed. R. E. Longton, pp. 135–68. Berlin: J. Cramer.

Rydin, H. & Clymo, R. S. (1989). Transport of carbon and phosphorus compounds about *Sphagnum*. *Proceedings of the Royal Society of London, B*, **237**, 63–84.

Satake, K., Shibata, K., & Bando, Y. (1990). Mercury sulphide (HgS) crystals in the cell walls of the acquatic bryophytes, *Jungermannia vulcanicola* Steph. and *Scapania undulata* (L.) Dum. *Aquatic Botany*, **36**, 325–41.

Schmitt, C. K. & Slack, N. G. (1990). Host specificity of epiphytic lichens and bryophytes: a comparison of the Adirondack Mountains (New York) and the Southern Blue Ridge Mountains (North Carolina). *Bryologist*, **93**, 257–74.

Sentenac, H. & Grignon, C. (1981). A model for predicting ionic equilibrium concentrations in cell walls. *Plant Physiology*, **68**, 415–19.

Sérgio, C. (1987). Epiphytic bryophytes and air quality in the Tejo Estuary. *Symposia Biologica Hungarica*, **35**, 795–814.

Shacklette, H. T. (1965). Element content of bryophytes. *Geological Survey Bulletin*, **1198-D**, D1–D21.

Shaw, A. J. (1994). Adaptation to metals in widespread and endemic plants. *Environmental Health Perspectives*, **102**, 105–8.

Sjögren, E. (1975). Epiphyllous bryophytes of Madeira. *Svensk Botaniska Tidskrift*, **69**, 217–88.

Sjögren, E. (1995). Changes in the epilithic and epiphytic moss cover in two deciduous forest areas on the island of Öland (Sweden) – a comparison between 1958–1962 and 1988–1990. *Studies in Plant Ecology*, **19**, 1–108.

Slack, N. G. (1977). Species diversity and community structure in bryophytes: New York State studies. *New York State Museum Bulletin*, **428**, 1–70.

  (1988). The ecological importance of lichens and bryophytes. *Bibliotheca Lichenologia*, **30**, 23–53.

Smith, A. J. E. (ed.) (1982a). *Bryophyte Ecology*. London: Chapman & Hall.

Smith, A. J. E. (1982b). Epiphytes and epiliths. In *Bryophyte Ecology*, ed. A. J. E. Smith, pp. 191–227. London: Chapman & Hall.

Söderlund, S., Forsberg, A., & Pedersén, M. (1988). Concentrations of cadmium and other metals in *Fucus vesciculosus* L. and *Fontinalis dalecarlica* Br. Eur. from the northern Baltic Sea and the southern Bothnian Sea. *Environmental Pollution*, **51**, 197–212.

Söderström, L. (1988). Sequence of bryophytes and lichens in relation to substrate variables of decaying coniferous wood in Northern Sweden. *Nordic Journal of Botany*, **8**, 89–97.

Southorn, A. L. D. (1976). Bryophyte recolonization of burnt ground with particular reference to *Funaria hygrometrica*. I. Factors affecting the pattern of recolonization. *Journal of Bryology*, **9**, 63–80.

  (1977). Bryophyte recolonization of burnt ground with particular reference to *Funaria hygrometrica*. II. The nutrient requirements of *Funaria hygrometrica*. *Journal of Bryology*, **9**, 361–73.

Stone, D. F. (1989). Epiphyte succession on *Quercus ganyana* branches in the Willamette Valley of Western Oregon. *Bryologist*, **92**, 81–94.

Streeter, D. T. (1970). Bryophyte ecology. *Science Progress*, **58**, 419–34.

Studlar, S. M. (1982). Succession of epiphytic bryophytes near Mountain Lake, Virginia. *Bryologist*, **85**, 51–63.

Sumerling, T. J. (1984). The use of mosses as indicators of airborne radionuclides near a major nuclear installation. *Science of the Total Environment*, **35**, 251–65.

Svensson, G. K. & Lidén, K. (1965). The quantitative accumulation of $^{95}Zr + ^{95}Nb$ and $^{140}Ba + ^{140}La$ in carpets of forest moss. A field study. *Health Physics*, **11**, 1033–42.

Syratt, W. J. & Wanstall, P. J. (1969). The effects of sulphur dioxide on epiphytic bryophytes. In *Air Pollution*, Proceedings of the First European Congress on the Influence of Air Pollution on Plants and Animals, pp. 79–85. Wageningen: Centre for Agricultural Publishing and Documentation.

Taoda, H. (1996). Studies on the effect of environmental acidification on bryophytes and lichens. *Global Environmental Research, Japan*, **2**, 63–8.

Tamm, C. O. (1953). Growth, yield and nutrition in carpets of a forest moss (*Hylocomium splendens*). *Meddelanden från Statens Skogsforskningsinstitut*, **43**, 1–140.

Tewari, M., Upreti, N., Pandey, P., & Singh, S. P. (1985). Epiphytic succession on tree trunks in a mixed oak–cedar forest, Kumaun Himalaya. *Vegetatio*, **63**, 105–12.

van Tooren, B. F. (1988). Decomposition of bryophyte material in two Dutch chalk grasslands. *Journal of Bryology*, **15**, 343–52.

van Tooren, B. F., den Hertog, J., & Verhaar, J. (1988). Cover, biomass and nutrient content of bryophytes in Dutch chalk grassland. *Lindbergia*, **14**, 47–58.

van Tooren, B. F., van Dam, D., & During, H. J. (1990). The relative importance of precipitation and soil as sources of nutrients for *Calliergonella cuspidata* in a chalk grassland. *Functional Ecology*, **4**, 101–7.

Trebacz, K., Simonis, W., & Schönknecht, G. (1994). Cytoplasmic Ca^{2+}, K^+, Cl^- and NO_3^- activities in the liverwort *Conocephalum conicum* L. at rest and during action potentials. *Plant Physiology*, **106**, 1073–84.

Trynoski, S. E. & Glime, J. M. (1982). Direction and height of bryophytes on four species of northern trees. *Bryologist*, **85**, 281–300.

Tuba, Z. & Csintalan, Z. (1993). The use of moss transplantation technique for bioindication of heavy metal pollution. In *Plants as Biomonitors: Indicators for Heavy Metals in the Terrestrial Environment*, ed. B. Markert, pp. 253–9. Weinheim: VCH.

Tyler, G. (1990). Bryophytes and heavy-metals: a literature review. *Botanical Journal of the Linnean Society*, **104**, 231–53.

UKCLAG, United Kingdom Critical Loads Advisory Group (1996). *Critical Levels of Air Pollutants for the United Kingdom*. Penicuik: Institute of Terrestrial Ecology.

Vanderpoorten, A. & Klein, J.-P. (1999). Variations of aquatic bryophyte assemblages in the Rhine Rift related to water quality. 2. The waterfalls of the Vosges and the Black Forest. *Journal of Bryology*, **21**, 109–15.

Voth, P. D. (1943). Effects of nutrient-solution concentration on the growth of *Marchantia polymorpha*. *Botanical Gazette*, **104**, 591–601.

Warner, F. & Harrison, R. M. (eds) (1993). *Radioecology after Chernobyl*. Chichester: Wiley.

Webster, H. J. & Sharp, A. J. (1973). Bryophytic succession on caribou dung in Arctic Alaska. *American Biological Society Bulletin*, **20**, 90.

Wells, J. M. & Boddy, L. (1995). Phosphorus translocation by saprotrophic basidiomycete mycelial cord systems on the floor of a mixed deciduous woodland. *Mycological Research*, **99**, 977–80.

Wells, J. M. & Brown, D. H. (1996). Mineral nutrient recycling within shoots of the moss *Rhytidiadelphus squarrosus* in relation to growth. *Journal of Bryology*, **19**, 1–17.

Wells, J. M. & Richardson, D. H. S. (1985). Anion accumulation by the moss *Hylocomium splendens*: uptake and competition studies involving arsenate, selenate, selenite, phosphate, sulphate and sulphite. *New Phytologist*, **101**, 571–83.

Winner, W. E. (1988). Responses of bryophytes to air pollution. In *Lichens, Bryophytes and Air Quality*, ed. T. H. Nash III & V. Wirth, pp. 141–73. Berlin: J. Cramer.

Winner, W. E. & Bewley, J. D. (1978). Terrestrial mosses as bioindicators of SO_2 pollution stress. *Oecologia*, **35**, 221–30.

(1983). Photosynthesis and respiration of feather mosses fumigated at different hydration levels with SO_2. *Canadian Journal of Botany*, **61**, 1456–61.

Winner, W. E. & Koch, G. W. (1982). Water relations and SO_2 resistance of mosses. *Journal of the Hattori Botanical Laboratory*, **52**, 431–40.

Woodin, S. & Lee, J. A. (1987). The fate of some components of acidic deposition in ombrotrophic mires. *Environmental Pollution*, **45**, 61–72.

Woodin, S., Press, M. C., & Lee, J. A. (1985). Nitrate reductase activity in *Sphagnum fuscum* in relation to wet deposition of nitrate from the atmosphere. *New Phytologist*, **99**, 381–8.

Woolgrove, C. E. & Woodin, S. J. (1996a). Effects of pollutants in snowmelt on *Kiaeria starkei*, a characteristic species of late snowbed bryophyte dominated vegetation. *New Phytologist*, **133**, 519–29.

(1996b). Current and historical relationships between the tissue nitrogen content of a snowbed bryophyte and nitrogenous air pollution. *Environmental Pollution*, **91**, 283–8.

Woollon, F. B. M. (1975). Mineral relationships and ecological distribution of *Fissidens cristatus* Wils. *Journal of Bryology*, **8**, 455–64.

10

Peatlands: ecosystems dominated by bryophytes

10.1 Introduction

Peatlands are unbalanced ecosystems where plant production exceeds decomposition of organic material. As a result, organic material, or peat, accumulates. This organic material is composed primarily of plant fragments remaining after partial decomposition of the plants that at one time lived on the surface of the peatland. Decomposition occurs through the action of microorganisms that have the ability to utilize dead plant components as sources of carbon for respiration (Thormann & Bayley 1997) in both the aerobic acrotelm and the anaerobic catotelm (Clymo 1984, Wieder *et al.* 1990, Kuhry & Vitt 1996). Labile cell contents, cellulose, and hemicellulose are more readily available sources of carbon than recalcitrant fractions that contain lignin-like compounds, with these latter compounds being concentrated in peat by decomposition (Turetsky *et al.* 2000). Vascular-plant-dominated, tree, shrub, and herb layers produce less biomass (Campbell *et al.* 2000) and decompose more readily than the bryophyte-dominated ground layer (Moore 1989). Surfaces of northern peatlands are almost always completely covered by a continuous mat of moss (National Wetlands Working Group 1988, Vitt 1990), and the large amount of biomass contained in this layer is composed of cell-wall material that decomposes slowly. This slow decomposition coupled with saturated, anaerobic conditions, cool climate, and/or short growing season allows organic matter to accumulate over large areas. Thus, most peat from northern peatlands is largely composed of a high percentage of material derived from bryophytes.

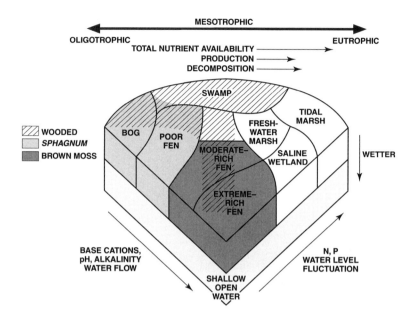

Fig. 10.1. Ternary diagram showing five wetland classes in relation to hydrology, chemistry, and vegetation. (Modified from Vitt 1994 and Zoltai & Vitt 1995.)

10.2 Classification

Wetlands are, in general, ecosystems that have accumulated some organic matter and have an abundance of hydrophytic vegetation. They can be divided into five basic types – three of which are non-peat-forming systems that are often defined as areas with less than 40 cm of peat (Zoltai & Vitt 1995). These three non-peat-forming wetland types may have a well-developed tree or shrub layer (swamps), be dominated by sedges and rushes without trees and shrubs (marshes), or contain emergent vegetation in less than a meter of water (shallow open water) (National Wetlands Working Group 1988, Zoltai & Vitt 1995). All of these wetland types have seasonally fluctuating water tables that are strongly influenced by surrounding surface and ground waters (Zoltai & Vitt 1995). Nitrogen mineralization rates are high (Bridgham *et al.* 1998), surface inflow often contains abundant nutrients, and these wetlands are often eutrophic (Mitch & Gosselink 1993) (Fig. 10.1). The lack of a well-developed bryophyte-dominated ground layer, coupled with abundant vascular

Fig. 10.2. Aerial view of two *Sphagnum*-dominated peatlands from north-eastern Alberta, Canada. The large peatland on the left is a continental bog wooded with scattered *Picea mariana*, surrounded by a narrow fen border (lagg) where inflow water to the basin is channeled. The peatland on the right represents a patterned fen, with strings and flarks oriented perpendicular to waterflow. Outflows for both peatlands are present at the top of the photo.

plant litter, allows relatively rapid decomposition and results in little peat accumulation (Thormann *et al.* 1999). Peat-forming wetlands (often termed 'mires' in Europe and 'peatlands' in North America) are ecosystems that accumulate organic matter. Although only two basic types of peatlands have generally been recognized (bogs and fens), peatlands can be more clearly defined by using hydrological, chemical, or vegetational criteria (Zoltai & Vitt 1995).

10.2.1 Hydrology

Peatlands that derive their water and nutrient supplies from precipitation and from water that has been in contact with upland soils are termed fens. Water flows through fens via one to several inflows and outflows (Fig. 10.2). The surrounding upland soil-water chemistry influences the water chemistry of fens (Siegel & Glaser 1987) often causing them to be relatively rich in base cations; however, if the surrounding uplands are relatively acidic the fens may be poor in base cations (Halsey *et al.* 1997*a*, *b*).

Peatlands that derive their water and nutrient supplies solely from precipitation are termed bogs. These peatlands are somewhat elevated above the surrounding area, and water flows from the raised bog surface onto the surrounding wetland or upland (Fig. 10.2). Therefore, bogs have relatively stagnant waters that do not reflect the surrounding soil conditions. For this reason, chemistry of the precipitation has the most important influence on the chemistry of the bog waters (Malmer 1962, Vitt *et al.* 1990, Malmer *et al.* 1992). If hydrology is considered the most significant criterion for peatland classification, then the primary division of these ecosystems is ombrogenous bogs and geogenous fens. Bogs can be viewed as ecosystems that are oligotrophic with ombrotrophic vegetation and fens are ecosystems that may be either oligotrophic or mesotrophic and are dominated by minerotrophic vegetation.

10.2.2 Chemistry

In the 1940s, Einar DuReitz recognized that Scandinavian peatlands could be divided into several types based on vegetation structure and floristic species composition (DuReitz 1949). He recognized that some peatlands have a large number of plant species with high fidelity to particular site conditions. The fens that were "rich" in site indicators he called rich fens. Other fens had fewer species with high fidelity and he called these poor fens. He recognized that ombrotrophic bogs had no plant species that were exclusive to bog conditions. In the 1950s, Hugo Sjörs published two classic papers that related pH and conductivity of the surface waters to the floristic types described by DuReitz. Sjörs (1950, 1952) described pH and conductivity gradients (as a surrogate for ionic composition) that ranged from acidic, lowly conductive bogs through somewhat less acidic poor fens to basic, highly conductive waters of rich fens. He proposed that, in fact, rich fens consisted of two types: moderate-(transitional) rich fens and extreme-rich fens. Further work by a number of researchers have carefully characterized these four peatland types in terms of both chemistry and vegetation (e.g., Gorham 1956, Slack *et al.* 1980, Malmer 1986, Vitt & Chee 1990, Gorham & Janssens 1992, Vitt *et al.* 1995a).

The range of pH in peatland surface waters of acidic bogs is from 3.0 to 4.5; acidic, poor fens is from 4.5 to 5.5; acidic to neutral, moderate-rich fens is from 5.5 to 7.0, and basic, extreme-rich fens with pH of 7.0 to about 8.5 or even higher (reviewed in Vitt 1990). Associated with this acidity gradient is one of alkalinity, with bogs and poor fens having no alkalinity; moderate-rich fens having some alkalinity (500–1000 μequivalents l^{-1}) and extreme-rich fens characterized by highly alkaline waters, often depositing Ca_2CO_3 as marl (Vitt *et al.* 1995b). Along these acidity (H^+) to

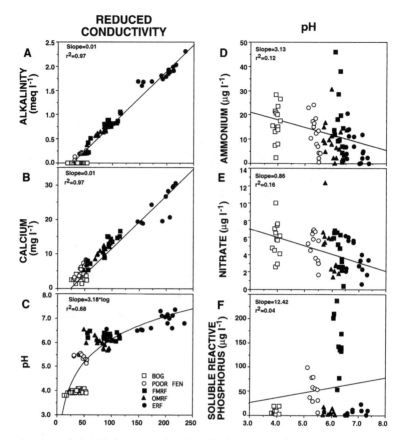

Fig. 10.3. Relationship between reduced conductivity (μS cm⁻¹) and A) alkalinity, B) calcium, and C) pH and between pH and the nutrients D) ammonium, E) nitrate, and F) soluble reactive phosphorus for surface waters along the bog–rich-fen gradient. Open symbols include bogs and poor fens, while solid symbols include several rich fens. Reduced conductivity (often termed corrected conductivity) is the total electrical conductivity minus that supplied by H⁺ (Sjörs 1952). (Modified from Vitt *et al.* 1995*a*.)

alkalinity (HCO₃⁻) gradients, base cations (Ca²⁺, Mg²⁺, Na⁺, K⁺) increase. However the limiting nutrients (NO₃⁻, NH₄⁺, P) are highly variable and show little correlation to these defining chemical components (Fig. 10.3). Additionally, bogs have surface waters with less than 3 mg l⁻¹ of calcium, whereas poor fens are somewhat higher with values of around 5 mg l⁻¹. In contrast, rich fens contain calcium concentrations ranging from 5 to 35 mg l⁻¹ or more (Fig. 10.3). Chemically, it seems that poor fens are more similar to bogs than they are to rich fens. Thus, when chemical char-

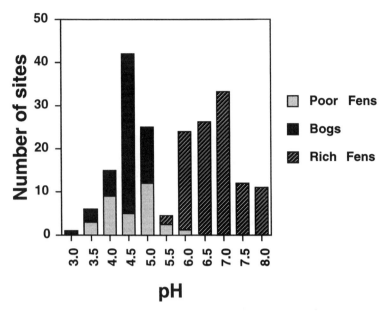

Fig. 10.4. Histogram of pH and peatland type. There are relatively few sites that have surface water pH between 5.2 and 5.7, when alkalinity values approach zero. n = 100 from continental western Canada.

acteristics are considered the critical division is between systems that are neutral to basic and alkaline (rich fens), compared to those that are acid and possess no alkalinity (poor fens and bogs).

At the regional scale, when numerous (>100) peatlands are systematically sampled, their abundance is clearly bimodal, with bogs and poor fens having pH less than ~5.0 forming one node, and rich fens with pH greater than ~6.0 forming a second node (Fig. 10.4). Fens with intermediate pH around 5.2–5.7 are rare on the northern landscape. This narrow area of pH is correlated closely with alkalinity values approaching zero (Vitt *et al.* 1995*a*).

10.2.3 Vegetation and flora

Physiognomically, peatlands vary considerably. Bogs are relatively dry, have a relatively deep acrotelm (aerobic layer), and have a large percentage of their area covered by hummocks; they may be wooded, shrub-dominated, or completely without trees (open) (Glaser & Janssens 1986, Belland & Vitt 1995). Generally, maritime bogs are open and contain pools of water (which may be arranged in reticulate or parallel patterns), while continental bogs are wooded and have no open water (Damman 1979, Glaser & Janssens 1986, Davis & Anderson 1991, Vitt *et al.* 1994). Bogs almost exclusively lack a sedge component, often have an abundance of ericaceous shrubs, and they are always dominated by *Sphagnum* moss or feather mosses and lichens (Glaser 1992, Belland & Vitt 1995). Poor and rich fens are relatively wet, have a shallow acrotelm, and have a higher percentage of their area covered by lawns and carpets (Vitt 1990). Pools are sometimes present (Vitt *et al.* 1975a). These fens may be arranged in reticulate or parallel patterns of raised, dry, elongate strings separated by pools (flarks) of water often filled with carpets of moss (Fig. 10.2) (Foster *et al.* 1983, Halsey *et al.* 1997b). Fens are usually sedge-dominated, and they may be wooded, shrubby, or open (Vitt & Chee 1990, Halsey *et al.* 1997b). Poor fens are dominated by *Sphagnum* moss and ericaceous shrubs may be present; moderate- and extreme-rich fens are dominated by "brown mosses" and ericaceous shrubs are lacking (Vitt 1990). Vascular-plant indicator species of these peatland types are regionally based, however the broad ranges of bryophytes coupled with their sensitivity to nutrient, acidity/alkalinity gradients, and water levels makes bryophytes nearly perfect sensitive indicators of peatland conditions (Table 10.1).

In conclusion, when bogs, poor fens, and rich fens are compared floristically through multivariate techniques (e.g., Nicholson *et al.* 1996, Gignac *et al.* 1998), bogs and poor fens are generally more closely related than poor fens and rich fens. Vegetationally, all three of these peatland types may be wooded, shrubby, and totally without trees (Glaser 1992, Belland & Vitt 1995, Halsey *et al.* 1997b). Bogs differ from fens in a general lack of sedges (Glaser 1992, Belland & Vitt 1995). Bogs and poor fens are generally *Sphagnum*-dominated, while rich fens are brown-moss-dominated. In oceanic areas, bogs may be patterned, while fens tend to be patterned more frequently in more continental areas. Bogs have water flowing away from their centers, while fens have water flowing through the system. Bogs have a well-developed acrotelm and are drier, while fens of all types have a poorly developed acrotelm and are wetter (Vitt *et al.* 1994). The dominance of *Sphagnum* in poor fens and bogs and lack of it in rich fens correlates well with acidity/alkalinity criteria (compare open symbols: *Sphagnum*-dominated peatlands, to solid symbols: brown-moss-dominated peatlands of Fig. 10.3). Thus if chemical, vegetational, and floristic criteria are considered important, then poor fens and bogs should be classified together as "*Sphagnum*-dominated peatlands" versus rich fens that should be called "brown-moss-dominated peatlands."

Table 10.1. Sequence of bryophyte species along the pool–hummock microtopographic gradient for the bog-rich fen vegetation–chemical gradients; species are the dominant ones found in central and western Canada (species found in oceanic peatlands of the east and west coast are not included)

Abbreviations C. = Calliergonella, D. = Drepanocladus, S. = Sphagnum, and T. = Tomentypnum.

	Permafrost bog	Bog	Poor fen	Moderate-rich fen	Extreme-rich fen
Hummock: top	S. fuscum S. lenense	S. austinii S. capillifolium S. fuscum	S. fuscum T. falcifolium	S. fuscum S. warnstorfi T. nitens	S. fuscum S. warnstorfi T. nitens
Hummock: side	S. magellanicum	S. magellanicum	S. magellanicum	S. warnstorfi T. nitens	S. warnstorfi T. nitens
Lawn	S. angustifolium S. balticum	S. angustifolium S. rubellum	S. angustifolium S. papillosum	D. vernicosus S. teres	Campylium stellatum
Carpet	S. jensenii S. majus S. riparium	S. lindbergii S. tenellum	S. jensenii S. riparium	C. cuspidata S. subsecundum	D. revolvens
Pool	none	D. fluitans S. cuspidatum	D. exannulatus	D. aduncus D. lapponicus	Scorpidium scorpioides

Source: Data derived from numerous sources, including Vitt et al. (1975b), Slack et al. (1980), Crum (1988), Gignac and Vitt (1990), Gignac et al. (1991a,b). Modified from Vitt (1994).

However if hydrological criteria are considered most important, than geogenous rich and poor fens are more similar to one another than ombrogenous bogs.

10.3 Ecological importance of the moss layer

The ground layer in peatlands is dominated by a 90–100% cover of mosses. Functioning of the peatland ecosystem is highly dependent on this moss layer, and both production and decomposition, as well as community development, are all influenced by this layer of mosses. In particular, the moss layer influences peatland function in four major ways: 1) nutrient sequestration (oligotrophication), 2) water-holding abilities, 3) decomposition, and 4) acidification.

10.3.1 Nutrient sequestration

Increasing nitrogen input to bogs (Rochefort *et al.* 1990, Aerts *et al.* 1992) and rich fens (Rochefort & Vitt 1988) results in increased moss growth, suggesting that nitrogen is limiting in most cases. However, in a few experiments, other factors such as precipitation (Bayley 1993) and phosphorus (Aerts *et al.* 1992), may influence plant growth as well.

When isotope-labeled nitrogen was added to peatlands (a bog and a rich fen) through precipitation, 98% of the nitrogen recovered after the first year was found in the top 12 cm of peat (Table 10.2). After two years, less than 2% of the added nitrogen was found in the vascular plants. In both cases, the mosses increased in growth while the vascular plants did not (Li & Vitt 1997). Although the study divided the ground layer and upper peat column into several 2–5-cm layers it did not differentiate whether the added nitrogen was situated in microorganisms, within moss cell walls, or contained in the moss cells. It appears that nitrogen is quickly sequestered by the moss layer and subsequent movement to vascular plants is dependent on release of the nitrogen from the moss and mineralization rates within the moss layer. *Sphagnum* tightly holds much of this added nitrogen and in the second year 19% of the first year's nitrogen was found in new second year's growth indicating that *Sphagnum* can translocate nitrogen upward (Li & Vitt 1997).

Recent analysis of net primary production shows that in continental western Canada at the regional scale, the four peatland types do not differ in annual production (Campbell *et al.* 2000). Overall, variability in ground production is high both spatially and temporally, but in general, the

Table 10.2. *Summary of recovery of applied nitrogen at the end of the first and second growing seasons*

	End of first growing season		End of second growing season	
	% Recovery of applied [15]N	% Distribution of [15]N	% Recovery of applied [15]N	% Distribution of [15]N
Moss-peat layer – Bog (n = 5)				
0–5 cm	44.66	65	28.84	34
5–12 cm	22.89	33	13.61	16
12–20 cm	–	–	17.73	21
20–30 cm	–	–	12.79	15
30–45 cm	–	–	11.29	12
Subtotal	67.56	98	84.26	98
Above-ground shrubs	0.99	1	1.59	2
Total	68.55	100[a]	85.85	100
Moss-peat layer – Rich fen (n = 5)				
0–5 cm	29.34	48	19.83	25
5–12 cm	31.54	51	28.29	36
12–20 cm	–	–	22.50	28
20–30 cm	–	–	5.53	7
30–45 cm	–	–	2.23	3
Subtotal	60.88	99	78.38	99
Above-ground shrubs	+	<1	1.16	1
Total	61.24	100	79.54	100

Notes: [a] does not = 100% because of rounding off of numbers.
Source: Data are from Li and Vitt (1997).

ground layer produces about 41% of the total annual plant production (Table 10.3). Thus, mosses sequester nutrients efficiently, and through release and mineralization, the moss layer effectively controls subsequent nutrient movement, and hence plant production, in peatlands (Rochefort *et al.* 1990, Li & Vitt 1997).

10.3.2 Water-holding capacity

In comparison to almost all vascular plants that are drought-avoidant, mosses are drought-tolerant. They are active when they are wet, but they also have the ability to become inactive when dry and can revitalize when re-wetted (Bewley 1979, Proctor 1979, 1984). Whereas mosses occurring in dry habitats (e.g., *Grimmia, Orthotrichum, Tortula*) dry and re-wet on daily cycles, peatland mosses have evolved these re-wetting abilities to a much

Table 10.3. *Summary of net primary production (in g m⁻²yr⁻¹); pooled state/
province means by layer for wetland types*

	Tree	Shrub	Herb	Moss	Total
Bog	106 (192)	247 (104)	13	156 (157)	449 (215)
Wooded fen	44	108	64	74	358
Shrubby fen	x	63	125	118	263
Open fen	x	x	365 (458)	163	268 (34)
Peatlands	88 (68)	210 (136)	166 (298)	139 (106)	337 (142)
Wooded swamp	542 (279)	31 (29)	62	x	654 (197)
Shrubby swamp	x	480 (260)	727 (667)	x	1232 (405)
Marsh	x	x	999 (529)	x	1034 (156)
Northern wetlands	542 (279)	255 (296)	820 (592)	x	970 (467)

Notes: Peatland and northern wetland means are pooled by wetland type and location. Standard deviations are shown in brackets for those layers where original published data did not include pooled means. For layers containing pooled means no standard deviations could be calculated. X = layers that are not present for the particular peatland type. For details see Campbell *et al.* (2000).

lesser extent (Glime & Vitt 1984). Instead, these peatland inhabitants have developed morphological adaptations to remain moist and active as long as possible, allowing longer photosynthetically active periods and thus greater growth (Stålfelt 1937). Pool species such as *Scorpidium scorpioides, Drepanocladus lapponicus, Drepanocladus exannulatus,* and *Sphagnum cuspidatum* live submerged or form poorly consolidated, emergent carpets (Vitt & Chee 1990). These species appear to have limited physiological abilities to live for extended time after drying out (but little information is available on this topic). Among the rich-fen species, hummock species such as *Tomentypnum nitens* have abundant stem tomentum and numerous side branches supposedly allowing for water uptake. Dense canopy structure of *Tomentypnum* also decreases the evaporative rates. These adaptations can also be seen in other peatland species such as *Aulacomnium palustre, Catoscopium nigritum, Dicranum undulatum, Polytrichum strictum,* and *Tomentypnum falcifolium.*

Although dead hyaline leaf cells that contain pores and hold large amounts of water occur sporadically throughout mosses (Proctor 1984), their best development is in the genus *Sphagnum.* There the leaves consist of alternating large, hyaline cells enclosing smaller, living, green cells, in a 1:2 ratio that is unique among plants. Additionally, the stems and branches are often encased in one or more layers of dead enlarged cells.

Fig. 10.5. Photograph of *Sphagnum fimbriatum* (right) and *S. palustre* (left). Note the complex canopy of dense capitulum and associated spreading and hanging branches.

These hyaline cells have lost their living cell contents very early on in stem development and as a result, the ratio of carbon to nitrogen remaining in the dead cell walls increases. This large increase in relatively high carbon:nitrogen ratio influences decomposition rates. However, the large hyaline cells, reinforced with internal cell-wall thickenings (termed fibrils), partially to entirely enclose photosynthetically active green cells. Thus, through internal hyaline cell water-holding capacity and through a complex canopy of dense capitula (Rydin & McDonald 1985), spreading and hanging branches, and concave leaves, *Sphagnum* produces a plant morphology suggestive of drought avoidance, not drought toleration (Fig. 10.5). Hummock species such as *S. fuscum* and *S. capillifolium* have the most highly developed canopies that protect the living green cells from drying out, while lawn species such as *S. angustifolium* have more poorly developed water-retention abilities. Research done by Titus and Wagner (1984) and Wagner and Titus (1984) has shown that species occurring on the highest hummocks occur there not because they are drought-tolerant, but because they are drought-avoidant. Lawn species, however, occurring in wetter, but more variable conditions, are in fact physiologically more drought-insensitive and lack the highly developed canopy modifications of the true hummock formers. This evidence is made more convincing through establishment experiments where Li and Vitt (1994) showed that the lawn species *S. angustifolium* establishes efficiently on bare peat without the protection of a developed canopy, whereas *S. fuscum* and *S. magellanicum,* both hummock formers, do not establish efficiently on bare peat.

These water-holding capacities allow *Sphagnum* to raise local water tables and increase available peatland habitats in areas where upland-adapted plants once occurred. This swamping or paludification of neighboring habitats is a major factor in increasing the amount of organic terrain in northern landscapes (Vitt & Kuhry 1992).

The presence of dead hyaline cells in *Sphagnum* allows these species to enclose living cells with water and conserve temperature and evaporative stress. Hyaline-cell modifications allow *Sphagnum* species to retain hydrated conditions longer in order to increase photosynthetic active periods and hence growth. In addition, hyaline cells allow *Sphagnum* species to move water away from the water table, thus increasing water table height and expanse, creating greater amounts of anaerobic conditions; and through complex canopy development allow *Sphagnum* species to expand habitat space.

10.3.3 Decomposition

Bryophytes are small organisms that appear to have difficulty tolerating large water-level fluctuations (Zoltai & Vitt 1990). In wetlands where water-level fluctuations are seasonally quite variable, such as in swamps and marshes, bryophytes are not dominant (Vitt 1994). These non-peat-forming wetlands, dominated by vascular plants that produce copious litter, have relatively high rates of decomposition (Mitsch & Gosselink 1993). Although it has been commonly argued that one of the factors attributing to slow decomposition rates in peatlands is acidity, in fact both acidic and basic peatlands accumulate large amounts of peat. In continental boreal Canada, an analysis of 341 peatland cores clearly indicates that rich fens accumulate peat to depths similar to those found in poor fens and bogs (Fig. 10.6). In all these cores, the major component of the accumulated peat is bryophytic: *Sphagnum* in poor fens and bogs, and brown mosses (including species of the genera *Aulacomnium, Catoscopium, Campylium, Drepanocladus* sensu lato, *Meesia, Scorpidium,* and *Tomentypnum*) in rich fens.

Comparison of hummock decomposition using litter bags in rich fens (*Tomentypnum nitens*) compared to bogs (*Sphagnum fuscum*) indicates that over the short term (16 months) significantly more decomposition

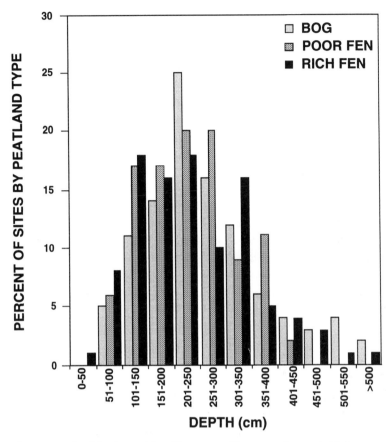

Fig. 10.6. Peatland depths partitioned by percent of bogs, poor fens, and rich fens. Number of sites include: bogs = 129, poor fen = 66, and rich fen = 146. All sites are from continental western Canada (Alberta, Saskatchewan, and Manitoba). (Data from Zoltai *et al.* 2000.)

occurred in hummocks composed of *Tomentypnum nitens* (Vitt 1990, Li & Vitt 1997). Within bogs along the hummock–hollow gradient, hummock species retain about 86% of their initial dry mass after three years, while hollow species retain only 74% after a similar time period (Rochefort *et al.* 1990). Additionally, moss net primary production (NPP) in hummocks is about half that in bog hollows (carpets and lawns), while moss NPP of fen hummocks is greater than or equal to that of hollows (reviewed in Vitt 1990). Thus, hummocks can develop higher than hollows above the water table in both bogs and rich fens due to either enhanced rates of moss NPP (fens) or less decomposition in hollows compared to hummocks (bogs).

326 DALE H. VITT

When *Sphagnum*-dominated hummocks are compared to brown-moss-dominated hummocks, both with well-developed acrotelms, the *Sphagnum* system decomposes 11% less than the brown moss hummocks after two years (Li & Vitt 1997). These data seem to suggest that there should be a fundamental difference in peat quality upon entry to the catotelm; however, there have been no comparative studies of peat quality in rich fens and bogs to my knowledge. Some ideas that need to be tested are that the slower rates of decomposition in *Sphagnum*-dominated peatlands are due to 1) the high carbon:nitrogen ratio of *Sphagnum* thus reducing the nitrogen needed for microbial growth, and 2) the chemical composition of *Sphagnum* cell walls that inhibits decomposition. In the case of *Sphagnum*-dominated bogs, the anaerobic catotelm receives from the acrotelm a relatively large amount (although probably less than 20% of the original mass) of undecomposed material with low bulk density. Once in the catotelm, deposition is considerably less, and has been modeled following simple exponential decay functions (Clymo 1984). In contrast, assuming similar catotelm functions, brown-moss-dominated fens have catotelms receiving material with lower carbon:nitrogen ratios and higher bulk densities. Despite these differences in acrotelms, brown-moss fens accumulate peat as efficiently as *Sphagnum*-dominated bogs (Fig. 10.6).

10.3.4 Acidification

In 1963, Clymo argued that peatland acidity is produced by *Sphagnum* cell walls where hydrogen ions produced by uronic acid molecules held on the cell wall are exchanged for base cations contained in pore waters. Thus, the base cations are absorbed into the cell walls while hydrogen ions are released into pore waters. This cation exchange ability of *Sphagnum* can easily be demonstrated by immersing some living or dead *Sphagnum* into doubly distilled water and measuring the pH change and then by adding common table salt to the same solution and again measuring the pH change. In the former case, no pH change is evident, while in the latter, the pH will immediately decrease 2–3 pH units. In 1980, Harry Hemond argued that although this process of cation exchange by *Sphagnum* undoubtedly occurs, it is not sufficient to produce the acidity that is present in bogs (i.e., pH 3.0–3.7). He concluded that bog acidity is largely due to decomposition and the production of humic acids present in pore water as dissolved organic carbon (DOC), and that acidity in bogs is a result of hydrogen release through decomposition of organic molecules that in turn are dissolved in the pore water as organic carbon. In the former case, the anions are attached to the *Sphagnum* cell walls (Richter &

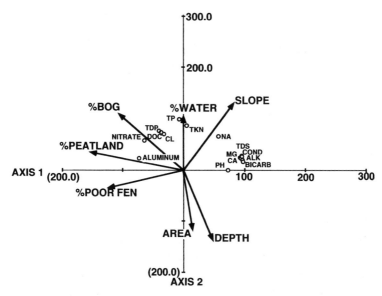

Fig. 10.7. Biplot of the chemical parameters measured in water from 29 boreal lakes in north-eastern Alberta, Canada. Watershed variables that explain a significant amount of the variation on the first two axes are shown by directed arrows. %WATER = percent of open water in the watershed. Depth and area refers to lake and area depth, while slope is the regional watershed slope. (Reprinted with permission from Kluwer Academic Publishers) *Water, Air, and Soil Pollution*, vol. **96**, 1997, Influence of peatlands on the acidity of lakes in northeastern Alberta, Canada, by Linda A. Halsey, Dale H. Vitt, and David Trew, Figure 5, page 33.

Dainty 1989), while in the later case the anions are dissolved in the pore water as DOC (Hemond 1980) and thus create the predominant "brown water" characteristic of bogs.

Regional landscape analyses of natural lake acidity support this pattern. Halsey *et al.* (1997*a*), showed that within 29 watersheds, bog cover and high DOC concentrations are positively correlated. In comparison, watersheds with a high percentage of poor fen cover are found in association with acidic lakes with low pH (Fig. 10.7). Importantly, this study indicates that lakes occurring in watersheds with greater than 30% cover of poor fens and bogs are nearly always acidic, while those without extensive acidic peatland cover have higher pH (all lakes of this study were situated on acidic shales). In addition, flow through poor fens appears to have more influence on downstream acidity than through stagnant bogs.

Sphagnum cation exchange activity is dependent on the presence of free cations in the surface waters of a peatland. Rich fens have large amounts of free cations (Vitt *et al.* 1995*a*), however these free base cations – largely calcium – are associated with HCO$_3^-$ (Vitt & Chee 1990). Gignac (1994) has shown that *Sphagnum* quickly dies when grown in water having any amount of alkalinity. Bogs, on the other hand, have few free base cations (Vitt *et al.* 1995*a*); and the lack of base cations under ombrotrophic conditions severely hinders the efficiency of the cation exchange. Thus, under bog conditions one would expect that cation exchange capacity would not be base-saturated, and that Hemond is correct: *Sphagnum*-produced acidity in bogs is not sufficient to produce the acidity. However, the question then becomes whether there are conditions where the alkalinity is low enough to allow *Sphagnum* to live and yet have sufficient base cations to allow acidification to occur.

The answer may actually lie in the fact that *Sphagnum* acidity is critical for successional transitions between moderate-rich fens and poor fens and may continue to dominate in flow-through-poor fens where base cations are more abundant then in bogs. The switch from a brown-moss-dominated system to one dominated by *Sphagnum* is critical in the evolution of acidic peatlands and many bogs. The macrofossil record clearly shows that the change from rich fen to poor fen occurs very rapidly (Fig. 10.8). This species change is associated with changes in acidity, generally from around pH 7 in moderate-rich fens to 5 in poor fens. Acidity produced via exchange of base cations (readily available in the minerotrophic fens) for hydrogen ions via the *Sphagnum* ion exchange system will be relatively more important in accounting for the acidity at higher pH (4–7). At low pH (3–4) where much higher concentrations of hydrogen ions are required and where base cations are low due to ombrotrophic conditions, acidity is due more to humic acid decomposition. This acidity would in many cases produce waters with high concentrations of DOC and these would be more heavily stained. Thus, a corollary is that *Sphagnum*-produced acidity may more strongly influence external downstream chemistry, while decomposition acidity may be more internally influential, especially in stagnant bogs. The relative importance of these two types of acidity needs further study, especially in poor fens.

The available evidence suggests that rich fens may persist without change for thousands of years, with fens being just as deep as bogs (Fig. 10.6). However, if alkalinity concentrations allow establishment of *Sphagnum*, then rapid acidification via *Sphagnum* cation exchange may result in the development of poor fen vegetation within 100–200 years (Fig. 10.8) (Vitt & Kuhry 1992). This rapid change at pH around 5.5 appears to be

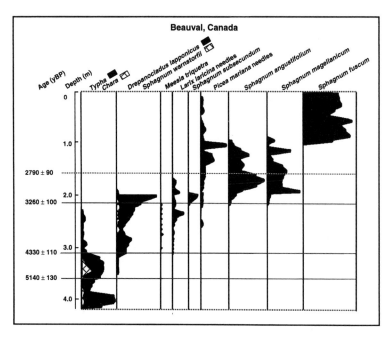

Fig. 10.8. Macrofossil analysis from Beauval, Saskatchewan, Canada showing abundance changes of the dominant moss species as well as other important indicators of local conditions. Note the transition period between brown-moss- and *Sphagnum*-dominated records is marked by the transient presence of *Sphagnum subsecundum*. Radiocarbon years are shown along the side. (Modified from Kuhry *et al.* 1993.)

responsible for the rarity of peatlands of this transitional nature on the boreal landscape (cf. Fig. 10.4).

10.4 Species occurrence

Bryophytes, especially mosses, have long been utilized as indicator species for floristically based classifications of peatlands. Bryophytes are small plants closely tied to their substrate. They are especially sensitive to acidity and alkalinity gradients that are both critical in understanding peatland structure and function. Additionally, and despite having narrow habitat ranges, they have broad geographical ranges. Many species are circumboreal in distribution, and these broad ranges allow a uniform set of indicator species in both Eurasia and North America. The high abundance of bryophytes in peatlands, their narrow substrate, chemical, and nutrient tolerances, as well as their sensitivity to water levels, have made them ideal candidates for research in plant-community structure. Early studies on ground layer structure followed from floristically based classification systems. Through these early studies, it was soon discovered that both *Sphagnum* and the dominant brown mosses had clearly defined habitats (DuReitz 1949, Eurola 1962, 1968, Ruuhijärvi 1960, 1962, 1963). With the advent of more quantitative techniques these habitat preferences became well known (e.g., Gignac *et al.* 1991*a*, *b*). For *Sphagnum* in North America, both Andrus in the east and Vitt in the west carefully quantified habitats for the major peatland species. Vitt and Slack advocated the view that although most bryophytes are non-competitive, in peatlands the critical driving force for species positions along water table and chemical gradients was competition (Vitt & Slack 1975). More recently Li and Vitt (1994) have argued that much of the mature species pattern present in peatlands is a reflection of past events and has resulted from establishment regimes and early interactions among species. Håkin Rydin's careful experimental approach (Rydin & McDonald 1985, Rydin 1986) has been instrumental in helping our understanding of the habitat tolerances of *Sphagnum* species.

Local habitat limitations of peatland bryophytes appear to be primarily controlled by chemical conditions, thus partitioning habitats along an acidity/alkalinity gradient that is reflected in our currently recognized rich-fen–bog classification (Gignac & Vitt 1990). Secondly, water level that is reflected in a classification of surface heights of pool, carpet, lawn, and hummock (Fig. 10.9; Andrus *et al.* 1983, Gignac *et al.* 1991*a*, *b*, Nordbakken 1996) is a controlling factor. All of these physical, limiting factors are then modified through species interactions at the microhabitat scale (Rydin 1993*b*). Rydin (1993*a*) has shown through transplants that *Sphagnum fuscum*, normally a hummock grower, will successfully occupy much lower (and wetter) conditions if transplanted there. Also, lawn species such as *S. angustifolium* that seemingly cannot form large hummocks due to drought limitation occurs as individuals in well-formed *S. fuscum* hummocks. In this latter case, *Sphagnum angustifolium* has expanded its habitat within populations of *S. fuscum*, occurring there in situations where it could not occur by itself. Thus it appears that interactions between *Sphagnum* species serve both to limit and expand habitats along the water-height gradient.

At the regional scale, both peatland occurrence and extent (Vitt *et al.* 1995*a*, Halsey *et al.* 1997*a*) and bryophyte ranges are correlated with

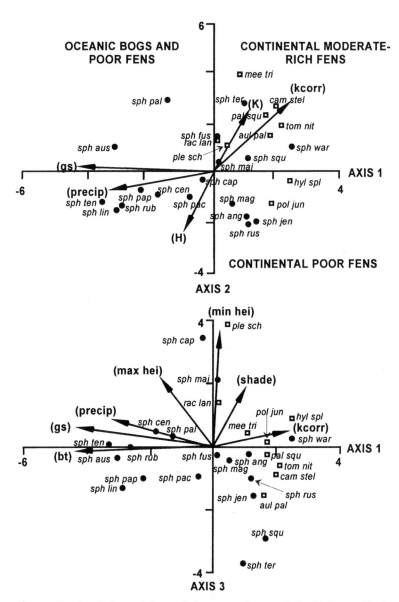

Fig. 10.9. Results of a detrended canonical correspondence analysis of *Sphagnum* (dots) and other bryophyte (squares) species and climatic and environmental variables. The climatic and environmental variables are shown by arrows and their abbreviations are: min hei = minimum height; max hei = maximum height; kcorr = corrected conductivity; bt = biotemperature; gs = growing season; precip = total precipitation falling as rain; ai = aridity index; pdays = number of precipitation days. (From Gignac & Vitt 1990.)

climatic factors (Gignac & Vitt 1990). Through the use of response surface modeling, Gignac has produced a series of papers documenting where *Sphagnum* and brown moss species occur in climatic space (Gignac & Vitt 1990, Gignac *et al.* 1991*a*, *b*, Gignac 1993). He then used these responses to develop models of ecosystem responses to climatic changes (Gignac & Vitt 1994, Gignac *et al.* 1998). Through a series of climatically based response surfaces, he and his co-authors have demonstrated that habitat space changes in response to climate. For example, as temperatures drop, permafrost forms and increases the height of hummocks in northern bogs (Vitt *et al.* 1994). These higher hummocks essentially create new habitats not present farther south and functionally expand the habitat response for such species as *Dicranum undulatum* (Nicholson & Gignac 1995). Furthermore, when individual species response surfaces are combined with General Circulation Models (GCM), predictions of how peatland species and peatland ecosystems will respond to both past and future climatic change can be made. Scenarios using $2 \times CO_2$ climate in north-western Canada clearly predict that the southern limit of peatlands will shift northwards about 700 km, assuming an equilibrium response (Gignac *et al.* 1998)

10.5 Species richness

Sphagnum dominates bogs and poor fens in boreal and northern regions of the world. These acidic peatlands may cover substantial areas where climate, substrate, and topography are suitable. Ingrained in our thinking is the idea that peatlands are most dominant in oceanic climates such as in Ireland, Britain, and Norway in Europe, and Newfoundland, coastal British Columbia, and Alaska in North America. While acidic peatlands are abundant in these area, the truly large expanses of peatland are in interior, continental boreal areas such as across boreal Canada, interior Alaska, and Siberia. It is in these areas that peatlands may dominate the landscape. Inventories from the interior of continental Canada (Alberta, Saskatchewan, and Manitoba) indicate that peatlands cover 21% of the landscape and dominate the landscape over hundreds of square kilometers (Vitt *et al.* 2000). Discriminant canonical correspondence analysis on wetland distributions in this region suggest that climate is the primary controlling factor on wetland distribution, while physiographic factors such as substrate topography and texture play secondary, but significant roles (Fig. 10.10). In addition, substrate acidity is correlated to peatland acidity. Areas dominated by acidic, Precambrian rocks have greater abun-

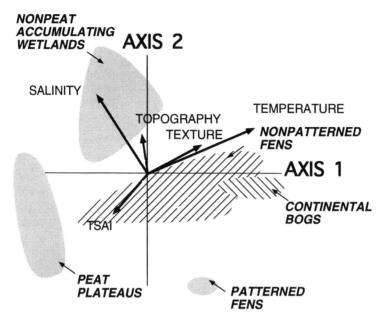

Fig. 10.10. Distribution of wetland types in Alberta, Canada using detrended canonical correspondence analysis. Environmental variables that significantly explain the first two axes included mean annual temperature, TSAI = thermal seasonal aridity index, salinity, topography, and texture. (See Halsey *et al.* (1997*b*) for methodology.)

dance of bogs than areas underlain by calcareous bedrock within similar climatic conditions (Halsey *et al.* 1997*b*).

Clymo (1970) has estimated that there is more biomass in *Sphagnum* then in any other plant in the world. However, the number of species of *Sphagnum* that occupy these large expanses of acidic peatland is relatively low when compared to the species richness of the genus. For example in North America, there are 84 of the approximately 250 species in the genus *Sphagnum* (Andrus & McQueen 1999), with about 57 of these species (23% of world diversity) found in peatlands. In continental western Canada, peatlands dominate the landscape in the Hudson Bay Lowlands, north of Lake Winnipeg and in northern Alberta's uplands (Fig. 10.11). However, these large expanses of peatland have only 20 species of *Sphagnum* (Vitt *et al.* 1995*c*), of which less than 10 have any appreciable abundance. Clearly the patterns of *Sphagnum* species richness do not parallel the abundance of peatlands. It is habitat diversity, not habitat abundance that controls species richness in peatlands.

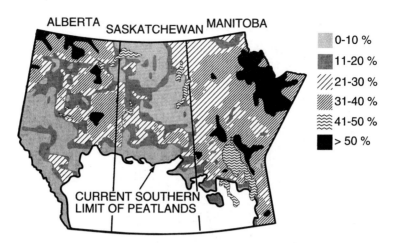

Fig. 10.11. Distribution of peatlands in continental western Canada (Alberta, Saskatchewan, and Manitoba). (Modified from Vitt *et al.* 2000.)

Species richness of *Sphagnum* along the rich-fen–bog gradient is surprisingly even (Fig. 10.12). Of the 20 species of *Sphagnum* found in a study of peatland bryophyte diversity by Vitt and Belland (1995) in western Canada, 12 species were found in bogs and 19 in fens. Even more interesting was that, within fens, 16 species occurred in poor fens, seven in moderate-rich fens, and nine in extreme-rich fens. Although the genus *Sphagnum* appears restricted to acidic (non-alkaline) habitats, a majority of species occur in habitats marginal to peatlands, in acidic microhabitats in rich fens, on cliff faces, along stream banks, on wet soils, and in non-peat-forming wetlands (cf. Vitt & Andrus 1977, Crum 1984, Andrus & McQueen 1999). Only a small number is actually involved in creating the large carbon-sequestering peatland expanses so abundant across continental regions of the northern hemisphere.

In peatlands the number of bryophyte species is similar to (or outnumbers) that of vascular plants. Since patterns of bryophyte richness often do not parallel those of other organisms (Vitt 1991), it is of some interest to examine richness patterns over a chemical gradient that defines the bog–rich-fen series of peatland types. Studying the South Pacific Islands, Vitt (1991) demonstrated that locally rare species (defined as species collected at only one site) of mosses compose 17–29% of the island floras. Likewise in peatlands studied in continental western Canada Vitt and Belland (1995) found that 20% of the moss flora of the 100 peatlands studied was locally rare. These percentages appear to be remarkably

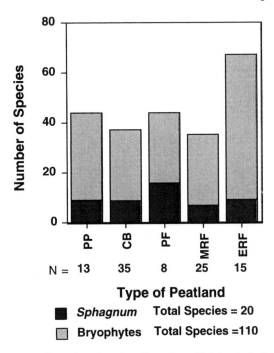

Peatland Gamma Diversity

Fig. 10.12. Total number of species of bryophytes (light) and of *Sphagnum* (dark) found in each of five peatland types in continental western Canada. PP = peat plateau, CB = continental bog, PF = poor fen, MRF = moderate-rich fen, and ERF = extreme-rich fen. (Data from Vitt *et al.* 1995*c*.)

similar when widespread dissimilar moss floras are compared. However, when peatlands are divided into five major types a pattern of uneven distribution emerges as locally rare species varied from 0% in moderate-rich fens to 5% in bogs, 9% in peat plateaus (bogs with permafrost), 11% in poor fens, and 22% in extreme-rich fens (Fig. 10.13).

When species richness of individual site diversity (alpha diversity) is compared among these five peatland types, considerable variation exists for all types. Alpha species richness values indicate that any of the five peatland types may be species-poor or species-rich (Table 10.4). Bogs that have been traditionally viewed as species-depauperate ecosystems at the individual site level can have fewer, the same, or three times as many species as any of the three fen types (Vitt *et al.* 1995*c*). When mean alpha diversities are compared among the five peatland types (Table 10.4), there

Locally Rare Species Richness

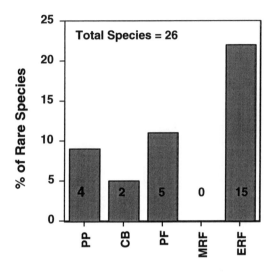

Type of Peatland

Fig. 10.13. Percent of locally rare bryophyte species within each of five peatland types in continental western Canada. PP = peat plateau, CB = continental bog, PF = poor fen, MRF = moderate-rich fen, and ERF = extreme-rich fen. Numbers within bars are numbers of locally rare species. (Data from Vitt *et al.* 1995*c*.)

are highly significant differences (analysis of variance, $P = 0.0001$). Bogs are not significantly different from moderate-rich fens, and, in fact, peat plateaus, poor fens, and extreme-rich fens form a group with the greatest alpha diversity. When comparing the standard deviations of alpha diversity for the five different peatlands types, the two bog types do not have as large of a standard deviation as do the three fen types (Table 10.4).

An analysis of species turnover, or beta diversity, yields a different conclusion. Beta diversities for the two bogs types as well as for poor fens and moderate-rich fens are quite similar (range: 2.73–3.25; Table 10.4). However beta diversity for extreme-rich fens is much higher (4.27). Likewise, when gamma diversity (total regional diversity) is compared among the peatlands, values range from 35 to 44 for bog types, poor fens, and moderate-rich fens while gamma diversity for extreme-rich fens is 67 (Fig. 10.12; Vitt *et al.* 1995*c*). Species richness suggests several patterns: 1) individual bog and fen sites are extremely variable; mean alpha diversity tells us very little, with the five peatland types divided into three significant, yet

Table 10.4. *Bryophyte species richness for peatland types of continental western Canada*

Peatland type	n	Range	Median, alpha	Mean, alpha	Standard deviation	Student–Newman Keuls groupings	Beta diversity
Peat plateau	13	12–23	17	17.4	3.3	A	2.53
Bog	35	6–18	12	11.4	2.9	C	3.25
Poor fen	8	8–24	17	16.1	5.6	A/B	2.73
Moderate-rich fen	25	6–23	11	12.7	4.3	C/B	3.04
Extreme-rich fen	15	8–28	14	15.7	6.8	A/B	4.27
Mean–median totals	96	8–23	14	14.4	–	–	3.16

Source: Data abstracted from Vitt *et al.* 1995c.

heterogenous groups, 2) the high beta diversity value for extreme-rich fens indicates that individual sites are much more different from each other than individual sites of the four other peatlands types, 3) The relatively high number of rare species (22%) (Fig. 10.13) found in extreme-rich fens is directly relatable to the high gamma diversity; this high regional species richness comes about because individual extreme-rich fens are more different from one another than are individual sites of the other four peatland types. The significance of this finding is that management of ecosystem types with high beta/gamma diversities must be done by careful selection of a relatively high number of sites for protection. Peatland types with low beta/gamma diversities can be managed with fewer sites, but the wide range of alpha diversities indicates that which sites are chosen is very important. Overall, extreme-rich fens are the most critical peatland types if local rarity of bryophytes is considered a high priority.

10.6 Conclusions

In many northern areas of the world with cool climates and short growing seasons, bryophyte-dominated peatlands form a substantial part of the landscape. These ecosystems have expanded over the past 6000 to 8000 years, and have sequestered large amounts of carbon. In continental western Canada, where data are available, peatlands contain 50 petagrams (10^{15} g) of carbon – this predominately composed of bryophyte material. Fifty percent of this carbon has been sequestered in the past 4000 years (Vitt *et al.* 2000). The functioning of these northern peatlands is strongly influenced by bryophytes, and our understanding of the nutrient flow, diversity, and carbon sequestering of these ecosystems can only be advanced by through knowledge of the bryophytes that dominate in these boreal and subarctic ecosystems.

Acknowledgments

This paper is based on research made possible through Natural Science and Engineering Research Grants (Canada). Support for individual projects include funding from Sun Gro Horticulture (through Tony Cable); Network of Centers of Excellence in Sustainable Forest Management; Canadian Forest Service; and the University of Alberta. I am especially grateful to the efforts and expertise of Linda Halsey, Laureen Snook, and Sandi Vitt for making this synthesis possible. These ideas and the data on which they are founded have accumulated through the years by interactions with my students and colleagues; in particular R. Andrus, S. Bayley, R. Belland, L. D. Gignac, J. Glime, L. Halsey, D. Horton, P. Kuhry, Y. Li, N. Malmer, B. Nicholson, L. Rochefort, N. Slack, M. Turetsky, K. Wieder, and S. Zoltai. A thorough review by Phil Camill resulted in a much better manuscript and is greatly appreciated.

REFERENCES

Aerts, R., Wallén, B. & Malmer, N. (1992). Growth-limiting nutrients in *Sphagnum*-dominated bogs subject to low and high atmospheric nitrogen supply. *Journal of Ecology*, **80**, 131–40.

Andrus, R. E., & McQueen, C. B. (1999). Sphagnaceae, *Sphagnum*. In *Bryophyte Flora of North America*, unpublished manuscript, Edmonton, Canada.

Andrus, R. E., Wagner, D. J., & Titus, J. E. (1983). Vertical zonation of *Sphagnum* mosses along hummock-hollow gradients. *Canadian Journal of Botany*, **61**, 3128–39.

Bayley, S. E. (1993). Mineralization of nitrogen in bogs and fens of western Canada. In *Proceedings of the American Society of Limnology and Oceanography and the Society of Wetland Scientists Combined Annual Meetings*, p. 88. Edmonton, Canada: University of Alberta.

Belland, R. J. & Vitt, D. H. (1995). Bryophyte vegetation patterns along environmental gradients in continental bogs. *Ecoscience*, **2**, 395–407.

Bewley, J. D. (1979). Physiological aspects of desiccation tolerance. *Annual Review of Plant Physiology*, **30**, 195–238.

Bridgham, S. D., Updegraff, K., & Pastor, J. (1998). Carbon, nitrogen, and phosphorus mineralization in northern wetlands. Ecology, **79**, 1545–61.

Campbell, C., Vitt, D. H., Halsey, L. A., Campbell, I., Thormann, M. N., & Bayley, S. E. (2000). *Net Primary Production and Standing Biomass in Northern Continental Wetlands*. Canadian Forest Service NOX Publication Series. Edmonton: Northern Forestry Centre.

Clymo, R. S. (1963). Ion exchange in *Sphagnum* and its relation to bog ecology. *Annnals of Botany*, **27**, 309–24.

(1970). The growth of *Sphagnum*: methods of measurement. *Journal of Ecology*, **58**, 13–49.

(1984). The limits of peat growth. *Proceedings of the Royal Society of London B,* **303**, 605–54.

Crum, H. (1984). *Sphagnopsida; Sphagnaceae*. North American Flora Series II, Part II. New York: New York Botanical Garden.

(1988). *A Focus on Peatlands and Peat Mosses*. Ann Arbor: University of Michigan Press.

Damman, A. W. H. (1979). Geographic patterns in peatland development in eastern North America. In *Classification of Peat and Peatlands,* Proceedings of the International Peat Society Symposium, Hyytälä, Finland, ed. E. Kivinen, L. Heikurainen, & P. Pakarinen, pp. 42–57. Helsinki: International Peat Society.

Davis, R. B. & Anderson, D. S. (1991). The eccentric bogs of Maine: a rare wetland type in the United States. *Maine State Planning Office Critical Areas Program Planning Report* **93**.

DuReitz, G. E. (1949). Huvidenheter och huvidgranser i Svensk myrvegetation. Summary: Main units and main limits in Swedish mire vegetation. *Svensk Botanisk Tidskrift*, **43**, 274–309.

Eurola, S. (1962). Über die regionale Einteilung der südfinnischen Moore. *Annales Botanici Societatis Vanamo*, **33**, 1–243.

Eurola, S. (1968). On the mire vegetation zones in northwestern Europe and their correlation to field and forest vegetation zones. *Luonnon Tutkija*, **72**, 83–97. [in Finnish]

Foster, D. R., King, G. A., Glaser, P. H., & Wright, H. E., Jr. (1983). Origin of string patterns in boreal peatlands. *Nature*, **306**, 256–8.

Gignac, L. D. (1993). Distribution of *Sphagnum* species, communities, and habitats in relation to climate. *Advances in Bryology*, **5**, 187–222.

(1994). Habitat limitations and ecotope structure of mire *Sphagnum* in western Canada. PhD dissertation, University of Alberta, Edmonton, Canada.

Gignac, L. D. & Vitt, D. H. (1990). Habitat limitations of *Sphagnum* along climatic, chemical and physical gradients in mires of western Canada. *Bryologist*, **93**, 7–22.

(1994). Responses of northern peatlands to climate change: effects on bryophytes. *Journal of the Hattori Botanical Laboratory*, **75**, 119–32.

Gignac, L. D., Vitt, D. H.,& Bayley, S. E. (1991*a*). Bryophyte response surfaces along ecological and climatic gradients. *Vegetatio*, **93**, 24–45.

Gignac, L. D., Vitt, D. H., Zoltai, S. C., & Bayley, S. E. (1991*b*). Bryophyte response surfaces along climatic, chemical and physical gradients in peatlands of western Canada. *Nova Hedwigia*, **53**, 27–71.

Gignac, L. D., Nicholson, B. J., & Bayley, S. E. (1998). The utilization of bryophytes in bioclimatic modeling: predicted northward migration of peatlands in the MacKenzie River basin, Canada, as a result of global warming. *Bryologist*, **101**, 572–87.

Glaser, P. H. (1992). Raised bogs in eastern North America – regional controls for species richness and floristic assemblages. *Journal of Ecology*, **64**, 535–54.

Glaser, P. H. & Janssens, J. A. (1986). Raised bogs in eastern North America: transitions in landforms and gross stratigraphy. *Canadian Journal of Botany*, **64**, 395–415.

Glime, J. M. & Vitt, D. H. (1984). The physiological adaptations of aquatic Musci. *Lindbergia*, **10**, 41–52.

Gorham, E. (1956). The ionic composition of some bog and fen waters in the English lake district. *Journal of Ecology*, **44**, 142–52.

Gorham, E. & Janssens, J. A. (1992). Concepts of fen and bog re-examined in relation to bryophyte cover and the acidity of surface waters. *Acta Societatis Botanicorum Poloniae*, **61**, 7–20.

Halsey, L. A., Vitt, D. H., & Trew, D. (1997*a*). Influence of peatlands on the acidity of lakes in northeastern Alberta, Canada. *Water, Air and Soil Pollution*, **96**, 17–38.

Halsey, L. A., Vitt, D. H., & Zoltai, S. C. (1997*b*). Climatic and physiographic controls on wetland type and distribution in Manitoba, Canada. *Wetlands*, **17**, 243–62.

Hemond, H. F. (1980). Biogeochemistry of Thoreau's Bog, Concord, Massachusetts. *Ecological Monographs*, **50**, 507–26.

Kuhry, P. & Vitt, D. H. (1996). Fossil carbon/nitrogen ratios as a measure of peat decomposition. *Ecology*, **77**, 271–5.

Kuhry, P., Nicholson, B. J., Gignac, L. D., Vitt, D. H., & Bayley, S. E. (1993). Development of *Sphagnum*-dominated peatlands in boreal continental Canada. *Canadian Journal of Botany*, **71**, 10–22.

Li, Y. & Vitt, D. H. (1994). The dynamics of moss establishment: temporal responses to nutrient gradients. *Bryologist*, **97**, 357–64.

(1997). Patterns of retention and utilization of aerially deposited nitrogen in boreal peatlands. *Ecoscience*, **4**, 106–16.

Malmer, N. (1962). Studies of mire vegetation in the archaean area of southwestern Götaland (south Sweden). II. Distribution and seasonal variation in elementary constituents on some mire sites. *Opera Botanica*, **7**, 1–67.

(1986). Vegetational gradients in relation to environmental conditions in northwestern European mires. *Canadian Journal of Botany*, **82**, 899–910.

Malmer, N., Horton, D. G., & Vitt, D. H. (1992). Elemental concentrations in mosses and surface waters of western Canadian mires relative to precipitation chemistry and hydrology. *Ecography*, **15**, 114–28.

Mitch, W. J. & Gosselink, J. G. (1993). *Wetlands*, 2nd edn. New York: Van Nostrand Reinhold.

Moore, T. R. (1989). Plant production, decomposition, and carbon efflux in a subarctic patterned fen. *Arctic and Alpine Research*, **21**, 156–62.

National Wetlands Working Group (1988). *Wetlands of Canada*. Ecological Land Classification Series, no. 24. Ottawa: Sustainable Development Branch, Environment Canada, and Montreal: Polyscience Publications Inc.

Nicholson, B. & Gignac, L. D. (1995). Ecotope dimensions of peatland bryophyte indicator species along gradients in the Mackenzie River Basin, Canada. *Bryologist*, **98**, 437–51.

Nicholson, B., Gignac, L. D., & Bayley, S. E. (1996). Peatland distribution along a north–south transect in the MacKenzie River Basin in relation to climatic and environmental gradients. *Vegetatio*, **126**, 119–33.

Nordbakken, J. F. (1996). Plant niches along the water-table gradient on an ombrotrophic mire expanse. *Ecography*, **19**, 114–21.

Proctor, M. C. F. (1979). Structure and eco-physiological adaptation in bryophytes. In *Bryophyte Systematics*, ed. G. C. S. Clarke & J. G. Duckett, pp. 479–509. London: Academic Press.

(1984). Structure and ecological adaptation. In *The Experimental Biology of Bryophytes*, ed. A. F. Dyer & J. G. Duckett, pp. 9–37. London: Academic Press.

Richter, C. & Dainty, J. (1989). Ion behaviour in plant cell walls. I. Characterization of the *Sphagnum russowii* cell wall ion exchanger. *Canadian Journal of Botany*, **67**, 451–9.

Rochefort, L. & Vitt, D. H. (1988). Effects of simulated acid rain on *Tomenthypnum nitens* and *Scorpidium scorpioides* in a rich fen. *Bryologist*, **91**, 121–9.

Rochefort, L., Vitt, D. H., & Bayley, S. E. (1990). Growth, production and decomposition dynamics of *Sphagnum* under natural and experimentally acidified conditions. *Ecology*, **71**, 1986–2000.

Ruuhijärvi, R. (1960). Über die regionale Einteilung der nordfinnischen Moore. *Annales Botanici Societatis Zoologicae Botanicae Fennici "Vanamo"*, **31**, 1–360.

(1962). Palsasoista ja niiden morfologiasta siitepolyanalyysin valossa [On the palsa mires and their morphology in the light of pollen analysis]. *Terra*, **74**, 58–68.

(1963). Zur Entwicklungsgeschichte der nordfinnischen Hockmoore. *Annales Botanici Societatis Zoologicae Botanicae Fennici "Vanamo"*, **34**, 1–40.

Rydin, H. (1986). Competition and niche separation in *Sphagnum*. *Canadian Journal of Botany*, **64**, 1817–24.

(1993*a*). Interspecific competition among *Sphagnum* mosses on a raised bog. *Oikos*, **66**, 413–23.

(1993*b*). Mechanisms of interactions among *Sphagnum* species along water-level gradients. *Advances in Bryology*, **5**, 153–85.

Rydin, H. & McDonald, A. J. S. (1985). Tolerance of *Sphagnum* to water level. *Journal of Bryology*, **13**, 571–8.

Siegel, D. I. & Glaser, P. H. (1987) Groundwater flow in a bog/fen complex, Lost River peatland, northern Minnesota. *Journal of Ecology*, **75**, 743–54.

Sjörs, H. (1950). Regional studies in north Swedish mire vegetation. *Botaniska Notiser*, **1950**, 174–221.

(1952). On the relation between vegetation and electrolytes in north Swedish mire waters. *Oikos*, **2**, 242–58.

Slack, N. G., Vitt, D. H., & Horton, D. G. (1980). Vegetation gradients of minerotrophically rich fens in western Alberta. *Canadian Journal of Botany*, **58**, 330–50.

Stålfelt, M. G. (1937). Der Gasaustausch der Moose. *Planta*, **27**, 30–60.

Thormann, M. N. & Bayley, S. E., (1997). Decomposition along a moderate-rich fen-marsh peatland gradient in Boreal Alberta, Canada. *Wetlands*, **17**, 123–37.

Thormann, M. N., Szumigalski, A. R., & Bayley, S. E. (1999). Aboveground peat and carbon accumulation potentials along a bog–fen–marsh peatland gradient in southern boreal Alberta, Canada. *Wetlands*, **19**, 305–17.

Titus, J. E. & Wagner, D. J. (1984). Carbon balance for two *Sphagnum* mosses: water balance resolves a physiological paradox. *Ecology*, **65**, 1765–74.

Turetsky, M. R., Wieder, R. K., Williams, C., & Vitt, D. H. (2000). Organic accumulation, peat chemistry, and permafrost melting in peatlands of boreal Alberta. *Ecoscience*, in press.

Vitt, D. H. (1990). Growth and production dynamics of boreal mosses over climatic, chemical, and topographic gradients. *Botanical Journal of the Linnean Society*, **104**, 35–59.

(1991). Distribution patterns, adaptive strategies and morphological changes of mosses along elevational and latitudinal gradients on South Pacific islands. In *Quantitative Approaches to Phytogeography*, ed. P. L. Nimis & T. J. Crovello, pp. 205–31. Dordrecht: Kluwer.

(1994). An overview of factors that influence the development of Canadian peatlands. *Memoirs of the Entomological Society of Canada*, **169**, 7–20.

Vitt, D. H. & Andrus, R. E. (1977) The genus *Sphagnum* in Alberta. *Canadian Journal of Botany*, **55**, 331–57.

Vitt, D. H. & Belland, R. (1995). The bryophytes of peatlands in continental western Canada. *Fragmenta Floristica et Geobotanica*, **40**, 339–48.

Vitt, D. H. & Chee, W. L. (1990). The relationships of vegetation to surface water chemistry and peat chemistry in fens of Alberta, Canada. *Vegetatio*, **89**, 87–106.

Vitt, D. H. & Kuhry, P. (1992). Changes in moss-dominated wetland ecosystems. In *Bryophytes and Lichens in a Changing Environment*, ed. J. W. Bates & A. M. Farmer, pp. 178–210. Oxford: Clarendon Press.

Vitt, D. H. & Slack, N. G. (1975). An analysis of the vegetation of *Sphagnum*-dominated kettle-hole bogs in relation to environmental gradients. *Canadian Journal of Botany*, **53**, 332–59.

Vitt, D. H., Achuff, P., & Andrus, R. E. (1975*a*). The vegetation and chemical properties of patterned fens in the Swan Hills, north central Alberta. *Canadian Journal of Botany*, **53**, 2776–95.

Vitt, D. H., Crum, H. A., & Snider, J. A. (1975*b*). The vertical zonation of *Sphagnum* species in hummock–hollow complexes in northern Michigan. *Michigan Botanist*, **14**, 190–200.

Vitt, D. H., Horton, D. G, Slack, N. G., & Malmer, N. (1990). *Sphagnum*-dominated peatlands of the hyperoceanic British Columbia coast: patterns in surface water chemistry and vegetation. *Canadian Journal of Forest Research*, **20**, 696–711.

Vitt, D. H., Halsey, L. A., & Zoltai, S. C. (1994). The bog landforms of continental western Canada, relative to climate and permafrost patterns. *Arctic and Alpine Research*, **26**, 1–13.

Vitt, D. H., Bayley, S. E., & Jin, T.-L. (1995*a*). Seasonal variation in water chemistry over a bog-rich fen gradient in continental western Canada. *Canadian Journal of Fisheries and Aquatic Sciences*, **52**, 587–606.

Vitt, D. H., Halsey, L. A., Thormann, M. N., & Martin, T. (1995*b*). *Peatland Inventory of Alberta*. Edmonton: Alberta Peat Task Force, University of Alberta.

Vitt, D. H., Li, Y., & Belland, R. J. (1995*c*). Patterns of bryophyte diversity in peatlands of continental western Canada. *Bryologist*, **98**, 218–27.

Vitt, D. H., Halsey, L. A., Bauer, I. E., & Campbell, C. (2000). Spatial and temporal trends of carbon sequestration in peatlands of continental western Canada through the Holocene. *Canadian Journal of Earth Sciences*, in press.

Wagner, D. J. & Titus, J. E. (1984). Comparative desiccation tolerance of two *Sphagnum* mosses. *Oecologia*, **62**, 182–7.

Wieder, R. K., Yavitt, J. B., & Lang, G. E. (1990). Methane production and sulfate reduction in two Appalachian peatlands. *Biogeochemistry*, **10**, 81–104.

Zoltai, S. C. & Vitt, D. H. (1990). Holocene climatic change and the distribution of peatlands in western interior Canada. *Quaternary Research*, **33**, 231–40.

(1995). Canadian wetlands: environmental gradients and classification. *Vegetatio*, **118**, 131–7.

Zoltai, S. C., Siltanen, R. M., & Johnson, J. D. (2000). *A Wetland Environmental Data Base*. Canadian Forest Service NOX Publication Series. Edmonton: Northern Forestry Centre.

11
_____

Role of bryophyte-dominated ecosystems in the global carbon budget

11.1 Introduction

There is growing consensus within the scientific community that increases in atmospheric methane (CH_4) and carbon dioxide (CO_2) are enhancing the earth's natural greenhouse effect. Because of the potential effects of these gases on the global energy budget and future climate, there is an urgent need to quantify terrestrial sources and sinks of carbon. Bryophytes are the primary form of carbon storage in many northern ecosystems. There is more carbon stored in *Sphagnum* and *Sphagnum* litter (150×10^{12} g) than in any other genus of plants, vascular or non-vascular (Clymo & Hayward 1982). Since the end of the last glacial period (~18 000 y.b.p.), the soils of the northern latitudes have served as a reservoir for terrestrial carbon (Harden *et al.* 1992). Northern peatlands alone may contain two to three times the amount of carbon stored in tropical rainforests (Post *et al.* 1982, Gorham 1991). The majority of this carbon has been frozen in permafrost soils and sequestered from atmospheric circulation for thousands of years. On a warming planet, this carbon represents a "ticking time bomb" that could rapidly decompose and increase the amount of CO_2 in the atmosphere by as much as 50% (Billings 1997, Goulden *et al.* 1998). Predicting how the vast stores of soil carbon in moss-dominated ecosystems will be affected by anthropogenic disturbance is critical for models of global climate change. This chapter will review the importance of bryophytes in determining patterns and rates of global carbon flux in northern ecosystems with particular attention given to how these systems might respond under different disturbance regimes.

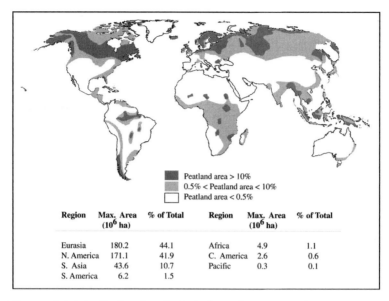

Region	Max. Area (10^6 ha)	% of Total	Region	Max. Area (10^6 ha)	% of Total
Eurasia	180.2	44.1	Africa	4.9	1.1
N. America	171.1	41.9	C. America	2.6	0.6
S. Asia	43.6	10.7	Pacific	0.3	0.1
S. America	6.2	1.5			

Peatland area > 10%
0.5% < Peatland area < 10%
Peatland area < 0.5%

Fig. 11.1. Global distribution of peatlands. (Redrawn from Gore 1986.)

11.1.1 Carbon storage in moss-dominated soils

Moss-dominated organic soils occur worldwide in regions where precipitation exceeds potential evapotranspiration and drainage is limited. The boreal and northern deciduous zones of Europe, Siberia, and North America contain expanses of flat terrain with a high groundwater table, making these areas ideal for peatland formation. More than 85% of all peatlands are located in northern temperate, boreal, and arctic ecosystems (Fig. 11.1: Gore 1983).

Tundra

Much of the arctic tundra has been underlain by permafrost since the last ice age. Even in late summer, the active layer (the part of the permafrost that thaws annually) may be less than 20 cm thick. All plant rooting and nutrient uptake is confined to this thin, cold, active layer, which severely limits the activity of vascular plants. For this reason, net annual production of bryophytes frequently exceeds that of vascular plants.

The largest stores of carbon in arctic ecosystems are in tussock tundra (29.1 Pg C; 1 Pg = 10^{15} g) and wet sedge tundra (14.4 Pg C) (Miller *et al.* 1983). As much as 95% of the carbon in tundra ecosystems is bound in dead organic matter as peat (Chapin *et al.* 1980). Until recently, tundra ecosystems were considered to be net sinks for carbon on the order of 0.1 to 0.3 Pg C yr^{-1} (Miller et al. 1983, Oechel & Billings, 1992, Marion & Oechel 1993). However, recent work by Oechel et al. (1993, 1995) suggests that increases in air and soil temperature may have already converted large regions of tundra from net sinks to net sources of CO_2.

Although most studies of CO_2 exchange from tundra soils focus only on emissions during the growing season, a thick snowpack may insulate soil microbial populations from cold temperatures and allow for continued decomposition of stored carbon throughout the winter (Fahnestock et al. 1998). As much as 22–37% of annual decomposition may occur beneath the snowpack (Havas & Mäenpää 1972, Pajari 1995), resulting in a dramatic short-term release of CO_2 during spring melt (Zimov et al. 1993). Annual estimates of net ecosystem exchange based solely on warm-season measurements may significantly underestimate the actual magnitude of CO_2 efflux (Oechel et al. 1997).

Boreal forests

As in tundra ecosystems, mosses form a nearly continuous understory throughout much of the boreal forest. In mature coniferous stands, cool temperatures and shaded soils favor the growth of weft-forming, pleurocarpous mosses such as Hylocomium splendens and Pleurozium schreberi. Sphagnum spp. are abundant in wetter locations. Due to their dominant role on the forest floor, mosses of the boreal forest modify soil conditions (e.g., temperature, moisture content, and pH) in ways that provide them with a competitive advantage over vascular plants. Mosses are often able to effectively compete with forest tree species for nutrient resources (Oechel & Van Cleve 1986) resulting in the low decomposition rates, reduced soil nutrient supply, and greatly reduced tree growth typical of taiga ecosystems.

Although boreal forests cover a little less than 17% of the earth's land surface, they contain more than one-third of all soil carbon (625 Pg) and currently function as a net carbon sink of 0.7 Gt C yr^{-1} (Apps et al. 1993). The majority of this carbon is stored in soils of moss-dominated, black spruce forests. In the interior of Alaska, Van Cleve et al. (1983) found that moss production was four times greater than annual black spruce foliage production. In central Canada, Sphagnum- and "feathermoss"-dominated areas of the landscape contained two to five times the amount of carbon stored in living and dead black spruce trees (Goulden et al. 1998). Although direct measurements of net ecosystem production at this site showed a net loss of 0.3 ± 0.5 t C ha^{-1} yr^{-1} from 1994 to 1997 (Goulden et al. 1998), ^{14}C dating of soil profiles suggests that over century and millennial time-scales, these soils have functioned as a net sink of 0.3 to 0.5 t C $ha^{-1} yr^{-1}$ (Harden *et al.* 1997).

Bogs and fens

Northern peatlands play a unique role in the terrestrial carbon cycle because they sequester the major greenhouse gas, CO_2, as peat while emitting the second most important greenhouse gas, CH_4, to the atmosphere. Boreal and subarctic peatlands cover nearly 3.5 million km^2 of the earth's land surface and store 300–455 Pg C (Sjörs 1980, Gorham 1991). However, most estimates of peatland carbon do not include the carbon stored in underlying mineral soils, which may contain an additional 2–5% of the carbon stored in the peat itself (Turunen *et al.* 1999). Based on records of peat accumulation, Gorham (1991) estimates a long-term carbon accumulation rate of 29 g C $m^{-2} yr^{-1}$, suggesting that boreal peatlands function as a net sink of 0.1 Pg C yr^{-1} (Gorham 1991).

Annual estimates for CH_4 emissions from northern peatlands (45–80° N) are on the order of 25 to 36 Tg yr^{-1} (Bartlett & Harriss, 1993, Vourlitis & Oechel 1997). Emissions from boreal wetlands (40–60° N) range from 23 mg CH_4 m^{-2} day^{-1} (forested bogs) to 128 mg CH_4 m^{-2} day^{-1} (open bogs). In arctic and subarctic systems, methane emissions range from 12 mg CH_4 m^{-2} day^{-1} (moist tundra) to 100 mg CH_4 m^{-2} day^{-1} (wet sedge tundra) (Vourlitis & Oechel 1997). Although the spatial and temporal variability associated with methane fluxes is large, as much as 60% of methane emissions from natural wetlands come from moss-dominated bogs located between 50–70° N (Fig. 11.2; Matthews & Fung 1987).

11.2 Factors controlling carbon storage in northern ecosystems

The carbon balance of an ecosystem is the difference between carbon uptake as CO_2 during photosynthesis and the respiratory loss of carbon as CO_2 by plants, animals, and decomposer organisms. In general, the accumulation of carbon in a newly developing soil follows a logarithmic pattern with an initial period of rapid carbon accumulation that declines slowly over time until the system reaches an equilibrium level of carbon storage. This relationship may be represented as:

$$\text{carbon accumulation} = \text{net primary production} - (\text{decomposition} * \text{stored carbon}) \tag{1}$$

In mature systems, the amount of carbon added to the system each year is balanced by the amount lost during decomposition. Carbon accumulation

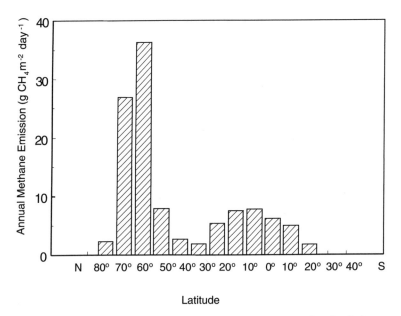

Fig. 11.2. Latitudinal distribution of methane emissions from natural wetlands. (Data from Matthews & Fung 1987.)

is then said to have reached "steady-state." Carbon storage in northern soils is controlled by rates of decomposition that are several orders of magnitude smaller than those in temperate or tropical systems (Table 11.1). Carbon storage and flux terms for a boreal black spruce forest in the former Soviet Union are provided in Fig. 11.3.

11.2.1 Carbon inputs
Northern ecosystems are a net sink for carbon because decomposition of organic matter does not keep pace with carbon deposition from plant biomass. Net primary production (NPP), or the rate at which carbon is stored by plants in excess of the amount required for respiration, is an essential link in the soil carbon cycle because it represents the main flux of carbon from the atmosphere to the soil. The primary factors governing the low carbon inputs to the soil are: temperature, moisture, nutrient availability, and moss productivity.

Temperature and moisture
As mosses develop in a boreal forest, they give rise to a cold, wet, nutrient-poor soil system that supports progressively less productive forests. In

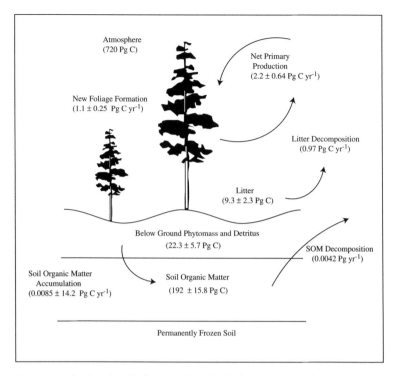

Fig. 11.3. CO₂ budget for a black spruce forest in the former Soviet Union. (Data from Kolchugina *et al.* 1995.)

turn, these same temperature and moisture conditions favor the growth of more mosses (Viereck 1983). Over time, mosses can succeed a forest community through a process known as *paludification* (Reiners *et al.* 1971). The original forest vegetation is replaced by species, particularly mosses, that are tolerant of a high water table and cold soil temperatures. Carbon concentrations in mineral soils underlying peatlands formed by paludification may be 1.2 to 1.8 times that in adjacent forest soils (Turunen *et al.* 1999).

Nutrient availability and moss productivity

Bryophytes are a highly effective competitor for nutrients in a nutrient-limited environment. Unlike most vascular plants, mosses obtain all of their nutrients directly from the atmosphere. In systems where mosses comprise a large portion of the forest floor, the moss layer is able to effectively intercept and assimilate nutrients derived from wet and dry deposition. Due to the relatively slow decomposition of moss litter (1–10% of the

Table 11.1. *Carbon inputs, decomposition rates, and storage in soils*

	Area (10^6 km)	Mean inputs (NPP) (g C m^{-2} yr^{-1})	Carbon storage (plants) (kg C m^{-2})	Carbon storage (soil) (kg C m^{-2})	Decomposition (yr^{-1})	Soil storage (Pg C)
Tundra and alpine meadow	8	65	0.3	21.6	–	173
Boreal forest	12	360	9	14.9	0.0028	179
Temperate forest	12	559	15	11.8	0.0115	142
Tropical forest	24.5	831	19	10.4	2.5000	255

Sources: NPP and carbon storage in plants from Whitakker and Likens (1973); soil carbon from Schlesinger (1977); decomposition from Cole and Rapp (1981) as modified by Schlesinger (1991).

rates found in vascular plant litter), these nutrients are released to the soil only over very long time-scales (Flanagan & Van Cleve 1977). As the forest floor matures, increasing amounts of nutrients are stored in the organic layers where they are unavailable for plant growth. Over time, the ability of the soil to support vascular plants declines. In mature black spruce stands, the high nutrient retention capacity of mosses may more than account for all of the elements (with the exception of calcium) input to the system via litterfall and throughfall (Oechel & Van Cleve 1986).

11.2.2 Decomposition

Moss-dominated soils differ from those of temperate and tropical regions in several important respects that contribute to their low decomposition rates: 1) soils may be frozen for much or all of the year; 2) the presence of a permanently frozen soil layer impedes the downward flow of water, resulting in saturated soil conditions; 3) the chemical composition of moss litter makes it more resistant to decomposition than that of many vascular plants; and 4) many mosses, particularly those in the genus *Sphagnum,* tend to acidify their surroundings, often lowering the pH of pore water below the tolerance of many decomposer organisms.

Temperature and moisture

Temperature is perhaps the most important environmental factor influencing the growth and activity of soil microorganisms. In general, for every 10^0 increase in temperature, the rate of a reaction increases by a factor of two to three ($Q_{10} = 2.0$–3.0). However, the response of soils to changes in temperature is not constant across all temperature ranges. In a review of laboratory data, Kirschbaum (1995) found significantly higher Q_{10} values at lower temperatures, with values decreasing from about 8 at 0 °C to approximately 2.5 at 20 °C. Because of this large increase in Q_{10} at low temperatures, Kätterer *et al.* (1998) suggest that Q_{10} values are not adequate for describing temperature relationships in high latitude systems. Regardless of the temperature function used, in northern ecosystems, where mean annual soil temperatures can be below 0 °C and soil temperatures during the summer may rise to only a few degrees above zero, even a one or two degree increase in soil temperature could have a significant effect on rates of CO_2 efflux. The availability of moisture in the soil is also a critical determinant of microbial activity and decomposition rates; maximum rates of decomposition are attained at intermediate moisture contents.

Bryophytes play a critical role in mediating both soil temperature and moisture. The high porosity and water-holding capacity of moss-derived organic matter make it an effective insulator; soils under bryophyte mats tend to be cold and wet. Weft- and hummock-forming mosses promote the development of permafrost by shielding soils from heat during the summer when air temperatures are above freezing and facilitating heat loss in the fall and winter (Bonan 1992). The combination of an impenetrable permafrost layer and a high moisture-holding capacity maintains the water table at shallow depths and results in the slow and incomplete decomposition of organic matter. Field transects in the Eurasian and Greenlandic arctic confirm the importance of water table and temperature in determining rates of soil respiration (Christensen *et al.* 1998).

Soil acidity

The pH of the soil affects nearly all soil processes – physical, chemical, and biological. In general, the greater the acidity of a soil, the fewer plant-available nutrients it contains, because hydrogen ions replace the nutrient cations on soil exchange sites. In addition, most decomposer organisms prefer a soil reaction that is close to neutral. Both the productivity of vascular plants and rates of decomposition tend to decline at lower pH values (Stewart & Wheatley 1990).

Mosses, in particular *Sphagnum* spp., actively influence soil pH. *Sphagnum* has a very high cation-exchange capacity resulting from an abundance of surface functional groups (e.g., $-COOH$). These functional groups can exchange their hydrogen ions for metal cations in solution and buffer the soil solution against the alkalinity of metallic cations brought in by precipitation. Although this process is not unique to mosses (all soil organic matter and most clay particles exhibit cation exchange), the vertical growth form of *Sphagnum* provides it with a unique ability to maintain acidic conditions despite constant flushing with dilute concentrations of salts and metal cations in rainwater (Clymo & Hayward, 1982). The ability to maintain a low pH not only confers a competitive advantage to acidophilic mosses, but also reduces rates of decomposition and promotes carbon storage.

11.3 Factors controlling methane dynamics in northern ecosystems

Quantifying the sources and sinks of atmospheric methane is of particular concern in global change research. Although the total amount of CH_4

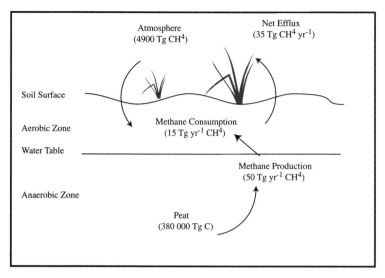

Fig. 11.4. Methane budget for northern peatlands. (Redrawn from Vourlitis & Oechel (1997); data for organic matter pool from Schlesinger (1991); CH_4 production and consumption from Reeburgh *et al.* (1994); atmospheric CH_4 from Bolin *et al.* (1994).)

present in the atmosphere is several orders of magnitude lower than the amount of CO_2, a single molecule of CH_4 can absorb 21 times more long-wave radiation (Ramanthan *et al.* 1985). Even small increases in atmospheric methane can have a significant impact on greenhouse warming.

Methane is released during the anaerobic decomposition of organic matter. Production occurs when soil bacteria use CO_2 or a methyl group as an electron acceptor and, in the process, produce gaseous methane:

$$CH_3COOH \rightarrow CO_2 + CH_4 \tag{2}$$

$$CO_2 + 4H_2 \rightarrow CH_4 + 2 H_2O \tag{3}$$

Methane exchange between the soil and the atmosphere is the net result of CH_4 production (methanogenesis) and CH_4 consumption (methano-trophy). In *Sphagnum* peatlands, photosynthesis by submerged mosses may provide a source of oxygen for methanotrophic activity and decrease net CH_4 emissions to the atmosphere (Verville *et al.* 1998). On average, approximately one-third of the methane produced in the anaerobic portion of the profile is consumed before reaching the surface (Fig. 11.4).

Methanogenic bacteria require highly reducing, anaerobic conditions ($E_h \leq -300$ mV) to produce CH_4, while methanotrophic bacteria tend to be most active along the boundary between aerobic and anaerobic portions of the soil profile (Reeburgh *et al.* 1994). Since much of the CH4 produced below the water table (up to 90%) can be consumed by methanotrophic bacteria in the aerated portion of the peat, atmospheric emissions are largely governed by the average position of the water table (Roulet *et al.*, 1992, Bubier *et al.* 1993, Moore & Roulet 1993). Bryophyte communities are often useful predictors of CH4 flux because their distribution in wetlands is closely related to the long-term average water table position (Bubier *et al.* 1995). Peat temperature (Crill *et al.* 1992, Dise *et al.* 1993) and substrate type (Crill *et al.* 1991) also play important roles in determining rates of CH4 emission. However, the relationship between these environmental variables is complex, since production and consumption may respond differently to the same environmental conditions.

11.4 Disturbance regimes

Because of the potential effects of increased atmospheric CO_2 and CH_4 on the global energy budget, the effect of a changing climate and anthropogenic disturbance on these critical soil systems is a topic of great interest in the scientific community. Excellent discussions of recent research may be found in Gorham (1991), Oechel *et al.* (1995), and Oechel and Vourlitis (1997).

11.4.1 Climate change

Most global circulation models suggest that increases in greenhouse gases will result in a warmer and wetter planet. Climate effects are expected to be most pronounced in boreal and arctic ecosystems, with a predicted surface temperature increase of 2–4 °C in the summer and as much as 6–15 °C in the winter and spring (Gates *et al.* 1992, Meehl *et al.* 1993). Arctic regions may also receive an additional 5–10% increase in precipitation (Maxwell 1992). There is compelling evidence to suggest that surface warming may already be occurring in arctic Alaska. Thermal profiles of permafrost and surface temperature records reveal that temperatures across the North Slope of Alaska have increased by 2–4 °C within the last century (Lachenbruch & Marshall 1986). Similar warming has also been observed in the Canadian arctic and boreal forest (Hengeveld 1991) and in the former Soviet Union (Karl *et al.* 1991).

Given the great uncertainty in global climate models, the ultimate effect of climate change on carbon cycling in moss-dominated ecosystems is difficult to quantify. Interactions between temperature, precipitation, soil moisture, evapotranspiration, vegetation, and gas exchange are complex and only partially understood. Regardless of the magnitude of climate change, high-latitude systems are particularly vulnerable to changes in temperature and moisture. Increased soil decomposition caused by warmer and drier soils and thawing permafrost could transform a region that is now a major sink for carbon into a significant source.

Direct effects of increased CO_2 on plant productivity
Higher concentrations of atmospheric CO_2 have the potential to increase both photosynthesis and biomass production. Over short time-scales, exposure to elevated CO_2 significantly increases rates of photosynthesis and carbon accumulation in many arctic species. However, both chamber and *in situ* studies of northern species show that this response may be short-lived (Billings *et al.* 1984, Oechel *et al.* 1994). Sustaining elevated rates of photosynthesis over long periods of time requires either the availability of sufficient nutrients (Chapin *et al.*, 1980) or improved nutrient-use efficiency (Oechel & Billings 1992). Because most northern ecosystems are nutrient-limited, the direct effects of CO_2 enhancements may be minimal (Oechel *et al.* 1994, Oechel & Vourlitis 1997). However, warmer and drier soil conditions may relieve nutrient limitations by increasing decomposition and release of stored nutrients in the forest floor (Nadelhoffer *et al.* 1991), thus stimulating ecosystem productivity and carbon accumulation (Shaver *et al.* 1992). In an experimental manipulation of air temperature in tussock tundra, Hobbie and Chapin (1998) found that increased air temperature stimulated rates of carbon turnover in soils (photosynthesis and decomposition) but had no significant effect on net carbon storage. These results suggest that the major effects of climate warming will occur through increased soil temperature and the consequent effects on decomposition and nutrient mineralization.

Enhanced CO_2 may benefit moss species more than vascular plants. Because mosses obtain their nitrogen from atmospheric deposition and cyanobacterial symbionts, they tend to suffer less from nutrient limitations than vascular plants. In addition, moss growth and photosynthesis is largely opportunistic with no period of senescence, allowing mosses to take greater advantage of an extended growing season (Oechel & Sveinbjörnsson 1978). In general, a warmer, moister, and CO_2-rich future environment may enhance bryophyte growth and result in thicker and denser bryophyte mats on the forest floor. However, the net effect of an extended growing season on carbon dynamics is uncertain, since both decomposition and carbon uptake are expected to increase (Oberbauer *et al.* 1998).

Effects of increased CO_2 on patterns of plant migration

Given the current temperature limitations on productivity in boreal ecosystems, it seems certain that soil warming will change the distribution of vegetation in northern latitudes. Differential responses to changing environmental conditions may change the competitive relationships between species. Experimental data from Chapin *et al.* (1983) suggest that increased temperature and nutrient availability in tundra soils provide a competitive advantage for vascular plants and may cause a decline in moss populations.

In a warmer climate, both the southern boundary of discontinuous permafrost and the treeline are expected to shift northward. In far northern ecosystems, this shift may increase rates of carbon storage by as much as 17.5% (0.070 Gt yr^{-1}) (Marion & Oechel 1993). The net effect on the boreal region is less certain since global warming is likely to result in increased storage along its northern boundary and decreased storage along the southern boundary (Ovendon 1990). At present, the best information about the rates of northward migration of plants in a warming climate is derived from pollen and microfossil records from the end of the last ice age (Overpeck *et al.* 1991). However, the rate of climate change currently predicted for the northern latitudes is much more rapid than ever previously experienced by terrestrial vegetation. There is no geologic evidence to suggest whether plant migration and related soil-forming processes could keep pace with expected changes in climate (Billings 1992).

Indirect effects of increased CO_2 on carbon storage in tundra and boreal forests

The most probable effects of higher CO_2 concentrations on tundra and boreal forests will be through their indirect effect on soil temperature and moisture. In forested soils, a warmer climate (with no change in current moisture conditions) is likely to result in increased decomposition and CO_2 efflux. In a black spruce stand in northern Manitoba, a net loss of 0.3 to 1.3 t C ha^{-1} yr^{-1} carbon from the subsoil was positively correlated with increases in air temperature (Goulden *et al.* 1998). Carbon isotopes ($^{14}C/^{12}C$) in soil CO_2 gas from this same stand indicated that much of the carbon released during the winter and spring was derived from older organic matter that had been sequestered at depth in the soil profile

(Winston *et al.* 1997). Significant increases in winter temperatures may result in greatly enhanced rates of decomposition and a net loss of stored carbon from the ecosystem.

In both tundra and forested systems, higher air and soil temperatures are associated with increased rates of evaporation, earlier spring thaw, thawing of permafrost layers, and lowered depth to water table, all of which could cause a significant reduction in soil moisture during the summer (Hinzman & Kane 1992). Unless these reductions in soil moisture are matched by an equivalent increase in precipitation, the net effect could range from a reduction in sequestration potential to a loss of stored carbon from the ecosystem. A series of controlled experiments by Billings *et al.* (1982, 1983) suggests that variations in water table have a greater effect on CO_2 flux from tundra soils than either warming the air temperature by 4 °C or doubling the atmospheric CO_2 concentration. Similar experiments on boreal soils showed that a reduction in water table increased CO_2 emissions and converted the soils from a net source to a net sink of CH_4 (Funk *et al.* 1994).

Ultimately, future carbon source–sink potential of moss-dominated soils will be determined by the temperature response of net primary productivity compared to that of organic matter decomposition (Kirschbaum 1995). Over short time-scales, soil decomposition is expected to be more responsive to changes in climate (Pastor & Post 1993, Oechel & Vourlitis 1997). However, over longer time-scales, changes in plant species composition (Smith & Shugart 1993) and increased nutrient availability via decomposition (Ratstetter *et al.* 1992) could improve ecosystem productivity and increase the amount of carbon stored in tree and shrub biomass. This increased carbon storage could balance or potentially exceed the short-term loss of carbon lost to enhanced soil respiration.

Effects of increased CO_2 on carbon dynamics in bogs and fens

Assuming that the position of the water table remains constant, warmer temperatures and greater plant productivity may cause an increase in CH_4 production (Whiting & Chanton 1993). However, as in forested ecosystems, warmer temperatures are also associated with increased evapotranspiration and permafrost thaw. As the water table drops, the thickness of aerobic zone increases and a greater proportion of CH_4 produced at depth may be oxidized before it reaches the surface. Even a relatively small decrease in the level of the water table (0.10–0.22 m) may cause a significant decline or elimination of CH_4 emissions to the atmosphere (Roulet *et al.* 1993) and increase CO_2 emissions by 50–100% (Silvola *et al.* 1996). On a landscape scale, dryer soils and warmer temperatures could convert large regions from a net source of CH_4 to a sink. However, climate warming may also increase the length of time that the soils are unfrozen and promote additional CH_4 emissions during the autumn and winter.

Warmer temperatures and increased aerobic decomposition are expected to cause a reduction in carbon accumulation. If warmer temperatures are also accompanied by a drop in the water table, enhanced decomposition of peat would result in a loss of stored carbon from the system (Shurpali *et al.* 1995, Carroll & Crill 1997, Bellisario *et al.* 1998). Based on the average Q_{10} for peatland soils, a 2–4 °C increase in temperature would cause a 30–60% increase in CO_2 efflux (Silvola *et al.* 1996). However, these carbon losses may be more than offset by increased moss and vascular plant productivity and a longer growing season. In the short term, increased carbon storage in tree biomass may balance losses of carbon resulting from peat decomposition. Furthermore, warmer air and soil temperatures may shift peatland landscapes further north and reactivate carbon accumulation in subarctic peatlands that are currently dormant (Gorham 1991).

11.4.2 Habitat destruction – agroforestry and peat harvesting

More than half of all peatlands are concentrated near industrialized areas in the northern temperate and boreal zones. As human populations in these regions grow, there is increasing pressure to utilize peatland resources. Peatlands can often be converted into productive agricultural or forest lands through ditching and draining. Peat itself is a valuable natural resource that has been mined for centuries as a source of fuel as well as for a variety of industrial applications.

As a result of these practices, the world area of peatlands in populated regions is declining rapidly. In the UK, more than 90% of blanket bogs and 98% of raised bogs have been lost, leaving only 125 000 ha and 1170 ha respectively (Foss 1997). At current rates of destruction, all unprotected raised bogs in Ireland will be gone by the turn of the century (Ryan & Cross 1984). No natural peatlands remain in the Netherlands and Poland; Switzerland and Germany each have less than 500 ha remaining (Foss 1997). In the former Soviet Union, carbon losses from peat harvesting currently exceed the amount of C stored in peatlands, resulting in a net source to the atmosphere of 70 Tg C yr^{-1} (Botch *et al.* 1995). Despite these high rates of CO_2 emission, the impact on the atmosphere is at least partially offset by the simultaneous reduction in methane production. In essence, drainage of peatlands converts large areas of the landscape from a net CH_4-source/CO_2-sink to a CO_2-source/CH_4-sink (Francez & Vasander 1995).

The ultimate effect of peatland conversion on carbon storage depends largely on the land-use practice. For example, ditching forested peatlands to increase tree productivity results in decomposition of stored carbon (Armentano & Menges 1986). However, if the carbon stored in tree biomass is not immediately removed from the site, the loss of carbon from peat decomposition may be balanced or exceeded by the amount of additional carbon sequestered in new tree biomass (Fowler *et al.* 1995). In the short term (<100 years), the net effect of drainage on the ecosystem may actually be an increase in carbon storage. On the other hand, peatlands drained for agricultural purposes may continue to release carbon as the remaining organic material is oxidized during subsequent tillage and harvesting.

11.4.3 Fire

Wildfire is a major control on patterns of carbon storage and release in northern landscapes. Between 5 and 12 million hectares of boreal forest burn every year releasing an estimated 10 to 70 t C ha^{-1} yr^{-1} (Stocks & Kauffman 1997, Kasischke *et al.* 2000). Due to the high organic content of the surface soils, fires in moss-dominated ecosystems can be difficult to control and may smolder for months. Global climate models suggest that a doubling of atmospheric CO_2 and the consequent change in global climate could increase both fire frequency and severity in the northern latitudes, releasing vast quantities of stored carbon to the atmosphere (Flannigan & Van Wagner 1991).

In addition to the immediate release of carbon during combustion, fire can also change the longer-term patterns of carbon storage and emission from northern soils. Following fire, the loss of insulating moss layers and darkened soil surface raise soil temperatures and cause permafrost to thaw. Over time, these changes result in a deeper rooting zone, improved drainage, and enhanced decomposition of stored carbon (MacLean *et al.* 1983, Richter *et al.* in press). Post-fire decomposition of carbon may equal or exceed the amount of carbon lost during combustion itself (Auclair & Carter 1993, Richter *et al.* 1999). The net effect of fire on the carbon cycle depends upon the balance between short-term losses of carbon during and immediately following burning and the gains of carbon in tree biomass during forest regrowth.

Bryophytes play a critical role in determining the pathway of secondary succession and rates of carbon storage. Even in fires where the moss layer is not entirely consumed, the reduction in forest canopy can result in die-back of the moss floor and the release of the large amounts of nitrogen and other nutrients stored there (Oechel & Van Cleve 1986). After fire, the burned ground surface experiences large diurnal fluctuations in temperature which may be greater than those tolerated by most seedlings. Mosses such as *Polytrichum* spp. and *Ceratodon purpureus* ("fire moss") can withstand these harsh conditions and rapidly colonize the surface. In the process, they create a more favorable environment for vascular plants by decreasing soil temperature and increasing moisture retention at the soil surface. Once a forest canopy has been established (30–40 years), *Hylocomium splendens* and *Pleurozium schreberi* become dominant. These mosses cool the soil, encourage the development of permafrost, and promote carbon accumulation.

11.4.4 Anthropogenic nutrient inputs

Bryophytes are particularly sensitive to atmospheric pollution. The growth form and high surface area of bryophytes make them well adapted for trapping and absorbing atmospheric nutrients. In addition, the leaves of many moss species are only one cell thick with the photosynthetic cells directly and continuously exposed to the atmosphere. Although bryophyte communities are exposed to a wide range of atmospheric pollutants, the oxides of nitrogen (NO_x) and sulfur (SO_x) are among the most important due to their role in soil acidification and fertilization.

Nitrogen

Like all plants, bryophytes require nitrogen to create proteins and enzymes. At the same time, many northern systems are severely nitrogen-limited. For this reason, a modest increase in atmospheric nitrogen inputs is not likely to be detrimental to peatland mosses. *Sphagnum* spp. are particularly effective at absorbing and assimilating nitrogen (Malmer 1990). However, at high levels, nitrogen deposition can exceed the assimilation capacity of *Sphagnum*. The excess nitrogen then enters the soil solution where it is taken up by vascular-plants. Under these conditions, vascular-plant productivity may increase and disrupt the balance between moss and shrub species (Lee & Studholme 1992). Of particular importance to the global carbon cycle is the fact that increased concentrations of nitrogen in moss tissues makes them more susceptible to decomposition (Clymo & Hayward 1982). Enhanced decomposition could reduce the thickness of the bryophyte mat and change the thermal, moisture, and nutrient status of forested, tundra, and peatland soils.

Sulfate

Unlike nitrogen, the deposition of SO_x compounds has a decidedly negative effect on the health of moss communities. Deposition of sulfates increases the acidity of soil water. Natural variation in the acidity of soil substrates and solutions is a major factor governing the composition of bryophyte communities. Moss communities that prefer a neutral reaction and are poorly buffered against changes in acidity, such as minerotrophic fens, are the most susceptible to the effects of acid deposition (Gorham *et al.* 1984).

Strong evidence for the detrimental effect of atmospheric sulfate deposition comes from the Sudbury region of Canada where metal smelting has released heavy metals and SO_2 into the environment. In peatlands closest to the smelter, all *Sphagnum* species were destroyed. Peatlands at a greater distance showed a significant reduction in the number of *Sphagnum* species (Gignac & Beckett 1986). Similarly, in the southern Pennine region of the UK, nearly all *Sphagnum* species have been lost or replaced since the Industrial Revolution (Lee & Studholme 1992). However, in cases where the sulfate percolates into the anaerobic zone, sulfate reduction may effectively buffer the peatland against further acidification (Gorham *et al.* 1984).

11.5 Conclusions

Since the end of the last glacial period, northern soils have served as a reservoir for terrestrial carbon. Due to temperature and moisture limitations, carbon dynamics in these systems are sensitive to even moderate changes in temperature and moisture conditions. At the same time, the effects of global warming are expected to be most pronounced in northern latitudes. Although the ultimate effect of climate change in moss-dominated ecosystems is difficult to quantify, should rates of soil decomposition increase more rapidly than net primary production, these systems might well cease to function as a net carbon sink and serve instead as a carbon source.

Northern ecosystems remain one of the last regions of the earth largely untouched by human activities. As populations in the northern hemisphere increase, greater human contact with boreal, tundra, and peatland systems can only result in increased rates of disturbance. Habitat destruction, fire, and deposition of pollutants all have detrimental and long-lived effects. The long time-scales over which these carbon-rich systems have developed means that recovery following major disturbance may require centuries or millennia. Predicting how the vast stores of soil carbon in moss-dominated ecosystems will be affected by anthropogenic disturbance and climate change is critical to models of global climate change.

REFERENCES

Apps, M. J., Kurz, W. A., Luxmoore, R. J., Nilsson, L. O., Sedjo, R. A., Schmidt, R., Simpson, L. G., & Vinson, T. S. (1993). Boreal forests and tundra. *Water, Air and Soil Pollution*, **70**, 39–53.

Armentano, T. V. & Menges, E. S. (1986). Patterns of change in the carbon balance of organic soil wetlands of the temperate zone. *Journal of Ecology*, **74**, 755–74.

Auclair, A. N. D. & Carter, T. B (1993). Forest wildfires as a recent source of CO_2 at northern latitudes. *Canadian Journal of Forest Research*, **23**, 1530–6.

Bartlett, K. & Harriss, R. C. (1993). Review and assessment of methane emissions from wetlands. *Chemosphere*, **26**, 261–320.

Bellisario, L. M., Moore, T. R., & Bubier, J. L. (1998). Net ecosystem CO_2 exchange in a boreal peatland, northern Manitoba. *Ecoscience*, **5**, 534–41.

Billings, W. D. (1992). Phytogeographic and evolutionary potential of the arctic flora and vegetation in a changing climate. In *Arctic Ecosystems in a Changing Climate, an Ecophysiological Perspective*, ed. F. S. Chapin III, R. L. Jeffries, J. F. Reynolds, G. R. Shaver, & J. Svoboda, pp. 91–109. New York: Academic Press.

(1997). Challenges for the future: arctic and alpine ecosystems in a changing world. In *Global Change and Arctic Terrestrial Ecosystems*, ed. W. C. Oechel, T. Callaghan, T. Gilmanov, J. I. Holten, B. Maxwell, U. Molau, & B. Sveinbjörnsson, pp. 1–18. New York: Springer-Verlag.

Billings, W. D., Luken, J. O., Mortensen, D. A., & Peterson, K. M. (1982). Arctic tundra: a source or sink for atmospheric carbon dioxide in a changing environment. *Oecologia*, **53**, 7–11.

(1983). Increasing atmospheric carbon dioxide: possible effects on arctic tundra. *Oecologia*, **58**, 286–9.

Billings, W. D., Peterson, K. M., Luken, J. O., & Mortensen, D. A. (1984). Interactions of increasing atmospheric carbon dioxide and soil nitrogen on the carbon balance of tundra microcosms. *Oecologia*, **65**, 26–9.

Bolin, B., Houghton, J., Meira-Filho, L. G. (1994). *Radiative Forcing of Climate Change: The 1994 Report of the Scientific Assessment Working Group of the Intergovernmental Panel on Climate Change*. Geneva: World Meteorological Organization.

Bonan, G. B. (1992). Soil temperature as an ecological factor in boreal forests. In *A Systems Analysis of the Global Boreal Forest*, ed. H. H. Shugart, R. Leemans, & G. B. Bonan, pp. 126–43. New York: Cambridge University Press.

Botch, M. S., Kobak, K. I., Vinson, T. S., & Kolchugina, T. P. (1993). Carbon pools and accumulation in peatlands of the former Soviet Union. *Global Biogeochemical Cycles*, 9, 37–46.

Bubier, J. L., Moore, T. R., & Roulet, N. T. (1993). Methane emissions from wetlands in the midboreal region of northern Ontario, Canada. *Ecology*, 74, 2240–54.

Bubier, J. L., Moore, T. R., & Juggins, S. (1995). Predicting methane emission from bryophyte distribution in northern peatlands. *Ecology*, 76, 677–93.

Carroll, P. & Crill, P. M. (1997). Carbon balance of a temperate poor fen. *Global Biogeochemical Cycles*, 11, 349–56.

Chapin, F. S., III (1983). Direct and indirect effects of temperature or arctic plants. *Polar Biology*, 2, 47–52.

Chapin, F. S., III, Miller, P. C., Billings, W. D., & Coyne, P. I. (1980). Carbon and nutrient budgets and their control in coastal tundra. In *An Arctic Ecosystem: The Coastal Tundra at Barrow, Alaska*, ed. J. Brown, P. C. Miller, L. L. Tieszen, & F. L. Bunnell, pp. 458–82. Stroudsburg, PA: Dowden, Hutchinson, Ross.

Christensen, T. R., Jonasson, S., Michelsen, A., Calaghan, T. V., & Hastrom, M. (1998). Environmental control on soil respiration in the Eurasian and reenlandic Arctic. *Journal of Geophysical Research*, 103, 29 015–21.

Clymo, R. S. & Hayward, P. M. (1982). The ecology of *Sphagnum*. In *Bryophyte Ecology*, ed. A. J. E. Smith, pp. 229–89. London: Chapman & Hall.

Cole, D. W. & Rapp, M. (1981). Element cycling in forest ecosystems. In *Dynamic Properties of Forest Ecosystems*, ed. D. E. Reichle, pp. 341–409. Cambridge: Cambridge University Press.

Crill, P. M., Harriss, R. C., & Bartlett, K. B. (1991). Methane fluxes from terrestrial wetland environments. In *Microbial Production and Consumption of Greenhouse Gases: Methane, Nitrogen Oxides, and Halomethanes*, ed. J. E. Rodgers & W. B. Whitman, pp. 91–110. Washington: American Society for Microbiology.

Crill, P. M., Bartlett, K. B., & Roulet, N. (1992). Methane flux from boreal peatlands. *Suo*, 43, 173–82.

Dise, N. B., Gorham, E., & Verry, E. S. (1993). Environmental factors controlling methane emissions from peatlands in northern Minnesota. *Journal of Geophysical Research*, 98, 10583–94.

Fahnestock, J. T., Jones, M. H., Brooks, P. D., Walker, D. A., & Welker, J. M. (1998). Winter and early spring CO_2 efflux from tundra communities of northern Alaska. *Journal of Geophysical Research*, 103, 29 023–7.

Flanagan, P. W. & Van Cleve, K. (1977). Microbial biomass, respiration, and nutrient cycling in a black spruce taiga ecosystem. In *Soil Organisms as Components of Ecosystems: International Soil Zoology Colloquium*, Ecological Bulletins 25, ed. U. Lohm & T. Persson, pp. 261–73. Stockholm: Swedish Natural Science Research Council.

Flannigan, M. D. & Van Wagner, C. E. (1991). Climate change and wildfire in Canada. *Canadian Journal of Forest Research*, 21, 61–72.

Foss, P. (1997). Ten years of the Save the Bogs Campaign. In *Conserving Peatlands*, ed. L. Parkyn, R. E. Stoneman, & H. A. P. Ingram, pp. 391–8. New York: CAB International.

Fowler, D., Hargreaves, K. J., MacDonald, J. A., & Gardiner, B. (1995). Methane and CO_2 exchange over peatland and the effects of afforestation. *Forestry*, **68**, 327–34.

Francez, A. J. & Vasander, H. (1995). Peat accumulation and peat decomposition after human disturbance in French and Finnish mires. *Acta Oecologia*, **16**, 599–608.

Funk, D. W., Pullman, E. R., Peterson, K. M., Crill, P. M., & Billings, W. D. (1994). Influence of water table on carbon dioxide, carbon monoxide, and methane fluxes from taiga bog microcosms. *Global Biogeochemical Cycles*, **8(3)**, 271–278.

Gates, W. L., Mitchell, W. F. B., Boer, G. J., Cubasch, U., & Meleshko, V. P. (1992). Climate modeling, climate prediction and model validation. In *Climate Change 1992: The Supplemental Report to the IPCC Scientific Assessment*, ed. J. T. Houghton, B. A. Callander, & S. K. Varney, pp. 97–135. Cambridge: Cambridge University Press.

Gignac, L. D. & Beckett, P. (1986). The effect of smelting operations on peatlands near Sudbury, Ontario, Canada. *Canadian Journal of Botany*, **64**, 1138–47.

Gore, A. J. P. (ed.) (1983). *Mires – Swamp, Bog, Fen, and Moor*. New York: Elsevier.

Gorham, E. (1991). Northern peatlands: role in the carbon cycle and probable responses to climate warming. *Ecological Applications*, **1**, 182–95.

Gorham, E., Bayley, S. E., & Schindler, D. W. (1984). Ecological effects of acid deposition upon peatlands: a neglected field in "acid-rain" research. *Canadian Journal of Fisheries and Aquatic Science*, **41**, 1256–68.

Goulden, M. L., Wofsy, S. C., Harden, J. W., Trumbore, S. E., Crill, P. M., Gower, S. T., Fries, T., Daube, B. C., Fan, S. M., Sutton, D. J., Bazzaz, A., & Munger, J. W. (1998). Sensitivity of boreal forest carbon to soil thaw. *Science*, **279**, 214–17.

Harden, J. W., Sundquist, E. T., Stallard, R. F., & Mark, R. K. (1992). Dynamics of soil carbon during deglaciation of the Laurentide Ice Sheet. *Science*, **258**, 1921–4.

Harden, J. W., O'Neill, K. P., Trumbore, S. E., Veldhuis, H., & Stocks, B. J (1997). Moss and soil contributions to the annual net carbon flux of a maturing boreal forest. *Journal of Geophysical Research*, **102**, 28805–16.

Havas, P. & Mäenpää, E. (1972). Evolution of carbon dioxide at the floor of a *Hylocomium myrtillus* type spruce forest. *Aquilo Series Botanique*, **11**, 22–40.

Hengeveld, H. (1991). *A State of the Environment Report*. Atmospheric Environment Service, Environment Canada.

Hinzman, L. D. & Kane, D. L. (1992). Potential response of an arctic watershed during a period of global warming. *Journal of Geophysical Research*, **97**, 2811–20.

Hobbie, S. E. & Chapin, F. S., III. (1998). The response of tundra plant biomass, aboveground production, nitrogen, and CO_2 flux to experimental warming. *Ecology*, **79**, 1526–44.

Karl, T. R., Kukla, G., Razuvayev, V. N, Changery, M. G., Quayle, R. G., Heim, R. R., Easterling, D. R., & Fu, C. B. (1991). Global warming: evidence for assymetric diurnal temperature change. *Geophysical Research Letters*, **18**, 2253–6.

Kasischke, E. S., O'Neill, K. P., French, N. H. F., & Borgeau-Chavez, L. L. (2000). Controls on patterns of biomass burning in Alaskan boreal forests. In *Fire, Climate Change, and Carbon Cycling in the North American Boreal Forest*, ed. E. S. Kasischke & B. J. Stocks, pp. 173–96. New York: Springer-Verlag.

Kätterer, T., Reichstein, M., Andrén, O., & Lomander, A. (1998). Temperature dependence of organic matter decomposition: a critical review using literature data analyzed with different models. *Biology and Fertility of Soils*, **27**, 258–62.

Kirschbaum, M. U. (1995). The temperature dependence of soil organic matter decomposition, and the effect of global warming on soil organic C storage. *Soil Biology and Biochemistry*, **27**, 753–60.

Kolchugina, T. P. & Vinson, T. S. (1993). Carbon sources and sinks in forest biomes of the former Soviet Union. *Global Biogeochemical Cycles*, **7**, 291–304.

Kolchugina, T. P., Vinson, T. S., Gaston, G. G., Rozhkov, V. A., & Schwidenko, A. Z. (1995). Carbon pools, fluxes, and sequestration potential in soils of the former Soviet Union. In *Advances in Soil Science, Soil Management, and the Greenhouse Effect*, ed. J. Kimball, E. Levine, B. A. Stewart, & R. Lal, pp. 25–40. Boca Raton: CRC Press.

Lachenbruch, A. H.& Marshall, B. V. (1986). Changing climate: geothermal evidence from permafrost in the Alaskan arctic. *Science*, **234**, 689–96.

Lee, J. A. & Studholme, C. J. (1992). Responses of *Sphagnum* to polluted environments. In *Bryophytes and Lichens in a Changing Environment*, ed. J. W. Bates & A. M. Farmer, pp. 314–32. New York: Oxford University Press.

MacLean, D. A., Woodley, S. J., Weber, M. G., & Wein, R. W. (1983). Fire and nutrient cycling. In *The Role of Fire in the Northern Circumpolar Ecosystem*, ed. R. W. Wein & D. A. MacLean, pp. 11–132. Chichester: Wiley.

Malmer, N. (1990). Constant or increasing nitrogen concentrations in *Sphagnum* mosses on mires in Southern Sweden during the last few decades. *Aquilo Series Botanique*, **28**, 57–65.

Marion, G. M. & Oechel, W. C. (1993). Mid- to late-Holocene carbon balance in Arctic Alaska and its implications for future global warming. *Holocene*, **3**, 193–200.

Matthews, E. & Fung, I. (1987). Methane emissions from natural wetlands: global distribution, area, and environmental character of sources. *Global Biogeochemical Cycles*, **1**, 61–86.

Maxwell, B. (1992). Arctic climate: potential for change under global warming. In *Arctic Ecosystems in a Changing Climate, an Ecophysiological Perspective*, ed. F. S. Chapin III., R. L. Jeffries, J. F. Reynolds, G. R. Shaver, & J. Svoboda, pp. 11–34. New York: Academic Press.

Meehl, G. A., Washington, W. M., & Karl, T. R. (1993). Low-frequency variability and CO_2 transient climate change: Part 1. Time averaged differences. *Climate Dynamics*, **8**, 117–33.

Miller, P. C., Kendall, R., & Oechel, W. C. (1983). Simulating carbon accumulation in northern ecosystems. *Simulation*, **40**, 119–31.

Moore, T. R. & Roulet, N. T. (1993). Methane flux: water table relations in northern wetlands. *Geophysical Research Letters*, **20**, 587–90.

Nadelhoffer, K. J., Giblin, A. E. Shaver, G. R., & Laundre, J. A. (1991). Effects of temperature and substrate quality on element mineralization in six arctic soils. *Ecology*, **72**, 242–53.

Oberbauer, S. F., Starr, G., & Pop, E. W. (1998). Effects of extended rowing season and soil warming on carbon dioxide and methane exchange of tussock tundra in Alaska. *Journal of Geophysical Research*, **103**, 29 075–82.

Oechel, W. C. & Billings, D. W. (1992). Anticipated effects of global change on the carbon balance of arctic plants and ecosystems. In *Physiological Ecology of Arctic Plants: Implications for Climate Change*, ed. F. S. Chapin III, R. L. Jefferies, G. R. Shaver, J. Reynolds, & J. Svoboda, pp. 139–68. New York: Academic Press.

Oechel, W. C. & Sveinbjörnsson, B. (1978). Primary production processes in arctic bryophytes at Barrow, Alaska. In *Vegetation and Production Ecology of the Alaskan Arctic Tundra*, ed. L. L. Tieszen, pp. 269–98. New York: Springer-Verlag.

Oechel, W. C. & Van Cleve, K. (1986). The role of bryophytes in nutrient cycling in the taiga. In *Forest Ecosystems in the Alaska Taiga*, ed. K. Van Cleve, F. S. Chapin III, P. W. Flanagan, L. A. Viereck, & C. T. Dyrness, pp. 121–37. New York: Springer-Verlag.

Oechel, W. C. & Vourlitis, G. (1997). Climate change in northern latitudes: alterations in ecosystem structure and function and effects on carbon sequestration. In *Global Change and Arctic Terrestrial Ecosystems*, ed. W. C. Oechel, T. Callaghan, T. Gilmanov, J. I. Holten, B. Maxwell, U. Molau, & B. Sveinbjörnsson, pp. 381–401. New York: Springer-Verlag.

Oechel, W. C., Hastings, S. J., Vourlitis, G., Jenkins, M., Riechers, G., & Grulke, N. (1993). Recent change of arctic tundra ecosystems from a net carbon dioxide sink to a source. *Nature*, **361**, 520–3.

Oechel, W. C., Cowles, S., Grulke, N., Hastings, S. J., Lawrence, B., Prudhomme, T., Riechers, G., Strain, B., Tissue, D., & Vourlitis, G. (1994). Transient nature of CO_2 fertilization in arctic tundra. *Nature*, **371**, 500–3.

Oechel, W. C., Vourlitis, G. L., Hastings, S. J., & Bochkarev, S. A. (1995). Change in arctic CO_2 flux over two decades: effects of climate change at Barrow, Alaska. *Ecological Applications*, **5**, 846–55.

Oechel, W. C., Vourlitis, G., & Hastings, S. J. (1997). Cold season CO_2 emissions from arctic soils. *Global Biogeochemical Cycles*, **11**, 163–72.

Ovendon, L. (1990). Peat accumulation in northern peatlands. *Quaternary Research*, **33**, 377–86.

Overpeck, J. T., Bartlein, P. J., & Webb, T., III (1991). Potential magnitude of future vegetation changes in eastern North America: comparisons with the past. *Science*, **254**, 692–5.

Pajari, B. (1995). Soil respiration in a poor upland site of Scots pine stand subjected to elevated soil temperatures and atmospheric carbon concentration. *Plant and Soil*, **168–9**, 563–70.

Pastor, J. & Post, W. M. (1993). Linear regressions do not predict the transient response of eastern North Amercan forests to CO_2 induced climate change. *Climate Change*, **23**, 111–119.

Post, W. M., Emmanuel, W. R., Zinke, P. J., & Stangenberger, A. G. (1982). Soil carbon pools and world life zones. *Nature*, **298**, 156–9.

Ramanthan, V., Cicerone, H. B., Singh, H. B., & Kiehl, J. T. (1985). Trace gas trends and their potential role in climate change. *Journal of Geophysical Research*, **90**, 5547–66.

Ratstetter, E. B., McKane, R. B., Shaver, G. R., & Melillo, J. M. (1992). Changes in C storage by terrestrial ecosystems: how C–N interactions restrict responses to CO_2 and temperature. *Water, Air and Soil Pollution*, **64**, 327–44.

Reeburgh, W. S., Roulet, N. T., & Svensson, B. H. (1994). Terrestrial biosphere–atmosphere exchange in high latitudes. In *Global Atmospheric–Biospheric Chemistry*, ed. R. G. Prinn, pp. 165–78. New York: Plenum Press.

Reiners, W. A., Worley, I. A., & Lawrence, D. B. (1971). Plant diversity in a chronosequence at Glacier Bay, Alaska. *Ecology*, **52**, 55–69.

Richter, D. D., O'Neill, K. P., & Kasischke, E. S. (2000). Stimulation of soil respiration in burned black spruce (*Picea mariana* L.) forest ecosystems: a hypothesis. In *Fire, Climate Change, and Carbon Cycling in the North American Boreal Forest*, ed. E. S. Kasischke, & B. J. Stocks, pp. 197–213. New York: Springer-Verlag.

Roulet, N. T., Ash, R., & Moore, T. R. (1992). Low boreal wetlands as a source of atmospheric methane. *Journal of Geophysical Research*, **97**, 3739–49.

Roulet, N. T., Ash, R., Quinton, W., & Moore, T. (1993). Methane flux from drained northern peatlands: effect of a persistent water table lowering on flux. *Global Biogeochemical Cycles*, **7**, 749–69.

Ryan, J. B. & Cross, J. C. (1984). Conservation of peatlands in Ireland. In *Proceedings of the 7th International Peat Conference, Dublin*, pp. 265–96. Jyväskyä, Finland: International Peat Society.

Schlesinger, W. H. (1991). *Biogeochemistry: An Analysis of Global Change*. New York: Academic Press.

Shaver, G. R., Billings, W. D., Chapin, F. S., III, Giblin, A. E., Nadelhoffer, K. J., Oechel, W. B. & Ratstetter, E. B. (1992). Global change and the carbon balance of arctic ecosystems. *BioScience*, **42**, 443–41.

Shurpali, N. J., Verma, S. B., Kim, J., & Arkebauer, T. J. (1995). Carbon dioxide exchange in a peatland ecosystem. *Journal of Geophysical Research*, **100**, 14319–26.

Silvola, J., Alm, J., Ahlo, U., Nykanen, H., & Martikainen, P. J. (1996). CO_2 fluxes from peat in boreal mires under varying temperature and moisture conditions. *Journal of Ecology*, **84**, 219–28.

Sjörs, H. (1980). Peat on earth: multiple use or conservation? *Ambio*, **9**, 303–8.

Smith, T. M. & Shugart, H. H. (1993). The transient response of terrestrial carbon storage to a perturbed climate *Nature*, **361**, 523–6.

Stewart, J. M. & Wheatley, R. E. (1990). Estimates of CO_2 production from eroding peat surfaces. *Soil Biology and Biochemistry* , **22** , 65–8.

Stocks, B. J. & Kauffman, J. B. (1997). Biomass consumption and behavior of wildland fires in boreal, temperate, and tropical ecosystems: parameters necessary to interpret historic fire regimes and future fire scenarios. In *Sediment Records of Biomass Burning and Global Change*, ed. J. S. Clark, H. Cachier, J. G. Goldhammer, & B. J. Stocks, pp. 169–88. Berlin: Springer-Verlag.

Turunen, J., Tolonen, K., Tolvanen, S., Remes, M., Ronkainen, J., & Jungner, H. (1999). Carbon accumulation in the mineral subsoil of boreal mires. *Global Biogeochemical Cycles*, **13**, 71–9.

Van Cleve, K., Oliver, L. K., Schlentner, P., Viereck, L. A., & Dyrness, C. T. (1983). Productivity and nutrient cycling in taiga forest ecosystems. *Canadian Journal of Forest Research*, **13**, 703–20.

Verville, J. H., Hobbie, S. E., Chapin, F. S., III, & Hooper, D. U. (1998). Response of tundra CH_4 and CO_2 flux to manipulation of temperature and vegetation. *Biogeochemistry*, **41**, 215–35.

Viereck, L. A., Dyrness, C. T., Van Cleve, K., & Foote, M. J. (1983). Vegetation, soils, and forest productivity in selected forest types in interior Alaska. *Canadian Journal of Forest Research*, **13**, 703–20.

Vourlitis, G. L. & Oechel, W. C. (1997). The role of northern ecosystems in the global methane budget. In *Global Change and Arctic Terrestrial Ecosystems*, ed. W. C. Oechel, T. Callaghan, T. Gilmanov, J. I. Holten, B. Maxwell, U. Molau, & B. Sveinbjörnsson, pp. 266–90. New York: Springer-Verlag.

Whiting, G. J. & Chanton, J. P. (1993). Primary production control of methane emission from wetlands. *Nature,* **364**, 794–5.

Whitakker, R. H. & Likens, G. E. (1973). Carbon in the biota. In *Carbon and the Biosphere,* ed. G. M. Woodwell & E. V. Pecan, pp. 281–302. Washington: National Technical Information Service.

Winston, G. C., Sundquist, E. T., Stephens, B. B., & Trumbore, S. E. (1997). Winter CO_2 fluxes in a boreal forest. *Journal of Geophysical Research,* **102**, 28 795–804.

Zimov, S. A., Semiletov, I. P., Daviodov, S. P., Voropaev, Y. V., Prosyannikov, S. F., Wong, C. S., & Chan, Y. H. (1993). Wintertime carbon dioxide emission from soils of northeastern Siberia. *Arctic,* **46**, 197–204.

12

Population ecology, population genetics, and microevolution

Bryophytes have a number of life history characteristics that make them interesting and tractable organisms for population studies. Their life cycle is unique among land plants in having a free-living (usually) perennial gametophyte and an annual sporophyte that remains attached to the gametophyte throughout its life. The sporophyte obtains a significant proportion of its nutrients from the maternal gametophyte to which it is attached, although in mosses it is green and photosynthetic when immature and premeiotic (Proctor 1977). A specialized tissue comprised of transfer cells occurs on either the sporophyte or gametophyte side, or both, of the junction of the two generations, and facilitates the movement of metabolites between the gametophyte and sporophyte (Ligrone *et al.* 1993). This physiological connection, in addition to the obvious genetic relationship between gametophyte and sporophyte, must contribute to complex evolutionary patterns into which bryophytes offer unique opportunities for investigation (Shaw & Beer 1997).

Although it is often said that it is the gametophyte rather than the sporophyte that is exposed to the external environment, there is no empirical evidence to suggest that the sporophyte is completely or even mostly shielded from the forces of natural selection. Indeed, the fact that "life history strategies" can be recognized and include both gametophytic and sporophytic traits (During 1979, Longton 1997) implies a history of selection on traits in both generations. Evolution of the whole bryophyte plant must involve interactions between each generation and the external environment, as well as interactions, genetic and physiological, between the sporophyte and gametophyte generations of a single plant.

This chapter provides a brief overview of ecological and evolutionary patterns and processes at the population level in bryophytes. It focuses on population processes directly relevant to evolutionary change while some ecological aspects of population biology are barely covered. In particular, demographic processes at the population (e.g., Watson 1974, Økland 1995) and metapopulation (e.g., Herben 1994) levels are given less consideration, although these processes obviously impinge upon genetic patterns.

Early genetic studies utilizing bryophytes include Allen's (1919) important work on the sex chromosomes of *Sphaerocarpos*. Allen subsequently conducted a series of crossing experiments examining segregation of gametophytic life history and morphological traits (reviewed in Allen 1935, 1945), work that has never been repeated or expanded. Other reviews of innovative earlier experimental studies on the genetics of bryophytes include Marchal (1907, 1911), Wettstein (1924, 1932), and Lewis (1961). McLetchie (1992) has recently taken up *Sphaerocarpos texanus* Aust. for experimental studies of life history evolution. *Sphaerocarpos* provides an especially convenient experimental system for genetic studies because the spores remain in tetrads.

Many typically soil-growing, "weedy" species of mosses also make good experimental organisms because they often have high spore germination rates, grow rapidly, form gametangia and sporophytes under controlled conditions, and can be cloned easily. In the absence of crossing studies and persistent spore tetrads, a lot of genetic information can be gleaned from growing single spore isolates from different sporophytes. Variation within and among so-called haploid sib families (gametophytes produced by the same sporophyte; Shaw 1991a) is useful for the analysis of quantitative characters (Shaw & Beer 1997, Shaw *et al.* 1997). In particular, because of the absence of dominance variation as a contributing and complicating factor, nested sib analyses can be used to separate additive and epistatic components of genetic variation for quantitative traits (Shaw *et al.* 1997). Epistatic variance is much harder to estimate in diploid organisms and its evolutionary significance has scarcely been investigated. Although the chromosome numbers of many bryophytes suggest that there has been polyploidization at some time in their history, all but demonstrably allopolyploid species seem to be functionally haploid, at least with regard to isozyme loci (Wyatt *et al.* 1989).

Species that have been used most frequently for experimental approaches to evolutionary processes include *Bryum argenteum* Hedw. (Longton & MacIver 1977, Longton 1981, Shaw & Albright 1990), *Funaria hygrometrica* Hedw. (Wettstein 1924, Dietert 1980, Shaw & Bartow 1992), *Ceratodon purpureus* Brid. (Jules & Shaw 1994, Shaw & Beer 1997, 1999), and

Physcomitrella patens (Hedw.) B.S.G. (Cove 1983). The latter has been used more for physiological and morphogenetic work (see chapter 6) than for investigations of natural populations, but it could provide a valuable system for such work. Species with unisexual gametophytes are best for work in which crossing studies are planned.

12.1 An overview of moss, liverwort, and hornwort life histories

A series of life history strategies, that is, covarying patterns of life history variation, have been defined subjectively for the mosses (During 1979) and liverworts (Longton & Schuster 1983). Hedderson and Longton (1995) used multivariate ordination techniques to show that recurring patterns indeed occur in three orders of mosses (Polytrichales, Funariales, Pottiales). Strategies are defined primarily by gametophyte sexuality, size and longevity, the occurrence of asexual propagules, and spore size. Longton (1988a) hypothesized multiple origins of specialized life histories from long-lived mosses of stable, mesic habitats.

Mosses, liverworts, and hornworts all go through a juvenile protonemal stage from which mature vegetative gametophytes, thallose or leafy, are produced. The vegetative gametophytes exist for a short to more extended duration before they form gametangia. The sporophytes are short-lived in all taxa, and in most species persist after maturity from a few days (many liverworts) to several months (most mosses). In a minority of taxa, sporophytes may persist for several years. Thus, the life cycle consists of several stages: juvenile (protonema), mature vegetative, a period of gametangial formation and display during which fertilization occurs, and sporophytic.

Spore longevity varies tremendously from species to species (Crum 1976). A period of dormancy or delayed germination has been reported in a few bryophytes (Longton 1997). Spores have been germinated after more than 15 years of storage, and the maximum age for retention of some viability is unknown. Spore-to-spore variation in longevity suggests an opportunity for selection to act, especially in species that contribute to the bank of ungerminated but viable spores that are present in many habitats (During 1997).

The protonemata of most mosses form a more or less extensive filamentous mat (often inconspicuous in established populations) from which many leafy gametophores are produced clonally. Such species obviously have the potential to form extensive clones from a single spore. *Sphagnum* forms only a single mature gametophore from each thallose protonema, although some thalli do form secondary thalli from which another gametophore may be produced (Anderson & Crosby 1965). Similarly, species of *Andreaea* form a small, essentially thallose immature stage from which a single leafy stem is typically produced. Liverworts also form just one or a few gametophores from each protonema (Schofield 1985).

Although most moss, liverwort, and hornwort gametophytes are perennial, short-lived or even annual life histories appear to have evolved in multiple unrelated lineages. So-called annuals are especially common in haplolepideous mosses of the Pottiales and Dicranales. Important temperate genera include *Astomum* in the Pottiaceae, *Pleuridium* and *Bruchia* in the Ditrichaceae, *Ephemerum* and *Micromitrium* in the Ephemeraceae. Species in the latter two genera form persistent protonemata whereas most "annuals" spend the bulk of their gametophytic life cycle as vegetative and reproductive gametophores and the protonemal stage is short-lived, as is typical of mosses. Ephemeral taxa are generally winter and/or spring annuals in the sense that the mature gametophores are visible for only a few weeks to a month or two. The sporophytes of such species are typically reduced and consist of small, round or oval capsules that are cleistocarpous (i.e., without differentiated and dehiscent operculum) and release their spores through irregular disintegration.

Some other mosses, especially in the Funariaceae, occupy disturbed habitats and typically disappear after a relatively short duration, at least in some geographic areas. Whether these, and also the more extremely reduced cleistocarpous species, actually have annual life cycles, is, however, open to question. Indeed, one has to be clear about the meaning of "annual" in the bryophytic context. It may be that even very short-lived species such as *Pleuridium*, *Ephemerum*, *Micromitrium*, and *Astomum* persist from year to year at least as underground rhizoids or other types of gametophytic diaspores (During 1997). This is difficult to determine from field observations, and indeed even experimentally. Populations of *Funaria*, *Physcomitrium*, and *Entosthodon*, although they disappear after sporophytes are produced and spores are shed, often reappear in successive years in precisely the same location. Succeeding generations may be formed from spores, but in most cases the possibility that they could also persist as rhizoids cannot be eliminated. It is noteworthy that *F. hygrometrica* appears to be generally rare in most diaspore banks that have been investigated (During & ter Horst 1983).

Diaspore banks consisting of spores, vegetative fragments, rhizoids, and subterranean tubers appear to be common and extensive in many soil-inhabiting bryophyte communities (During 1997). Some evidence exists for both persistent rhizoids and even perenniating protonema in a few mosses (Duckett & Ligrone 1992, Duckett & Matchum 1995). Some species of mosses consistently form multicellular tubers on their rhizoids that presumably serve a regenerative function. Newton and Mishler (1994) summarized the types of asexual diaspores in mosses and discussed their ecological and evolutionary significance. Correns (1899) literally wrote the book on asexual reproductive structures in mosses.

Experimental studies in which substrate samples are cultivated to see what species emerge often reveal diverse communities of diaspores, although it is generally difficult or impossible to determine if regenerating gametophytes are produced from gametophyte fragments or spores (During 1997). Indirect evidence suggests that ephemeral species that form spores abundantly regenerate from spores rather than vegetative fragments. Regeneration of the boreal liverwort *Barbilophozia lycopodioides* (Wall.) Loeske from the diaspore bank is an example of vegetative persistence since this species did not form spores or specialized gemmae in the area in which a study took place (Jonsson 1993). The liverwort *Blasia pusilla* L. forms two types of gemmae, one including symbiotic *Nostoc* filaments and the other without. The two gemma types differ in their germination patterns and appear to serve different ecological functions (Duckett & Renzaglia 1993). During (1997) notes that short-lived opportunistic rather than dominant perennial species are most abundantly represented in the diaspore bank.

Mature gametophytes of most temperate species form gametangia once each year, typically during a characteristic season, which might vary geographically (Dietert 1980). A multitude of environmental factors including substrate nutrient levels and chemical form, light quantity and quality, and temperature, have been shown to affect gametangial induction (Chopra 1984). Many liverworts are long-day plants with regard to gametangial formation, although a few short-day species are known as well (e.g., in the genus *Riccia*). Anthocerotes are said to form gametangia under short-day conditions (Bensen-Evans 1964). Most mosses appear to be day-neutral. The period of time between gametangial formation and sporophyte maturation may be as short as a month (e.g., *Physcomitrella patens*; Nakosteen & Hughes 1978) to well over a year (Wyatt & Derda 1997). Most species take about 7–15 months (Crum 1976). Many temperate species that have mature sporophytes in the spring undergo gametangial formation and fertilization the previous autumn. Many such species overwinter with sporophytes in the "spear stage." In some species (e.g., *Funaria hygrometrica, Ceratodon purpureus*), the degree to which sporophytes grow in the autumn depends solely on temperature; if transplanted into a greenhouse sporophytes of these species mature without a period of dormancy. Reproductive phenology has been described for a number of species in nature by Arnell (1905), Greene (1960), Briggs (1965), Odu (1981), and Miles *et al.* (1989) among others.

In general, spores are shed quickly in liverworts, slowly in mosses, and continuously in hornworts. The spores of cleistocarpous mosses, presumably, are shed over a short duration, more like liverworts than other mosses. Mosses with peristome teeth that are reflexed in wet weather, such as some Orthotrichaceae, release their spores under these conditions whereas most mosses release spores in drier weather when their peristome teeth are spreading or reflexed.

12.2 Estimating fitness in bryophytes

In order to assess any possible effects of natural selection on variation in natural populations, a genetic basis for these traits must be demonstrated and the fitness of different phenotypes must be estimated. Life history components such as gametangial formation, fertilization rates, sporophyte survival, and spore number are critical to plant fitness. Some of these parameters have been estimated for a few species, mainly mosses.

The proportion of gametophytic stems in a population that expresses sexuality (i.e., form gametangia) is often low in mosses. A few species of bryophytes are known from only one sex, or even no sexually expressive plants at all (see below). Otherwise, estimates range from less than 5% of stems in the dioicous species *Scopelophila cataractae* (Mitt.) Broth. (Shaw 1993) up to 80% or more in some autoicous species (Anderson & Lemmon 1974, Longton & Miles 1982, Shaw 1991b). McQueen (1985) found that 44% of the stems sampled from a population of *Sphagnum subtile* (Russ.) Warnst. were sterile. It takes an experimental approach to further interpret the relatively high frequency of sterile gametophores in populations. It may be that some individuals (i.e., genets; plants derived from different spores) do not form gametangia, or that environmental cues for gametangial formation are especially stringent in these individuals. On the other hand, since moss spores give rise to a protonemal mat that in turn gives rise to hundreds or thousands of genetically identical stems (ramets), it may be that all genets form gametangia and that sterile plants observed in populations represent ramets of individuals that also form gametangia in the population. Both of these factors may contribute to the occurrence of sterile stems in natural populations (Shaw & Beer 1999). The frequencies of steriles, male and female gametophytes, and gemma-producing plants of *Tetraphis pellucida* Hedw. vary with population density (Kimmerer 1991).

Little data exist on fertilization rates (the proportion of egg-bearing archegonia that are fertilized) in bryophytes. The percentage of female plants with sporophytes was highly variable (i.e., zero to 100%) among microsites within a population of the unisexual species *Rhytidiadelphus triquetrus* (Hedw.) Warnst. (Riemann 1972). McQueen (1985) found that over 90% of the female gametophytes detected in a population of *Sphagnum subtile* carried sporophytes. Longton and Miles (1982) tallied up the number of female and bisexual gametophores that carried sporophytes in two autoicous species, *Tortula muralis* Hedw. and *Atrichum undulatum* (Hedw.) P. Beauv., and found high fertilization rates in both bisexual and female gametophores. Stark (1983) likewise documented high fertilization rates in the autoicous species *Entodon cladorrhizans* (Hedw.) C. Müll. Fertilization rates in dioicous species are often much lower. In the British moss flora, sporophyte formation is much less frequent in dioicous than bisexual species (Longton & Miles 1982). Unisexual liverworts are also less likely to form sporophytes (Longton & Schuster 1983). More detailed studies of individual species are not consistent with regard to fertilization success in bisexual versus unisexual stems. Autoicous species are useful in this regard because many such species form both unisexual and bisexual stems. There was no difference in *Atrichum undulatum* and *Tortula muralis* (Longton & Miles 1982), but sporophytes were more frequent on bisexual stems in *E. cladorrhizans* (Stark 1983). Intensive studies of fertilization rates, possibly utilizing genetic markers to assess parentage of sporophytes that are formed, are needed to determine if some gametophytes in a population have higher reproductive success than others. The effects of other life history traits such as plant size, growth rates, and abundance of gametangial formation on fertilization success could then be determined.

Huge variation among bryophytes exists in the number of spores formed per capsule. Longton (1997) summarized estimates of spore output per sporophyte in 28 bryophyte species. The numbers range from a low of 16 in *Archidium alternifolium* (Hedw.) Schimp. to 80 million in

Dawsonia lativaginata Wijk. Combining such estimates with field observations of sporophyte density in natural populations, spore output per dm^2 ranges from about 300 000 to 383 million (Longton 1997). Data on infraspecific variation in spore number per capsule are more limited, although of more direct relevance to the possible significance in terms of the potential for infraspecific evolutionary change in this trait. Two British populations of *Atrichum undulatum* exhibited a 160% difference in mean number of spores per capsule (145 000 vs. 381 000; Longton & Miles 1982). This difference in reproductive capacity was greatly amplified when the number of sporophytes per unit area was also taken into consideration. One population was estimated to produce some 445 million spores per m^2 of moss cover while the other produced about 116 million. Although both numbers are huge, these differences suggest that there are ample opportunities for selection on reproductive capacity, assuming that at least some of the variation reflects genetic rather than just environmental differences between the populations. Year-to-year variation in spore output was recorded for the liverwort *Ptilidium pulcherrimum* (C. Web.) Hampe (Jonsson & Söderstöm 1988).

Common garden experiments are the only way to separate genetic and environmental contributions to variation in life history traits. Experimental studies of life history variation in *Ceratodon purpureus* indicate the presence of genetic variation within populations for virtually every component of gametophytic fitness (Shaw & Beer 1999). Differences among haploid sib families were observed for length (in days) of the juvenile protonemal stage, length of the mature vegetative stage, total biomass per genet, individual size, the proportion of gametophytes (genets) that formed gametangia during the experiment, the proportion of ramets per genet that formed gametangia, and the mean number of gametangial buds formed per reproductive ramet. These traits are presumed to affect the fitness of individual gametophytes and therefore be subject to evolutionary change by natural selection.

It is noteworthy, however, that no one has actually assessed the impact of differences in these traits on reproductive success of gametophytes in a mixed experimental population. For example, we know that gametophytes within populations of *C. purpureus* are highly variable in growth rates, but does growth rate translate into differential reproductive success in natural populations? In addition, although such common garden experiments demonstrate genetic variation in life history traits, the experiments are designed to minimize the effects of environmental (i.e., non-genetic) contributions to variation. High levels of non-genetic variation in nature could limit the efficacy of natural selection in effecting evolutionary change. In a statistical sense it is the genotype–environment interaction, the extent to which genotypes differ in their responses to environmental variation, that is most relevant to evolutionary inferences. Differences among species in response to environmental variation are suggestive, but still a step removed from evolutionary processes, which occur at the population level within species.

12.3 Mating systems in bryophytes

The bryophyte life cycle allows for diverse mating systems. Approximately 57% of mosses and 68% of liverworts have unisexual gametophytes; i.e., they form either archegonia or antheridia, but not both. There has been some discussion about whether such species should be called dioecious, as in unisexual seed plants, or dioicous, a term unique to bryophytes, to emphasize the nonhomology between unisexual sporophytes (as in seed plants) and unisexual gametophytes (as in bryophytes). Regardless of the terminology, it is important to recognize that unisexuality in bryophytes is not comparable, morphologically or genetically, to unisexuality in seed plants.

It is often stated that unisexual bryophytes are obligate outbreeders. However, mating between gametophytes produced by the same sporophyte is genetically comparable to "selfing" as the term is applied to heterosporous seed plants. Male and female gametes produced by a single sporophyte are formed via meiosis and are more genetically comparable to brothers and sisters than to a single individual (Shaw *et al.* 1997).

Bisexual species of bryophytes are at least potentially capable of true self-fertilization, i.e., fusion of egg and sperm produced mitotically from a single gametophyte. Such matings involve genetically identical gametes (barring mutation) and yield completely homozygous sporophytes in a single generation. True self-fertilization (also referred to as "intragametophytic selfing" [Klekowski 1972], or autogamy [Cruden & Lloyd 1995]) really constitutes a form of asexual reproduction, since all the spores will be genetically identical, as well as identical to the parental gametophyte (Mishler 1988, Newton & Mishler 1994). Crossing among brothers and sisters produced by the same sporophyte (haploid sibs), that is, selfing in the diploid sense, reduces heterozygosity by 50% each generation, and is expected to yield over 90% homozygosity in under 10

Table 12.1. *Sexual systems in bryophytes*

Dioicous (dioecious)	Separate male and female gametophytes
Phyllodioicy	Separate male and female gametophytes; male plants dwarfed and epiphytic on females
Monoicous (monoecious)	Both sexes (at least potentially) on one gametophyte
Synoicous	Archegonia and antheridia mixed in one inflorescence
Paroicous	Archegonia and antheridia in one inflorescence but separated by one or more bracts
Autoicous	Separate male (perigonia) and female (perichaetia) inflorescences on the same stem
Rhizautoicous	Separate perigonia and perichaetia on stems connected by rhizoids

Source: Simplified from Wyatt and Anderson (1984).

generations. If bryophyte colonies consist primarily of haploid sibs and more distant relatives from previous generations, mating patterns in nature probably involve various degrees of inbreeding. True outcrossing, mating between unrelated gametophytes, depends on the extent to which populations are established or subsequently colonized by propagules from more distant sources.

Bryophytes have variable arrangements of archegonia and antheridia (Table 12.1). The archegonia-bearing inflorescences are perichaetia and the antheridia-bearing inflorescences are perigonia. Unisexual species may have archegonia and antheridia mixed together in the same inflorescence (synoicous and paroicous) or in different inflorescences (autoicous). Autoicous species may have male and female inflorescences on a single branch, on different branches of the same stem, or on different stems that are connected only by rhizoids (rhizautoicous). The latter can be difficult to distinguish from dioicy in the absence of experimental work.

In unisexual species, male gametophytes are often smaller than female gametophytes. Such patterns can be difficult to detect in natural populations if the degree of sexual dimorphism is subtle (Shaw & Gaughan 1993). Nevertheless, male and female gametophytes sometimes differ not only in morphology but also in life history characteristics such as spore germination patterns (McLetchie 1992) and patterns of clonal proliferation (Wilson 1993, Shaw & Beer 1999).

Extremes in sexual dimorphism are attained in so-called phyllodioicous mosses that have small, budlike, "dwarf males" that grow epiphytically on much larger females (Loveland 1956, Ramsay 1979). Phyllodioicy has apparently evolved multiple times as it is known in diverse families including the Dicranaceae, Leucobryaceae, Orthotrichaceae, and Brachythecaceae. Over 100 species of mosses are reported to produce dwarf males (Ando 1977). Phyllodioicy is especially well-represented among epiphytic mosses. The recurrence of this phenomenon in diverse groups strongly suggests a role for natural selection. One possible adaptive explanation is that phyllodioicy increases sporophyte formation in species with unisexual gametophytes. This might be especially important in epiphytic species. However, estimates of sporophyte production by female plants of *Dicranoloma* with or without epiphytic dwarf males did not reveal any relationship between fertilization success and the presence of epiphytic males (Ramsay 1985). In *Holomitrium perichaetiale* (Hook.) Brid., abundant sporophytes were observed in a population within which no dwarf males were found (Ramsay 1986).

Phyllodioicy is a ripe area for additional research. It is still not clear in many cases to what extent male size is genetically fixed within a species, as opposed to environmental determination of male size. Both mechanisms appear to exist. In *Camptothecium megaptilum* Sull., male plants grown in culture, isolated from females, attain full size whereas most male plants in nature growing epiphytically on females are dwarf (Wallace 1970, cited in Crum 1976). The proximate basis for male dwarfism in this case appears to be environmental. The males of *Holomitrium perichaetiale* are consistently dwarfed (Ramsay 1986). In *Leucobryum glaucum* (Hedw.) Ångstr. and *L. juniperoideum* (Brid.) C. Müll. male plants vary from dwarf to full size (Blackstock 1987), but they have not been grown experimentally. In *Dicranum*, some species have males that are always dwarf (e.g., *D. polysetum* Sw.), others always have isomorphic males and females (e.g., *D. viride* (Sull. & Lesq. in Sull.) Lindb.), and still others have males that are sometimes large and sometimes dwarfed to varying degrees (e.g., *D. bonjeanii* De Not., *D. scoparium* Hedw.). Briggs (1965) found that species of *Dicranum* that always have dwarf males form only dwarf males in culture whereas species with males that vary from small to large form dwarf males only when grown in proximity to a female. Male plants of *Dicranoloma dicarpum* (Nees) Par. grown in isolation from spores are strongly dwarfed, essentially like males observed in natural populations (Ramsay 1985). Both genetic and environmental determination of sexual dimorphism has been documented in *Macromitrium* (Ramsay 1979), as in *Dicranum*.

The evolution of dwarf males may involve some form of neotenic development. When grown from spores, male plants of phyllodioicous species form a protonemal mat as in other species without extreme sexual dimorphism, but when the first leafy gametophores differentiate they almost immediately form gametangia. That is, they attain reproductive maturity at a stage that would be juvenile in normal mosses. Dwarf males in some taxa, at least, have persistent protonema (*Dicranum*; Briggs 1965, *Dicranoloma*; Ramsay 1985), another indication of evolutionary changes in the timing of development.

12.4 Sex ratios

A chromosomal mechanism of sex determination implies that male and female gametophytes should segregate in a 1:1 ratio at meiosis and, barring other complicating factors, the sexes should occur in roughly equal numbers in natural populations (Ramsay & Berrie 1982). Numerous studies have shown that this is not the case. Female-biased sex ratios are so commonly observed that this can be considered a characteristic of unisexual mosses. A general rarity of male plants has been implicated as a cause of infrequent sporophyte production in a number of species (see reviews by Longton 1976, Wyatt 1982). A few exceptions exist; male-biased sex ratios have been reported in *Racomitrium lanuginosum* (Hedw.) Brid. (Tallis 1959), *Atrichum crispum* (James) Sull. (Dixon 1924), *Bryum ruderale* Crundw. & Nyh., and *B. gemmiparum* De Not. (Whitehouse, quoted in Longton & Greene 1969). Many populations of *Ceratodon purpureus* have more females than males but the reverse is also sometimes observed (Shaw & Gaughan 1993). Some populations of this colonizing species are unisexual and in this case it appears that the sexual makeup of populations reflects stochastic colonizing events. Many of the mosses for which biased sex ratios are common are characteristic of more stable communities, however, and appear to have more complex explanations. It is less clear if similar patterns occur generally in liverworts (Longton & Schuster 1983), although female-biased sex ratios are reported for *Cryptothallus mirabilis* Malmb. (Lewis & Benson-Evans 1960) and *Sphaerocarpos texanus* (McLetchie 1992).

Female function appears to be emphasized even in autoicous species where each plant has the genetic potential to form both antheridia and archegonia. Four populations of *Tortula muralis* from Wales and England exhibited similar patterns: about 71% of the stems formed archegonia only, 8% were male, 3% were bisexual, and the remaining expressed no sex. Four populations of another autoicous species, *Atrichum angustatum* (Brid.) Bruch & Schimp, had 44% females, 11% males, 5% were bisexual, and about 37% were sterile. Anderson and Lemmon (1974) found that only about 10% of the plants in a population of *Weissia controversa* Hedw. formed both antheridia and archegonia, and a similar though less extreme pattern was observed in the autoicous species *Funaria hygrometrica* by Shaw (1991*b*). These patterns in which populations consist primarily of female and bisexual plants are analogous to gynodioecy in flowering plants.

Estimating sex ratios in natural populations is fraught with uncertainties. Steriles are obviously either male or female and it is not clear whether phenotypic sex ratios accurately reflect the true ratios of genetically male and female gametophytes. For example, an increasing proportion of sterile shoots in a population of *Funaria hygrometrica* followed over a growing season reflects continual branching of established plants. Thus, sampling shoots late in the growing season leads to an increasing percentage of steriles.

It may also be that males more frequently fail to form gametangia (Mishler & Oliver 1991). Sexual dimorphism has been observed for many morphological and life history traits, including sensitivity to environmental conditions that promote gametangial formation (Une 1985). Longton (1988*b*) observed that male plants of *Bryum argenteum* appear to be restricted to protected habitats in the subantarctic regions whereas female plants are widespread and, unlike males, occur on continental Antarctica. Stark *et al.* (1998) found no males in a sample of 481 stems of *Syntrichia caninervis* Mitt.; there were 146 females and 335 steriles. Only 9% of the stems, on average, formed archegonia in any one year. Poorly understood environmental factors clearly affect gametangial formation in males and females. Shaw (1993) found no differences between unisexual male and female populations of the rare "copper moss," *Scopelophila cataractae*, in tolerance of soils contaminated with heavy metals.

Utilizing a cytological difference to distinguish between prereproductive males and females in *Mnium undulatum* Sw., Newton (1972) showed that the sex ratio is female-biased at the time of germination (4:1). The sex ratio also is female-biased at germination in *Sphaerocarpos texanus* (McLetchie 1992). Shaw and Gaughan (1993) detected a 3:2 ratio of females to males among progeny from a population of *Ceratodon purpureus* grown to reproductive maturity as single spore isolates. In a more detailed analysis of two populations, Shaw and Beer (1999) found that gametophytic sex ratios varied significantly among haploid sib families with some families producing predominantly females and others predominantly males. Interestingly, there were roughly equivalent numbers of families biased in both directions within the populations so that neither population overall exhibited a biased sex ratio. McLetchie (1992) suggested local mate competition to explain the evolution of female-biased sex ratios in *Sphaerocarpos texanus*. This is a plausible (but untested) mechanism for *Sphaerocarpos*, which has spores persistent as tetrads, as well as for other mosses if most spores land in close proximity to the mother plant.

Some bryophytes exhibit truly bizarre sexual patterns. Only female gametophytes of the moss, *Tortula pagorum* (Milde) De Not. occur in North America and only male plants occur in Europe (Anderson 1948). Both sexes, and sporophytes, are recorded from Australia. Similar patterns of differing distributions and disjunctions between male and female conspecifics have been reported for liverworts in the genus *Plagiochila* (Longton & Schuster 1983). Among British mosses, 30 species are not known to form sporophytes anywhere (in Britain or elsewhere). Three such species are known only as males, four as females, and 10 have never been observed with gametangia of either sex.

12.5 Empirical studies of mating patterns in nature

Mating patterns in natural populations of mosses and liverworts are scarcely known, and this fundamental aspect of population biology remains uninvestigated in hornworts. Some of the earliest work using isozyme markers to estimate genetic variation in natural populations showed that colonies can be genetically heterogeneous over short distances (Cummins & Wyatt 1981). In general, levels of electrophoretically detectable genetic variation are higher than predicted for haploid organisms (Wyatt *et al.* 1989a), comparable in some species to the levels found in outcrossing seed plants (see below). Such measures do not, however, give a direct measure of outcrossing rates since populations may be genetically heterogeneous through recruitment of new genotypes from adjacent sites, even in the absence of effective crossing within populations.

Anderson and Lemmon (1974) used a cytological marker to investigate crossing rates in the autoicous species *Weissia controversa* in the southeastern United States. They located a site that had multiple cushions of this species differing in the number of accessory or m-chromosomes and conducted reciprocal transplants. Crosses could be identified by the numbers of m-chromosomes observed during meiosis. Although out-crossing was detected, even gaps of a few centimeters between colonies were sufficient to block gamete dispersal and cross-fertilization. If individual colonies are typically clonal, effective crossing in natural populations is probably minimal (Anderson & Lemmon 1974).

Innes (1990) found a similar pattern of restricted gene flow using isozyme markers to estimate crossing patterns in the unisexual species *Polytrichum juniperinum* Hedw. The number of multilocus isozyme genotypes within and among 16 proximate sites in Newfoundland was consistent with sexual reproduction and associated segregation of isozyme alleles, but analyses of sporophytes indicated that most matings were between gametophytes growing in the same or nearby population. Only two of 137 matings implicated male gametophytes that were not growing in the same small population. Nevertheless, one mating implicated a male that was more than 5 m from the fertilized female, and several others implicated males that were separated by 1–2 m. Fewer multilocus genotypes were detected in established populations than among spore progeny, suggesting differential mortality during the establishment phase. Isozyme markers have been used to detect outcrossing in the liverworts *Pellia* (Zielinski 1984, 1986) and *Conocephalum* (Odrzykoski & Szweykowski 1991), in *Sphagnum* (Cronberg 1994), and in the moss *Plagiothecium* (Hofman *et al.* 1991). All of six sampled sporophytes of the liverwort *Plagiochasma rupestre* (Forst.) Steph. had isozyme genotypes (for two loci) that were identical to the maternal gametophytes (Bischler & Boisselier-Dubayle 1997).

Demonstrations of gametophytic self-compatibility in bisexual species requires experimental culture of single spore isolates, and the ability to form completely homozygous offspring through selfing has been demonstrated in only a few species: *Physcomitrella patens* (Cove 1983), *Funaria hygrometrica* (Shaw 1990b), and *Desmatodon randii* (Kenn.) Laz. (Lazarenko 1974). Self-incompatibility has not been demonstrated in any bryophyte.

Although it takes considerable experimental effort to demonstrate self-compatibility in bisexual species (i.e., single spore cultivation), self-compatibility in the sense that seed-plant reproductive biologists use the term involves mating between haploid sibs produced by the same sporophyte. This is a much easier issue to investigate experimentally. Individual sporophytes can be opened onto soil or some other appropriate substrate, and the spore progeny permitted to grow and mate. If sporophytes are formed in such cultures, the plants are self-compatible in the diploid sense. This is an important issue in bryophyte reproductive biology and such simple experiments are encouraged.

12.6 Gene flow distances and clonal structure

Gene flow resulting from gamete dispersal appears to be quite restricted in bryophytes (Longton 1976, 1994, Wyatt 1982). The most common method for estimating gamete dispersal distances (in the absence of genetic markers) has been to measure the distance from a sporophyte to the nearest male plant. This provides a minimum estimate, since the sperm could have come from a more distant individual. Most estimates made in this manner suggest that female gametophytes are fertilized by males that are within about 10 cm for species without specialized mechanisms to promote dispersal (Longton 1976). Average sperm dispersal distances may be an order of magnitude greater in species, such as some members of the Polytrichaceae, that have the perigonial leaves forming a discoid splash cup (Longton 1976, Wyatt & Derda 1997). It is harder to get estimates for spore dispersal distances, but, as expected, it appears that spore dispersal distances are strongly leptokurtic. Most spores fall within a meter or two of the parental sporophyte, but some small percentage, potentially quite important, travel longer distances. Taking into account estimates of both sperm and spore dispersal, Wyatt (1982) estimated neighborhood size of *Atrichum angustatum* to be 8.08 m². Spore and perhaps gamete dispersal distances, and thus neighborhood sizes, may be much larger in epiphytic mosses. Potential interbreeding might also occur over much larger distances in taxa such as *Fontinalis*, which release their gametes into flowing water.

The physical coverage of single genets, i.e., clonal structure, has a large effect on mating patterns in bryophytes. Many colonies of such species as *Climacium americanum* Brid. appear to be clonal (Meagher & Shaw 1990), and the physical separation of male and female clones prohibits formation of sporophytes in most populations. Bedford (1938) found that sporophytes were formed in *C. dendroides* (Hedw.) Web. & Mohr., a species that is almost always without sporophytes in nature, when male plants were transplanted into what appeared to be a female clone. Longton and Greene (1969) conducted a similar experiment in *Pleurozium schreberi* (Brid.) Mitt., with comparable results. Large carpets of *Polytrichum* species are often obviously formed of extensive contiguous male or female sectors, and sporophytes are limited to those areas where both sexes occur in proximity (Longton & Greene 1967). At the extreme, in the rare copper moss, *Scopelophila cataractae*, both male and female populations occur in North America, but never together in the same population, and sporophytes do not occur here (Shaw 1993).

Although genetic diversity in many bryophytes is substantial, the efficacy of clonal growth is evident. Based on studies in which at least 10 isozyme loci were sampled, a few estimates of clonality are as follows: 86% of the colonies of *Plagiomnium ciliare* (C. Müll.) T. Kop. were fixed for a single multilocus isozyme genotype (Wyatt *et al.* 1989*a*), 70% in *Climacium americanum* (Meagher & Shaw 1990), 50% in *Mielichhoferia elongata* (Funck) Loeske (Shaw & Schneider 1995), 100% in *M. mielichhoferiana* (Hoppe & Hornsch.) Loeske (excluding populations where the two species hybridize), and 97% in *Scopelophila cataractae* (Shaw 1995). In the liverworts, 94% of *Pellia borealis* Lorb. populations appear to be clonal (Zielinski 1986), 63% in *Porella platyphylla* (L.) Pfeiff., 75% in *P. baueri* (Schiffn.) C. Jens., and 94% in *P. cordaeana* (Hüb.) Moore (Boisselier-Dubayle *et al.* 1998*a*).

The extent of clonal proliferation is especially interesting in peat bogs, where a particular species of *Sphagnum* may cover hectares. Shaw and Srodon (1995) detected seven multilocus isozyme genotypes among 100 plants of *S. rubellum* Wils. sampled from one bog. Two of the genotypes accounted for 66% of the plants. However, no single cross could account for the genotypic diversity observed in the bog, which required a minimum of four colonizations by different genotypes. In a more detailed study of clonal structure, Cronberg (1994) found four genotypes of *S. rubellum* and 18 of *S. capillifolium* (Ehrh.) Hedw. in a Swedish bog. Based on transects, the mean linear extent of clones was 35 ± 8 (SE) cm. The largest clone extended for 160 cm. Most genotypes were found in only one transect, and no genotype was encountered in more than four transects.

12.7 Genetic diversity and population structure

Cummins and Wyatt (1981) described the "traditional view" of bryophytes as slowly evolving relics from past geological eras, predominantly asexual and low in genetic variation. Numerous studies of natural populations using isozyme markers have shown that many or even most species harbor significant amounts of allelic diversity for isozymes, in some cases

Table 12.2. *Genetic diversity in populations of bryophytes and tracheophytes*

	Mosses	Liverworts	Tracheophytes
Percentage polymorphic loci within populations (P)	35.6	15.4	34.2
Mean number of alleles per polymorphic locus (A)	1.51	1.28	1.53
Gene diversity within populations (H_S)	0.134	0.044	0.113

Source: From Stoneburner *et al.* (1991).

comparable to the levels of variation typically found in outcrossing angiosperms (Wyatt *et al.* 1989*b*). Species of mosses, on average, contain almost identical levels of genetic diversity to the mean levels observed in angiosperms and gymnosperms (Table 12.2). Liverworts, in general, characteristically contain lower levels of diversity. Reviews of isozyme studies on mosses and liverworts can be found in Szwekowski (1982), Wyatt *et al.* (1989*b*), Ennos (1990), Newton (1990), Stoneburner *et al.* (1991), Wyatt (1992, 1994), Krzakowa (1996), and Bischler and Boisselier-Dubayle (1997).

Knowledge about how genetic diversity is "packaged" within species can suggest hypotheses about population processes. A common measure of population structure is G_{ST}, which quantifies that portion of the total genetic diversity within a species that can be attributed to differentiation among populations. (G_{ST} is defined as D_{ST} / H_T, where D_{ST} is the genetic diversity among populations and H_T is the total genetic diversity.) Estimated values for G_{ST} in bryophytes range from close to zero (no differentiation among populations) to unity (no variation within populations; all genetic variation among populations) (Table 12.3). Low values (e.g., < 0.2) approach those characteristic of wind-pollinated seed plants, but even moderate values (e.g., 0.2–0.5) imply significant differentiation. Very high values recorded in some species of mosses and liverworts suggest founding of many populations by one or a few propagules and little subsequent recruitment of genetically divergent individuals. G_{ST} values varied regionally in *Meesia triquetra* (Richt.) Ångstr. (Montagnes *et al.* 1993); no such geographic variation was detected in *Mielichhoferia elongata* although regional variation in total genetic diversity was observed (Shaw & Schneider 1995).

Estimates of effective gene flow among populations can be made from G_{ST} values (Wright 1951). $Nm = (1 - F_{ST}) 4F_{ST}$, where N is the recipient population size and m is the number of migrants per generation; F_{ST} is the

Table 12.3. G_{ST} values reported in isozyme studies of selected bryophytes

	G_{ST}	Source
Liverworts		
Conocephalum conicum "A"	0.231	Odrzykoski (1986) in Bischler & Boisselier-Dubayle (1997)
Conocephalum conicum "L"	0.232	Odrzykoski (1986) in Bischler & Boisselier-Dubayle (1997)
Conocephalum conicum "S"	0.276	Bischler & Boisselier-Dubayle (1997)
Conocephalum conicum "FS"	0.423	Akiyama & Hiraoka (1994)
Marchantia chenopoda	0.244	Moya (1993) in Bischler & Boisselier-Dubayle (1997)
Pellia epiphylla "S"	0.503	Bischler & Boisselier-Dubayle (1997)
Plagiochila asplenioides	0.714	Bischler & Boisselier-Dubayle (1997)
Porella platyphylla	0.893	Bischler & Boisselier-Dubayle (1997)
Preissia quadrata	0.928	Bischler & Boisselier-Dubayle (1997)
Reboulia hemisphaerica (group 1)	0.896	Bischler & Boisselier-Dubayle (1997)
Riccia dictyospora	0.692	Dewey (1989)
Mosses		
Hylocomium splendens	0.073	Cronberg et al. (1997)
Leucodon temperatus	0.042	Akiyama (1994)
Leucodon nipponicus	0.208	Akiyama (1994)
Meesia triquetra	0.454	Montagnes et al. (1993)
Mielichhoferia elongata	0.925	Shaw & Schneider (1995)
Mielichhoferia mielichhoferiana	1.000	Shaw & Schneider (1995)
Plagiomnium affine	0.231	Wyatt et al. (1993a)
Plagiomnium curvatulum	0.198	Wyatt et al. (1993a)
Plagiomnium elatum	0.198	Wyatt et al. (1993a)
Plagiomnium ellipticum	0.166	Wyatt et al. (1993a)
Plagiomnium medium	0.128	Wyatt (1992)
Plagiomnium tezukae	0.093	Wyatt et al. (1993b)

single locus equivalent of G_{ST}. The few estimates that have been made suggest that gene flow among populations is sufficiently restricted that genetic drift could play a significant role in population differentiation. Derda and Wyatt (1990), for example, calculated $N\,m$ as between 0.4 and 0.7 for *Polytrichum commune* Hedw. Values below 1.0 are thought to be low enough for drift to be a major determinant of population structure.

A problem with using statistics such as G_{ST} or other measures of current genetic structure to infer levels of gene flow is that it is impossible to separate the effects of current, ongoing gene flow from the effects of history (Templeton et al. 1995). The use of G_{ST} values assumes that patterns of population subdivision reflect current, and recurring, processes. Many bryophytes in the northern hemisphere have geographic distributions that include more than one continent, and several studies using isozymes have shown that North American and European populations are little or not at all differentiated (e.g., Derda & Wyatt 1990, Shaw & Schneider 1995). Does this lack of differentiation really result from current levels of intercontinental gene flow, or from previous mixing and a subsequent lack of differentiation? Templeton *et al.* (1995) have developed phylogenetic methods based on coalescence theory that attempt to separate the effects of history and current processes on population genetic structure. The approach has not yet been applied to bryophytes, but especially considering their intriguingly widespread distributions, Templeton's methods could offer new insights into bryophyte biogeography and population genetics.

Boisselier-Dubayle *et al.* (1995) used RAPDs to investigate population variation in *Marchantia* and observed patterns similar to those obtained from isozyme studies. Patterson *et al.* (1998) used a combination of PCR amplification and restriction enzymes to investigate genetic diversity and differences between *Leucobryum glaucum* and *L. albidum* (Brid. ex P. Beauv.) Lindb. Direct sequencing of the nuclear ribosomal internal transcribed spacer region (ITS) revealed high levels of variation among populations of *Mielichhoferia elongata* and *M. mielichhoferiana* (Shaw 2000). This variation, which includes both nucleotide substitutions and insertion–deletion events, is in contrast to the typical uniformity of the ITS region within species of angiosperms (Baldwin *et al.* 1992). Similar high levels of variation in the ITS region occurs in *Fontinalis antipyretica* Hedw. (Shaw & Allen, in review). Twenty-five New World populations of *Sphagnum magellanicum* Brid. in contrast, from Alaska to Chile, are virtually identical in ITS sequence (Shaw, unpublished data). Clearly there is much to be learned about molecular evolution in bryophytes, but analyses at the DNA level hold great promise for future studies in bryophyte population genetics, systematics, and biogeography.

12.8 Ecotypic differentiation

Ecotypic differentiation refers to population differentiation for ecological characters that are correlated with differences in habitat (Turesson 1922). Reviews of ecotypic differentiation in bryophytes were provided by Longton (1974) and Shaw (1990a, 1991a). If the existence of genetic variation is a good measure of "evolutionary potential," then ecotypic differentiation provides evidence that such genetic variation can be molded by natural selection for adaptation to heterogeneous environments. The proof (of evolutionary potential) is in the pudding.

Habitats contaminated by heavy metals provide natural laboratories for studying ecotypic differentiation, because the major selective pressure (toxic metals) is clear, tolerance to metals is easily measured, and many contaminated habitats are of recent origin (so a time-frame for evolution can be inferred). Metal-tolerant ecotypes have been demonstrated in both liverworts (Briggs 1972, Brown & House 1978) and mosses (Shaw 1991*a*, Jules & Shaw 1994). An absence of ecotypic differentiation was observed for *Bryum argenteum* populations growing in metal-contaminated habitats (Shaw & Albright 1990). Longton (1981) also found that widespread (polar to tropical) populations of *B. argenteum* showed no evidence of differential adaptation to temperature (although genetically based morphological differences were observed), a surprising observation except that it matches the results from work on metal tolerance in this species.

12.9 Hybridization

The occurrence of interspecific hybridization in mosses was suggested in the last century (Venturi 1881). The term "hybrid" should be restricted to the sporophyte generation since only it actually contains a diploid (or polyploid) genome from two different species. Offspring produced by hybrid sporophytes are best considered "recombinants" (Shaw 1994*b*). Gametophytic plants that are morphologically intermediate between two species have been described in a number of genera. Morphological evidence is always ambiguous, however, because many hybrids (Rieseberg 1995) and recombinants (Shaw 1998) are not morphologically intermediate, and many intermediates are not hybrids.

The best-known hybrids in mosses are those between cleistocarpous and phanerocarpous taxa (Table 12.4). Such hybrids have been reported in the Ditrichaceae, Funariaceae, and Pottiaceae. The reason that these are best known is obvious; the sporophytes of the parental species are very different and hybrids are easy to recognize. Closely related species in most genera have similar sporophytes and hybrids, if they occur, would not be recognized as such. Cleistocarpous and phanerocarpous species are sufficiently divergent that all known hybrid sporophytes appear to be sterile (Table 12.4). Anderson and Snider (1982) studied meiosis in hybrids between *Weissia* and *Astomum* (Pottiaceae) and *Ditrichum* and *Pleuridium*

Table 12.4. *Naturally occurring hybrids reported between cleistocarpous and phanerocarpous mosses*

Parents	Source
Astomum crispatum ♀ × *Weissia crispum* ♂	Nicholson (1905)
Astomum ludovicianum ♀ × *Weissia controversa* ♂	Reese & Lemmon (1965)
Astomum muhlenbergianum ♀ × *Weissia controversa* ♂	Anderson & Lemmon (1972)
Bruchia microspora ♀ × *Trematodon longicollis* ♂	Rushing & Snider (1985)
Pleuridium subulatum ♀ × *Ditrichum palludum* ♂	Andrews & Hermann (1959)
Ditrichum pallidum ♀ × *Pleuridium acuminatum* ♂	Limpricht (1887)

(Ditrichaceae), both of which produce no viable spores. In *Astomum–Weissia* hybrids, sterility was chromosomal in the sense that pairing during meiosis was irregular, whereas sterility in *Ditrichum–Pleuridium* hybrids was segregational in the sense that chromosome pairing appeared normal but spores aborted thereafter.

Molecular markers, including isozymes, have provided valuable sources of new information about hybridization in bryophytes, although even here inferences are not always without ambiguity. Cronberg (1989, 1994) suggested hybridization between two closely related species of *Sphagnum*, *S. rubellum* and *S. capillifolium*, as the explanation for individuals that combined alleles characteristic of the two species. However, the two species are not fixed for alternative alleles at any locus, and so it is impossible to know if such individuals represent rare variants of one or the other species, or true recombinants. On a methodological level, the occurrence of fixed allelic differences between species permits an unambiguous interpretation of hybridization for those rare individuals that do combine alleles. The occurrence of recombinants only when the two putative parental species are sympatric is added evidence.

By fortuitous chance, such a situation is found in the genus *Mielichhoferia*. *Mielichhoferia elongata* and *M. mielichhoferiana* are fixed for different alleles at five isozyme loci in all allopatric populations sampled from North America and Europe (Shaw & Rooks 1994, Shaw & Schneider 1995). However, recombinants occur in all of five populations sampled where the two species grow sympatrically. In fact, no sympatric populations have been identified where the two species do not hybridize. The frequency of recombinants, relative to the parental species, varies from population to population, but the recombinants are always less frequent than the parents. The relative frequencies of the parental species also varies, but the allelic profiles of recombinants can be accounted for by the profiles of parental gametophytes growing in the same patch.

An intensive analysis of one hybrid zone in the Rocky Mountains revealed that M. *mielichhoferiana* is most common (57% of plants sampled), then M. *elongata* (23%) and recombinants (19%) (Shaw 1998). Recombinants are intermediate between the two parental species for morphological characters that distinguish them, but are consistently closer to the M. *mielichhoferiana* parental type. Similarly, based on their isozyme profiles, the recombinants are on average closer to M. *mielichhoferiana* than to M. *elongata*. This pattern could result from backcrossing to M. *mielichhoferiana*, which is most common in the population, or from a selective advantage at this site for plants that approach M. *mielichhoferiana*. Estimates of linkage disequilibrium among isozyme alleles indicate that despite hybridization, the two parental species are maintaining their integrity, at least for the present. This is significant because the hybrid zone site is the only known locality for M. *mielichhoferiana* in the lower 48 states of the USA (except for two small populations in the same valley as the hybrid zone). Analyses of over 20 sporophytes produced in the hybrid zone in 1996, based on species-specific differences in nuclear ribosomal DNA sequences, showed that all were pure M. *elongata* (Shaw, unpublished data). Molecular studies of hybrid zones in angiosperms indicate that even in zones where hybrids are abundant, the actual formation of F_1 hybrids appears to be a rare event (Arnold 1997).

12.10 Allopolyploidy

Chromosome doubling following interspecific hybridization is known to be an important mechanism of evolution and speciation in angiosperms (Grant 1981). Although polyploidy is widespread in mosses, and to a lesser extent in liverworts and hornworts (Smith 1978, Newton 1984, Anderson 1980), most polyploids were until recently thought to be autopolyploids. Allopolyploid origins have been proposed from morphological considerations (e.g., *Weissia exserta* (Broth.) Chen.; Khanna 1960), but solid evidence for hybrid origins has come from genetic studies. Strong even if inconclusive evidence for an allopolyploid origin for *Plagiochila britannica* Paton came from cytological studies of it and its putative parents, P. *porelloides* (Torrey ex Nees) Lindenb. and P. *asplenioides* (L.) Dum. (Newton 1986). More recently, isozymes have provided strong evidence for allopolyploidy in bryophytes. See reviews by Wyatt *et al.* (1989*b*), Stoneburner *et al.* (1991), and Bischler and Boisselier-Dubayle (1997).

The strongest evidence for allopolyploid origins of liverwort species comes from work on *Pellia* in the Metzgeriales (Odrzykoski *et al.* 1996), and *Porella* in the Jungermanniales (Boisselier-Dubayle *et al.* 1998*a*). The polyploid *Pellia borealis* appears to be an allopolyploid derived from hybridization between two morphologically similar but genetically distinct sibling species within *Pellia epiphylla* (L.) Corda. *Porella baueri* is most likely derived from hybridization between two other haploid European species, *P. platyphylla* and *P. cordaeana*. The evidence for allopolyploid origins comes from patterns of fixed heterozygosity in isozyme profiles, in which alleles from the putative parents are combined in polyploid gametophytes.

Wyatt and co-workers (Wyatt *et al.* 1988, 1992, 1993*a*, *b*) have demonstrated allopolyploid origins for *Plagiomnium medium* (B.S.G.) T. Kop., *P. curvatulum* (Lindb.) Schljakov, and *Rhizomnium pseudopunctatum* (B.S.G.) T. Kop. in the Mniaceae. *Plagiomnium curvatulum* has three orphan alleles, not found in either parent, whereas *P. medium* has only one such unique allele, prompting the suggestion that *P. medium* originated more recently. Divergent patterns of fixed heterozygosity in different populations of *P. medium* indicate that the allopolyploid has arisen multiple times from independent hybridization events involving the same two parental species (Wyatt *et al.* 1988). Multiple origins were also proposed for the allopolyploid liverwort species *Porella baueri* (Boisselier-Dubayle *et al.* 1998*a*).

A polyploid origin for *Sphagnum russowii* Warnst. has been proposed on the basis of isozyme profiles, with putative parents being *S. girghensohnii* Russ. and *S. rubellum* or some closely related species (Cronberg 1994). This is consistent with the morphological characteristics of *S. russowii*. Flatberg (1988) has also suggested an allopolyploid origin for *S. skyense* Flatb. involving *S. quinquefarium* (Lindb. ex Braithw.) Warnst. and *S. subnitens* Russ. & Warnst. in Russ. as parents, but this hypothesis, based on morphological features, has not been tested using genetic markers.

12.11 Cryptic speciation

Morphologically defined species of bryophytes tend to have broad, often intercontinental geographic distributions. Whether such broad ranges reflect very effective dispersal of tiny wind-blown spores, or ancient distributions without subsequent morphological differentiation, is a topic that has received extensive discussion during the last century (e.g., Herzog 1926, Crum 1972, van Zanten & Pócs 1981, van Zanten & Gradstein 1988). On average, congeneric bryophyte species tend to be more distinct than are congeneric angiosperms (Wyatt *et al.* 1989*b*), at least in terms of isozyme loci. For example, the mean genetic identity (*I*; Nei 1972) for pairs of congeneric seed plants is 0.67, whereas most pairs of bryophyte species have values of *I* less than 0.6, and often much lower (Wyatt *et al.* 1989, Shaw & Rooks 1994). Higher levels of differentiation are consistent with a generally older age for extant bryophyte species; in terms of genetic isolation, bryophytes are comparable to homosporous pteridophytes (Soltis & Soltis 1989).

Although electrophoretic studies have shown than in some species of mosses there is little or no genetic differentiation between North American and European populations, other work, especially on liverworts, has revealed that morphologically uniform, widespread species sometimes actually consist of two or more cryptic or sibling species (Bischler & Boisselier-Dubayle 1997). Genetic identity values among such cryptic species are often very low (<0.5) although they are morphologically indistinguishable, or nearly so.

Conocephalum conicum (L.) Lindb., widespread around the northern hemisphere, consists of at least seven cryptic species. Their ranges are largely though not completely allopatric, and have pairwise genetic identities ranging from 0.2 to 0.7 (Szweykowski & Krzakowa 1979, Szweykowski *et al.* 1980, Odrzykoski 1987, 1995, Odrzykoski & Szweykowski 1991, Akiyama & Hiraoka 1994). *Riccia dictyospora* M. A. Howe consists of at least three cryptic species with genetic identities ranging from 0.2 to 0.5. *Pellia endiviifolia* (Dicks.) Dum. consists of at least four cryptic species; two in Europe, one in Asia, and one in North America (Krzakowa 1981, Zielinski 1986, 1987). Cryptic species have also been documented in the thallose genera *Reboulia* (Boisselier-Dubayle *et al.* 1998*b*), *Marchantia* (Boisselier-Dubayle *et al.* 1995), and *Aneura* (Szweykoski & Odrzykoski 1990), and in the leafy liverwort genus *Porella* (Therrien *et al.* 1998).

In the mosses, isozyme analyses indicate that *Mielichhoferia elongata* consists of two groups of populations. One group is known only from North America whereas the other is represented in both North America and Europe (Shaw & Schneider 1995). The groups differ in allele frequencies but do not exhibit any fixed allelic differences. The same two groups are, however, fixed for different nucleotide substitutions in nuclear ribosomal DNA, and phylogenetic analyses suggest that they represent two mutually monophyletic clades within *M. elongata*, i.e., cryptic species (Shaw 2000). A quantitative analysis of morphological traits did not reveal any differences between the two cryptic species, nor geographically within either taxon (Shaw 1994a). Both chloroplast and nuclear DNA sequences suggest that European and North American populations of *Fontinalis antipyretica* represent different cryptic species that do not seem to differ morphologically (Shaw & Allen 2000). Cryptic or nearly cryptic species of mosses have also been documented in *Neckera* (Appelgren & Cronberg 1999), *Mnium* (Wyatt *et al.* 1997), and *Plagiomnium* (Wyatt & Odrzykoski 1998). Other cases of cryptic speciation will undoubtedly be uncovered in the bryophytes.

REFERENCES

Akiyama, H. (1994). Allozyme variability within and among populations of the epiphytic moss *Leucodon* (Leucodontaceae: Musci). *American Journal of Botany*, **81**, 1280–7.

Akiyama, H. & Hiraoka, T. (1994). Allozyme variability within and divergence among populations of the liverwort *Conocephalum conicum* (Marchantiales: Hepaticae) in Japan. *Journal of Plant Research*, **107**, 307–20.

Allen, C. E. (1919). The basis of sex inheritance in *Sphaerocarpos*. *Proceedings of the Philosophical Society*, **58**, 289–316.

(1935). Genetics of bryophytes. I. *Botanical Review*, **1**, 269–91.

(1945). Genetics of bryophytes. II. *Botanical Review*, **11**, 260–87.

Anderson, L. E. (1948). The distribution of *Tortula pagorum* (Milde) De Not. in North America. *Bryologist*, **46**, 47–66.

(1980). Cytology and reproductive biology of mosses. In *The Mosses of North America*, ed. R. J. Taylor & A. E. Leviton, pp. 37–76. San Francisco: American Association for the Advancement of Science.

Anderson, L. E. & Crosby, M. R. (1965). The protonema of *Sphagnum meridense* (Hampe) C. Muell. *Bryologist*, **68**, 47–54.

Anderson, L. E. & Lemmon, B. E. (1972). Cytological studies of natural hybrids between species of the moss genera, *Astomum* and *Weissia*. *Annals of the Missouri Botanical Garden*, **59**, 382–416.

(1974). Gene flow distances in the moss, *Weissia controversa* Hedw. *Journal of the Hattori Botanical Laboratory*, **38**, 67–90.

Anderson, L. E. & Snider, J. A. (1982). Cytological and genetic barriers in mosses. *Journal of the Hattori Botanical Laboratory*, **52**, 241–54.

Ando, H. (1977). Topics in the sexuality of bryophytes, *Journal of the Japanese Botanical Society*, **2**, 30–2.

Andrews, A. L. & Hermann, F. J. (1959). A natural hybrid in the Ditrichaceae. *Bryologist*, **62**, 119–22.

Appelgren, L. & Cronberg, N. (1999). Genetic and morphological variation in the rare epiphytic moss *Neckera pennata* Hedw. *Journal of Bryology*, **21**, 97–107.

Arnell, H. W. (1905). Phenological observations on mosses. *Bryologist*, **8**, 41–4.

Arnold, M. L. (1997). *Natural Hybridization and Evolution*. New York: Oxford University Press.

Baldwin, B. G. (1992). Phylogenetic utility of the internal transcribed spacers of nuclear ribosomal DNA in plants. *Molecular Phylogenetics and Evolution*, 1, 3–16.

Bedford, T. H. B. (1938). Sex distribution in colonies of *Climacium dendroides* W. & M. *North Western Naturalist*, 13, 213–21.

Benson-Evans, K. (1964). Physiology of reproduction of bryophytes. *Bryologist*, 64, 430–45.

Bischler, H. & Boisselier-Dubayle, M. C. (1997). Population genetics and variation in liverworts. *Advances in Bryology*, 6, 1–34.

Blackstock, T. H. (1987). The male gametophores of *Leucobryum glaucum* (Hedw.) Ångstr. and *L. juniperoideum* (Brid.) C. Muell. in two Welsh woodlands. *Journal of Bryology*, 14, 535–41.

Boisselier-Dubayle, M. C., Jubier, M. F., Lejeune, B., & Bischler, H. (1995). Genetic variability in the three subspecies of *Marchantia polymorpha* (Hepaticae): isozymes, RFLP and RAPD markers. *Taxon*, 44, 363–76.

Boisselier-Dubayle, M. C., Lambourdiere, J., & Bischler, H. (1998a). The leafy liverwort *Porella baueri* (Porellaceae) is an allopolyploid. *Plant Systematics and Evolution*, 210, 175–97.

(1998b). Taxa delimitation in *Reboulia* investigated with morphological, cytological, and isozyme markers. *Bryologist*, 101, 61–9.

Briggs, D. A. (1965). Experimental taxonomy of some British species of genus *Dicranum*. *New Phytologist*, 64, 366–86.

(1972). Population differentiation in *Marchantia polymorpha* L. in various lead pollution levels. *Nature*, 238, 166–7.

Brown, D. H. & House, K. L. (1978). Evidence of a copper tolerant ecotype of the hepatic, *Solenostoma crenulata*. *Annals of Botany*, 42, 1383–92.

Chopra, R. N. (1984). Environmental factors affecting gametangial induction in bryophytes. *Journal of the Hattori Botanical Laboratory*, 55, 99–104.

Correns, C. (1899). *Untersuchungen über die Vermehrung der Laubmoose durch Brutorgane und Stecklinge*. Jena: Fisher.

Cove, D. J. (1983). Genetics of Bryophyta. In *New Manual of Bryology*, vol. 1, ed. R. M. Schuster, pp. 222–31. Nichinan: Hattori Botanical Laboratory.

Cronberg, N. (1989). Patterns of variation in morphological characters and isozymes in populations of *Sphagnum capillifolium* (Ehrh.) Hedw. and *S. rubellum* Wils. from two bogs in southern Sweden. *Journal of Bryology*, 15, 683–96.

(1994). Genetic diversity and reproduction in *Sphagnum* (Bryopsida): isozyme studies in *S. capillifolium* and related species. PhD dissertation, University of Lund.

Cronberg, N., Molau, U., & Sonesson, M. (1997). Genetic variation in the clonal bryophyte *Hylocomium splendens* at hierarchical geographic scales in Scandinavia. *Heredity*, 78, 293–301.

Cruden, R. W. & Lloyd, R. M. (1995). Embryophytes have equivalent sexual phenotypes and breeding systems: why not a common terminology to describe them? *American Journal of Botany*, 82, 816–25.

Crum, H. A. (1972). The geographic origins of the mosses of eastern North America's eastern deciduous forest. *Journal of the Hattori Botanical Laboratory*, 35, 269–98.

(1976). *Mosses of the Great Lakes Forest*. Ann Arbor: University of Michigan Herbarium.

Cummins, H. & Wyatt, R. (1981). Genetic variability in natural populations of the moss *Atrichum angustatum*. *Bryologist*, 84, 30–8.

Derda, G. S. & Wyatt, R. (1990). Genetic variation in the common hair-cap moss, *Polytrichum commune*. *Systematic Botany*, **15**, 592–605.

Dewey, R. M. (1989). Genetic variation in the liverwort *Riccia dictyospora* (Ricciaceae, Hepaticopsida). *Systematic Botany*, **14**, 155–67.

Dietert, M. F. (1980). The effect of temperature and photoperiod on the development of geographically isolated populations of *Funaria hygrometrica* and *Weissia controversa*. *American Journal of Botany*, **67**, 369–80.

Dixon, H. N. (1924). *The Student's Handbook of British Mosses*, ed. 3. Eastbourne: Sumfield & Day.

Duckett, J. G. & Ligrone, R. (1992). A survey of diaspore liberation mechanisms and germination patterns in mosses. *Journal of Bryology*, **17**, 335–54.

Duckett, J. G. & Matchum, H. W. (1995). Studies on protonemal morphogenesis in mosses. III. The perennial gemmiferous protonema of *Dicranella heteromalla* (Hedw.) Schimp. *Journal of Bryology*, **18**, 407–24.

Duckett, J. G. & Renzaglia, K. S. (1993). The reproductive biology of the liverwort *Blasia pusilla* L. *Journal of Bryology*, **17**, 541–52.

During, H. J. (1979). Life strategies of bryophytes: a preliminary review. *Lindbergia*, **5**, 2–18.

(1997). Bryophyte diaspore banks. *Advances in Bryology*, **6**, 103–34.

During, H. J. & ter Horst, B. (1983). The diaspore banks of bryophytes and ferns in chalk grassland. *Lindbergia*, **9**, 57–64.

Ennos, R. A. (1990). Population genetics of bryophytes. *Trends in Ecology and Evolution*, **5**, 38–9.

Flatberg, K. I. (1988). *Sphagnum skyense* sp. nov. *Journal of Bryology*, **15**, 101–7.

Grant, V. (1981). *Plant Speciation* 2nd edn. New York: Columbia University Press.

Greene, S. W. (1960). The maturation cycle, of the stages of development of gametangia and capsules in mosses. *Transactions of the British Bryological Society*, **3**, 736–45.

Hedderson, T. A. & Longton, R. E. (1995). Patterns of life history variation in the Funariales, Polytrichales, and Pottiales. *Journal of Bryology*, **18**, 639–75.

Herben, T. (1994). The role of reproduction for persistence of bryophyte populations in transient and stable habitats. *Journal of the Hattori Botanical Laboratory*, **76**, 115–26.

Herzog, T. (1926). *Geographie der Moose*. Jena: Fisher.

Hofman, A., van Delden, W., & van Zanten, B. O. (1991). Population genetics of the moss *Plagiothecium undulatum* (Hedw.) Schimp. I. Inheritance of allozymes. *Heredity*, **67**, 13–18.

Innes, D. J. (1990). Microgeographic genetic variation in the haploid and diploid stages of the moss *Polytrichum juniperinum* Hedw. *Heredity*, **64**, 331–40.

Jonsson, B. G. (1993). The bryophyte diaspore bank and its role after small-scale disturbance in a boreal forest. *Journal of Vegetation Science*, **4**, 819–26.

Jonsson, B. G. & Söderström, L. (1988). Growth and reproduction in the leafy hepatic *Ptilidium pulcherrimum* (G. Web.) Vainio during a 4-year period. *Journal of Bryology*, **15**, 315–25.

Jules, E. S. & Shaw, A. J. (1994). Adaptation to metal-contaminated soils in populations of the moss, *Ceratodon purpureus*: vegetative growth and reproductive expression. *American Journal of Botany*, **81**, 791–7.

Khanna, K. R. (1960). Studies in natural hybridization in the genus *Weissia*. *Bryologist*, **63**, 1–16.

Kimmerer, R. W. (1991). Reproductive ecology of *Tetraphis pellucida* I. Population density and reproductive mode. *Bryologist*, **94**, 255–60.

Klekoswski, E. J., Jr. (1972). Genetical features of ferns as contrasted to seed plants. *Annals of the Missouri Botanical Garden*, **59**, 138–51.

Krzakowa, M. (1981). Evolution and speciation in *Pellia*, with special reference to the *Pellia megaspora-endiviifolia* complex (Metzgeriales), IV. Isozyme investigations. *Journal of Bryology*, **11**, 447–50.

Krzakowa, M. (1996). Review of genetic investigations on bryophytes in Poland. *Cryptogamie, Bryologie et Lichénologie*, **17**, 237–43.

Lazarenko, A. S. (1974). Some considerations on the nature and behavior of the relic moss, *Desmatodon randii*. *Bryologist*, **77**, 474–7.

Lewis, K. R. (1961). The genetics of bryophytes. *Transactions of the British Bryological Society*, **4**, 111–30.

Lewis, K. R. & Benson-Evans, K. (1960). The chromosomes of *Cryptothallus mirabilis* (Hepaticae: Riccardiaceae). *Phyton*, **14**, 21–35.

Ligrone, R., Duckett, J. G., & Renzaglia, K. S. (1993). The gametophyte–sporophyte junction in land plants. *Advances in Botanical Research*, **19**, 231–317.

Limpricht, K. G. (1887). *Die Laubmoose Deuschlands, Oesterreiches und der Schweiz*. Leipzig.

Longton, R. E. (1974). Genecological differentiation in bryophytes. *Journal of the Hattori Botanical Laboratory*, **38**, 49–65.

(1976). Reproductive biology and evolutionary potential in bryophytes. *Journal of the Hattori Botanical Laboratory*, **41**, 205–23.

(1981). Inter-population variation in morphology and physiology in the cosmopolitan moss *Bryum argenteum*. *Journal of Bryology*, **11**, 501–20.

(1988a). Life-history strategies among bryophytes of arid regions. *Journal of the Hattori Botanical Laboratory*, **64**, 15–28.

(1988b). *The Biology of Polar Bryophytes and Lichens*. Cambridge: Cambridge University Press.

(1994). Reproductive biology in bryophytes. The challenges and the opportunities. *Journal of the Hattori Botanical Laboratory*, **76**, 159–72.

(1997). Reproductive biology and life history strategies. *Advances in Bryology*, **6**, 65–101.

Longton, R. E. & Greene, S. W. (1967). The growth and reproduction of *Polytrichum alpestre* Hoppe on South Georgia Island. *Philosophical Transactions of the Royal Society B***252**, 295–327.

(1969). Relationship between sex distribution and sporophyte production in *Pleurozium schreberi* (Brid.) Mitt. *Annals of Botany*, **33**, 83–105.

Longton, R. E. & MacIver, M. A. (1977). Climatic adaptation in Antarctic and Northern Hemisphere populations of a cosmopolitan moss, *Bryum argenteum*. In *Adaptations within Antarctic Ecosystems*, ed. G. A. Llano, pp. 899–919. Washington: Smithsonian Institution.

Longton, R. E. & Miles, C. J. (1982). Studies on the reproductive biology of mosses. *Journal of the Hattori Botanical Laboratory*, **52**, 219–40.

Longton, R. E. & Schuster, R. M. (1983). Reproductive biology. In *New Manual of Bryology*, vol. 1, ed. R. M. Schuster, pp. 386–462. Nichinan: Hattori Botanical Laboratory.

Loveland, H. F. (1956). Sexual dimorphism in the moss genus *Dicranum* Hedw. PhD dissertation, University of Michigan.

Marchal, E. (1907). Aposporie et sexualité les mousses. II. *Bulletin de l'Académie royale de Belgique, Classe des sciences,* 1–50.

(1911). Aposporie et sexualité les mousses. III. *Bulletin de l'Académie royale de Belgique. Classe des sciences,* 750–78.

McLetchie, D. N. (1992). Sex ratio from germination through maturity and its reproductive consequences in the liverwort *Sphaerocarpos texanus. Oecologia,* **92,** 273–8.

McQueen, C. B. (1985). Spatial patterns and gene flow distances in *Sphagnum subtile. Bryologist,* **88,** 333–6.

Meagher, T. R. & Shaw, A. J. (1990). Clonal structure in the moss, *Climacium americanum* Brid. *Heredity,* **64,** 233–8.

Miles, C. J., Odu, E. A., & Longton, R. E. (1989). Phenological studies on British mosses. *British Bryological Society,* **15,** 607–21.

Mishler, B. D. (1988). Reproductive ecology of Bryophytes. In *Plant Reproductive Ecology: Patterns and Processes,* ed. J. Lovett Doust & L. Lovett Doust, pp. 285–306. New York: Oxford University Press.

Mishler, B. D. & Oliver, M. J. (1991). Gametophytic phenology of *Tortula ruralis,* a desiccation-tolerant moss, in the Organ Mountains of southern New Mexico. *Bryologist,* **94,** 143–53.

Montagnes, R. J. S., Bayer, R. J., & Vitt, D. H. (1993). Isozyme variation in the moss *Meesia triquetra* (Meesiaceae). *Journal of the Hattori Botanical Laboratory,* **74,** 155–70.

Nakosteen, P. C. & Hughes, K. W. (1978). Sexual life cycle of three species of Funariaceae in culture. *Bryologist,* **81,** 307–14.

Nei, M. (1972). Genetic distance between populations. *American Naturalist,* **106,** 283–92.

Newton, A. E. & Mishler, B. D. (1994). The evolutionary significance of asexual reproduction in mosses. *Journal of the Hattori Botanical Laboratory,* **76,** 127–45.

Newton, M. E. (1972). Sex-ratio differences in *Mnium hornum* Hedw. and *M. undulatum* Sw. in relation to spore germination and vegetative regeneration. *Annals of Botany,* **36,** 163–78.

(1984). The cytogenetics of bryophytes. In *The Experimental Biology of Bryophytes,* ed. A. F. Dyer & J. G. Duckett, pp. 65–96. London: Academic Press.

(1986). Bryophyte phylogeny in terms of chromsome cytology. *Journal of Bryology,* **14,** 43–57.

(1990). Genetic structure of hepatic species. *Botanical Journal of the Linnean Society,* **104,** 215–29.

Nicholson, W. E. (1905). Notes on two forms of hybrid *Weissia. Revue Bryologique,* **32,** 19–25.

Odrzykoski, I. J. (1987). Genetic evidence for reproductive isolation between two European "forms" of *Conocephalum conicum. Symposia Biologica Hungarica,* **35,** 577–87.

(1995). Sibling species of the *Conocephalum conicum* complex (Hepaticae: Conocephalaceae) in Poland: distribution maps and description of an identification method based on isozyme markers. *Fragmenta Floristica et Geobotanica,* **40,** 393–404.

Odrzykoski, I. J. & Szweykowski, J. (1991). Genetic differentiation without concordant morphological differgence in the thallose liverwort *Conocephalum conicum. Plant Systematics and Evolution,* **178,** 135–51.

Odrzykoski, I. J., Chudszinska, E., & Szweykowski, J. (1996). The hybrid origin of the polyploid liverwort *Pellia borealis. Genetica,* **98,** 75–86.

Odu, E. A. (1981). Reproductive phenology of some tropical African mosses. *Cryptogamie, Bryologie et Lichénologie*, **2**, 91–9.

Økland, R. H. (1995). Population biology of the clonal moss *Hylocomium splendens* in Norwegian spruce forests. I. Demography. *Journal of Ecology*, **83**, 697–712.

Patterson, E., Boles, S. B., & Shaw, A. J. (1998). Nuclear ribosomal DNA variation in *Leucobryum glaucum* and *L. albidum* (Leucobryaceae): a preliminary investigation. *Bryologist*, **101**, 272–7.

Proctor, M. C. F. (1977). Evidence on the carbon nutrition of moss sporophytes from $^{14}CO_2$ uptake and subsequent movement of labeled assimilate. *Journal of Bryology*, **9**, 375–86.

Ramsay, H. P. (1979). Anisospory and sexual dimorphism in the Musci. In *Bryophyte Systematics*, ed. G. C. E. Clarke & J. G. Duckett, pp. 281–316. London: Academic Press.

(1985). Cytological and sexual characteristics of the moss *Dicranoloma* Ren. *Monographs in Systematic Botany from the Missouri Botanical Garden*, **11**, 93–110.

(1986). Studies on *Holomitrium perichaetiale* (Hook.) Brid. (Dicranaceae: Bryopsida). *Hikobia*, **9**, 307–14.

Ramsay, H. P. & Berrie, G. K. (1982). Sex determination in bryophytes. *Journal of the Hattori Botanical Laboratory*, **52**, 255–74.

Reese, W. D. & Lemmon, B. E. (1965). A natural hybrid between *Weissia* and *Astomum* and notes on the nomenclature of North American species of *Astomum*. *Bryologist*, **68**, 73–7.

Riemann, B. (1972). On the sex distribution and occurrence of sporophytes in *Rhytidiadelphis triquetrus* (Hedw.) Warnst. In Scandinavia. *Lindbergia*, **1**, 219–24.

Rieseberg, L. H. (1995). The role of hybridization in evolution: old wine in new skins. *American Journal of Botany*, **82**, 944–53.

Rushing, A. E. & Snider, J. A. (1985). A natural hybrid between *Bruchia microspora* Nog. and *Trematodon longicollis* Michx. *Monographs in Systematic Botany from the Missouri Botanical Garden*, **11**, 121–32.

Schofield, W. B. (1985). *Introduction to Bryology*. New York: Macmillan.

Shaw, A. J. (1990a). Metal tolerance in bryophytes. In *Heavy Metal Tolerance in Plants: Evolutionary Aspects*, ed. A. J. Shaw, pp. 133–52. Boca Raton: CRC Press.

(1990b). Intraclonal variation in morphology, growth rate, and copper tolerance in the moss, *Funaria hygrometrica*. *Evolution*, **44**, 441–7.

(1991a). Ecological genetics, evolutionary constraints, and the systematics of bryophytes. *Advances in Bryology*, **4**, 29–74.

(1991b). The genetic structure of sporophytic and gametophytic populations of the moss, *Funaria hygrometrica* Hedw. *Evolution*, **45**, 1260–74.

(1993). Population biology of the rare copper moss, *Scopelophila cataractae*. *American Journal of Botany*, **80**, 1034–41.

(1994a). Systematics of *Mielichhoferia* (Bryaceae: Musci). II. Morphological variation among disjunct populations of *M. elongata* and *M. mielichhoferiana*. *Bryologist*, **97**, 47–55.

(1994b). Systematics of *Mielichhoferia* (Bryaceae: Musci). III. Hybridization between *M. elongata* and *M. mielichhoferiana*. *American Journal of Botany*, **81**, 782–90.

(1995). Genetic biogeography of the rare "copper moss," *Scopelophila cataractae*. *Plant Systematics and Evolution*, **197**, 43–58.

(1998). Genetic analysis of a hybrid zone in *Mielichhoferia* (Musci). In *Bryology for the Twenty-first Century*, ed. J. W. Bates, N. W. Ashton, & J. G. Duckett, pp. 161–74. Leeds: Maney and British Bryological Society.

Shaw, A. J. (2000). Molecular phylogeography and cryptic speciation in the mosses *Mielichhoferia elongata* and *M. mielichhoferiana* (Bryaceae). *Molecular Ecology*, in press.

Shaw, A. J. & Albright, D. L. (1990). Potential for the evolution of metal tolerance in *Bryum argenteum*, a moss. II. Generalized tolerances among diverse populations. *Bryologist*, **93**, 187–92.

Shaw, A. J. & Allen, B. H. (2000). Phylogenetic relationships, morphological incongruence, and geographic speciation in the Fontinalaceae (Bryophyta). *Plant Systematics and Evolution*, in press.

Shaw, A. J. & Bartow, S. M. (1992). Genetic structure and phenotypic plasticity in proximate populations of the moss, *Funaria hygrometrica*. *Systematic Botany*, **17**, 257–71.

Shaw, A. J. & Beer, S. C. (1997). Gametophyte–sporophyte variation and covariation in mosses. *Advances in Bryology*, **6**, 35–63.

(1999). Life history variation in gametophyte populations of the moss *Ceratodon purpureus* (Ditrichaceae). *American Journal of Botany*, **86**, 512–21.

Shaw, A. J. & Gaughan, J. (1993). Control of sex ratios in haploid populations of the moss, *Ceratodon purpureus*. *American Journal of Botany*, **80**, 584–91.

Shaw, A. J. & Rooks, P. R. (1994). Systematics of *Mielichhoferia* (Bryaceae: Musci). I. Morphological and genetic analyses of *M. elongata* and *M. mielichhoferiana*. *Bryologist*, **97**, 1–12.

Shaw, A. J. & Schneider, R. E. (1995). Genetic biogeography of the rare copper moss, *Mielichhoferia elongata* (Bryaceae). *American Journal of Botany*, **82**, 8–17.

Shaw, A. J. & Srodon, M. (1995). Clonal diversity in *Sphagnum rubellum* Wils. *Bryologist*, **98**, 261–4.

Shaw, A. J., Weir, B. S., & Shaw, H. (1997). The occurrence and significance of epistatic variance for quantitative characters and its measurement in haploids. *Evolution*, **51**, 348–53.

Smith, A. J. E. (1978). Cytogenetics, biosystematics and evolution of Bryophyta. *Advances in Botanical Research*, **6**, 195–276.

Soltis, D. E. & Soltis, P. S. (1989). Polyploidy, breeding systems, and genetic differentiation in homosporous pteridophytes. In *Isozymes in Plant Biology*, ed. D. E. Soltis & P. S. Soltis, pp. 241–58. Portland: Dioscorides Press.

Stark, R. L. (1983). Reproductive biology of *Entodon cladorrhizans*. 1. Reproductive cycle and frequency of fertilization. *Systematic Botany*, **8**, 381–8.

Stark, R. L., Mishler, B. D., & McLetchie, D. N. (1998). Sex expression and growth rates in natural populations of the desert soil crustal moss *Synthrichia caninervis*. *Journal of Arid Environments*, **40**, 401–16.

Stoneburner, A., Wyatt, R., & Odrzykoski, I. J. (1991). Applications of enzyme electrophoresis to bryophyte systematics and population biology. *Advances in Bryology*, **4**, 1–27.

Szweykowski, J. (1982). Genetic differentiation of liverwort populations and its significance for bryotaxonomy and bryogeography. *Journal of the Hattori Botanical Laboratory*, **53**, 21–8.

Szweykowski, J. & Krzakowa, M. (1979). Variation of four enzyme systems in Polish populations of *Conocephalum conicum* (L.) Dum. (Hepaticae, Marchantiales). *Bulletin de l'Académie Polonaise des Sciences Biologiques, Classe II*, **27**, 37–41.

Szweykowski, J. & Odrzykoski, I. J. (1990) Chemical differentiation of *Aneura pinquis* (L.) Dum. (Hepaticae, Aneuraceae) in Poland and some comments on the application of enzymatic markers in bryology. In *Bryophytes: Their Chemistry and Chemical Taxonomy*, ed. H. D. Zinmeister & R. Mues, pp. 437–48. Oxford: Clarendon Press.

Szweykowski, J., Odrzykoski, I. J., & Zielinski, R. (1980). Further data on the geographic distribution of two genetically different forms of the liverwort *Conocephalum conicum* (L.) Dum.: the sympatric and allopatric regions. *Bulletin de l'Académie Polonaise des Sciences Biologiques, Classe II*, **28**, 437–49.

Tallis, J. H. (1959). Studies in the biology and ecology of *Racomitrium lanuginosum* Brid. II. Growth, reproduction and physiology. *Journal of Ecology*, **47**, 325–50.

Templeton, A. R., Routmanm, E., & Phillips, C. A. (1995). Separating population structure from population history: a cladistic analysis of the geographical distribution of mitochondrial DNA haplotypes in the tiger salamander, *Amblystoma tigrinum*. *Genetics*, **140**, 767–82.

Therrien, J. P., Crandall-Stotler, B. J., & Stotler, R. E. (1998). Morphological and genetic variation in *Porella platyphylla* and *P. platyphylloidea* and their systematic implications. *Bryologist*, **101**, 1–19.

Turesson, G. (1922). The genotypical response of the plant species to the habitat. *Hereditas*, **3**, 211–350.

Une, K. (1985). Factors restricting the formation of normal male plants in isosporous species of *Macromitrium* (Musci: Orthotrichaceae) in Japan. *Journal of the Hattori Botanical Laboratory*, **59**, 523–9.

Venturi, G. (1881). Une mousse hybride. *Revue Bryologique*, **8**, 20–2.

Watson, M. (1974). The population biology of six closely related species of Bryophytes (Musci). PhD dissertation, Yale University.

Wettstein, F. von (1924). Morphologie und Physiologie des Formswecksels der Moose auf genetischer Grundlage. I. *Zeitschrift für induktive Abstammungs- und Vererbungslehre*, **33**, 1–236.

(1932). Genetik. In *Manual of Bryology*, ed. F. Verdoorn, pp. 233–72. The Hague: Martinus Nijdhoff.

Wilson, C. E. (1993). Sexually dimorphic traits affecting clonal architecture and life history in *Polytrichum commune*. PhD dissertation, State University of New York at Stony Brook.

Wright, S. (1951). The genetical structure of populations. *Annals of Genetics*, **15**, 323–54.

Wyatt, R. (1982). Population ecology of bryophytes. *Journal of the Hattori Botanical Laboratory*, **52**, 179–98.

(1992). Conservation of rare and endangered bryophytes: input from population genetics. *Biological Conservation*, **59**, 99–107.

(1994). Population genetics of bryophytes in relation to their reproductive biology. *Journal of the Hattori Botanical Laboratory*, **76**, 147–57.

Wyatt, R. & Anderson, L. E. (1984). Breeding systems of bryophytes. In *The Experimental Biology of Bryophytes*, ed. A. F. Dyer & J. G. Duckett, pp. 39–64. London: Academic Press.

Wyatt, R. & Derda, G. S. (1997). Population biology of the Polytrichaceae. *Advances in Bryology*, **6**, 265–96.

Wyatt, R. & Odrzykoski, I. J. (1998). On the origins of the allopolyploid moss *Plagiomnium cuspidatum*. *Bryologist*, **101**, 263–71.

Wyatt, R., Odrzykoski, I. J., & Stoneburner, A., Bass, H. W., & Galau, G. (1988). Allopolyploidy in bryophytes: multiple origins of *Plagiomnium medium*. *Proceedings of the National Academy of Sciences USA*, **85**, 5601–4.

Wyatt, R., Odrzykoski, I. J., & Stoneburner, A. (1989*a*). High levels of genetic variability in the haploid moss *Plagiomnium ciliare*. *Evolution*, **43**, 1085–96.

Wyatt, R., Stoneburner, A., & Odrzykoski, I. J. (1989*b*). Bryophyte isozymes: systematic and evolutionary implications. In *Isozymes in Plant Biology*, ed. D. E. Soltis & P. M. Soltis, pp. 221–34. Portland: Dioscorides Press.

Wyatt, R., Odrzykoski, I. J., & Stoneburner, A. (1992). Isozyme evidence of reticulate evolution in the mosses: *Plagiomnium medium* is an allopolyploid of *P. ellipticum* × *P. insigne*. *Systematic Botany*, **17**, 532–50.

(1993*a*). Isozyme evidence regarding the origins of the allopolyploid moss *Plagiomnium curvatulum*. *Lindbergia*, **18**, 49–58.

(1993*b*). Isozyme evidence proves that the moss *Rhizomnium pseudopunctatum* is an allopolyploid of *R. gracile* × *R. magnifolium*. *Memoirs of the Torrey Botanical Club*, **25**, 20–34.

Wyatt, R., Odrzykoski, I. J., & Koponen, T. (1997). *Mnium orientale* sp. nov. from Japan is morphologically and genetically distinct from *M. hornum* in Europe and North America. *Bryologist*, **100**, 226–36.

Zanten, B. O. van & Gradstein, S. R. (1988). Experimental dispersal geography of neotropical liverworts. *Beihefte zur Nova Hedwigia*, **90**, 41–94.

Zanten, B. O. van & Pócs, T. (1981). Distribution and dispersal of bryophytes. *Advances in Bryology*, **1**, 479–562.

Zielinski, R. (1984). Electrophoretic evidence of cross-fertilization in the monoecious *Pellia epiphylla*, n = 9. *Journal of the Hattori Botanical Laboratory*, **56**, 255–62.

(1986). Cross-fertilization in the monoecious *Pellia borealis*, n = 18, and spatial distribution of two peroxidase genotypes. *Heredity*, **56**, 299–304.

(1987). Interpretation of electrophoretic patterns in population genetics of bryophytes. VI. Genetic variation and evolution of the liverwort genus *Pellia* with special reference to Central European territory. *Lindbergia*, **12**, 87–96.

13

Bryogeography and conservation of bryophytes

13.1 Introduction

Except for the marine environment, bryophytes, as a group, are nearly cosmopolitan in distribution. Because of the small size of spores and the frequent occurrence of vegetative propagules, bryophytes are easily dispersed across the landscape. Consequently, many show much wider distribution patterns than the seed plants (Watson 1974). This is especially true at the family and generic levels. The purported greater age of bryophytes may have contributed further to the wider range of bryophytes (Schofield 1992).

The ecological and biological factors shaping the distribution patterns of seed plants have also affected the bryophytes. Indeed, like the seed plants, many bryophyte species found in the arctic and boreal zones are the same in North America, Asia, and Europe. In the tropics, there is a significant number of moss taxa that are pantropical, but the number is far lower when compared to the circumboreal taxa.

Although a great number of bryophyte genera and species are cosmopolitan and distributed throughout different climatic regions, many do exhibit a disrupted or narrow range. The latter represent the uncommon, rare and endemic bryophytes. Their ultimate survival is critically dependent on the preservation of their natural habitats.

13.2 Factors affecting the dispersal of bryophyte diaspores

Diaspores are defined as any propagative parts of bryophytes, be they spores or gemmae, capable of giving rise to a new individual. Among the bryophytes, few members, like the moss family Splachnaceae, produce sticky spores that are dependent on flies for their dispersal. Instead, many bryophytes are dispersed over great distances by air currents. A few bryophytes that display a bipolar distribution in arctic and subantarctic regions have been alleged to be the outcome of transequatorial dispersal brought by migratory birds. Others are thought to travel over long distances via the jet stream in the upper atmosphere. However, the majority of bryophytes probably disperse across the land by slow expansion through forest and by island or mountain hopping. This hypothesis is supported by the fact that spores of many bryophytes are ill-equipped against desiccation, extreme temperature, and strong UV radiation (van Zanten 1976, 1978). Furthermore, many epiphytic mosses and epiphyllous liverworts produce green spores that are rather short-lived and will not survive long-distance travel (cf. Schuster 1983b).

According to van Zanten and Pócs (1981), Hepaticae, in general, is less effectively dispersed by air currents than Musci. One of the reasons is the small spores of many tropical rain forest hepatics do not seem to tolerate desiccation. Accordingly, a spore size less than 25 μm is best suited for air transport over long distances within a climatological belt.

Imura (1994) reviewed the vegetative diaspores of 186 out of 1183 species of Japanese mosses. He concluded that specialized vegetative diaspores are found more frequently in dioicous than monoicous mosses.

After dispersal, the next critical factor is the establishment of a viable population at the new site. This is where the ecological requirement and adaptation of the bryophyte come into play. A few on-site studies showed that diaspores of many more species of bryophytes had landed but failed to germinate and mature (Miller & Ambrose 1976).

Interestingly, more than half of the bryophytes in the world are dioicous. This poses a problem in sexual reproduction when the gametophyte of only one sex arrives at a distant locality. The subsequent survival and expansion of the population must depend on asexual propagation. Such is the situation of the present-day widely disjunctive female populations of *Takakia lepidozioides* found in Alaska, British Columbia, Japan, Himalayas, and on Mount Kinabalu of Borneo.

The changing geographical position of land masses brought about by plate tectonics and the climatic vicissitudes through Tertiary and Quaternary periods are two other factors shaping the modern distribution of bryophytes. Fossil records, such as corticolous mosses and liverworts preserved in ambers, have revealed the existence of subtropical to tropical floras in many parts of Europe during the Tertiary period (Grolle 1988, Frahm 1996).

In modern times, humans have become the unwitting vector of bryophyte diaspores. A number of bryophytes have been introduced well beyond their natural range and have become weedy in foreign lands. Well-known examples of alien bryophytes are the European *Pseudoscleropodium purum* in the southern hemisphere and North America, and the southern hemisphere *Orthodontium lineare* and *Campylopus introflexus* in Europe, especially in the UK (Söderström 1992). Furthermore, several bryophytes have become adapted to man-made habitats and are found more commonly around human settlements than in their natural habitats. Examples are *Barbula unguiculata, Bryum argenteum, Hyophila propagulifera, Leptobryum pyriforme, Tortula pagorum,* and *T. muralis.*

In summary, it is important to emphasize that no regional bryoflora is developed from a single event or two to three diaspore introductions. In all probability, various components of the flora have reached the locality successively in different ways (Tan 1992). A good case in point is the bryogeography of the Greater and Lesser Antillean mosses which was shown by Buck (1990) and Delgadillo (1993*b*) to be the result of complex processes of long-distance dispersal, vicariance events, extinction, and changes of sea level brought by the changing climate.

13.3 Major patterns of bryophyte distributions

In many parts of the world the diversity and number of taxa are greater in mosses than in liverworts and hornworts. The exceptions are the extreme wet or dry environments, e.g., the exposed forest canopy, alpine/tundra habitats, super-humid montane forest, and wet coastal vegetation, where the increasing trend of diversity is in favor of liverworts.

Modern distribution of plants is affected primarily by the prevailing climatic pattern and, secondarily, by the available habitat. Herzog, in his monumental work *Geographie der Moose* (1926), outlined several distribution patterns of bryophytes that are still recognized today. To understand the bryogeography of a region, it is imperative to dissect the flora into its floristic components by grouping together the species exhibiting similar distribution patterns. From such a floristic analysis, the origin and evolution of the flora can be deduced.

The worldwide distributions of bryophytes can be grouped into five major patterns following closely the prevailing climatic zones of the world. Minor patterns of distribution such as bipolar and other disjunctions are discussed separately below. Within each major category, the range of a species can be continuous or discontinuous depending on other factors such as substrate and moisture. Because of the presence of microhabitats that are often different from the macroenvironment in the region, a large number of taxa, especially those with broad ecological amplitude, can be observed to have a total distribution combining two to three of the five patterns outlined below.

1. Arctic–alpine distribution
There is a good number of bryophytes that are widespread across the Arctic Circle. Many have spilled south towards the Equator along high mountain summits that support alpine vegetation. Examples are species of *Andreaeobryum, Anthelia, Arnellia, Arctoa, Conostomum, Kiaeria, Oedipodium, Oreas, Stegonia, Tetralophozia, Voitia,* and *Plagiothecium berggrenianum* (Fig. 13.1). Because of the proximity of lands in the Arctic and the availability of ice-free areas serving as refugia during the ice ages, the arctic flora today has an impressive number of bryophytes close to 600 (Longton 1988).

2. Boreal distribution
These are the species found abundantly across the northern coniferous forest belts in all major continents. Many have extended their ranges northward into the colder arctic zone or southward into the warmer temperate zone, especially in the mountains. Most have continuous ranges across the three continents, namely Europe, Asia, and North America; hence, the term circumboreal taxa. Examples are species of *Abietinella, Aneura, Blasia, Catascopium, Cinclidium, Pellia, Pleurozium, Ptilidium, Ptilium, Tortella, Schistostega, Splachnum,* and *Warnstorfia.*

3. North and south temperate distributions
These are the warmth-loving bryophytes found mainly at low latitudes in northern and southern hemispheres. They are important plant components of semi-evergreen, broadleaved, and mixed deciduous temperate forests. They may show a continuous distribution confined to one mountain range or a disruptive range between two continents.
3(a). North temperate distribution
   Examples of bryophytes showing a north temperate distribution are species of *Acrobolbus, Anastrophyllum, Brotherella, Entodon, Gollania, Hookeria, Leucodon, Neckera, Schwetschkeopsis,* and *Tritomaria.*
   In historical bryogeographical parlance, bryophytes that exhibit boreal and temperate ranges in the northern hemisphere are considered Laurasian elements. This is because many have their alleged origin from the ancient Mesozoic landmass called Laurasia.

Gradstein and Vána (1987) mapped the world distributions of several Laurasian hepatics with broad ranges in the northern hemisphere reaching the tropics at high altitudes. Examples are *Blepharostoma trichophyllum*, *Lepidozia reptans* (Fig. 13.2), and *Nowellia curvifolia*.

3(b). South temperate distribution

Because of the land configuration, the distributions of bryophytes in the southern hemisphere are discontinuous. Owing to their alleged origin from another Mesozoic supercontinent south of the equator, Gondwana, the bryophytes of the southern hemisphere are called Gondwanic taxa. Examples are species of *Balantiopsis, Calomnium, Catagonium, Carrpos, Echinodium, Gigaspermum, Herzogiaria, Hydropogon, Hygrolembidium, Lepyrodon, Marsupidium, Monoclea, Neomeesia, Pleurophascum, Rhaphidorrhynchium, Verdoornia, Vetaforma,* and *Wardia*. Like their counterparts in the northern hemisphere, many Gondwanic taxa also reach the tropics in their distributions. Some even successfully penetrated into the northern hemisphere. Examples of the latter group are species of *Bescherellia, Dawsonia, Leptostomum, Plagiotheciopsis,* and members of Phyllogoniaceae.

4. Tropical distribution

There are as many bryophytes found in the tropics as in the temperate regions of the world. The magnitude of diversity of temperate bryophytes rivals well the acclaimed high diversity of tropical bryophytes. This evidence has been taken to support a temperate origin of bryophytes.

About 80% of bryophytes of all the tropical rainforests belong to tropical families such as Calymperaceae, Fissidentaceae, Meteoriaceae, Neckeraceae, Pterobryaceae, and Sematophyllaceae among mosses, and Frullaniaceae, Lejeuneaceae, Lepidoziaceae, Plagiochilaceae, and Radulaceae among liverworts (Gradstein & Pócs 1989). Many of them are best represented in the tropics but are also fairly widespread in temperate zones.

Less than 20% of tropical bryophytes are pantropical like *Erpodium biseriatum* (Fig. 13.3). The majority are either confined to the Old World tropics (paleotropical) or the New World tropics (neotropical). Examples of paleotropical bryophytes are *Cephaloziella kiaerii* (Fig. 13.4) and genera such as *Cladopodanthus, Conoscyphus, Cuspidatula, Desmotheca, Macvicaria, Notoscyphus, Papillidiopsis, Ptychanthus,* and *Trichocoleopsis,* whereas examples of genera of neotropical bryophytes are *Aureolejeunea, Caribaeohypnum, Cyclolejeunea, Curviramea, Hildebrandtiella, Puiggariella, Ruizanthus, Vanaea,* and *Vitalianthus*.

In the Old World, equatorial Africa has both a lower diversity and a smaller number of bryophytes than equatorial Asia. Recent

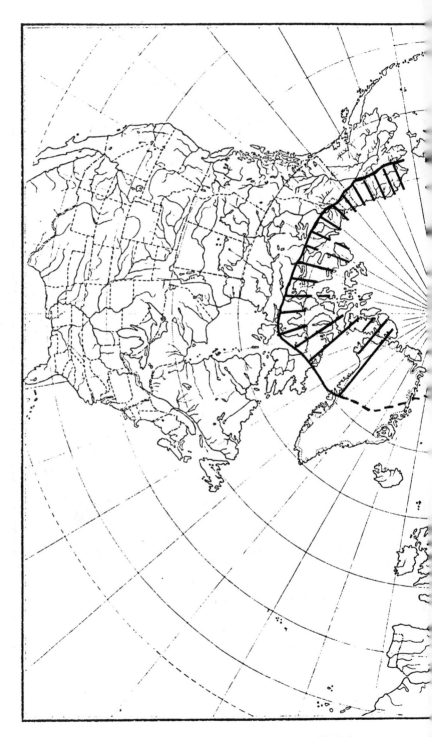

Fig. 13.1. Circumarctic distribution of *Plagiothecium berggrenianum*. (Modified after Ignatov 1993.)

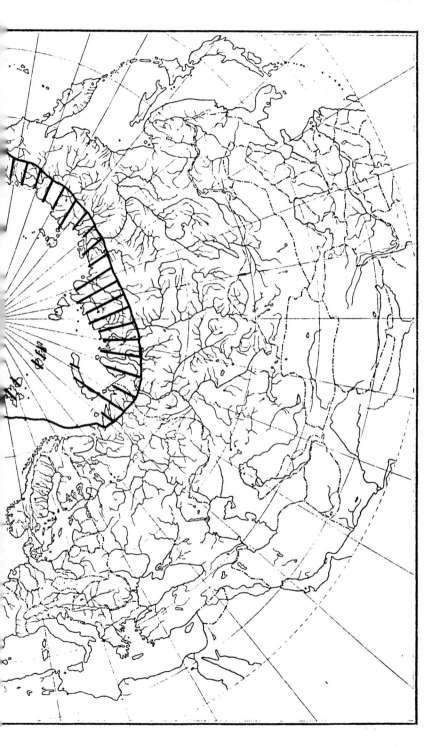

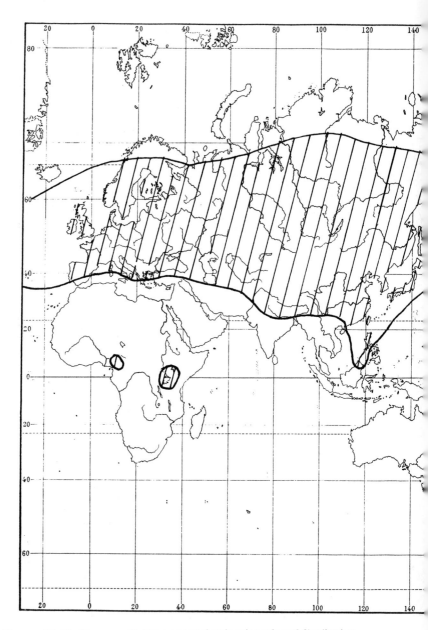

Fig. 13.2. Worldwide range of *Lepidozia reptans* showing circumboreal distribution
with extended ranges in high mountains in the tropics. (Modified after
Gradstein & Vána 1987.)

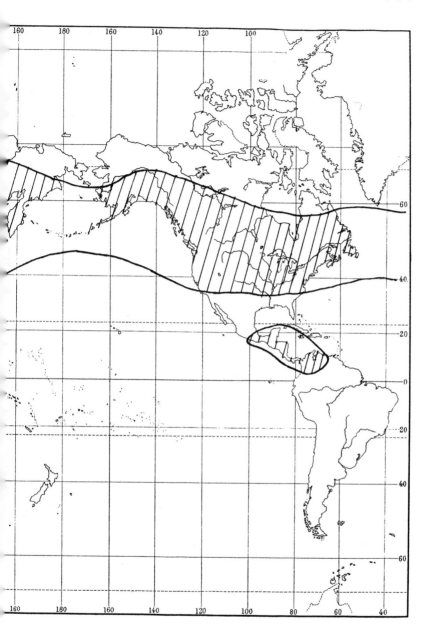

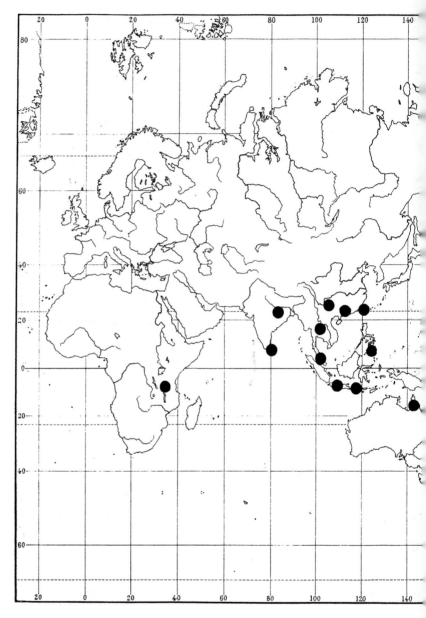

Fig. 13.3. Nearly pantropical, albeit discontinuous, distribution of *Erpodium biseriatum* (syn. *Solmsiella biseriata*).

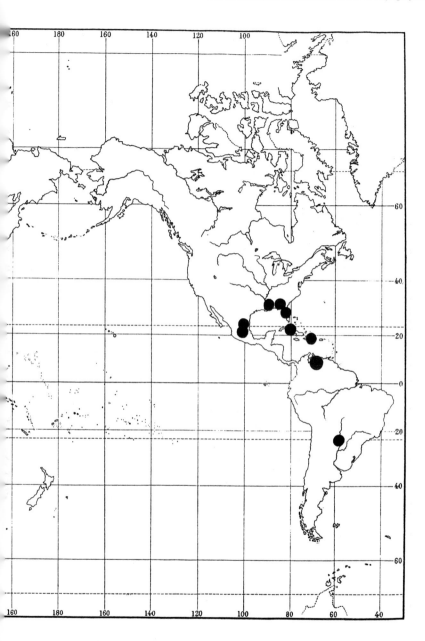

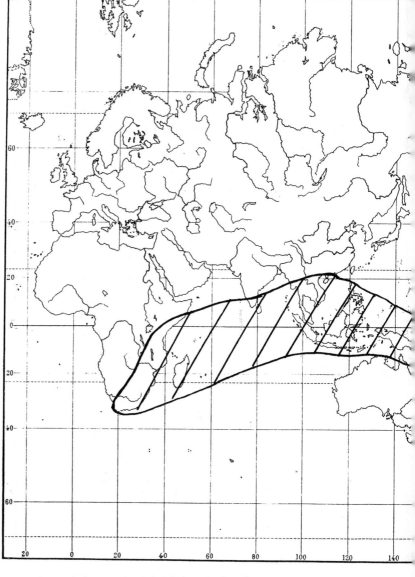

Fig. 13.4. Paleotropical range of *Cephaloziella kiaerii*. (After Pócs 1992.)

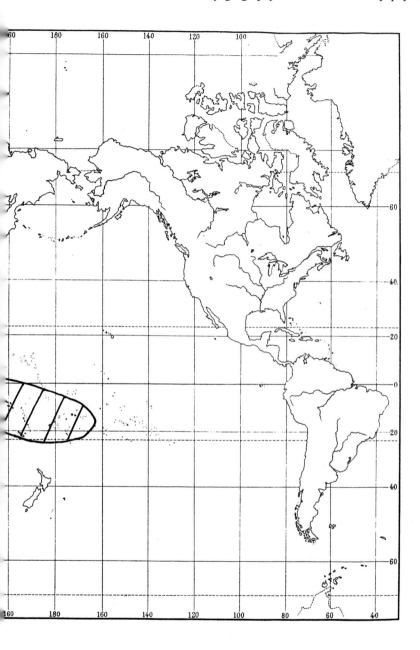

investigation shows that a handful of paleotropical mosses, especially species of taxonomically difficult genera, such as *Acroporium* and *Trichosteleum*, has been treated under different names in tropical Africa and tropical Asia. There is thus a need for critical taxonomic comparisons of moss floras of the regions under study before an accurate plant geographical analysis can be made.

5. Subantarctic and antarctic distributions

To date, some 160 genera and 380 species of bryophytes have been reported from the subantarctic and antarctic regions. The bryoflora of the Antarctic south of 60th latitude in the southern hemisphere has around 100 mosses and 25–30 hepatics (Seppelt *et al.* 1998).

The rather limited diversity of subantarctic and antarctic mosses includes several taxa of acrocarpous mosses such as species of *Andreaea, Bartramia, Bryum, Campylopus, Ceratodon, Distichium, Ditrichum, Grimmia, Racomitrium, Polytrichum, Pottia, Sphagnum*, and *Tortula,* in addition to a few pleurocarpous mosses. The latter belong mostly to the widespread genera *Amblystegium, Brachythecium, Calliergon,* and *Drepanocladus.* The liverworts with restricted subantarctic and antarctic ranges are mostly members of Gymnomitriaceae, Jungermanniaceae, Lepidoziaceae, Lophoziaceae, Lophocoleaceae, and Cephaloziaceae. A good example is the genus *Herzogobryum* of Gymnomitriaceae (Fig. 13.5).

Many species of bryophytes found in the Subantarctic and Antarctic are locally endemic. In his discussion on the possible origin of the antarctic moss flora, Robinson (1972) doubted the antiquity and validity of the many local endemics. True enough, later revisions have shown that a number of the local endemics are but incorrectly determined morphotypes of much more widespread taxa (Seppelt *et al.* 1998).

Surprisingly, there are two endemic families of liverworts described from the subantarctic region. They are Vetaformaceae (monotypic) and Phyllothalliaceae (with one genus and two species). The latter reaches as far as Patagonia in South America.

13.4 Disjunctive distributions

This is the most fascinating and much-studied pattern of plant distribution. The causes of disjunction in plant distributions are many and are difficult to ascertain because of the incomplete pictures of past and present distributions of plants around the world. Most likely, the causal events leading to range disruption vary from species to species. Usually, plate tectonics and climatic changes in the region are the two principal causes. Allen (1996) suggested that some disjunctions were better explained by long-distance dispersal followed by short land and air dispersals across the continent.

Plant disjunction can be local between two nearby mountains or thousands of kilometers apart across an ocean. Some bryophytes even show tri- to quadricentric disjunctions across the continents or oceans.

The well-known transcontinental or transoceanic disjunctions are the bipolar (Fig. 13.6), amphi-Atlantic (Fig. 13.7), Mediterranean–Californian–Australian, East Asiatic–Eastern North American, circum-Northern Pacific, and South American–Australasian disjuncts. These disjunctive patterns have been reviewed repeatedly by many, including Iwatsuki and Sharp (1967), Schofield and Crum (1972), and Schofield (1974, 1984, 1985, 1988). Other less dramatic disjunctions, such as the Eastern–Western North American (Schofield 1980, Belland 1987), North–South American, and intra-Malesian disjunctions (Touw 1992, Tan 1998), are of regional interest.

At present, fewer than 60 bryophytes show an amphi-Pacific disjunct between the paleotropics and neotropics. In contrast, there are more bryophytes, such as species of *Arachniopsis, Bryopteris, Renauldia*, and *Symbiezidium*, that have a disjunctive range between tropical Africa and tropical America (cf. Fig. 13.7). Frahm (1982), Gradstein *et al.* (1983), Buck and Griffin (1984), Reese (1985), Ochyra *et al.* (1992), and Allen (1996) contributed much to the understanding of African and South American disjunctions. According to Delgadillo (1993a), the number of mosses exhibiting a broad neotropical–African disjunction pattern total more than 334 species. This is a reflection of the long history of juxtaposition of the two continents before the breaking-up of Gondwana in early Cretaceous time.

Pócs (1976, 1992) has made significant contributions to the knowledge on tropical African and tropical Asiatic disjunction of bryophytes. A total of 70 liverworts and 108 mosses were identified to have Afro-Asian bicontinental disjunctions.

Recently, Schofield (1994) identified the Pacific coast of North America as the more likely place of origin for xerophytic bryophytes now inhabiting disjunctively the Mediterranean climatic regions in the world. His interpretation is based on the wider ecological tolerance and the morphological variability of a high proportion of so-called Mediterranean taxa in Pacific North America than in Eurasia and Africa. His viewpoint on the origin of Mediterranean flora differs from that of Frey and Kürschner (1988) who suggested an ancient Tethyan Sea origin. This underscores the

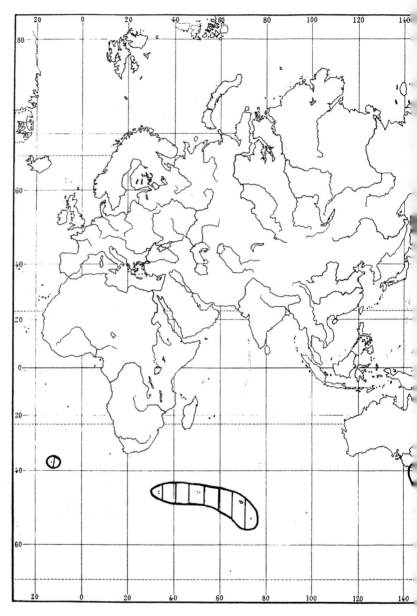

Fig. 13.5. The disjunctive subantarctic range of the genus *Herzogobryum*.

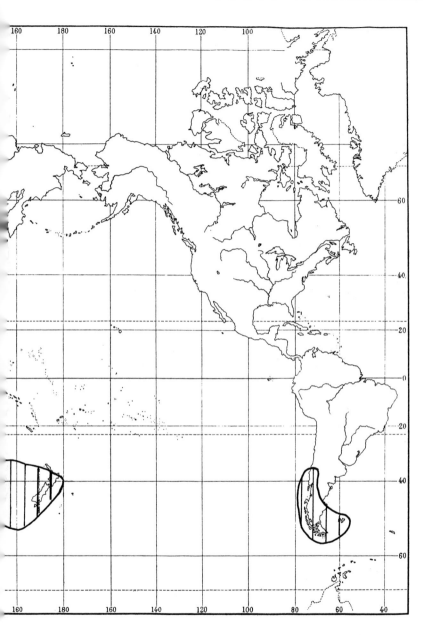

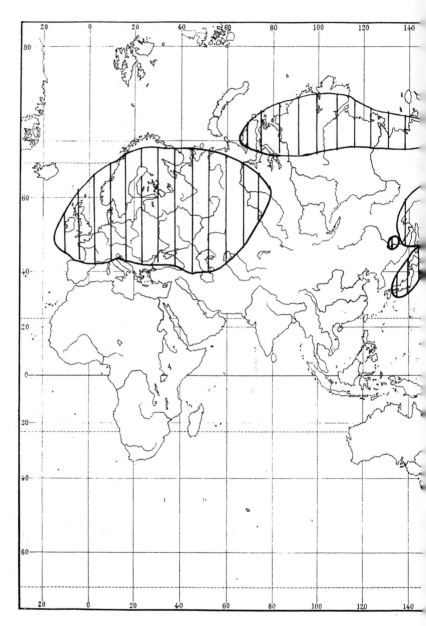

Fig. 13.6. Bipolar distribution of *Buxbaumia aphylla*. According to Schofield (personal communication, 1999), there is a doubtful record from Tasmania.

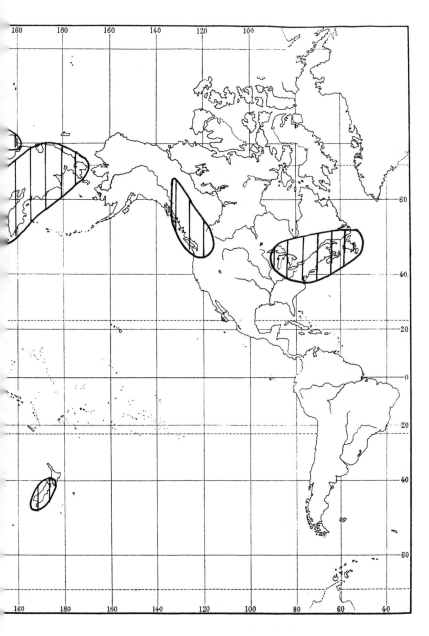

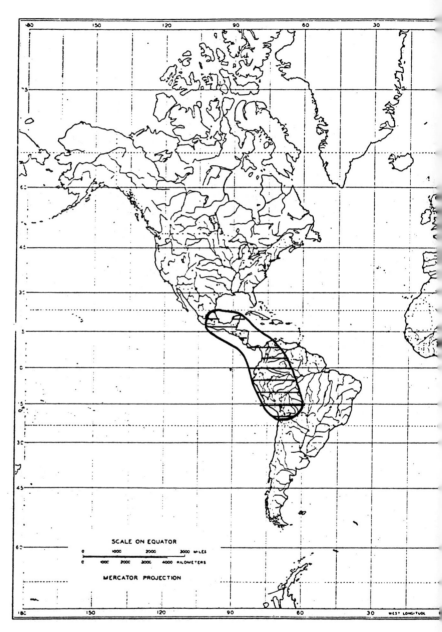

Fig. 13.7. Tricentric disjunctive range of *Campylopus nivalis* in Central and South America and Africa. (Modified after Frahm 1982.)

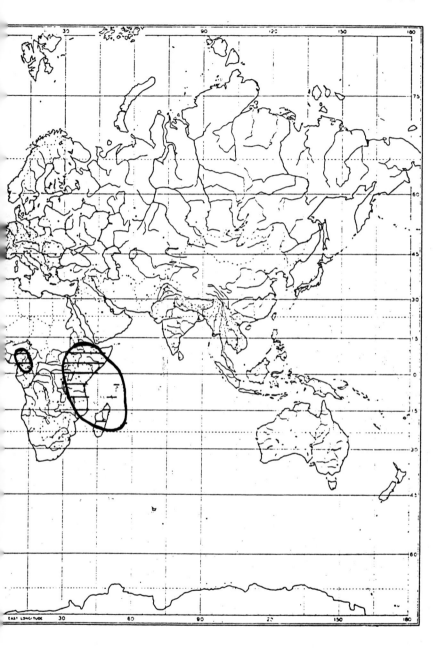

fact that, with the accumulation of distribution data, there will be new insight into and re-interpretation of existing hypotheses and principles governing the study of plant geography.

13.5 Vicariance and vicariads

In plant geography, two closely related forms or species found in widely separated places are called vicarious species, or simply, vicariads. Like plant disjunction, the causes of vicariance in plants are many. Theoretically, vicariads are formed by the fragmentation of a single parental taxon. The process involves: 1) formation of geographical, climatic, or ecological barriers leading to the fragmentation of a parental population, and 2) the speciation of fragmented populations in isolation.

Vicariance has been extensively investigated in seed plants but not among the bryophyte taxa. Schofield (1980) and Iwatsuki (1992) provided 12 examples of moss vicariads in Eastern Asia and Eastern North America, e.g., *Bruchia microspora* and *B. sullivantii, Climacium japonicum* and *C. americanum, Drummondia sinensis* and *D. prorepens, Diphyscium involutum* and *D. cumberlandianum,* and *Syrrhopodon kiiensis* and *S. texanus.* Gradstein *et al.* (1983) listed 16 pairs of vicariad taxa of liverworts between tropical America and Africa. Some of their examples are *Calypogeia peruviana* and *C. afrocoerulea, Frullania riojaneirensis* and *F. africana, Plagiochila columbica* and *P. integerrima,* and *Symbiezidium barbiflorum* and *S. madagascariense* (Fig. 13.8).

13.6 Plate tectonics and bryophyte distribution

The theory of continental drift published by Alfred Wegener in 1915 which influenced the study of sea-floor spreading and plate tectonics is perhaps the most important geological concept to have revolutionized the study of modern biogeography. Indeed, the reconstruction of the history of continental drift through geologic times has provided a solid basis for plant geographers to understand and explain the taxic similarity and floristic affinity between regions that are widely separated today.

Raven and Axelrod (1974) presented a useful review of the continental drift events and their significance in phytogeography. According to them, the breaking-up of the supercontinent called Pangaea into two large landmasses, Laurasia and Gondwanaland, some 180 million years ago, and the final drifting of the continents to their present positions some 5 million years ago, constitute a critical sequence of events in the long history of the earth. The intermittent collision of continental and oceanic plates not only generated massive mountain ranges serving either as geographical barriers or corridors to plant migration, but also brought together disparate floras from various geographical sources to mix freely and co-exist.

Recently, Frahm (1994) compared the *Campylopus* floras of Sri Lanka and Madagascar. Although the two islands are widely separated today by the Indian Ocean, both had a common origin as part of the former Gondwana. It is no wonder that two-thirds of the *Campylopus* taxa of these two islands were found to be similar or very closely related to the African flora.

On a worldwide basis, Schuster (1982, 1983a) reviewed the family and generic distributions of bryophytes in relation to the plate tectonics. Earlier, Schuster (1972) presented a good discussion of the historical migration of Gondwanic elements into the rainforests of south-east Asia via the northward drifting of the Indian and Australian plates. His studies revealed further that many antipodal liverworts possess "primitive" characters and widely disjunctive ranges in the southern hemisphere.

However, not all of the currently observed patterns of plant distributions can be attributed to the influences of plate tectonics. In a number of cases, long-distance dispersal of diaspores may be the primary causal mechanism.

13.7 Bryophyte endemism

Bryophytes having small or narrow ranges, be they continuous or discontinuous, are recognized as endemics. Distinction should be made between two kinds of endemics: neo-endemic and paleo-endemic. The former refers to recently evolved species that have not had time to achieve a wider distribution, whereas the latter refers to ancient, surviving species with a shrinking range. In practice, it is difficult to establish the true nature of an endemic unless the fossil records can provide unequivocal support. As a general statement, the endemics described from young geological formations, such as the Andes, Himalayas, or Mount Kinabalu, can be referred to as neo-endemics.

Schofield (1985) stated that endemism in the bryoflora is positively correlated to the following three conditions: 1) the length of time during which the locality was available for colonization, 2) the environmental diversity, especially in the availability of moisture, and 3) the length of time in isolation.

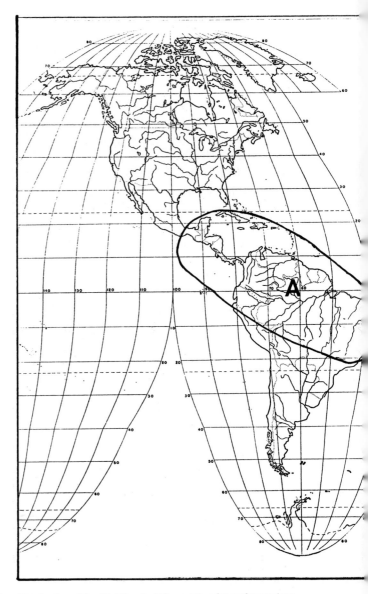

Fig. 13.8. Vicariant distribution of *Symbiezidium barbiflorum* (A) and *S. madagascariense* (B).

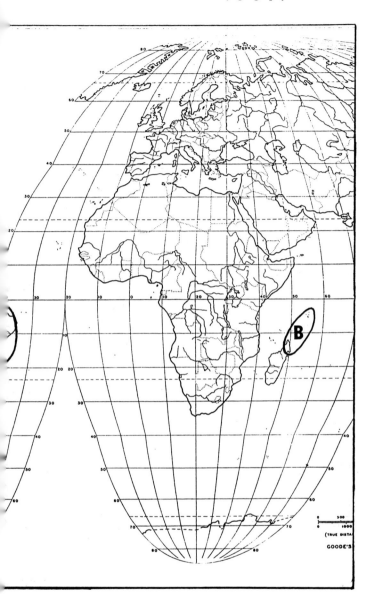

Interestingly, the percentages of generic and species endemism of bryophytes are lower than those calculated for the seed plants in many parts of the world. Furthermore, the estimated percentages of bryophyte endemism for the temperate and tropical rainforests appear to be similar.

Areas in the world with reportedly high bryophyte endemism and diversity are shown in Fig. 13.9. Several of them are large islands with a long history of isolation. Others are topographically diverse, mountainous areas. Considering the alarming rate of destruction of natural habitats worldwide, efforts should be organized to protect as many endemics in these areas as possible.

Although more than 75% of the families of bryophytes are widespread in both northern and southern hemispheres, there are a few endemic families with rather narrow ranges. These endemic families are either stenotypic or monotypic. They are: Pseudoditrichaceae, restricted to arctic Canada; Gyrothyraceae, restricted to the Pacific coast of North America; Rutenbergiaceae, restricted to coastal Tanzania, Madagascar, and adjacent islands, and Hydropogonaceae, restricted to tropical and subtropical South America.

In Western Melanesia, Piippo (1992) calculated a high endemism of 38% for the entire hepatic flora. Percentages of endemism, however, vary from family to family, such that 55% in Plagiochilaceae, 61% in Frullaniaceae, 27% in Lepidoziaceae, and none in Herbertaceae and Marchantiaceae.

13.8 Island biogeography and bryophytes

Many islands are known to produce endemics in spite of the fact that some of the recognized endemics may simply be the result of insufficient survey of the floras of adjacent continents. Two islands well known for their high endemism of seed plants and bryophytes are Madagascar and New Caledonia. In the case of New Caledonian *Fissidens*, Iwatsuki (1990) reduced significantly the percentage of species endemism from 80.6% to 10.7% after a careful revision. However, the percentage of species endemism in New Caledonian *Frullania* remained high even after a modern taxonomic revision (Iwatsuki 1990).

There are two kinds of islands from the viewpoint of plant geography, namely, continental and oceanic islands. The former became part of a continent when the sea level lowered during the ice ages. The past land connections permitted movement of plants from the continent to the island in small numbers. Predictably, the flora of a continental island is like a reduced version of the large continental flora. Such is the case of the moss flora of Sri Lanka vis-à-vis the flora of continental southern India. For similar reason, the bryoflora of Taiwan is closely related to the much larger flora of continental China.

Oceanic islands, on the other hand, lack any land connection with a continent. Their origin from the ocean floor precluded any direct landward migration of plants into the island. Most of their diaspores arrived via long-distance dispersal. The flora of oceanic islands therefore is a mixture of chancy and adventive plant taxa from various distant sources. A good example is the moss flora of the Hawaiian Islands.

The theory of island biogeography made famous by MacArthur and Wilson (1967) has not been tested with bryophytes. There has been no experimental study on rate of migration and rate of extinction of bryophytes under the island conditions. Based on limited data of bryophyte distributions, the alleged positive correlation between island size and species diversity appears to be valid (cf. Pócs 1988). However, Frahm *et al.* (1996) found that the topography and elevational variation, more than the island size, affect the number of bryophyte taxa present on an island. As to the relationship between the distance of the island from the continent and the degree of diversity of its bryofloras, the few recent studies (Menzel & Passow-Schindhelm 1990, Dirkse *et al.* 1993, Frahm *et al.* 1996) unfortunately do not show any clear correlation. Perhaps the barrier created by the distance of the island from the continent has been overcome by the efficacy of dispersability of air-borne diaspores of many bryophytes over the years.

Piippo (1992) and Piippo and Koponen (1997) reported that the floristic affinities of mosses of the island groups in Western Melanesia and Oceania are more closely related to Asia than to Australia and South America. Equally interestingly, Tan and Engel (1990) observed some differences in the patterns of distributions of mosses and liverworts in Borneo, New Guinea, and the Philippine archipelago. For large moss genera, they reported many species shared by the three large island groups. However, the same pattern is not seen among large liverwort genera. Instead, each of the large liverwort genera has its own set of endemics present on each of the island groups. According to them, these differences in distribution patterns between mosses and liverworts probably resulted from the differences in their ecological adaptation and reproductive biology evolved under island conditions.

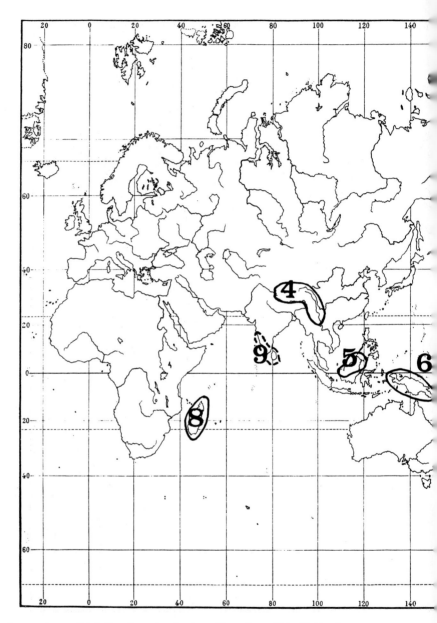

Fig. 13.9. Areas of high diversity and endemism of bryophytes: (1) Pacific North-west of North America; (2) Central America; (3) Northern Andes; (4) Himalayas and vicinity; (5) Borneo; (6) New Guinea; (7) New Caledonia; and (8) Madagascar. Broken line indicates area with unconfirmed report of high endemism and diversity of bryophytes: (9) Western Ghats and Sri Lanka.

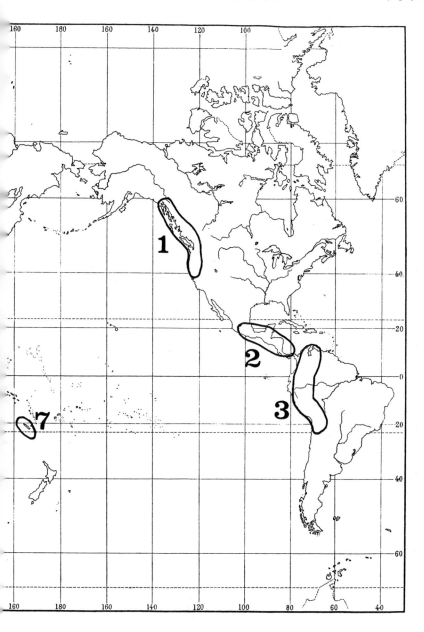

As mentioned above, island habitats tend to promote endemic taxa. Based on epiphyllous bryophytes, Pócs (1996) showed that the percentage of species endemism is highest in islands. The percentages of species endemism are about 10–16% for Cuba and the Galapagos (Gradstein & Weber 1982, Pócs 1988), and about 15–25% for larger islands like Madagascar and Sri Lanka.

Recently, Söderström (1996) compared the number of hepatic taxa of Samoa, Fiji, Vanuatu, and the Solomon Islands in the Pacific Ocean and several countries in central Africa. He found that the number of taxa in general is higher and the number of endemic taxa is significantly higher on the Pacific islands compared with the African countries.

13.9 Bryofloristic regions of the world

Schofield (1992) presented a discussion of six broad bryophyte floristic kingdoms, namely, Holarctic, Paleotropical, Neotropical, Australian, Cape or South African, and Holantarctic. In practice, however, many bryologists have come to accept the 20 floristic regions of the world published by van der Wijk *et al.* in the Index Muscorum (1959–69). Although the delineation of the 20 regions is supported by past and present diversity research and floristic analyses, the boundary between adjacent regions in some parts of the world can be refined further. For example, the Far East of Russia can be placed under Asia 2, instead of Asia 1. Likewise, the Tibetan plateau and portions of western Yunnan and western Sichuan of China should form part of Asia 3, instead of Asia 2. The inclusion of the Indochina floristic region in Asia 3 also needs reconsideration. Lastly, the Bismarck, New Britain, and Solomon Island groups should constitute part of Asia 4, instead of Oceania, and Lord Howe Island and New Caledonia should also be included in Australia 1.

Owing to space limitations, only selected floristic regions of the world with better known bryodiversity are briefly reviewed below.

13.9.1 North America
This is the Am 1 region of the *Index Muscorum*'s bryogeographical scheme. The region commences from north of Mexico and extends all the way into the Arctic Ocean, including Greenland. Because of heavy glaciation during the Pleistocene, the total number of known bryophytes today is not great in relation to its expansive land area. Nonetheless, it is one of

the two regions in the world where the bryoflora is best inventoried and analyzed. The other region is Europe.

There are about 315 genera and 1325 species of mosses and 120 genera and 570 species of hepatics and hornworts in North America (Schofield & Thiers, personal communication, 1999). The overall percentage of species endemism of bryophytes is about 18%, although the figure can be higher or lower in different parts of North America. Sixteen endemic genera and 25 endemic species of mosses are known from this region. Genera endemic to eastern North America are *Brachelyma*, *Bryocrumia*, *Bryoandersonia*, *Donrichardsia*, *Neomacounia*, and *Ozobryum* and while genera endemic to western North America are *Alsia*, *Andreaeobryum*, *Bestia*, *Bryolawtonia*, *Dendroalsia*, *Geothallus*, *Gyrothyra*, *Leucolepis*, *Pseudobraunia*, *Rhytidiopsis*, *Roellia*, *Schofieldia*, and *Trachybryum*. Many of these are monotypic genera.

The bryoflora of North America consists of a major component of boreal taxa and a small south-eastern component with strong tropical and subtropical affinities. In addition, the drier south-western portion has a "Mediterranean" flora. By and large, the floristic affinity of western North America is closer to eastern Asia, while the eastern North American flora has a closer relation with the Atlantic European flora.

Recently Frahm and Vitt (1993) made a floristic comparison between the moss floras of North America and Europe. They calculated that about 43% of the North American species are not found in Europe. They also concluded that the Pleistocene extinction was less severe in North America than in Europe. For them, the North American moss flora is richer because of the presence of neotropical taxa in the south-eastern United States and the circumpacific taxa in western North America.

According to Schofield (personal communication, 1999), there are five centers of bryophyte diversity and endemism in North America: 1) the southern Appalachian Mountains, 2) the mixed deciduous forest in eastern North America, 3) the Pacific coast of western North America, 4) the unglaciated western North American Arctic, and 5) California.

13.9.2 Tropical America

This floristic region is known also as the Neotropics. It includes the areas of Am 2, Am 3, Am 4, and Am 5 of the *Index Muscorum*'s bryogeographical system. The region stretches from Mexico and the West Indies southward to Bolivia and Brazil. The presence of the north–south-oriented Cordilleran mountain range in Central America and the Andean mountain range in South America have provided, no doubt, important migratory pathways for North American and Patagonian bryophytes to enter and enrich the floral diversity of this region. Although smaller than North America in land area, this region exhibits a much more diverse bryophyte flora. In fact, the total diversity and taxic number of bryophytes of the tropical American region are probably the highest in the world.

There are approximately 3900 species of bryophytes, of which about 2600 are mosses and about 1300 are liverworts and hornworts. The species endemism has been estimated to be as high as 48% (Delgadillo 1994). Of the 390 genera of mosses and 200 genera of liverworts and hornworts, 78 and 50, respectively, are endemic. Examples of endemic moss genera are *Acritodon, Actinodontium, Amblytropis, Diploneuron, Hydropogon, Hydropogonella, Kingiobryum, Phyllodrepanium, Schliephackea,* and *Stenodictyon,* and examples of endemic liverworts are *Alobiella, Alobiellopsis, Bromeliophila, Cephaloziopsis, Fulfordianthus, Fuscocephaloziopsis, Haesselia, Luteolejeunea, Micropterygium, Omphalanthus, Paracromastigum, Physantholejeunea, Schusterolejeunea, Szweykowskia, Trabacellula,* and *Zoopsidella.* Schuster (1990) offered a variety of explanations for the rather high proportion of endemic genera of Jungermanniales in the neotropics, among which isolation and the antiquity of the continent are two important factors.

There are about 10 centers of high diversity and endemism of bryophytes in this region, namely, from north to south, 1) Mexico, 2) Central America, 3) West Indies, 4) Choco, 5) Northern Andes, 6) Central Andes, 7) Amazonia, 8) Guyana Highland, 9) Brazilian Planalto, and 10) Southeastern Brazil. Of these, the two areas of the Andes are the primary center of bryodiversity in the neotropics. Together they have 80% of the neotropical liverwort flora and about 55% of the neotropical moss flora. In terms of moss diversity, this is likely the richest of the world's tropical regions (Churchill *et al.* 1995).

13.9.3 Europe
This is the other region of the world where the bryoflora is well investigated and documented. The region is the same as the Eur of the *Index Muscorum* system. The total number of bryophyte taxa in Europe and Macaronesia consists of 340 genera and about 1690 species, of which about 40 genera are monotypic. For mosses alone, Europe has about 232 genera and 1084 species (Frahm & Vitt 1993).

Since a large part of continental Europe and the British Isles were glaciated at the end of Pleistocene, there is no endemic family known from this region. The endemic genera are *Alophosia, Andoa, Cryptothallus, Nobregaea, Ochyraea, Pictus, Saccogyna, Tetrastichium,* and *Trochobryum.* Because of the long history of cultivation and industrialization, Europe has a few endemic species induced anthropogenically. A good example is *Ditrichum cornubicum* found growing restrictively on mine tailings. The region also has documented records of 19 bryophytes introduced from regions outside Europe (Frahm & Vitt 1993).

Floristically, northern Europe has the largest concentration of circumboreal bryophytes and is especially rich in species of *Sphagnum* and Amblystegiaceae. The Atlantic European region has a strongly oceanic equable climate and supports a surprising number of taxa with subtropical affinity shared with the Himalayas, Japan, and the Pacific north-west of North America. Central Europe, with its many high mountains and a more continental climate, supports an equally diverse bryoflora consisting of a mixture of boreal and temperate taxa. The floristic affinity and diversity of mosses of the European part of Russia was summarized by Ignatov (1993) who reported high species diversity in oceanic parts and mountain areas of Russia, and low diversity in the continental lowlands. His discussion of the moss diversity and affinity for the six floristic divisions of the former Soviet Union, based on a total of 1157 species of mosses, is most informative.

The bryoflora of the British Isles, which comprises 716 species of mosses, 284 liverworts, and four hornworts, represents about two thirds of the European total bryodiversity. Hill and Preston (1998) did a remarkably detailed analysis of the floristic composition and affinity of the British and Irish bryophytes, especially in comparison with the North American bryoflora. They reported 17 endemic species of bryophytes and about 44 hyperoceanic taxa, mostly liverworts, confined to western part of the British Isles. Many of these are disjunctively found also in western North America in similar hyperoceanic habitats.

The Mediterranean part of Europe, and northern Africa as well, has many xerophytic taxa belonging to Ricciaceae, Pottiaceae, and Grimmiaceae that are shared disjunctively with California and south-western Australia where a similar climate of summer-dry–winter-wet prevails. Other Mediterranean taxa have developed a more or less discontinuous range extending into the drier parts of Central Asia.

The Macaronesian island groups, which include the Azores, Canary Islands, Madeira, and Cape Verde, have a rich bryoflora in their evergreen cloud forests that are linked phytogeographically to Europe, northern Africa, the neotropics, paleotropics, and Australia.

In a recent publication, Hodgetts (1996) identified Macaronesia and the Mediterranean as two centers of diverse bryophytes in Europe. The Alps, the Carpathians, and the Atlantic coast are three other centers of bryophyte diversity in continental Europe.

13.9.4 Africa

The floristic map of the African continent and vicinity in *Index Muscorum* comprises four parts (Afr 1–4). The Afr 1 or North Africa has little floristic individuality and represents the extension of xerophytic bryofloras of the Mediterranean and the Middle East. The higher mountains, though, have a large northern temperate bryoflora.

The three parts of sub-Saharan Africa (Afr 2–4) have an uneven pattern of bryodiversity distributed across the continent. In Afr 2, the tropical taxa dominate, with a rich representation of *Archilejeunea*, *Cololejeunea*, and *Lopholejeunea* among hepatics, and *Callicostella*, *Fissidens*, *Macromitrium*, *Syrrhopodon*, and among the mosses. The same tropical bryoflora is represented further south in the wetter mountain ranges of Transvaal and Natal. The high mountains of central Africa support distinctive afroalpine paramo vegetation with many endemics. In southern Africa (Afr 4), there is, increasingly, a higher proportion of southern temperate Gondwanic elements. Some of these have apparent Australasian and Andean links (e.g., *Carrpos*, *Eccremidium*).

The hepatic taxa of sub-Saharan Africa (Afr 2–4) were recently summarized by Wigginton and Grolle (1996), and those of mosses by O'Shea (1995). However, these two checklists reflect more the level of exploration in the different countries rather than their real species richness.

The total number of Hepaticae in sub-Saharan Africa is about 894 species, of which 713 occur on the continent and 436 on the Indian Ocean islands. According to O'Shea (1997), the number of known African mosses is 3048 species. This number will go down to probably 1300 taxa after thorough taxonomic revisions.

Two endemic moss families (Rutenbergiaceae and Serpotortellaceae) occur in continental Africa and the adjacent Indian Ocean islands. The endemic African hepatic genera are few (*Capillolejeunea*, *Cephalojonesia*, *Cladolejeunea*, and *Evansiolejeunea*). The number of endemic moss genera is somewhat higher, about 20, e.g., *Kleioweisiopsis*, *Husnotiella*, *Hypodontium*, *Loiseaubryum*, *Neorutenbergia*, *Perssonia*, *Pocsiella*, and *Quathlamba*. At the generic level the neotropical affinity of the African bryoflora seems strong (Gradstein *et al.* 1983). At the species level the African bryoflora has a stronger link with Asia and Oceania (Pócs 1992).

The Indian Ocean island groups, including Madagascar, have the highest percentage of endemism in Africa. The endemic moss family Serpotortellaceae, and the endemic hepatic genus *Capillolejeunea* occur here. The region is, no doubt, a center of evolution for several genera, such as *Diplasiolejeunea*, *Leucoloma*, and *Rutenbergia*. It has an interesting link also with the neotropics, sharing a number of species that do not occur in mainland Africa, e.g., *Phyllodrepanium falcifolium* and *Phyllogonium fulgens*. The Indian Ocean is also the westernmost limit of a number of Asiatic taxa.

The following are the seven centers of high endemism and bryodiversity in Africa: 1) Mount Cameroon, 2) Gabon, 3) East Zairean–Rwandan highlands, 4) the afro-alpine areas of Ruwenzori, Mount Elgon, Mount Kenya, and Mount Kilimanjaro, 5) the crystalline chain of the Eastern Arc mountains in Tanzania, to the Mulanje Mountains in Malawi, 6) Drakensberge in Natal–Lesotho, and 7) the Cape.

13.9.5 East Asia

This is another floristic region very high in bryophyte diversity. It is roughly equivalent to the As 2 of the *Index Muscorum*. Because the region experienced little glaciation during the Pleistocene, many rare and monotypic taxa have survived *in situ* as relicts of a Tertiary flora.

The rise of the Himalayas after the collision of the Indian plate against the Asian continent about 45 million years ago is the single most important geological event in Asia, producing radical changes in the climate, ecology, and evolution of biota in continental Asia. Its impact on the East Asiatic bryophyte flora can be seen today in the presence of a high diversity of taxa and great number of endemics around the Himalayas.

Approximately, East Asia has 450 genera and 2300 species of mosses. The percentage of species endemism is estimated to be 20%. The liverwort and hornwort floras comprise about 150 genera and 1100 species (Piippo & R.-L. Zhu, personal communication, 1999).

Bryophyte genera endemic to this region are numerous, including 50 moss taxa. Examples are *Actinothuidium*, *Bissetia*, *Boulaya*, *Brachymeniopsis*, *Bryonoguchia*, *Dolichomitra*, *Dolichomitriopsis*, *Dozya*, *Handeliobryum*, *Hondaella*, *Horikawaea*, *Mamillariella*, *Miyabea*, *Orthodontopsis*, *Orthomitrium*, *Palisadula*, *Podperaea*, *Reimersia*, *Sciaromiopsis*, *Struckia*, and *Taxiphyllopsis*. Examples of endemic liverwort genera are *Cavicularia*, *Cryptocoleopsis*, *Hattoria*, *Macvicaria*, *Neotrichocolea*, *Nipponolejeunea*, *Tuzibeanthus*, and *Xenochila*.

Because of its long tradition of forest protection and a cultural liking for bryophytes, Japan has a bryoflora that is the best preserved and most studied in East Asia. Although the country itself has no endemic genera and only a small number of endemic species, its flora contains 324 genera and 1180 species of mosses, about half of the moss diversity of the entire East Asia region. For the hepatic flora, Piippo (1992) reported 124 genera and 530 species.

Deguchi and Iwatsuki (1984) published a detailed floristic analysis of the Japanese bryophytes. According to them, the long chain of islands arranged in a north–south direction from the subarctic and boreal to subtropical regions has provided diverse habitats for many bryophytes to colonize and survive. Additionally, Iwatsuki (1992) discussed the floristic connection between Japan and North America, citing 34 Japanese mosses with an East Asiatic–Eastern North American disjunct pattern.

China, with its vast and diverse terrain, is the most important repository of bryodiversity in the East Asiatic region. Unfortunately, the destruction of forest and other natural habitats in China is also the most serious in the world. There are about 420 genera and 2200 species of mosses, and about 125 genera and 680 species of liverworts and hornworts known from China including Taiwan.

Briefly stated, the bryoflora of China, like its vascular counterparts, shows a strong Sino-Japanese floristic influence in the eastern and central parts of the country. The south-western and southern parts of China share a flora with the neighboring Himalayas and Malesia. The boreal taxa are confined to the north-eastern corner of China and the interior mountains of Inner Mongolia and Xinjiang Provinces. The arid Gobi interior and the frigid Tibetan plateau harbor a significant number of Irano-Turanian elements or bryophytes of Central Asiatic affinity. The lower and mid-elevations of the Himalayan range in China are particularly rich in oligotypic genera representing the neo-endemics and wafted elements of Gondwanic flora.

In conclusion, East Asia has the following centers of high endemism and bryodiversity: 1) Japan and eastern China, 2) central and south-western karst provinces of China, and 3) the Hengduan mountain ranges.

13.9.6 Malesia

Insular south-east Asia or Malesia (As 4) is geologically unique as the result of its complex history of plate tectonics. The long chain of island countries, which include Malaysia, Indonesia, Philippines, Brunei, and Papua New Guinea, is the result of collision of the two continental shelves, namely the Sahul and Sunda of the Asiatic and Australian plates, during the mid-Tertiary. It is here that the famous Wallace's Line is situated. This biogeographical barrier separating the Asiatic and Australian biota, however, has had only a moderate effect in preventing the mixing of Asiatic and Australasian mosses (Tan 1992). Geologically, several of the islands in this region, such as Sulawesi and the Philippine archipelago, have now been shown to be of composite nature formed by the accretion of microplates and terranes from both Laurasian and Gondwanic origins (Hall & Holloway 1998). The coming together of these various microplates also carried with them dissimilar floras that subsequently became integrated.

The region has about 330 genera and 1600 species of mosses and about 135 genera and 800 species of liverworts. Pócs (1996) counted 224 species of epiphyllous liverworts from Malesia making this region the highest in epiphyllous bryodiversity in the world.

Briefly stated, the lowland bryophyte flora of Malesia is primarily Asiatic in composition while its montane flora has a strong representation of Australian flora. Also, the western portion of Malesia is more Asiatic in floristic character while the eastern portion is strongly influenced by the Australasian taxa.

The estimated percentage of species endemism of bryophytes is about 25% because of the large number of endemics described from Borneo and New Guinea, the two main centers of bryodiversity and endemism in Malesia. Examples of endemic genera are *Aphanotropis, Brotherobryum, Calatholejeunea, Cladopodanthus, Dactylophorella, Hattoriolejeunea, Macgregorella, Merrilliobryum, Pachyneuropsis, Sclerohypnum,* and *Taxitheliella*.

13.9.7 Australasia

This region is defined here to include Australia, New Zealand, New Caledonia and the adjacent Norfolk, Cocos, and Lord Howe Islands. It is the equivalent of Aus 1 and 2 of the *Index Muscorum* system. A total of about 290 genera and 1000 species of mosses and about 160 genera and 850 species of liverworts and hornworts are known from this region. There is currently no endemic moss family in the region. The two endemic liverwort families, Vandiemeniaceae and Perssoniellaceae, are both monotypic. The former is endemic to Australia while the latter is endemic to New Caledonia.

The bryoflora of Australia is relatively diverse. The continent itself has about 275 genera and 950 species of mosses and about 125 genera and 670 species of liverworts and hornworts. In northern Queensland one can find a large number of tropical Malesian bryophytes. A large portion of the land west of the Great Dividing Range has little rainfall and supports a xeric scrub to subxeric savannah vegetation dominated by *Eucalyptus* and other myrtaceous trees. This is where the xeromorphic bryophytes of Australia are most diverse and the soil crust bryophytes most concentrated. The south-eastern coast of Australia has much rainfall, coupled with a mild winter. Here, as in New Zealand and Tasmania, the native *Nothofagus* forests are well developed offering a great diversity of bryophytes that are of Gondwanic origin.

Strongly influenced by the orographic factor, New Zealand has about 208 genera and 523 species of mosses and about 135 genera and 500 species of liverworts and hornworts (Fife 1995). Its bryoflora is strongly related to that of Australia in spite of its long history of separation from the latter.

Including the unquestionably rich bryoflora of New Caledonia, endemic species of bryophytes in Australasia are plentiful, especially in Pottiaceae, Ditrichaceae, and Archidiaceae. Many genera in Australasia, on the other hand, are broadly distributed also in temperate South America and southern Africa, as well as in New Guinea. There are about 17 endemic genera of mosses and 14 endemic genera of liverworts known from Australia and New Zealand. New Caledonia alone has four endemic moss genera, all monotypic. Some examples of endemic moss genera in Australasia are *Beeveria, Bryostreimannia, Crosbya, Cryptodendron, Franciella, Hypnobartlettia, Mesochaete, Mesotus, Parisia, Stonea, Tetracoscinodon, Touwia,* and *Wildia.* Examples of endemic genera of liverworts in Australasia are *Archeophylla, Chloranthelia, Eoisotachis, Lamellocolea, Megalembidium, Neogrollea, Perssoniella, Stolonivector, Vandiemenia, Verdoornia,* and *Xenothallus.*

Streimann (personal communication, 1999) has identified the following centers of high bryodiversity and endemism in Australasia: 1) semi-arid interior, 2) north-east Queensland, 3) Tasmania, 4) Lord Howe Island, and 5) New Caledonia.

13.10 Conservation

It is a well-established fact that the survival of many bryophytes depends on the preservation of their natural habitats. Habitats such as the primary forests provide good protection and the necessary microclimates and substrata needed for the bryophytes to colonize and diversify. Consequently, the elimination of forests leads to the extinction of many bryophyte residents. This is especially true for the lignicolous species inhabiting decaying wood and the epiphyllous taxa living on leaves.

All endangered bryophytes can be protected in two ways: 1) the habitat approach and 2) the species approach. The latter publicizes the species as a nationally endangered taxon justifying protection in its natural habitat. To date, only a few countries in Europe, North America, and Asia have produced a national red data book for bryophytes (Hallingbäck & Hodgetts 1999).

The International Association of Bryologists (IAB) and the International Union of Conservation of Nature and Natural Resources (IUCN) have recently joined together to identify the world's most endangered bryophytes. The first red list of the world's most endangered bryophytes, consisting of 91 species of which 40 are mosses and 51 liverworts and hornworts, was subsequently published in the *Bryological Times* (International Association of Bryologists 1997). The complete listing can be gleaned at the websites of the IUCN and the IAB.

The selection of red-listed bryophyte taxa is based on the following six criteria: 1) the species must be threatened worldwide, 2) the species must be confined to threatened habitat, 3) the species must have a narrow range, 4) the species is not overlooked or undercollected, 5) the species has a unique morphology/biology, and 6) the species occupies a special evolutionary position.

The same IAB committee monitoring the world's most endangered bryophytes also identified the following as serious threats to the long-term existence of bryophytes: 1) forest destruction, 2) modern agricultural practices, e.g., uses of herbicides and chemical fertilizers, 3) industrial pollution, e.g., acid rain, 4) introduction of weedy/alien species, 5) habitat degradation caused by tourism, and 6) excessive collection for commercial purposes (see Hallingbäck & Hodgetts 1999).

In order to preserve more effectively the regional bryodiversity with limited financial and land resources, Tan and Iwatsuki (1996, 1999) made attempts to identify hot spots of high diversity and endemism of bryophytes in East Asia for immediate protection. The selection criteria used in choosing the hot spots are: 1) high species number and taxonomic diversity, 2) presence of many endemics with narrow ranges, 3) great variation of habitats and plant communities, 4) complementarity of floristic compositions, 5) legislated status as parks and nature reserves, and 6) confirmed existence of the reported bryoflora. The eight selected East Asiatic

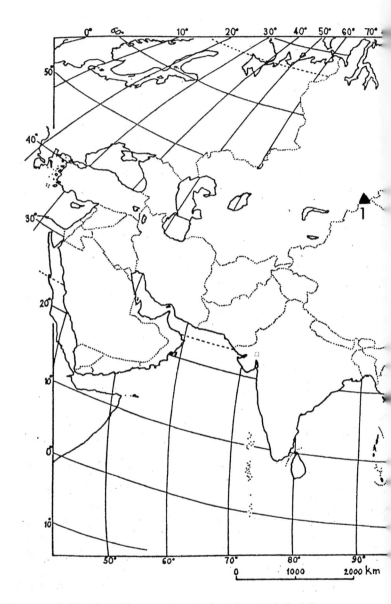

Fig. 13.10. Recommended bryological hot spots in east and south-east Asia for full protection (see Tan & Iwatsuki 1996, 1999): (1) Mount Altai Nature Reserve, Russia and China; (2) Mount West Tianmu Protected Nature, China; (3) Mount Fanjing Nature Reserve, China; (4) Yakushima Nature Reserve, Japan; (5) Mount Amuyaw Forest Reserve, the Philippines; (6) Mount Kinabalu Nature Park, Malaysia; (7) Gunung Rindjani Nature Reserve, Lombok, Indonesia; (8) Mount Wilhelm Forest Reserve, Papua New Guinea. Two additional nature reserves under study for hot spot recommendation: (9) Bureya Nature Reserve, Russia; and (10) Gunong Tahan National Park, peninsular Malaysia.

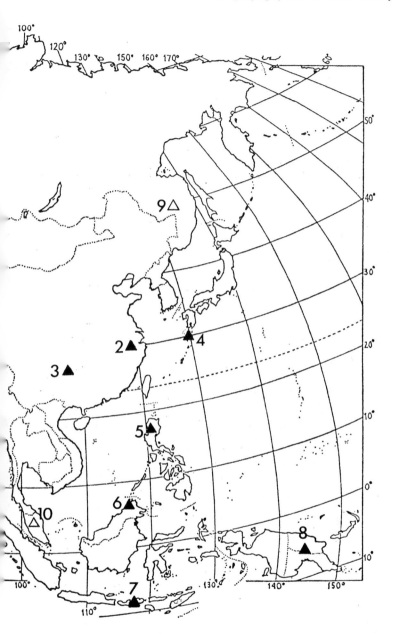

hot spots of bryophyte diversity are shown in Fig. 13.10. It is estimated that more than one-third of the bryophyte diversity and endemic taxa in East Asia will come under protection if the above eight hot spots are fully preserved.

Recently, there have been suggestions to give priority consideration in conservation to bryophyte taxa that rank high in genome diversity or phylogenetic character analysis (Vane-Wright *et al.* 1991, Hedenäs 1996). However, for a group of plants like bryophytes which has only incomplete information on its evolution and phylogeny, it is not practical at the moment to pursue such a strategy of conservation.

To evaluate the deteriorating situation of bryophyte diversity in the tropics, Ines-Sastre and Tan (1995) reviewed and compared the loss of bryophyte diversity in Puerto Rico and in the Philippines. Approximately 12% of the moss species reported from Puerto Rico have become extinct within the 20th century, with the island's cloud forest suffering the heaviest loss of diversity. In the Philippines, Mount Santo Tomas, a limestone mountain situated in northern Luzon, was shown to have lost to deforestation and agriculture 50% of its moss species over a period of 50 years. Because the moss diversity in many mountains in the tropics has not been fully investigated, the actual loss of bryodiversity brought by the deforestation and other environmental disturbances in many tropical countries will be difficult to assess.

Bryophytes are inherently small, inconspicuous plants that do not get attention from the general public or government bureaucrats. Their protection has to be carried out in connection with the conservation of attention-getting, "flagship" or "keystone" plant and animal species, such as the panda, tiger, elephant, and *Rafflesia*. By helping to establish sufficiently large nature reserves needed by these flagship or keystone species, the survival of bryophytes that share the same ecosystem can then be assured.

Acknowledgments

We are grateful to Drs H. Akiyama, T. Furuki, Z. Iwatsuki, L. Harrison, M. Ignatov, D. H. Murphy, S. Piippo, R. Pursell, W. B. Schofield, H. Sipman, H. Streimann, B. Thiers, I. Turner, and R.-L. Zhu and to Mr M.-S. Choi for discussion, reading drafts of the manuscript, and also providing information on the distributions of some bryophtyes.

REFERENCES

Allen, B. (1996). *Diphyscium pocsii* (Musci: Buxbaumiaceae), an African species new to Honduras. *Nova Hedwigia*, **62**, 371–5.

Belland, R. (1987). The disjunct moss element of the Gulf of St. Lawrence Region: glacial and postglacial dispersal and migration histories. *Journal of the Hattori Botanical Laboratory*, **63**, 1–76.

Buck, W. R. (1990). Biogeography of the Greater Antillean mosses. *Tropical Bryology*, **2**, 35–48.

Buck, W. R. & Griffin, D., III (1984). *Trachyphyllum*, a moss genus new to South America with notes on African-South American bryogeography. *Journal of Natural History*, **18**, 63–9.

Churchill, S. P., Griffin, D. III, & Lewis, M. (1995). Moss diversity of the tropical Andes. In *Biodiversity and Conservation of Neotropical Montane Forests*, ed. S. P. Churchill, D. Griffin III, & M. Lewis, pp. 335–46. New York: New York Botanical Garden.

Deguchi, H. & Iwatsuki, Z. (1984). Bryogeographical relationships in the moss flora of Japan. *Journal of the Hattori Botanical Laboratory*, **55**, 1–11.

Delgadillo, C. M. (1993*a*). The neotropical-African moss disjunction. *Bryologist*, **96**, 604–15.

(1993*b*). The Antillean Arc and the distribution of neotropical mosses. *Tropical Bryology*, **7**, 7–12.

(1994). Endemism in the neotropical moss flora. *Biotropica*, **26**, 12–16.

Dirkse, G. M., Bouman, A. C., & Losada-Lima, A. (1993). Bryophytes of the Canary Islands, an annotated checklist. *Cryptogamie: Bryologie, Lichénologie*, **14**, 1–47.

Fife, A. J. (1995). Checklist of the mosses of New Zealand. *Bryologist*, **98**, 313–37.

Frahm, J.-P. (1982). Grossdisjunktionen von Arealen südamerikanischer und afrikanischer *Campylopus*-Arten. *Lindbergia*, **8**, 45–52.

(1994). The affinities between the *Campylopus* floras of Sri Lanka and Madagascar – or: which species survived on Noah's ark? *Hikobia*, **11**, 371–6.

(1996). Mosses newly recorded from Saxonian amber. *Nova Hedwigia*, **63**, 525–7.

Frahm, J.-P. & Vitt, D. H. (1993). Comparisons between the moss floras of North America and Europe. *Nova Hedwigia*, **56**, 307–33.

Frahm, J.-P., Lindlar, A., Sollman, A. P., & Fischer, E. (1996). Bryophytes from the Cape Verde Islands. *Tropical Bryology*, **12**, 123–53.

Frey, W. & Kürschner, H. (1988). Re-evaluation of *Crossidium geheebii* (Broth.) Broth. (Pottiaceae) from Sinai, a xerothermic Pangean element. *Journal of Bryology*, **15**, 123–6.

Gradstein, S. R. & Pócs, T. (1989). Bryophytes. In *Tropical Rain Forest Ecosystems*, ed. H. Lieth & M. J. A. Werger, pp. 311–25. Amsterdam: Elsevier.

Gradstein, S. R. & Vána, J. (1987). On the occurrence of Laurasian liverworts in the tropics. *Memoirs of New York Botanical Garden*, **45**, 388–425.

Gradstein, S. R. & Weber, W. A. (1982). Bryogeography of the Galapagos Islands. *Journal of the Hattori Botanical Laboratory*, **52**, 127–52.

Gradstein, S. R., Pócs, T., & Vána, J. (1983). Disjunct hepaticae in tropical America and Africa. *Acta Botanica Hungarica*, **29**, 127–71.

Grolle, R. (1988). Bryophyte fossils in amber. *Bryological Times*, **47**, 4–5.

Hall, R. & Holloway, J. D. (1998). *Biogeography and Geological Evolution of Southeast Asia*. Leiden: Backhuys.

Hallingbäck, T. & Hodgetts, N. G. (eds) (1999). *Status Survey and Conservation Action Plan for Mosses, Liverworts and Hornworts*. Gland: International Union of Conservation of Nature and Natural Resources/Species Survival Commission Bryophyte Specialist Group Report.

Hedenäs, L. (1996). How do we select species for conservation? *Anales de Instituto de Biologia dela Universidad Nacional Autonomia Mexico, Serie Botanica*, **67**, 129–45.

Herzog, Th. (1926). *Geographie der Moose*. Jena: Fischer.

Hill, M. O. & Preston, C. D. (1998). The geographical relationships of British and Irish bryophytes. *Journal of Bryology*, **20**, 127–226.

Hodgetts, N. G. (1996). Threatened bryophytes in Europe. *Anales de Instituto de Biologia dela Universidad Nacional Autonomia Mexico, Serie Botanica*, **67**, 183–200.

Ignatov, M. S. (1993). Moss diversity patterns on the territory of the former USSR. *Arctoa*, **2**, 13–47.

Imura, S. (1994). Vegetative diaspores in Japanese mosses. *Journal of the Hattori Botanical Laboratory*, **77**, 177–232.

Ines-Sastre, J. & Tan, B. C. (1995). Problems of bryophyte conservation in the tropics: a discussion, with case example from Puerto Rico and the Philippines. *Caribbean Journal of Science*, **31**, 200–6.

International Association of Bryologists (1997). Red list of the world's most endangered bryophytes. *Bryological Times*, **93**, 1–7.

Iwatsuki, Z. (1990). Origin of the New Caledonian bryophytes. *Tropical Bryology*, **2**, 139–48.

    (1992). The moss flora of Japan and its North American connections. *Bryobrothera*, **1**, 1–7.

Iwatsuki, Z. & Sharp, A. J. (1967). The bryological relationships between Eastern Asia and North America I. *Journal of the Hattori Botanical Laboratory*, **30**, 152–70.

Longton, R. E. (1988). *The Biology of Polar Bryophtyes and Lichens*. Cambridge: Cambridge University Press.

MacArthur, R. H. & Wilson, E. O. (1967). *The Theory of Island Biogeography*. Princeton: Princeton University Press.

Menzel, M. & Passow-Schindhelm, R. (1990). The mosses of the Maldive Islands. *Cryptogamie: Bryologie Lichénologie*, **11**, 363–7.

Miller, N. G. & Ambrose, L. J. H. (1976). Growth in culture of wind-blown bryophyte gametophyte fragments from Arctic Canada. *Bryologist*, **79**, 55–63.

Ochyra, R., Bednarek-Ochyra, H., Pócs, T., & Crosby, M. R. (1992). The moss *Adelothecium bogotense* in continental Africa, with a review of its world range. *Bryologist*, **95**, 287–95.

O'Shea, B. (1995). Checklist of the mosses of sub-Saharan Africa. *Tropical Bryology*, **10**, 91–198.

    (1997). The mosses of sub-Saharan Africa. 2. Endemism and biodiversity. *Tropical Bryology*, **13**, 75–87.

Piippo, S. (1992). On the phytogeographical affinities of temperate and tropical Asiatic and Australasiatic hepatics. *Journal of the Hattori Botanical Laboratory*, **71**, 1–35.

Piippo, S. & Koponen, T. (1997). On the phytogeographic biodiversity of Western Melanesian mosses. *Journal of the Hattori Botanical Laboratory*, **82**, 191–201.

Pócs, T. (1976). Correlations between the tropical African and Asian bryofloras. I. *Journal of the Hattori Botanical Laboratory*, **41**, 95–106.

    (1988). Biogeography of the Cuban bryophyte flora. *Taxon*, **37**, 615–21.

(1992). Correlation between the tropical African and Asian bryofloras. II. *Bryobrothera*, **1**, 35–47.

(1996). Epiphyllous liverwort diversity at worldwide level and its threat and conservation. *Annales de Instituto de Biologia dela Universidad de Nacional Autonomia de Mexico, Serie Botanica*, **67**, 109–27.

Raven, P. H. & Axelrod, D. L. (1974). Angiosperm biogeography and past continental movements. *Annals of the Missouri Botanical Garden*, **61**, 539–673.

Reese, W. D. (1985 [1987]). Tropical lowland mosses disjunct between Africa and the Americas, including *Calyptothecium planifrons* (Ren. & Par.) Argent, new to the Western Hemisphere. *Acta Amazonica*, **15** (1–2, Suppl.), 115–21.

Robinson, H. E. (1972). Observations on the origin and taxonomy of the Antarctic moss flora. In *Antarctic Terrestrial Biology*, ed. G. A. Llano, pp. 163–77. Washington: American Geophysical Union.

Schofield, W. B. (1974). Bipolar disjunctive mosses in the Southern Hemisphere, with particular reference to New Zealand. *Journal of the Hattori Botanical Laboratory*, **38**, 13–32.

(1980). Phytogeography of the mosses of North America (north of Mexico). In *The Mosses of North America*, ed. R. J. Taylor & A. E. Leviton, pp. 131–70. San Francisco: Pacific Division, American Association for the Advancement of Science, California Academy of Sciences.

(1984). Bryogeography of the Pacific coast of North America. *Journal of the Hattori Botanical Laboratory*, **55**, 35–43.

(1985). *Introduction to Bryology*. New York: Macmillan.

(1988). Bryophyte disjunctions in the Northern Hemisphere: Europe and North America. *Botanical Journal of the Linnean Society*, **98**, 211–24.

(1992). Bryophyte distribution patterns. In *Bryophytes and Lichens in a Changing Environment*, ed. J. W. Bates & A. M. Farmer, pp. 103–30. Oxford: Clarendon Press.

(1994). Bryophytes of Mediterranean climates in British Columbia. *Hikobia*, **11**, 407–14.

Schofield, W. B. & Crum, H. A. (1972). Disjunctions in bryophytes. *Annals of the Missouri Botanical Garden*, **59**, 174–202.

Schuster, R. M. (1972). Continental movements, "Wallace's Line" and Indomalayan–Australasian dispersal of land plants: some eclectic concepts. *Botanical Review*, **38**, 3–86.

(1982). Generic and familial endemism in the hepatic flora of Gondwanaland: origins and causes. *Journal of the Hattori Botanical Laboratory*, **52**, 3–35.

(1983a). Phytogeography of Bryophyta. In *New Manual of Bryology*, vol. 1, ed. R. M. Schuster, pp. 463–626. Nichinan: Hattori Botanical Laboratory.

(1983b). Reproductive biology, dispersal mechanisms and distribution patterns in Hepaticae and Anthocerotae. *Sonderbildung naturwissenschaftlichen Verein Hamburg*, **7**, 119–62.

(1990). Origins of neotropical leafy hepaticae. *Tropical Bryology*, **2**, 239–64.

Seppelt, R. D., Lewis-Smith, R. I., & Kanda, H. (1998). Antarctic bryology: past achievements and new perspectives., *Journal of the Hattori Botanical Laboratory*, **84**, 203–39.

Söderström, L. (1992). Invasions and range expansions and contractions of bryophytes. In *Bryophytes and Lichens in a Changing Environment*, ed. J. W. Bates & A. M. Farmer, pp. 131–58. Oxford: Clarendon Press.

(1996). Islands – endemism and threatened bryophytes. *Anales de Instituto de Biologia dela Universidad de Nacional Autonomia de Mexico, Serie Botanica*, **67**, 201–11.

Tan, B. C. (1992). Philippine muscology (1979–1989). *Bryobrothera*, **1**, 137–41.

(1998). Noteworthy disjunctive patterns of Malesian mosses. In *Biogeography and Geological Evolution of Southeast Asia*, ed. R. Hall & J. D. Holloway, pp. 235–41. Leiden: Backhuys.

Tan, B. C. & Engel, J. J. (1990). A preliminary study on the affinities of Philippine, Bornean and New Guinean hepatics. *Tropical Bryology*, **2**, 265–72.

Tan, B. C. & Iwatsuki, Z. (1996). Hot spots of mossses in East Asia. *Anales de Instituto de Biologia dela Universidad de Nacional Autonomia de Mexico, Serie Botanica*, **67**, 159–67.

(1999). Four hot spots of moss diversity in Malesia. *Bryobrothera*, **5**, 247–52.

Touw, A. (1992). Biogeographical notes on the Musci of South Malesia, and of Lesser Sunda Islands in particulars. *Bryobrothera*, **1**, 143–55.

van der Wijk, R., Margadant, W. D., & Florschütz, P. A. (1959–1969). *Index Muscorum*, vols. 1–5. Utrecht: International Association for Plant Taxonomists.

Vane-Wright, R. I., Humphries, C. J., & Williams, P. H. (1991). What to protect? – Systematics and the agony of choice. *Biological Conservation*, **55**, 235–54.

Watson, E. V. (1974). *The Structure and Life of Bryophytes*. London: Hutchinson.

Wigginton, M. J. & Grolle, R. (supplemented by A. Gyarmati) (1996). Catalogue of the Hepaticae and Anthocerotae of Sub-Saharan Africa. *Bryophytorum Bibliotheca*, **50**, 1–267.

Zanten, B. O. van (1976). Preliminary report on germination experiments designed to estimate survival chances of moss spores during aerial trans-oceanic long-range dispersal in Southern Hemisphere, with particular reference to New Zealand. *Journal of the Hattori Botanical Laboratory*, **41**, 133–46.

(1978). Experimental studies on trans-oceanic long range dispersal of moss spores in the Southern Hemisphere. *Journal of the Hattori Botanical Laboratory*, **44**, 455–82.

Zanten, B. O. van & Pócs, T. (1981). Distribution and dispersal of bryophytes. *Advances in Bryology*, **1**, 479–562.

Index

Page references in *italics* refer to figures and tables

[449]